AF680151

Pathogenic Streptococci

From Genomics to Systems Biology and Control

Edited by:

Yuqing Li and Xuedong Zhou

State Key Laboratory of Oral Diseases
National Clinical Research Center for Oral Diseases
West China Hospital of Stomatology
Sichuan University
Chengdu 610041, China

Caister Academic Press **www.caister.com**

Copyright © 2019

Caister Academic Press
Norfolk, UK

www.caister.com

British Library Cataloguing-in-Publication Data
A catalogue record for this book is available from the British Library

ISBN: 978-1-912530-22-9 (book)
ISBN: 978-1-912530-23-6 (ebook)

Description or mention of instrumentation, software, or other products in this book does not imply endorsement by the author or publisher. The author and publisher do not assume responsibility for the validity of any products or procedures mentioned or described in this book or for the consequences of their use.

All rights reserved. No part of this publication may be reproduced, stored in a retrieval system, or transmitted, in any form or by any means, electronic, mechanical, photocopying, recording or otherwise, without the prior permission of the publisher. No claim to original Government works.

Contents

Current books of interest

- SUMOylation and Ubiquitination: Current and Emerging Concepts 2019
- Avian Virology: Current Research and Future Trends 2019
- Pathogenic Streptococci: From Genomics to Systems Biology and Control 2019
- Insect Molecular Virology: Advances and Emerging Trends 2019
- Methylotrophs and Methylotroph Communities 2019
- Prions: Current Progress in Advanced Research (Second Edition) 2019
- Microbiota: Current Research and Emerging Trends 2019
- Microbial Ecology: Current Advances from Genomics, Metagenomics and Other Omics 2019
- Porcine Viruses: From Pathogenesis to Strategies for Control 2019
- Lactobacillus Genomics and Metabolic Engineering 2019
- Cyanobacteria: Signaling and Regulation Systems 2018
- Viruses of Microorganisms 2018
- Protozoan Parasitism: From Omics to Prevention and Control 2018
- Genes, Genetics and Transgenics for Virus Resistance in Plants 2018
- Plant-Microbe Interactions in the Rhizosphere 2018
- DNA Tumour Viruses: Virology, Pathogenesis and Vaccines 2018
- Pathogenic Escherichia coli: Evolution, Omics, Detection and Control 2018
- Postgraduate Handbook: A Comprehensive Guide for PhD and Master's Students and their Supervisors 2018
- Enteroviruses: Omics, Molecular Biology, and Control 2018
- Molecular Biology of Kinetoplastid Parasites 2018
- Bacterial Evasion of the Host Immune System 2017
- Illustrated Dictionary of Parasitology in the Post-Genomic Era 2017
- Next-generation Sequencing and Bioinformatics for Plant Science 2017
- Brewing Microbiology: Current Research, Omics and Microbial Ecology 2017
- Metagenomics: Current Advances and Emerging Concepts 2017
- The CRISPR/Cas System: Emerging Technology and Application 2017
- Bacillus: Cellular and Molecular Biology (Third edition) 2017
- Cyanobacteria: Omics and Manipulation 2017
- Foot-and-Mouth Disease Virus: Current Research and Emerging Trends 2017
- Staphylococcus: Genetics and Physiology 2016
- Microbial Biodegradation: From Omics to Function and Application 2016
- Influenza: Current Research 2016
- Chloroplasts: Current Research and Future Trends 2016
- The Bacteriocins: Current Knowledge and Future Prospects 2016
- Brain-eating Amoebae: Biology and Pathogenesis of Naegleria fowleri 2016

Full details at: www.caister.com

Chapter 1

https://doi.org/10.21775/9781912530229.01

CRISPR-Cas Systems in Streptococci

Tao Gong[#], Miao Lu[#], Xuedong Zhou, Anqi Zhang, Boyu Tang, Jiamin Chen, Meiling Jing, Yuqing Li*

State Key Laboratory of Oral Diseases, National Clinical Research Center for Oral Diseases, West China Hospital of Stomatology, Sichuan University, Chengdu, China

Tao Gong[#] and Miao Lu[#] contributed equally to this work.

*Corresponding author: liyuqing@scu.edu.cn

Abstract

Streptococci are one of the most important and common constituents of the host's microbiota and can colonize and live in the upper respiratory and urogenital tract of humans and animals. The CRISPR-Cas systems (i.e., clustered regularly interspaced short palindromic repeat, with CRISPR-associated proteins) found in bacteria and archaea provide sequence-based adaptive immunity against mobile genetic elements, especially in the streptococci. Here, recent research progress on CRISPR-Cas systems in the streptococci is reviewed, including their classification (mainly type I, type II, and type III), physiological function, defense mechanism (CRISPR adaptation, crRNA biogenesis, and target interference) and applications, which are useful for a better understanding of the functions of such systems. Finally, the advances that have been made in streptococci may help in the discovery of further novel CRISPR-Cas systems for use in new technologies and applications in other species.

1. Introduction

Streptococci, a heterogeneous group of Gram-positive facultative anaerobic bacteria, commonly colonize and live in the upper respiratory and urogenital tract of humans and animals, but also are present in other hosts. The genus *Streptococcus* consists of more than 100 diverse species (Willenborg and Goethe, 2016) that tend to be adapted to their hosts, reflecting co-evolution. Thus, streptococci are very important components of the host's microbiota. Although streptococci are commonly regarded as commensals, streptococcal diseases range from moderately serious infections such as impetigo, cellulitis or pharyngitis to severe life-threatening diseases such as meningitis, pneumonia, endocarditis and septicemia. Of the streptococci only *Streptococcus thermophilus* does not have pathogenicity-related genes and is therefore regarded as nonpathogenic (Hols et al., 2005). However, *Streptococcus pyogenes*, *Streptococcus pneumoniae*, *Streptococcus mutans* and *Streptococcus agalactiae* are considered as important human pathogens. For instance, *S. pyogenes* usually colonizes skin epithelial surfaces or the throat and can cause a series of clinical diseases such as necrotizing fasciitis or streptococcal toxic shock syndrome in humans (Hamada et al., 2015); *S. pneumoniae* is the most common cause of community-acquired pneumonia and meningitis (Lynch and Zhanel, 2009); *S. mutans* is regarded as the major etiological agent of dental caries (Tanzer et al., 2001) and *S. agalactiae* is an emerging cause of septic arthritis in some countries (Louthrenoo et al., 2014). Nevertheless, other streptococcal species related to the human microbiota can also cause invasive or noninvasive diseases, including *S. sanguinis* (endocarditis) (Jung et al., 2012) and *S. gordonii* (septic arthritis) (Yombi et al., 2012).

Streptococcus species undergo frequent and extensive horizontal gene transfer (HGT), through transformation, transduction, or conjugation (Choi et al., 2012). HGT has played a major role in genomic evolution, helping in the acquisition of new traits that impact processes such as metabolic pathways, pathogenesis and antibiotic resistance. For example, prophages make up to 14% of the *S. pyogenes* genome and encode virulence factors including DNases, pyrogenic exotoxins and the phospholipase SlaA (Fischetti, 2007). However, some innate strategies have also evolved to protect against invading foreign genetic material, including ComX or SigX factors, CSP (competence stimulating peptide), restriction-modification (R-M) systems or recombination machinery (Fontaine et al., 2015).

Besides the innate immune strategies mentioned above, some bacteria also possess an alternative adaptive immune system. CRISPR-Cas systems (clustered regularly interspaced short palindromic repeat, with CRISPR-associated proteins) are found in 40% of bacteria and 90% of archaea and provide sequence-based adaptive immunity against mobile genetic elements such as phages, invasive conjugative plasmids and transposable elements (Marraffini, 2015). The two essential components, CRISPR loci and their associated *cas* genes, constitute a small RNA-guided system that helps prokaryotes ward off invasive genetic elements. A CRISPR locus contains arrays of 'spacers' derived from the genomes of previous invaders, usually viruses or plasmids, and those critical 'spacer' sequences are inserted between the repeat elements of natural CRISPR loci. A cluster of *cas* genes is located upstream of a CRISPR locus and different CRISPR-Cas systems possess diverse *cas* genes (Makarova et al., 2011). CRISPR-Cas systems were first identified as a specific sequence-based barrier against three major routes of foreign DNA introduction: plasmid conjugation, phage transduction and DNA transformation (Marraffini, 2015). To date, the CIRSPR-Cas system is the only known prokaryotic immune system that enables precise recognition and targeting of phages once a specific sequence of from its genome has been 'memorized'. Importantly, this 'memory' can be passed down to subsequent generations.

In this review, we discuss current knowledge on the distribution, physiological functions and mechanisms of CRISPR-Cas systems encoded in streptococci, their role in the biology of these organisms and their applications.

2. Distribution of CRISPR-Cas systems in streptococci

CRISPR-Cas systems are divided into two distinct classes, characterized by their effector module design (Amitai and Sorek, 2016). Class 1 CRISPR-Cas systems, containing the most common and diverse type I systems, type III predominantly in archaea, as well as type IV systems, are defined by the presence of a multi-subunit crRNA-effector complex. Class 2 CRISPR-Cas systems, including the most prevalent type II, type V, and a new type VI, are defined by the presence of a large, single, multi-domain crRNA-effector module. Each type can be further partitioned into several subtypes, up to 19 subtypes in all (Koonin et al., 2017). More and more CRISPR-Cas systems are being revealed in the streptococci. As shown in Table 1, streptococci mainly harbor three types of CRISPR-Cas systems: type I, type II and type III, in addition to the individual type V and unknown CRISPR loci. Of these, the number of type II systems, especially subtype II-A, is much larger than the other types, suggesting that streptococci mainly depend on type II systems to defend against invading foreign genetic material or participate in physiology and pathogenesis, whereas many types remain to be identified. The CRISPR loci of several representative types are shown in Figure 1.

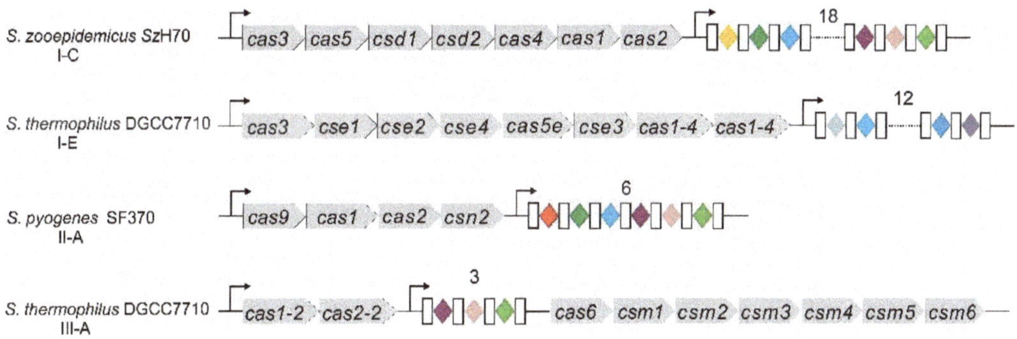

Figure 1. The characteristics of four representative CRISPR-Cas systems in streptococci.

Table 1 Types of streptococcal CRISPR-Cas systems covered in this review.

Species	Type I			Type II			Type III		Type V	Unknown	References
	C	E	Unknown	A	C	Unknown	A	Unknown	A		
S. agalactiae											Lier et al., 2015; Lopez-Sanchez et al., 2012
S. anginosus											Olson et al., 2013; Makarova et al., 2015
S. australis											Maeda et al., 2011
S. constellatus subsp.*constellatus*											Olson et al., 2013
S. dysgalactiae subsp. *equisimilis*											Shimomura et al., 2011; Makarova et al., 2015
S. gallolyticus											Louwen et al., 2014
S. gordonii											Louwen et al., 2014; Makarova et al., 2015
S. infantarius subsp. *infantarius*											Jans et al., 2013; Makarova et al., 2015
S. iniae											Zhang et al., 2014; Makarova et al., 2015
S. intermedius											Olson et al., 2013
S. lutetiensis											Makarova et al., 2015
S. macedonicus											Makarova et al., 2015
S. mitis											Kilian et al., 2014
S. mutans											Serbanescu et al., 2015; Louwen et al., 2014
S. oligofermentans											Makarova et al., 2015
S. oralis											Kilian et al., 2014
S. parasanguinis											Makarova et al., 2015
S. parauberis											Nho et al., 2013
S. pasteurianus											Makarova et al., 2015
S. phocae subsp. *phocae*											Bethke and Avendano-Herrera, 2017
S. phocae subsp. *salmonis*											Bethke and Avendano-Herrera, 2017
S. pyogenes											Hynes et al., 2017; Louwen et al., 2014
S. ruminantium											Tohya et al., 2018
S. salivarius											Makarova et al., 2015
S. sanguinis											Louwen et al., 2014
S. suis											Yao et al., 2015; Makarova et al., 2015
S. thermophilus											Makarova et al., 2015
S. troglodytae											Okamoto et al., 2016
S. uberis											Hossain et al., 2015
S. zooepidemicus											Waller and Robinson, 2013; Makarova et al., 2015

2.1 Type I CRISPR-Cas systems

As shown in Table 1, 30% (9/30) of analyzed streptococci harbor type I-C, 13.3% (4/30) have type I-E and there are two unknown CRISPR loci (2/30). All type I loci contain the signature gene *cas3* (or its variant *cas3′*), which encodes a helicase protein with a demonstrated capacity to unwind double-stranded DNA (dsDNA) and RNA-DNA duplexes (Huo et al., 2014). Type I systems are currently divided into seven sub-types, I-A to I-F and I-U (Makarova et al., 2011), of which subtype I-B, containing the complete gene locus (*cas1-cas2-cas3-cas4-cas5-cas6-cas7-cas8*) seems to be best defined by ancestral type I gene composition, whereas other subtypes are divergent derivatives of type I-B with loss of different genes or rearranged gene order (Makarova et al., 2013). For example, subtype I-C seems to be a derivative of subtype I-B that lacks *cas6*, which seems to be functionally replaced by *cas5*. Subtype I-E appears to lack *cas4*, suggesting that diverse types may undergo evolution in complex environmental conditions.

2.2 Type II CRISPR-Cas systems

63.3%(19/30) of the analyzed streptococci in Table 1 contain type II CRISPR-Cas systems (mostly type II-A and several type II-C), which are currently divided into three subtypes: II-A, II-B, and II-C (Chylinski et al., 2014). The signature gene for type II systems is *cas9,* which encodes a multi-domain protein that combines the functions of the crRNA-effector complex to cleave double-stranded DNA targets (Gasiunas et al., 2012), and also contributes to adaptation (Heler et al., 2015). The type II-A CRISPR-Cas system is regarded as the minimal CRISPR-Cas system and contains a repeat-spacer array, only four *cas* genes (*cas9*, *cas1*, *cas2 and csn2*), a partially complementary trans-acting RNA (tracrRNA) and RNase III (Figure 2). Subtype II-B lacks *csn2* but includes *cas4*, and the Cas4 protein belongs to the PD-(D/Æ)xK family of nucleases that have been shown to possess 5' to 3' single-stranded DNA (ssDNA) exonuclease activity or both directions (Makarova et al., 2015). While Csn2 functions in spacer integration, the role of Cas4 in the CRISPR-Cas system remains to be identified (Barrangou et al., 2007). Subtype II-C loci only have three protein-coding genes: *cas1*, *cas2,* and *cas9* (Chylinski et al., 2013).

2.3 Type III CRISPR-Cas systems

Of the analyzed streptococci, 26.7% (8/30) possess type III-A CRISPR-Cas systems, and an unknown CRISPR locus (1/30). All type III systems contain the signature gene *cas10* and the Cas10 proteins show significant sequence variation among the diverse type III CRISPR-Cas systems. In fact, there are four type III subtypes: III-A to III-D (Makarova et al., 2015). Type III CRISPR loci have more diverse genetic organization than type I systems due to gene duplications or deletions, and domain insertions or fusions. For instance, type III-C and type III-D system lack the well conserved *cas1* and *cas2* genes (Vestergaard et al., 2014). Moreover, subtype III-A loci usually include *cas1*, *cas2* and *cas6* genes, but most type III-B loci lack these genes and therefore depend on other CRISPR-Cas systems present in the same genome. Both subtype III-A and III-B CRISPR-Cas systems have been shown to co-transcriptionally target RNA and DNA (Samai et al., 2015).

2.4 Other putative CRISPRs in streptococci

About 53.3% (16/30) of the analyzed streptococci possess uncharacterized

CRISPR-Cas systems that have been identified bioinformatically (Table 1). With the rapid development of complete genome sequencing technology, bioinformatic tools have been used to identify CRISPR-Cas systems in many streptococci. These tools include CRISPRFinder (Grissa et al., 2007b), CRISPRdetect (Biswas et al., 2016), CRISPROne (http://omics.informatics.indiana.edu/CRISPRone) for detecting CRISPRs, CRISPRTarget (Biswas et al., 2013) and so forth. For example, prediction of CRISPRs in *S. iniae* SF1 (serotype I) was performed using the CRISPR recognition tool (http://crispr.u-psud.fr/crispr/). It has also been reported that *S. anginosus* contains a type V CRISPR-Cas system (Makarova et al., 2015). Moreover, CRISPRs were detected in the genomes of *Streptococcus infantarius* subsp. *infantarius* CJ18 and ATCC BAA-102T using CRISPRfinder and CRISPRdb (Grissa et al., 2007a, b), but CJ18 contains a smaller number of CRISPR spacers, which indicates a reduced defense capability against foreign DNA (Jans et al., 2013). *S. mitis, S. oralis, S. parauberis, S. trogolodytae* and *S. ubris* have also been found to possess CRISPR sequences via CRISPRfinder, but these have not been characterized (Kilian et al., 2014; Makarova et al., 2015; Okamoto et al., 2016).

3. Physiological functions of CRISPR-Cas systems

Recent studies have shown that CRISPR-Cas systems can possess alternative roles in pathogenesis or host cell physiology, including in biofilm formation, DNA repair, stress response, immune evasion, quorum sensing and interactions between bacteria (Louwen et al., 2014; Sampson and Weiss, 2014) in addition to their classical function in defense against foreign DNA. For example, the number of CRISPR sites in forty-five *S. mutans* strains isolated from plaque samples was associated with the clinical manifestations of early childhood caries (ECC). This study found that the biofilm formation and the ability to synthesize EPS by *S. mutans* strains with two CRISPR sites were significantly improved compared to strains without CRISPR loci, suggesting the CRISPR sites might contribute to the cariogenic potential of *S. mutans* (Chen et al., 2017). The CRISPR-Cas system in *Pseudomonas aeruginosa* plays an important role in modulating biofilm formation. Although the detailed mechanism remains unknown, the study showed that when a certain bacteriophage invades *Pseudomonas aeruginosa*, the CRISPR-Cas system can interact with a gene in the chromosomally integrated prophage to inhibit biofilm formation (Cady and O'Toole, 2011; Zegans et al., 2009). In *S. mutans* UA159, the CRISPR-Cas system was also found to be involved in the stress response, DNA repair and gene expression (Serbanescu et al., 2015). Moreover, Cas1, found in all CRISPR systems, and a crRNA array derived from *Escherichia coli* are essential for DNA repair (Babu et al., 2011). Cas2 also influences the ability of *Legionella pneumophila* to survive within amoebae (Gunderson and Cianciotto, 2013). As amoebae are essential for *Legionella pneumophila* to survive in a complex environment (Abu et al., 1998), the function of Cas2 is predicted to be essential for environmental persistence and subsequent transmission to other hosts.

In pathogens, such as the Gram-negative intracellular pathogen *Francisella novicida*, its type II CRISPR-Cas system plays an important role in bypassing detection by the host's recognition receptors. This system targets an endogenous transcript encoding an immunostimulatory bacterial lipoprotein

(BLP), resulting in mRNA degradation and decreased transcript levels. In contrast, in the absence of this CRISPR system, BLP levels increase slowly, which triggers the activation of a Toll-like Receptor 2 (TLR2)-dependent proinflammatory response, leading to destruction of the bacteria during infection (Jones et al., 2012). Based on the diverse roles of CRISPR systems in different organisms, we speculate that different bacteria may have co-opted distinct CRISPR-Cas components for diverse functional outcomes. As the physiological roles of CRISPR-Cas systems in streptococci are still poorly understood, other functions of these systems remain to be identified.

4. Defense mechanism of CRISPR-Cas systems in streptococci

The molecular mechanism of the CRISPR-Cas system mediates immunity in three phases: (1) CRISPR adaptation, (2) crRNA biogenesis and (3) target interference (see Figure 2). The adaptation phase involves the integration of new spacers derived from plasmids or viruses into the CRISPR array. In the second phase, the CRISPR array is transcribed as a precursor CRISPR RNA (pre-crRNA), which is then processed to form mature crRNAs including a part of the repeat and the spacer. During the interference phase, the mature crRNA in combination with single or multiple Cas proteins, functions as a guide to recognize and cleave invading nucleic acids (also known as protospacers), such as plasmids or viruses (Makarova et al., 2015). In some CRISPR-Cas systems (i.e., types I, II and V), a short protospacer adjacent motif (PAM) sequence found adjacent to the targeted protospacer sequence is required for both the adaptation and interference phases. For these CRISPR-Cas systems, the PAM sequence is located on the 3' or 5' end of the foreign targeted DNA but is absent in the CRISPR array, which functions in target recognition between its own and foreign DNA (Leenay and Beisel, 2017). However, CRISPR-Cas systems without PAM sequences possess strategies to avoid self-targeting, such as lacking complementarity between the 5' repeat handle of the crRNA and the 3' flanking region of the target RNA in some type III systems (Tamulaitis et al., 2017).

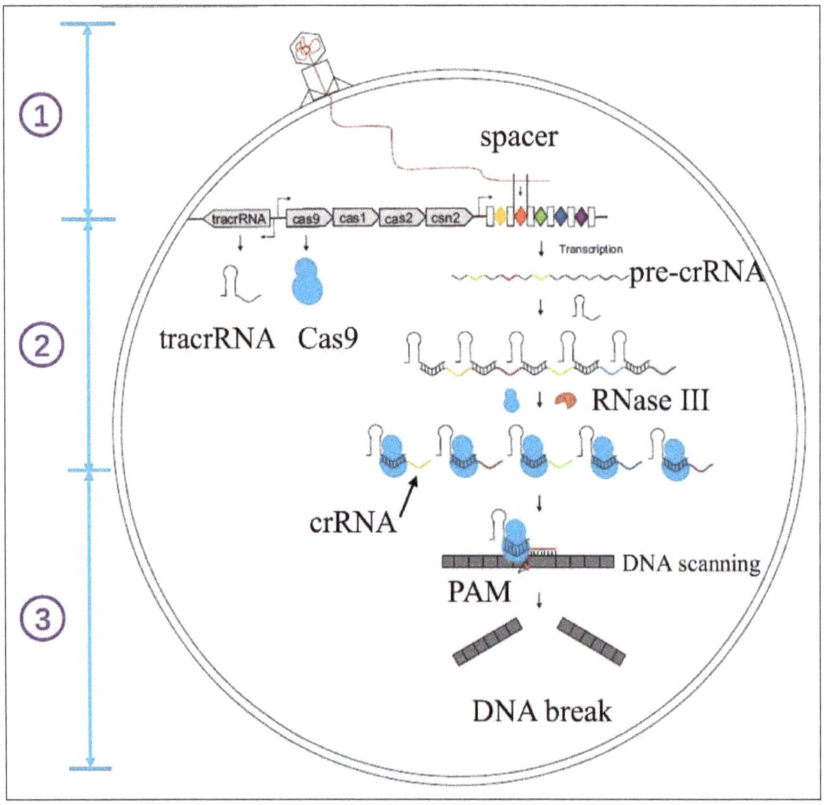

Figure 2. The three stages of CRISPR-Cas9 adaptive immunity: (1) CRISPR adaptation, (2) crRNA biogenesis and (3) target interference.

4.1 CRISPR adaptation

During the CRISPR adaptation phase, short fragments of exogenous DNA derived from plasmids or viruses are incorporated into the CRISPR array within the bacterial genome, and these protospacers act as new spacer sequences. In this way hosts can combat future attacks by similar invaders and keep a genetic record for similar encounters (Makarova et al., 2015). The adaptation mechanism is best characterized in *E. coli.* The Cas1 and Cas2 proteins, which form a $Cas1_4$-$Cas2_2$ complex (hereafter, Cas1-Cas2), are mainly responsible to spacer integration, whereas some elements such as at least one CRISPR repeat that is part of the leader sequence (Wei et al., 2015) and a few host factors for repair of the insertion sites (Ivancic-Bace et al., 2015) are also required. Spacer acquisition can be divided into three steps: (1) substrate capture, (2) recognition of the CRISPR locus and (3) spacer integration into CRISPR loci (Jackson et al., 2017) (Figure 3).

4.1.1 Cas1-Cas2 substrate capture

In the substrate capture phase, production of prespacers (partially duplexed dsDNA) from foreign DNA can occur by two pathways: naïve CRISPR adaptation and primed CRISPR adaptation (Heler et al., 2014). Cas1/Cas2-mediated acquisition can add a spacer-repeat unit to a CRISPR locus consisting of only one repeat sequence, which is referred to as naïve acquisition. However, primed acquisition occurs when the CRISPR array already possesses spacers matching the target DNA and the presence of spacers can increase the frequency of acquiring another spacer. The ends of the dsDNA prespacer combined with Cas1-Cas2 are separated by tyrosine wedges in each Cas1 dimer, which locks and opens the DNA branch points. Each 3' end of the prespacer spreads into the active subunits of each corresponding Cas1 dimer (Nunez et al., 2015; Wang et al., 2015) (Figure 3).

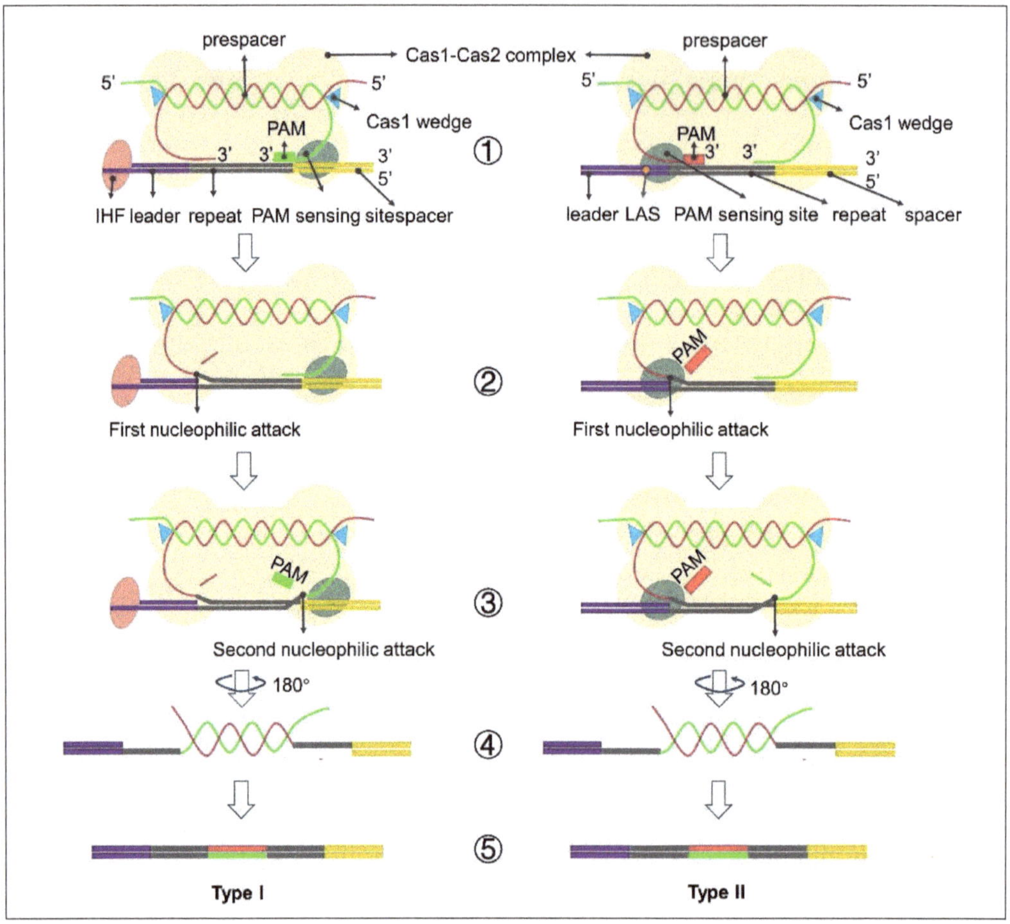

Figure 3. Spacer integration process: (1) Cas1-Cas2 complex load with prespacer into Cas1-Cas2-prespacer complex, which binds leader and first repeat. For type I systems (e.g., I-E in *E. coli*), Cas1-Cas2 binding to the leader-proximal repeat is assisted by integration host factor (IHF), whereas Cas1-Cas2 binding to the leader-proximal repeat in type II systems (e.g., II-A in *S. pyogenes*) need recognize the leader-anchoring site (LAS). (2) The first nucleophilic attack likely occurs at the leader-repeat junction and forms a half-site intermediate. (3) The second nucleophilic attack occurs at the existing repeat-spacer junction into full-site integration. (4) and (5) Host DNA repair mechanisms fill the integration site to acquire perfect integration.

4.1.2 Recognition of the CRISPR array

During recognition, the leader-repeat sequences upstream of the CRISPR arrays directly guide leader-polarized spacer integration and are assisted by direct Cas1-Cas2 complex recognition and host proteins (Wright and Doudna, 2016). For the type I-E system from *E. coli*, leader-repeat recognition is required by the integration host factor (IHF) heterodimer. IHF, which binds the CRISPR leader, helps to localize Cas1-Cas2 to the leader-repeat junction, which also contributes to stabilizing the interaction between Cas1-Cas2 and the leader-repeat, increasing the efficiency of spacer acquisition (Nunez et al., 2016). However, in type II systems from *S. pyogenes*, a short leader-anchoring site (LAS) adjacent to the first repeat has been found to be essential for CRISPR spacer adaptation (Wei et al., 2015) (Figure 3).

4.1.3 Integration into the CRISPR array

For CRISPR-Cas types that depend on PAM sequences for target recognition, prespacer substrates at specific positions are processed to produce suitable PAMs. While each of the four Cas1 proteins contains a PAM-sensing domain, one is sufficient (Wang et al., 2015). Furthermore, the presence of a PAM in the prespacer can also ensure integration of the CRISPR array in the correct orientation (Shmakov et al., 2014). The prespacer processed by Cas1 then creates the two 3'-OH ends necessary for nucleophilic attack on the junction between the repeat and other sequences. The first nucleophilic attack most likely occurs at the leader-repeat junction and a second attack at the existing repeat-spacer junction (Jackson et al., 2017). In brief, Cas1-Cas2 depending on

IHF or LAS binds to the leader-proximal repeat, which helps junction between the leader and the repeat, or repeat and spacer, finally, the DNA repair mechanisms fill in the integration site (Figure 3).

4.2 crRNA biogenesis

During the second phase, crRNA biogenesis pathways diverge among the different CRISPR-Cas types. For example, in most type I and type III systems, Cas6 is essential for the primary processing of pre-crRNA (Makarova et al., 2015). Cas6 is a metal-independent endoribonuclease that recognizes and cleaves a single phosphodiester bond in the repeat sequences of pre-crRNA transcripts (Marraffini and Sontheimer, 2010). Type I-C systems do not contain Cas6 but instead contain the metal-independent endoribonuclease Cas5. For example, Cas5c in *S. pyogenes* not only processes pre-crRNA but also binds to the crRNA 5' handle, which is similar in function to Cas6 and Cas5 from type I-E systems in *E. coli*. Cas5d from the type I-C system in *S. mutans* UA159 also has RNase activity. Although present in types I-A, I-B and I-E, Cas5 is only directly involved in the maturation process in type I-C systems. However, type II-A systems *in S. pyogenes* use a distinct mechanism for crRNA biogenesis and this process depends on a tracrRNA and a host RNase III. The tracrRNA contains extensive secondary structure that is recognized and bound by Cas9 and also harbors a sequence that is complementary to the repeat sequence of the pre-crRNA. Annealing of these complementary sequences results in the formation of double stranded RNAs that are cleaved at one end by RNase III (Jinek et al., 2012). This cleavage liberates the small crRNAs from the precursor, which remain bound to Cas9 via their association with the tracrRNA (Pennisi, 2013).

4.3 Target interference

In the final phase, target interference, each crRNA combines with Cas protein(s) to form an effector complex, which is guided to target specific sequences (DNA or RNA) for recognition and destruction. In most cases, selection of an appropriate target sequence isn't entirely complementary to the crRNA, but is complementary to a so-called seed sequence (about seven or eight base pairs in close proximity to the PAM) (Sternberg et al., 2014). Several base mismatches between the spacer and protospacer are often tolerated(Semenova et al., 2011).

In type I systems, the Cascade complex scans DNA for complementary target sites, firstly identifying an appropriate PAM motif, then by partial melting and base pairing with the guide's seed sequence and eventually the formation of a complete R-loop structure, resulting in DNA degradation by the Cas3 nuclease/helicase (Jore et al., 2011). While type III systems are structurally similar to the type I systems (Jackson and Wiedenheft, 2015), there are some substantial mechanistic variations. Recently, it has been demonstrated that type III complexes are transcription-dependent DNA nucleases (Tamulaitis et al., 2014) meaning that they initially recognize target sequences via a specific interaction between the crRNA and a complementary nascent mRNA, following which cleavage of the flanking DNA sequences occurs. However, class 2 systems need only a single protein for interference. In the type II systems, the crRNA complex, which is involved in target recognition and cleavage, consists of Cas9 bound to a crRNA guide base paired with the tracrRNA. The crystal structures of Cas9 show that the protein has two distinct roles in target recognition and nuclease activity. Finally, binding and cleavage of the target DNA by the Cas9-crRNA effector complex depends upon the recognition of an appropriate PAM located at the 3' end of the protospacer. Cas9 generates a blunt-ended double-strand break, typically located 3 nucleotides from the 3' end of the protospacer (Garneau et al., 2010).

5. Applications of CRISPR-Cas systems

5.1 Gene therapy

Genome editing using CRISPR-Cas9 technology is revolutionizing studies of gene functions and likely will give rise to an entirely new class of therapeutics for a wide range of diseases. One such study has successfully removed a mutation in the dystrophin protein, which is responsible for the most common form of Duchenne's muscular dystrophy (Dmd) (Nelson et al., 2016). Furthermore, CRISPR-Cas systems have been used in novel therapeutic approaches against human immunodeficiency virus type-1 (HIV-1), aiming to permanently inactivate all virus genomes or to prevent viral persistence in latent reservoirs (Wang et al., 2018). Although it remains to be seen which disease indications are most suited for CRISPR-Cas9-mediated genome editing, optimization of delivery and assessment of specificity are critical for the safe and effective translation of this technology to the clinical, especially in terms of off-target effects.

5.2 Transcriptional modulation

To repurpose the CRISPR-Cas system for genome regulation as opposed to genome editing, investigators have found that a catalytically inactive version of

Cas9 can be used as a platform for RNA-guided transcriptional regulation. A RNA-guided DNA recognition platform using co-expression of a mutant Cas9 with a small-guided (sgRNA) designed with a ~20 base-pair (bp) complementary region that yields specific silencing of a gene of interest without off-target effects has been developed and is known as CRISPR interference (CRISPRi) (Qi et al., 2013). This minimal CRISPRi system can not only effectively silence transcription initiation and elongation but is also highly specific and efficient with up to 1,000-fold repression. Moreover, the CRISPRi gene knockdown system is inducible and reversible and can be used to simultaneously regulate multiple genes, meaning that the CRISPRi system has the potential for genome-scale analysis and regulation of microbial genomes (Qi et al., 2012) and can be utilized as a flexible framework for engineering transcriptional regulatory networks. Finally, the CRISPRi system can also be used to knock down gene expression in mammalian cells (Qi et al., 2013).

5.3 Functional genomic screens

The high efficiency of gene editing based on the CRISPR-Cas system makes it possible to enable thorough genome-wide functional screens to identify essential or non-essential genes. Genome-wide loss-of-function screens using RNAi have many disadvantages, including partial knockdown of genes and serious off-target effects. However, Cas9-mediated sgRNA screens have high screening sensitivity and practicality and can be easily designed to target almost any DNA fragment. For example, high-throughput CRISPRi phenotyping has identified new essential genes in *S. pneumoniae* through the construction of a knockdown library targeting potentially essential genes by CRISPR interference and showing a growth phenotype for most of them (Liu et al., 2017). Future application should be involve using multiplex sgRNA libraries to perturb noncoding genetic elements, distant enhancers, general promoter architectures or other transcriptional regulatory factors (Shalem et al., 2014).

5.4 Rapid cell generation and animal models

In the past few years, studies have shown that gene editing based on the CRISPR-Cas system can be used to accelerate the generation of transgenic models. To generate new cellular models, the sgRNA is designed from the sequence of interest, which is cloned into a plasmid encoding Cas9 and this plasmid is introduced into the target cells through transient transfection. For the generation of transgenic animal models, a previous study showed that Cas9 and the desired sgRNA can be directly injected into fertilized zygotes to obtain heritable gene modification in animals such as monkeys (Yang et al., 2013). Double-stranded DNA breaks (DSB) produced by Cas9 at specific genomic loci are repaired by the error-prone non-homologous end-joining (NHEJ) pathway generating insertions or deletions, or by homology-directed repair (HDR) using a homologous DNA repair template to generate the desired mutants (Salsman and Dellaire, 2017). Furthermore, CRISPRi can also be used to repress gene expression to identify its functions, rather than produce DSB. Therefore, based on these strategies, new transgenic animal models can be constructed. Comparison with the traditional process of generating transgenic animals, the generation time for mutant models can be reduced from more than a year to only several weeks, saving costs and enlarging the scale at which these experiments can be performed. More interestingly, Cas9 can also be used for direct modification of somatic tissue to bypass embryonic manipulation (Hsu et al., 2014).

5.5 Using CRISPR-Cas systems as antimicrobials

Regulating the composition of microbial populations plays a critical role in medicine and biotechnology. While traditional strategies such as antibiotics and antimicrobial peptides provide partial solutions, these have many disadvantages, such as lack of specificity or high cost. A study demonstrated that RNA-directed CRISPR-Cas systems as antimicrobials can overcome those drawbacks (Gomaa et al., 2014). Although the CRISPR-Cas system naturally evolved to provide adaptive immunity against invasive genetic elements in bacteria and archaea, genome editing technologies using this system can be co-opted to target chromosomal sequences, also referred as 'self-targeting'. However, self-targeting by CRISPR-Cas systems is typically lethal in bacteria if DNA repair is absent (Bikard and Barrangou, 2017). For example, studies have found that crRNAs guide Cas9 to cleave all copies of the genome in *E. coli* so efficiently that it is impossible to make repair these breaks through recombination with a sister chromosome, resulting in cell death (Cui and Bikard, 2016). However, a major limit remains in the effective and steady delivery of a CRISPR-Cas9 system into host cells, especially in Gram-positive bacteria, although several strategies using phages, lentivirus (LV), adenovirus (AV) and adeno-associated virus (AAV) as vectors have been successful (Wang et al., 2017). Future research should focus on CRISPR-Cas and delivery system combinations to develop specific and programmable antimicrobials that will open new avenues for the application of CRISPR-Cas9 systems to effectively kill pathogens and precisely alter microbiome composition (Bikard and Barrangou, 2017; Citorik et al., 2014).

5.6 Anti-CRISPR

Like the human immune system, the CRISPR-Cas system can rapidly recognize and clear invading genetic elements in bacteria. However, some phages have also developed a way to escape recognition by CRISPR-Cas systems. These evasion strategies contain point mutations, large-scale deletions, DNA modifications or phage-encoded proteins known as anti-CRISPRs (Acrs) (Hynes et al., 2017). To date, 21 unique families of anti-CRISPR proteins against type I and type II CRISPR–Cas systems have been discovered. For example, AcrF1 and AcrF2 were shown to interact directly with type I-F Cascade complex in *P. aeruginosa*, inhibiting its ability to bind with DNA (Bondy-Denomy et al., 2013). Moreover, AcrIIA4 abolishes the type II-A CRISPR-Cas9 system in *S. pyogenes* by modeling DNA sequence and occupying the PAM site bound by Cas9, resulting in the failure of Cas9 to bind target DNA (Dong et al., 2017). However, hosts can also acquire more new spacers derived from viruses in response to these escape mutations (Fineran et al., 2014). Therefore, there is an evolutionary arms race between anti-CRISPR and CRISPR-Cas systems. The presence of anti-CRISPRs also contributes to CRISPR-Cas system diversity, which provides strong selective pressure on the evolution of anti-CRISPR proteins. Two anti-CRISPRs derived from *L. monocytogenes* have the ability to prevent genome editing commonly used by *S. pyogenes* Cas9 (SpyCas9) in human cells (Rauch et al., 2017), which may act as an 'off-switch' to control the activity of CRISPR-Cas systems. Moreover, as there are many human pathogens with CRISPR-Cas systems, phages used for gene therapy could be equipped with a variety of anti-CRISPRs to prevent the bacterial adaptive immune response. Finally, our understanding of anti-CRISPRs in its infancy, but these studies have opened up a new area of CRISPR-Cas research and subsequent studies will uncover more mechanisms to enrich CRISPR-Cas system diversity (Maxwell, 2017).

5.7 Other applications

To study real-time interactions between specific genes, robust strategies are required to visualize DNA in living cells. Traditional methods (such as fluorescence *in situ* hybridization, FISH) for labeling DNA require sample fixation and can't capture all live processes. Cas9 tagged with fluorescent labels for labeling specific DNA loci was recently developed as a powerful tool for live-cell-imaging and is known as DNA-FISH (Chen et al., 2013). Moreover, the application of CRISPR-Cas systems for molecular detection includes the detection of specific DNA or RNA sequences, detection of mutations and single nucleotide variants (SNVs), and the detection of pathogens (Zhou et al., 2018). To date, the majority of CRISPR-based applications have focused on the type II-A CRISPR-Cas9 system in *S. pyogenes*, whereas a significant diversity in CRISPR types remains to be utilized, expanding their potential applications.

6. Conclusion

This paper has reviewed current advances in our understanding of CRISPR-Cas systems in streptococci, and has looked at the potential for the application of this technology to other fields. Interestingly, those microbes that are responsible for some of the most prevalent infectious diseases in the world may also harbor a cure for a number of genetic diseases and may provide new therapies to combat diseases and new genetic tools that can revolutionize medicine. However, the applications of CRISPR-Cas systems face several challenges, including off-target effects, delivery efficiency and unwanted mutations. Thus, future work is required to discovering other novel CRISPR-Cas systems, overcome current limitations and address the functions of the CRISPR-Cas locus in ecology, evolution and pathogenesis.

7. Acknowledgements

This work was supported by grants from the National Natural Science Foundation of China (grant numbers 31200985, 81430011, and 31870065). The authors declare no potential conflicts of interest with regard to the authorship and/or publication of this article.

8. References

Abu Kwaik Y, Gao LY, Stone BJ, Venkataraman C, Harb OS. (1998). Invasion of protozoa by Legionella pneumophila and its role in bacterial ecology and pathogenesis. Appl Environ Microbiol. 64(9):3127-3133. PMID: 9726849

Amitai G, Sorek R. (2016). CRISPR-Cas adaptation: insights into the mechanism of action. Nat Rev Microbiol. 14(2):67-76. DOI: 10.1038/nrmicro.2015.14.

Babu M, Beloglazova N, Flick R, Graham C, Skarina T, Nocek B, Gagarinova A, Pogoutse O, Brown G, Binkowski A, et al. (2011). A dual function of the CRISPR-Cas system in bacterial antivirus immunity and DNA repair. Mol Microbiol. 79(2):484-502. DOI: 10.1111/j.1365-2958.2010.07465.x.

Barrangou R, Fremaux C, Deveau H, Richards M, Boyaval P, Moineau S,

Romero DA, Horvath P. (2007). CRISPR provides acquired resistance against viruses in prokaryotes. Science. 315(5819):1709-1712. DOI: 10.1126/science.1138140.

Bethke J, Avendaño-Herrera R. (2017). Comparative genome analysis of two *Streptococcus phocae* subspecies provides novel insights into pathogenicity. Mar Genomics. 31:53-61. DOI: 10.1016/j.margen.2016.09.005.

Bikard D, Barrangou R. (2017). Using CRISPR-Cas systems as antimicrobials. Curr Opin Microbiol. 37:155-160. DOI: 10.1016/j.mib.2017.08.005.

Biswas A, Gagnon JN, Brouns SJ, Fineran PC, Brown CM. (2013). CRISPRTarget: bioinformatic prediction and analysis of crRNA targets. RNA Biol. 10(5):817-827. DOI: 10.4161/rna.24046.

Biswas A, Staals RH, Morales SE, Fineran PC, Brown CM. (2016). CRISPRDetect: A flexible algorithm to define CRISPR arrays. BMC Genomics. 17:356. DOI: 10.1186/s12864-016-2627-0.

Bondy-Denomy J, Pawluk A, Maxwell KL, Davidson AR. (2013). Bacteriophage genes that inactivate the CRISPR/Cas bacterial immune system. Nature. 493(7432):429-432. DOI: 10.1038/nature11723.

Cady KC, O'Toole GA. (2011). Non-identity-mediated CRISPR-bacteriophage interaction mediated via the Csy and Cas3 proteins. J Bacteriol. 193(14):3433-3445. DOI: 10.1128/JB.01411-10.

Chen B, Gilbert LA, Cimini BA, Schnitzbauer J, Zhang W, Li GW, Park J, Blackburn EH, Weissman JS, Qi LS, et al. (2013). Dynamic imaging of genomic loci in living human cells by an optimized CRISPR/Cas system. Cell. 155(7):1479-1491. DOI: 10.1016/j.cell.2013.12.001.

Chen J, Li T, Zhou X, Cheng L, Huo Y, Zou J, Li Y. (2017). Characterization of the clustered regularly interspaced short palindromic repeats sites in Streptococcus mutans isolated from early childhood caries patients. Arch Oral Biol. 83:174-180. DOI: 10.1016/j.archoralbio.2017.07.023.

Choi SC, Rasmussen MD, Hubisz MJ, Gronau I, Stanhope MJ, Siepel A. (2012). Replacing and additive horizontal gene transfer in Streptococcus. Mol Biol Evol. 29(11):3309-3320. DOI: 10.1093/molbev/mss138.

Chylinski K, Le Rhun A, Charpentier E. (2013). The tracrRNA and Cas9 families of type II CRISPR-Cas immunity systems. RNA Biol. 10(5):726-737. DOI: 10.4161/rna.24321.

Chylinski K, Makarova KS, Charpentier E, Koonin EV. (2014). Classification and evolution of type II CRISPR-Cas systems. Nucleic Acids Res. 42(10):6091-6105. DOI: 10.1093/nar/gku241.

Citorik RJ, Mimee M, Lu TK. (2014). Sequence-specific antimicrobials using efficiently delivered RNA-guided nucleases. Nat Biotechnol. 32(11):1141-1145. DOI: 10.1038/nbt.3011.

Cui L, Bikard D. (2016). Consequences of Cas9 cleavage in the chromosome of Escherichia coli. Nucleic Acids Res. 44(9):4243-51. DOI: 10.1093/nar/gkw223.

Dong, Guo M, Wang S, Zhu Y, Wang S, Xiong Z, Yang J, Xu Z, Huang Z. (2017). Structural basis of CRISPR-SpyCas9 inhibition by an anti-CRISPR protein. Nature. 546(7658):436-439. DOI: 10.1038/nature22377.

Facklam R. (2002). What happened to the streptococci: overview of taxonomic and nomenclature changes. Clin Microbiol Rev. 15(4):613-630. DOI: 10.1128/CMR.15.4.613-630.2002.

Fineran PC, Gerritzen MJ, Suárez-Diez M, Künne T, Boekhorst J, van Hijum SA, Staals RH, Brouns SJ. (2014). Degenerate target sites mediate rapid primed

CRISPR adaptation. Proc Natl Acad Sci U S A. 111(16):E1629-38. DOI: 10.1073/pnas.1400071111.

Fischetti VA. (2007). In vivo acquisition of prophage in Streptococcus pyogenes. Trends Microbiol. 15(7):297-300. DOI: 10.1016/j.tim.2007.05.003.

Fontaine L, Wahl A, Fléchard M, Mignolet J, Hols P. (2015). Regulation of competence for natural transformation in streptococci. Infect Genet Evol. 33:343-360. DOI: 10.1016/j.meegid.2014.09.010.

Garneau JE, Dupuis MÈ, Villion M, Romero DA, Barrangou R, Boyaval P, Fremaux C, Horvath P, Magadán AH, Moineau S. (2010). The CRISPR/Cas bacterial immune system cleaves bacteriophage and plasmid DNA. Nature. 468(7320):67-71. DOI: 10.1038/nature09523.

Gasiunas G, Barrangou R, Horvath P, Siksnys V. (2012). Cas9-crRNA ribonucleoprotein complex mediates specific DNA cleavage for adaptive immunity in bacteria. Proc Natl Acad Sci U S A. 109(39):E2579-86. DOI: 10.1073/pnas.1208507109.

Gomaa AA, Klumpe HE, Luo ML, Selle K, Barrangou R, Beisel CL. (2014). Programmable removal of bacterial strains by use of genome-targeting CRISPR-Cas systems. MBio. 5(1):e00928-13. DOI: 10.1128/mBio.00928-13.

Grissa I, Vergnaud G, Pourcel C. (2007a). The CRISPRdb database and tools to display CRISPRs and to generate dictionaries of spacers and repeats. BMC Bioinformatics. 8:172. DOI: 10.1186/1471-2105-8-172.

Grissa I, Vergnaud G, Pourcel C. (2007b). CRISPRFinder: a web tool to identify clustered regularly interspaced short palindromic repeats. Nucleic Acids Res. 35(Web Server issue):W52-7. DOI: 10.1093/nar/gkm360.

Gunderson FF, Cianciotto NP. (2013). The CRISPR-associated gene cas2 of

Legionella pneumophila is required for intracellular infection of amoebae. MBio. 4(2):e00074-13. DOI: 10.1128/mBio.00074-13.

Hamada S, Kawabata S, Nakagawa I. (2015). Molecular and genomic characterization of pathogenic traits of group A *Streptococcus pyogenes*. Proc Jpn Acad Ser B Phys Biol Sci. 91(10):539-559. DOI: 10.2183/pjab.91.539.

Heler R, Marraffini LA, Bikard D. (2014). Adapting to new threats: the generation of memory by CRISPR-Cas immune systems. Mol Microbiol. 93(1):1-9. DOI: 10.1111/mmi.12640.

Heler R, Samai P, Modell JW, Weiner C, Goldberg GW, Bikard D, Marraffini LA. (2015). Cas9 specifies functional viral targets during CRISPR-Cas adaptation. Nature. 519(7542):199-202. DOI: 10.1038/nature14245.

Hols P, Hancy F, Fontaine L, Grossiord B, Prozzi D, Leblond-Bourget N, Decaris B, Bolotin A, Delorme C, Dusko Ehrlich S, et al. (2005). New insights in the molecular biology and physiology of Streptococcus thermophilus revealed by comparative genomics. FEMS Microbiol Rev. 29(3):435-463. DOI: 10.1016/j.fmrre.2005.04.008.

Hossain M, Egan SA, Coffey T, Ward PN, Wilson R, Leigh JA, Emes RD. (2015). Virulence related sequences; insights provided by comparative genomics of *Streptococcus uberis* of differing virulence. BMC Genomics. 16:334. DOI: 10.1186/s12864-015-1512-6.

Hsu PD, Lander ES, Zhang F. (2014). Development and applications of CRISPR-Cas9 for genome engineering. Cell. 157(6):1262-1278. DOI: 10.1016/j.cell.2014.05.010.

Huo Y, Nam KH, Ding F, Lee H, Wu L, Xiao Y, Farchione MD Jr, Zhou S, Rajashankar K, Kurinov I, et al. (2014). Structures of CRISPR Cas3 offer mechanistic insights into Cascade-activated DNA unwinding and degradation. Nat Struct Mol Biol. 21(9):771-777. DOI: 10.1038/nsmb.2875.

Hynes AP, Rousseau GM, Lemay ML, Horvath P, Romero DA, Fremaux C, Moineau S. (2017). An anti-CRISPR from a virulent streptococcal phage inhibits Streptococcus pyogenes Cas9. Nat Microbiol. 2(10):1374-1380. DOI: 10.1038/s41564-017-0004-7.

Ivančić-Baće I, Cass SD, Wearne SJ, Bolt EL. (2015). Different genome stability proteins underpin primed and naïve adaptation in *E. coli* CRISPR-Cas immunity. Nucleic Acids Res. 43(22):10821-10830. DOI: 10.1093/nar/gkv1213.

Jackson RN, Wiedenheft B. (2015). A conserved structural chassis for mounting versatile CRISPR RNA-Guided immune responses. Mol Cell. 58(5):722-728. DOI: 10.1016/j.molcel.2015.05.023.

Jackson SA, McKenzie RE, Fagerlund RD, Kieper SN, Fineran PC, Brouns SJ. (2017). CRISPR-Cas: Adapting to change. Science. 356(6333). pii: eaal5056. DOI: 10.1126/science.aal5056.

Jans C, Follador R, Hochstrasser M, Lacroix C, Meile L, Stevens MJ. (2013). Comparative genome analysis of Streptococcus infantarius subsp. infantarius CJ18, an African fermented camel milk isolate with adaptations to dairy environment. BMC Genomics. 14:200. DOI: 10.1186/1471-2164-14-200.

Jinek M, Chylinski K, Fonfara I, Hauer M, Doudna JA, Charpentier E. (2012). A programmable dual-RNA-guided DNA endonuclease in adaptive bacterial immunity. Science. 337(6096):816-821. DOI: 10.1126/science.1225829.

Jones CL, Sampson TR, Nakaya HI, Pulendran B, Weiss DS. (2012). Repression of bacterial lipoprotein production by Francisella novicida facilitates evasion of innate immune recognition. Cell Microbiol. 14(10):1531-1543. DOI:

10.1111/j.1462-5822.2012.01816.x.

Jore MM, Lundgren M, van Duijn E, Bultema JB, Westra ER, Waghmare SP, Wiedenheft B, Pul U, Wurm R, Wagner R, et al. (2011). Structural basis for CRISPR RNA-guided DNA recognition by Cascade. Nat Struct Mol Biol. 18(5):529-536. DOI: 10.1038/nsmb.2019.

Jung CJ, Yeh CY, Shun CT, Hsu RB, Cheng HW, Lin CS, Chia JS. (2012). Platelets enhance biofilm formation and resistance of endocarditis-inducing streptococci on the injured heart valve. J Infect Dis. 205(7):1066-1075. DOI: 10.1093/infdis/jis021.

Kilian M, Riley DR, Jensen A, Brüggemann H, Tettelin H. (2014). Parallel evolution of Streptococcus pneumoniae and Streptococcus mitis to pathogenic and mutualistic lifestyles. MBio. 5(4):e01490-14. DOI: 10.1128/mBio.01490-14.

Koonin EV, Makarova KS, Zhang F. (2017). Diversity, classification and evolution of CRISPR-Cas systems. Curr Opin Microbiol. 37:67-78. DOI: 10.1016/j.mib.2017.05.008.

Leenay RT, Beisel CL. (2017). Deciphering, Communicating, and Engineering the CRISPR PAM. J Mol Biol. 2017 Jan 20;429(2):177-191. DOI: 10.1016/j.jmb.2016.11.024.

Lier C, Baticle E, Horvath P, Haguenoer E, Valentin AS, Glaser P, Mereghetti L, Lanotte P. (2015). Analysis of the type II-A CRISPR-Cas system of *Streptococcus agalactiae* reveals distinctive features according to genetic lineages. Front Genet. 6:214. DOI: 10.3389/fgene.2015.00214.

Liu X, Gallay C, Kjos M, Domenech A, Slager J, van Kessel SP, Knoops K, Sorg RA, Zhang JR, Veening JW. (2017). High-throughput CRISPRi phenotyping identifies new essential genes in Streptococcus pneumoniae. Mol Syst Biol.

13(5):931. DOI: 10.15252/msb.20167449.

Lopez-Sanchez MJ, Sauvage E, Da Cunha V, Clermont D, Ratsima Hariniaina E, Gonzalez-Zorn B, Poyart C, Rosinski-Chupin I, Glaser P. (2012). The highly dynamic CRISPR1 system of *Streptococcus agalactiae* controls the diversity of its mobilome. Mol Microbiol. 85(6):1057-1071. DOI: 10.1111/j.1365-2958.2012.08172.x.

Louthrenoo W, Kasitanon N, Wangkaew S, Hongsongkiat S, Sukitawut W, Wichainun R. (2014). Streptococcus agalactiae: an emerging cause of septic arthritis. J Clin Rheumatol. 20(2):74-78. DOI: 10.1097/RHU.0000000000000071.

Louwen R, Staals RH, Endtz HP, van Baarlen P, van der Oost J. (2014). The role of CRISPR-Cas systems in virulence of pathogenic bacteria. Microbiol Mol Biol Rev. 78(1):74-88. DOI: 10.1128/MMBR.00039-13.

Lynch JP, Zhanel GG. (2009). Streptococcus pneumoniae: epidemiology, risk factors, and strategies for prevention. Semin Respir Crit Care Med. 30(2):189-209. DOI: 10.1055/s-0029-1202938.

Maeda Y, Goldsmith CE, Coulter WA, Mason C, Dooley JS, Lowery CJ, Snelling WJ, Moore JE. (2011). Prevalence of clustered regulatory interspaced short palindromic repeat (CRISPR)-like sequences in mitis-group streptococci. Br J Biomed Sci. 68(2):65-8. PMID: 21706916

Makarova KS, Haft DH, Barrangou R, Brouns SJ, Charpentier E, Horvath P, Moineau S, Mojica FJ, Wolf YI, Yakunin AF, et al. (2011). Evolution and classification of the CRISPR-Cas systems. Nat Rev Microbiol. 9(6):467-477. DOI: 10.1038/nrmicro2577.

Makarova KS, Wolf YI, Alkhnbashi OS, Costa F, Shah SA, Saunders SJ, Barrangou R, Brouns SJ, Charpentier E, Haft DH, et al. (2015). An updated evolutionary classification of CRISPR-Cas systems. Nat Rev Microbiol. 13(11):722-736. DOI: 10.1038/nrmicro3569.

Makarova KS, Wolf YI, Koonin EV. (2013). The basic building blocks and evolution of CRISPR-CAS systems. Biochem Soc Trans. 41(6):1392-1400. DOI: 10.1042/BST20130038.

Marraffini LA, Sontheimer EJ. (2010). CRISPR interference: RNA-directed adaptive immunity in bacteria and archaea. Nat Rev Genet. 11(3):181-190. DOI: 10.1038/nrg2749.

Marraffini LA. (2015). CRISPR-Cas immunity in prokaryotes. Nature. 526(7571):55-61. DOI: 10.1038/nature15386.

Maxwell KL. (2017). The Anti-CRISPR Story: A Battle for Survival. Mol Cell. 68(1):8-14. DOI: 10.1016/j.molcel.2017.09.002.

Nelson CE, Hakim CH, Ousterout DG, Thakore PI, Moreb EA, Castellanos Rivera RM, Madhavan S, Pan X, Ran FA, Yan WX, et al. (2016). In vivo genome editing improves muscle function in a mouse model of Duchenne muscular dystrophy. Science. 351(6271):403-407. DOI: 10.1126/science.aad5143.

Nho SW, Hikima J, Park SB, Jang HB, Cha IS, Yasuike M, Nakamura Y, Fujiwara A, Sano M, Kanai K, et al. (2013). Comparative genomic characterization of three *Streptococcus parauberis* strains in fish pathogen, as assessed by wide-genome analyses. PLoS One. 8(11):e80395. DOI: 10.1371/journal.pone.0080395.

Nuñez JK, Bai L, Harrington LB, Hinder TL, Doudna JA. (2016). CRISPR immunological memory requires a host factor for specificity. Mol Cell. 62(6):824-833. DOI: 10.1016/j.molcel.2016.04.027.

Nuñez JK, Harrington LB, Kranzusch PJ, Engelman AN, Doudna JA. (2015). Foreign DNA capture during CRISPR-Cas adaptive immunity. Nature.

527(7579):535-538. DOI: 10.1038/nature15760.

Okamoto M, Naito M, Miyanohara M, Imai S, Nomura Y, Saito W, Momoi Y, Takada K, Miyabe-Nishiwaki T, Tomonaga M, et al. (2016). Complete genome sequence of Streptococcus troglodytae TKU31 isolated from the oral cavity of a chimpanzee (Pan troglodytes). Microbiol Immunol. 60(12):811-816. DOI: 10.1111/1348-0421.12453.

Olson AB, Kent H, Sibley CD, Grinwis ME, Mabon P, Ouellette C, Tyson S, Graham M, Tyler SD, Van Domselaar G, et al. (2013). Phylogenetic relationship and virulence inference of *Streptococcus anginosus* Group: curated annotation and whole-genome comparative analysis support distinct species designation. BMC Genomics. 14:895. DOI: 10.1186/1471-2164-14-895.

Pennisi E. (2013). The CRISPR craze. Science. 341(6148):833-836. DOI: 10.1126/science.341.6148.833.

Qi L, Haurwitz RE, Shao W, Doudna JA, Arkin AP. (2012). RNA processing enables predictable programming of gene expression. Nat Biotechnol. 30(10):1002-1006. DOI: 10.1038/nbt.2355.

Qi LS, Larson MH, Gilbert LA, Doudna JA, Weissman JS, Arkin AP, Lim WA. (2013). Repurposing CRISPR as an RNA-guided platform for sequence-specific control of gene expression. Cell. 152(5):1173-1183. DOI: 10.1016/j.cell.2013.02.022.

Rauch BJ, Silvis MR, Hultquist JF, Waters CS, McGregor MJ, Krogan NJ, Bondy-Denomy J. (2017). Inhibition of CRISPR-Cas9 with bacteriophage proteins. Cell. 168(1-2):150-158.e10. DOI: 10.1016/j.cell.2016.12.009.

Salsman J, Dellaire G. (2017). Precision genome editing in the CRISPR era. Biochem Cell Biol. 95(2):187-201. DOI: 10.1139/bcb-2016-0137.

Samai P, Pyenson N, Jiang W, Goldberg GW, Hatoum-Aslan A, Marraffini LA. (2015). Co-transcriptional DNA and RNA cleavage during Type III CRISPR-Cas immunity. Cell. 161(5):1164-1174. DOI: 10.1016/j.cell.2015.04.027.

Sampson TR, Weiss DS. (2014). CRISPR-Cas systems: new players in gene regulation and bacterial physiology. Front Cell Infect Microbiol. 4:37. DOI: 10.3389/fcimb.2014.00037.

Semenova E, Jore MM, Datsenko KA, Semenova A, Westra ER, Wanner B, van der Oost J, Brouns SJ, Severinov K. (2011). Interference by clustered regularly interspaced short palindromic repeat (CRISPR) RNA is governed by a seed sequence. Proc Natl Acad Sci U S A. 108(25):10098-10103. DOI: 10.1073/pnas.1104144108.

Serbanescu MA, Cordova M, Krastel K, Flick R, Beloglazova N, Latos A, Yakunin AF, Senadheera DB, Cvitkovitch DG. Role of the Streptococcus mutans CRISPR-Cas systems in immunity and cell physiology. J Bacteriol. 197(4):749-761. DOI: 10.1128/JB.02333-14.

Shalem O, Sanjana NE, Hartenian E, Shi X, Scott DA, Mikkelson T, Heckl D, Ebert BL, Root DE, Doench JG, et al. (2014). Genome-scale CRISPR-Cas9 knockout screening in human cells. Science. 343(6166):84-87. DOI: 10.1126/science.1247005.

Shimomura Y, Okumura K, Murayama SY, Yagi J, Ubukata K, Kirikae T, Miyoshi-Akiyama T. (2011). Complete genome sequencing and analysis of a Lancefield group G *Streptococcus dysgalactiae* subsp. equisimilis strain causing streptococcal toxic shock syndrome (STSS). BMC Genomics. 12:17. DOI: 10.1186/1471-2164-12-17.

Shmakov S, Savitskaya E, Semenova E, Logacheva MD, Datsenko KA,

Severinov K. (2014). Pervasive generation of oppositely oriented spacers during CRISPR adaptation. Nucleic Acids Res. 42(9):5907-5916. DOI: 10.1093/nar/gku226.

Sternberg SH, Redding S, Jinek M, Greene EC, Doudna JA. (2014). DNA interrogation by the CRISPR RNA-guided endonuclease Cas9. Nature. 507(7490):62-67. DOI: 10.1038/nature13011.

Tamulaitis G, Kazlauskiene M, Manakova E, Venclovas Č, Nwokeoji AO, Dickman MJ, Horvath P, Siksnys V. (2014). Programmable RNA shredding by the type III-A CRISPR-Cas system of Streptococcus thermophilus. Mol Cell. 56(4):506-517. DOI: 10.1016/j.molcel.2014.09.027.

Tamulaitis G, Venclovas Č, Siksnys V. (2017). Type III CRISPR-Cas immunity: major differences brushed aside. Trends Microbiol. 25(1):49-61. DOI: 10.1016/j.tim.2016.09.012.

Tanzer JM, Livingston J, Thompson AM. (2001). The microbiology of primary dental caries in humans. J Dent Educ. 65(10):1028-1037. PMID: 11699974

Tohya M, Sekizaki T, Miyoshi-Akiyama T. (2018). Complete genome sequence of *Streptococcus ruminantium* sp. nov. GUT-187T (=DSM 104980T =JCM 31869T), the type strain of *S. ruminantium*, and comparison with genome sequences of *Streptococcus suis* strains. Genome Biol Evol. 10(4):1180-1184. DOI: 10.1093/gbe/evy078.

Vestergaard G, Garrett RA, Shah SA. (2014). CRISPR adaptive immune systems of Archaea. RNA Biol. 11(2):156-167. DOI: 10.4161/rna.27990.

Waller AS, Robinson C. (2013). *Streptococcus zooepidemicus* and *Streptococcus equi* evolution: the role of CRISPRs. Biochem Soc Trans. 2013 Dec;41(6):1437-1443. DOI: 10.1042/BST20130165.

Wang G, Zhao N, Berkhout B, Das AT. (2018). CRISPR-Cas based antiviral strategies against HIV-1. Virus Res. 244:321-332. DOI: 10.1016/j.virusres.2017.07.020.

Wang HX, Li M, Lee CM, Chakraborty S, Kim HW, Bao G, Leong KW. (2017). CRISPR/Cas9-Based genome editing for disease modeling and therapy: challenges and opportunities for nonviral delivery. Chem Rev. 117(15):9874-9906. DOI: 10.1021/acs.chemrev.6b00799.

Wang J, Li J, Zhao H, Sheng G, Wang M, Yin M, Wang Y. (2015). Structural and sechanistic basis of PAM-dependent spacer acquisition in CRISPR-Cas systems. Cell. 163(4):840-853. DOI: 10.1016/j.cell.2015.10.008.

Wei Y, Chesne MT, Terns RM, Terns MP. (2015). Sequences spanning the leader-repeat junction mediate CRISPR adaptation to phage in Streptococcus thermophilus. Nucleic Acids Res. 43(3):1749-1758. DOI: 10.1093/nar/gku1407

Wei Y, Chesne MT, Terns RM, Terns MP. (2015). Sequences spanning the leader-repeat junction mediate CRISPR adaptation to phage in *Streptococcus therm*ophilus. Nucleic Acids Res. 43(3):1749-1758. DOI:10.1093/nar/gku1407.

Willenborg J, Goethe R. (2016). Metabolic traits of pathogenic streptococci. FEBS Lett. 590(21):3905-3919. DOI: 10.1002/1873-3468.12317.

Wright AV, Doudna JA. (2016). Protecting genome integrity during CRISPR immune adaptation. Nat Struct Mol Biol. 23(10):876-883. DOI: 10.1038/nsmb.3289.

Yang H, Wang H, Shivalila CS, Cheng AW, Shi L, Jaenisch R. (2013). One-step generation of mice carrying reporter and conditional alleles by CRISPR/Cas-mediated genome engineering. Cell. 154(6):1370-1379. DOI: 10.1016/j.cell.2013.08.022.

Yao X, Li M, Wang J, Wang C, Hu D, Zheng F, Pan X, Tan Y, Zhao Y, Hu L, Tang J, Hu F. (2015). Isolation and characterization of a native avirulent strain of *Streptococcus suis* serotype 2: a perspective for vaccine development. Sci Rep. 5:9835. DOI: 10.1038/srep09835.

Yombi Jc, Belkhir L, Jonckheere S, Wilmes D, Cornu O, Vandercam B, Rodriguez-Villalobos H. (2012). Streptococcus gordonii septic arthritis: two cases and review of literature. BMC Infect Dis. 12:215. DOI: 10.1186/1471-2334-12-215.

Zegans ME, Wagner JC, Cady KC, Murphy DM, Hammond JH, O'Toole GA. (2009). Interaction between bacteriophage DMS3 and host CRISPR region inhibits group behaviors of Pseudomonas aeruginosa. J Bacteriol. 191(1):210-219. DOI: 10.1128/JB.00797-08.

Zhang BC, Zhang J, Sun L. (2014). *Streptococcus iniae* SF1: complete genome sequence, proteomic profile, and immunoprotective antigens. PLoS One. 9(3):e91324. DOI: 10.1371/journal.pone.0091324.

Zhou L, Peng R, Zhang R, Li J. (2018). The applications of CRISPR/Cas system in molecular detection. J Cell Mol Med. 22(12):5807-5815. DOI: 10.1111/jcmm.13925.

https://doi.org/10.21775/9781912530229.02

Genetic Regulation of Streptococci by Small RNAs

Ye Tu[1], Xiaoyue Jia[1], Ran Yang[2], Xian Peng[1,3], Xuedong Zhou[1,3], Xin Xu [1,3]*

1. State Key Laboratory of Oral Diseases & National Clinical Research Center for Oral Diseases & Department of Cariology and Endodontics, West China Hospital of Stomatology, Sichuan University, Chengdu, China
2. State Key Laboratory of Oral Diseases & National Clinical Research Center for Oral Diseases & Department of Pediatric Dentistry, West China Hospital of Stomatology, Sichuan University, Chengdu, China
3. Clinical Research Center for Oral Diseases, Sichuan Province

Corresponding author: Xin Xu, No. 14 Section 3, Renmin South Road, West China Hospital of Stomatology, Sichuan University, Chengdu, China 610041. Email: xin.xu@scu.edu.cn

Abstract

Streptococcal species constitute a large group of commensal and pathogenic Gram-positive bacteria that exist in a wide variety of habitats. The family of small RNAs is typically ranged in size from 50 to 300 nucleotides, and acts as regulators in bacteria. The last decade has witnessed the increasing findings of small RNAs (sRNAs), which play important regulatory roles in the variety of biological processes in streptococci. In this review, we summarized the recent achievements in the identification of streptococcal sRNAs, mainly in *Streptococcus pyogenes* and *Streptococcus pneumoniae*. In addition, we particularly focused on the functions that sRNAs exert in the regulatory networks of both phenotypical traits and pathogenicity. The fact that sRNAs act as a critical fine-tuning regulator of streptococci may not only reveal in-depth mechanisms of bacterial post-transcriptional regulations in response to environmental perturbance, but also provide promising approaches to the better management of streptococcal infections.

Keywords

Streptococci, small RNA, gene expression, pathogenicity, virulence

Highlights of Contents

1) sRNAs are recognized as an important regulatory component in both physiological phenotype and pathogenicity of streptococci.
2) Apart from best studied *S. pyogenes* and *S. pneumoniae,* sRNAs have also been detected in other streptococcal species including *S. suis*, *S. mutans*, *S. sanguinis, S. thermophiles*, *S. mitis* and *S. orais*.
3) sRNAs are involved in the physiological phenotypes of streptococci, and mostly in stress response and drug resistance.
4) sRNAs regulate bacterial invasion and cytotoxin secretion of streptococci.

1. Introduction

Streptococcal species are Gram-positive bacteria that possess the ability to colonize and invade into human and animals, whose interactions with host organisms can vary from commensal to life-threatening. The bacterial pathogen *Streptococcus pyogenes*, also designated as group A streptococcus (GAS), is the etiological agent of several human diseases, including pharyngitis, impetigo, acute rheumatic fever, streptococcal toxic-shock-like syndrome, and necrotizing fasciitis (Cunningham, 2000), leading to a substantial mortality and morbidity (Carapetis et al., 2005). *Streptococcus agalactiae*, commonly referred to as group B streptococcus (GBS) (Lancefield, 1933), colonizes the vaginal and gastrointestinal tracts of healthy adults asymptomatically, and it can cause fatal invasive disease such as neonatal meningitis in susceptible host (Edwards et al., 1985; Phares et al., 2008). *Streptococcus pneumoniae* is another human pathogen responsible for many human diseases, including pneumonia bacteremia and endocarditis in both children and adults (Picazo, 2009). Besides, there are large group of commensal streptococcal species including *Streptococcus mutans, Streptococcus thermophilus*, *Streptococcus sanguinis* and so on (Zorgani et al., 2016).

The polymicrobial nature of streptococcal infections in the most cases necessitates biofilm, a highly-structured community of microbial cells and extracellular matrix adhering to abiotic or biological surface (Costerton et al., 1999 Galante et al., 2015; Krzysciak et al., 2014; Rosini and Margarit, 2015) and inter-species/inter-kingdom interactions, which are mediated by complex signaling such as Quorum Sensing systems (QS) in GAS, *S. pneumoniae* as well as *S. mutans* (Bassler, 1999; Galante et al., 2015; Jimenez and Federle, 2014; Shanker and Federle, 2017). In addition, the bacterial two-component signal transduction systems (TCSTSs) including the CovRS/CsrRS system (Dalton and Scott, 2004; Heath et al., 1999; Kreikemeyer et al., 2003; Lamy et al., 2004; Levin and Wessels 1998; Pan et al., 2009; Park et al., 2012; Roberts and Scott, 2007; Trevino et al., 2009), the CiaHR system (Guenzi et al., 1994; Halfmann et al., 2007; Levesque et al., 2007; Liu and Burne, 2009; Patenge et al., 2013; Qi et al., 2004; Quach et al., 2009), the Ihk/Irr system (Han et al., 2012; Hertzen et al., 2012; Shelburne et al., 2005; Voyich et al., 2003), and the VicRK system (Li et al., 2010; Liu et al., 2006; Winkler and Hoch, 2008) have been identified to contribute to environmental fitness as well as host colonization and invasion of the streptococci Besides, eukaryotic-type serine/threonine kinases (Echenique et al., 2004; Hanks et al., 1988; Jin and Pancholi, 2006; Novakova et al., 2005; Rajagopal et al., 2003) is known to being related with biofilm formation. Surface proteins such as protein F(PrtF/Sfbl), Fba and SfbX (Roberts and Scott, 2007) have been identified as the major determinant of microbial binding to the mucosal receptor fibronection. The antiphagocytic M protein (Lancefield, 1962; Swanson et al., 1969), the hyaluronic acid capsule (Wessels et al., 1991), C5a peptidase (ScpA) (Ji et al., 1997; Lynskey et al., 2017) and streptococcal inhibitor of complement (Agarwal et al., 2015; Blom et al., 2014) function to help the streptococci escaping the host immunity (Roberts and Scott, 2007). Soluble cytolysins or secreted virulence factors, such as streptolysin S (SLS) (Datta et al., 2005; Nizet, 2002), streptococcal pyrogenic exotoxin B (SpeB) (Biswas et al., 2001; Chen et al., 2003; Chiang-Ni and Wu, 2008; Tsai et al., 1999), and streptokinase (Christensen, 1945 Steiner and Malke, 2002; Wang et al., 1998) are critical for the adherence to and internalization into host epithelial cells. Stand-alone transcriptional regulators, such as Mga (Caparon and Scott, 1987; Hemsley et al., 2003; Hondorp and McIver, 2007; Kreikemeyer et al., 2003; Ribardo and McIver, 2006; Solano-Collado et al., 2012; Vahling and McIver, 2006), Lux/AI-2 (Blehert et al., 2003; Chao and Vogel, 2010; Federle and Bassler, 2003; Joyce et al., 2004; Lyon et al., 2001; Marouni and Sela, 2003; Petersen et al., 2006; Schauder et al., 2001; Siller et al., 2008; Vendeville et al., 2005; Vidal et al., 2011; Wang et al., 2011; Yoshida et al., 2005), metabolic regulators like CcpA (Almengor et al., 2007; Giammarinaro and Paton, 2002; Iyer et al., 2005; Saier et al., 1996; Shelburne et al., 2008; Willenborg et al., 2011), CodY (Guedon et al., 2005; Hendriksen et al., 2008; Kreth et al., 2011; Malke and Ferretti, 2007; Malke et al., 2006; Sonenshein, 2005), and CRISPR/Cas system(Barrangou et al., 2007; Deveau et al., 2008; Deveau et al., 2010; Ishino et al., 1987; Marraffini and Sontheimer, 2010) has also been identified as strategy for the pathogen to modulate its fitness and virulence in response to the microenvironment perturbations, contributing to the pathogenicity and ultimate clinical outcome of streptococcal infections.

Recently, sRNAs, whose complex regulatory roles in bacterial virulence have been reported in multiple pathogens, are recognized as an important regulatory component of streptococci (Quereda and Cossart, 2017; Waters and Storz, 2009) Despite the limited number of sRNAs discovered in streptococci compare to other bacteria to date, their critical regulatory roles streptococcal species have drawn increasing attentions of both scientists and clinical professionals.

The family of small RNAs (sRNAs) is typically composed of 50 to 300 nucleotide (nt) long transcripts in bacteria (Le Rhun et al., 2016). The first sRNA, 6S RNA, was identified in 1967 in the fractionation of crude preparations of *E.coli* tRNA by disc electrophoresis on polyacrylamide gel (Hindley, 1967). In 1984, the first chromosomally encoded sRNAs, MicF RNA was reported, which inhibits translation of the OmpF mRNA of *E.coli* (Mizuno et al., 1984). Accredited to the technological advances, sRNAs have been so far found in almost all the bacterial species such as *Escherichia coli* (Gottesman, 2004), Vibrio cholera (Bardill and Hammer, 2012), *Listeria monocytogenes* (Mellin and Cossart, 2012), *Pseudomonas aeruginosa* (Sonnleitner et al., 2012), *Staphylococcus aureus* (Romilly et al., 2012), *Clostridium perfringens* (Mraheil et al., 2010). sRNAs regulate gene expression to allow bacteria responding more quickly to environmental cues, and consequently contribute to the fitness and microbial invasion to the host (Quereda and Cossart, 2017; Sharma and Heidrich, 2012). Both genome-wide computational and experimental approaches have been applied to predict or discover sRNAs (Le Rhun and Charpentier, 2012), and multiple regulatory mechanisms of sRNAs have been proposed (Figure 1). (Waters and Storz, 2009).

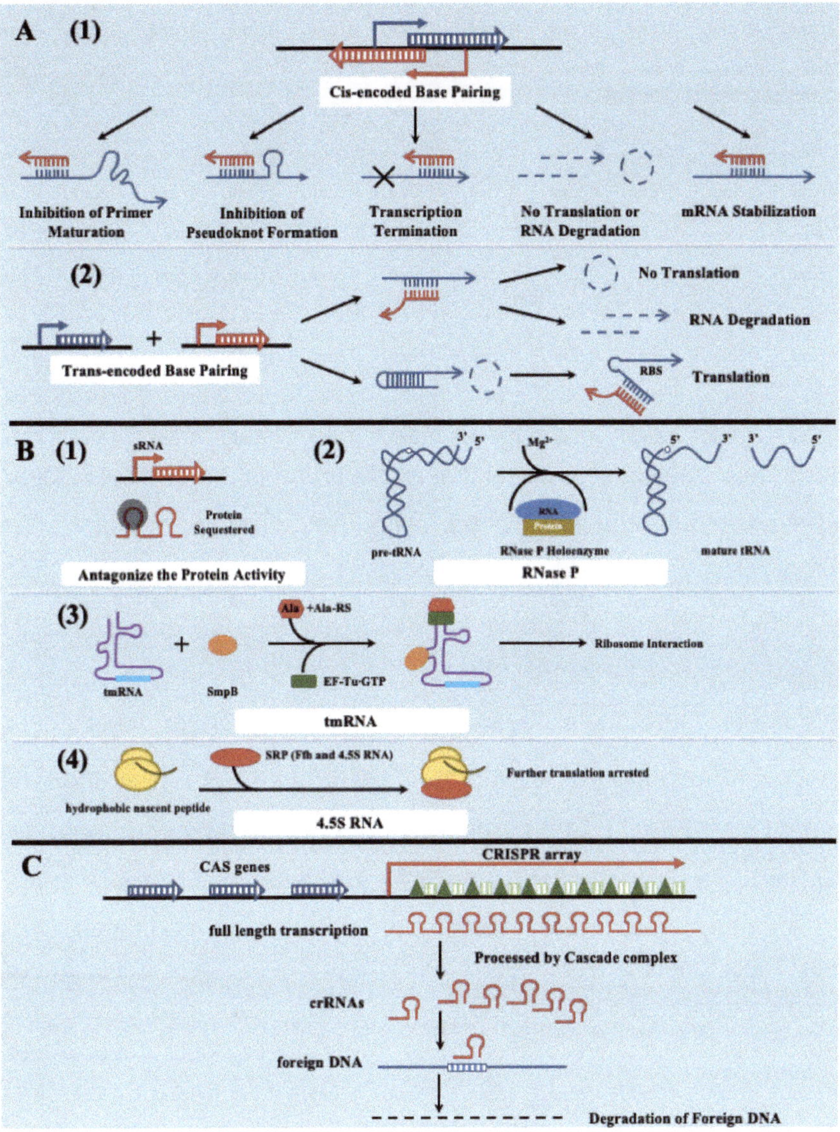

Figure 1. Regulatory Mechanism Exerted by sRNAs. A. Base-pairing Regulatory sRNAs. (1) cis-encoded base pairing sRNAs. cis-encoded base pairing sRNAs are encoded in cis on the same DNA locus and share extended regions of complete complementarity with their target mRNA, which result to the inhibition of primer maturation, inhibition of RNA pseudoknot formation, transcriptional attenuation, inhibition of translation, and promotion of RNA degradation or cleavage. (2) tran-encoded base pairing sRNAs. trans-encoded sRNAs share only limited complementarity with their target mRNAs. These sRNAs have been confirmed regulating the translation and/or stability of

target mRNAs. B. Protein-binding sRNAs. (1) sRNAs that antagonize the protein activity. Protein-binding sRNAs antagonize the activities of the cognate proteins by mimicking the structures of other nucleic acid. (2) RNase P. (3) tmRNA. (4) 4.5S RNA. C. CRISPR RNAs (crRNAs). crRNAs are a unique class of regulatory RNAs that provide resistance to bacteriophage and prevent plasmid conjugation.

There are several kinds of sRNAs. Firstly, cis-acting elements located on untranslated regions (UTRs) of a translated mRNA can control the expression of the enclosed gene(s) via modulation of the secondary structure or stability(Garst et al., 2011; Johansson et al., 2002). Another class of sRNAs is short noncoding RNA(ncRNA) which belongs to antisense RNAs (asRNAs) (Wahlestedt, 2013), and its regulatory activity occurs through base-pairing to the mRNA encoded by the opposite DNA strand molecule (Thomason and Storz, 2010). Beside, RNAs that target other RNAs, foreign DNA, or proteins is defined as trans-acting sRNAs (Gottesman and Storz, 2011). Sharing limited complementarity with their target, trans-acting sRNAs can bind to multiple mRNAs. However, this type of interaction requires the RNA chaperone protein Hfq for stabilization, and Hfq is absent in the Streptococcus species (Nielsen et al., 2010). Recently, the simplified nomenclatures and the classification of bacterial sRNAs were proposed and reemphasized (Zorgani et al., 2016), dividing sRNAs into two major categories. The first category is caRNAs (cis-acting RNAs) that act or are located at the on the untranslated regions (UTPs) of mRNA molecules. caRNAs usually refer to UTRs, which can be divided into 5' UTRs and 3' UTRs. The second category are the traRNAs (trans-acting RNAs), comprising the asRNAs (antisense RNAs also known as cis-encoded sRNAs) and treRNAs (trans-encoded sRNAs). asRNAs can be divided into 5' asRNAs and 3' asRNAs, and treRNAs consist of csRNAs (cia-dependent small RNAs), msRNAs (microRNA-size sRNAs), crRNAs (CRISPR RNAs) as well as tracrRNAs (trans-activating CRISPR RNAs).

2. sRNAs in Streptococci

In gram-positive bacterial pathogens, almost all reported biological functions of sRNAs are part of complex regulatory processes involved in adaptation or virulence processes. Most of the characterized sRNA genes are found in the intergenic regions (IGRs) of the core genome, while a few sRNAs are transcribed from the complementary strand of genes located in pathogenicity islands (PAI) (Pichon et al., 2012). Encoded on the genome, sRNAs of streptococci are associated with loci expressing TCSTSs (e.g, GAS fasX, rivX, *S. pneumoniae* csRNA) or transcriptional regulators (e.g, GAS rivX) (Romby and Charpentier, 2010). A large proportion of sRNAs act at the mRNA level, whereas a few sRNAs, such as Pel of GAS, possess the ability in encoding proteins with virulent functions (Biswas et al., 2001). The sRNAs that have been investigated to date are mainly expressed in a growth phase-dependent manner, and some of them have been discovered to be controlled by a few transcriptional regulators or TCSTSs (Romby and Charpentier, 2010).

2.1 sRNAs in S. pyogenes

GAS is the best studied streptococcal strain with sRNA. Several sRNAs including FasX, Pel, RivX, 4.5S RNA, tracrRNA and crRNA have been identified to regulate bacterial virulence in GAS (Deltcheva et al., 2011; Garneau et al., 2010; Kihlberg et al., 1995; Kreikemeyer et al., 2001; Perez et al., 2009; Roberts and Scott, 2007; Trevino et al., 2010). Pel (pleioteropic effect locus) was the first sRNA discovered in GAS in 1995 (Kihlberg et al., 1995), working as a bifunctional sRNA encoding protein and meanwhile regulating virulence factor expression. The region of *pel* gene locates at the site of Tn917 insertion and surrounding sequence on the chromosome (Li et al., 1999). *pel* has been demonstrated being identical to *sagA*, and the *sag/pel* locus is the structural gene for the key virulence factor of GAS, Streptolysin S (Li et al., 1999). Being regulated in a growth-phase-dependent manner (Mangold et al., 2004), Pel has been demonstrated exerting its effects on virulence gene expression at both transcriptional level (*emm* gene encoding M protein and *sic* gene encoding a streptococcal inhibitor of the complement system and post-transcriptional level (cysteine protease SpeB) (Biswas et al., 2001; Li et al., 1999; Mangold et al., 2004; Perez et al., 2009). Multiple regulators are coordinated with the expression of the *sag/pel* operon, among which CodY, Irr, SLS and SptS seem regulating positively, while CcpA, CovR, FasX, LuxS, Mga, RopB and members of the Ralp regulator family have repressing effects on it. In addition, the cell nutritional state also regulates *sag/pel* expression (Le Rhun and Charpentier, 2012). Although the *pel* locus is conserved in GAS, the *pel* deletion phenotypes vary among isolates, suggesting that other regulatory factors are involved in the network of Pel-mediated pleiotropic phenotype on GAS virulence (Biswas et al., 2001; Le Rhun and Charpentier, 2012; Li et al., 1999; Perez et al., 2009). *fasBCAX* operon is a polycystronic message containing genes encoding two histidine protein kinases (HPKs) (FasB and FasC) and one response regulator (RR) (FasA) (Kreikemeyer et al., 2001). FasX is a small 300 nucleotide monocistronic transcript, whose encoded gene (*fasX*) has been identified downstream of *fasBCA* operon(Kreikemeyer et al., 2001). Deletion of *fasX* results in a similar phenotype of *fasBCA* or *fasA* mutants, while a complementary expression in trans from plasmid restores the wild-type fasBCA regulation pattern, further suggesting the main role of *fasX* in the *fas* regulon (Kreikemeyer et al., 2001). FasX sRNA has been found conserved in GAS (Ramirez-Pena et al., 2010 while at the same time presents differently in several clinical isolates. In an M49 serotype (CS101), luciferase promoter fusion revealed a growth phase-associated transcription of *fasX* with peak activities during the late exponential phase, depending on the RR FasA (Kreikemeyer et al., 2001). In M49 serotype (NZ131), expression of FasX was influenced by amino acid starvation (Steiner and Malke, 2001), while enhanced by the luxS/AI2 signaling system in an M19 serotype (RDN02) (Siller et al., 2008). RivX (RofA like proteinIV regulator X) is a processed sRNA originating from a *rivRX* transcript that encoding the regulatory protein RivR (Ralp4 family of regulators) (Roberts and Scott, 2007). Being conserved in all 12 sequenced strains of GAS, *rivX* is mapped immediately down-stream of the *rivR* gene and contains a small ORF (Roberts and Scott, 2007). By analyzing the *rivX*-harbouring nonsense mutation, the activity of RivX sRNA was demonstrated being regulated by *rivX* transcript itself rather than the peptide encoded within it (Roberts and Scott, 2007). RivX sRNA promotes several mRNAs that encoding the virulence factors including C5a peptidase, cysteine protease SpeB and M protein, in the company of RivR (Roberts and Scott, 2007). *rivR* gene has been found directly repressed by CovR regulator (Roberts et al., 2007), and the products of the *rivRX* locus exert positive control over the transcription of Mga regulon (Roberts and Scott, 2007), indicating that RivR and RivX may link the CovRS system to the Mga regulon. 4.5S RNA (or scRNA for small cytoplasmic RNA) acts as a regulator in virulence factor expression and secretion. 4.5S RNA together with fifty-four homologue protein (Ffh) constitute SRP (signal recognition particle) from GAS, a ribonucleoprotein complex that targets polypeptides to the secretory apparatus in a co-translational manner (Trevino et al., 2010). Although not being essential, mutants affected in 4.5S RNA gene are attenuated in bacterial virulence (Trevino et al., 2010). In the SRP from *S. pneumoniae*, 4.5S RNA was demonstrated binding to the M-domain of Ffh, however the modulating ability to bacterial virulence has not been investigated yet (Zheng et al., 2002). crRNAs and tracrRNA play important roles in the CRISPR/Cas system of GAS. The precursor crRNA (pre-crRNA) is expressed in a repeat-spacer array of CRISPR/Cas system, and will be further processed into mature crRNAs. Mature crRNAs are regarded as unique RNAs conferring immunity against invading phages and plasmids by exerting as traRNAs targeting DNA and guiding Cas protein to the invading cognate nucleic sequences for Cas-executed cleavage (Deltcheva et al., 2011; Le Rhun and Charpentier, 2012; Sapranauskas et al., 2011). Discovered via dRNA-seq analysis, tracrRNA harbors 24 nucleotide complementarity to the repeat regions of the pre-crRNA transcript. By the activities of the widely conserved endogenous RNase III and the CRISPR-associated Csn1 protein， tracrRNA acts in trans to activate the maturation of pre-crRNA into crRNAs in the CRISPR/Cas system of Type II (Deltcheva et al., 2011). The presence of tracrRNA was also revealed by northern blot profiling in *S. thermophilus* LMD-9 genome (Karvelis et al., 2013). A genome-wide study using tiling microarray predicted 14 new sRNAs of GAS, which were further confirmed by Northern blot analysis (Perez et al., 2009). Recently, by using high-throughput sRNA sequencing in GAS SF370 clinical isolate, a novel asRNA has been discovered for the first in this pathogen (Le Rhun et al., 2016).

2.2 sRNAs in *S. pneumoniae*

Other than three highly conserved sRNAs (RNase P, tmRNA and scRNA) or previously known families of caRNAs like riboswitches or leader regions (TPP riboswitch, T-box), several so-called functional sRNAs have been identified in the genomes of *S. pneumoniae* (Tsui et al., 2010; Wilton et al., 2015). csRNAs, the first sRNAs described in *S. pneumoniae*, were initially discovered by mapping the genome of *S. pneumoniae* for promoters regulated by CiaR regulator (Halfmann et al., 2007). Being transcribed from the five strongest promoters of the pneumococcal CiaR regulon, csRNA1-5 (or CcnA-E) share a high degree of sequence similarity especially in the characteristic unpaired region between the two stem-loop structures (Halfmann et al., 2007; Marx et al., 2010). The complementarity to the Shine-Dalgarno (SD) sequence and the start codon AUG within this unpaired region suggest that csRNA could control translation initiation of mRNAs (Marx et al., 2010). An additional conserved sequence CCUC(6N)CAU located in the middle of the unpaired region was identified in all crRNA, resulting into the potential to base pair with RBS (Halfmann et al., 2007). In *S. pneumoniae* strain R6, csRNA4 and csRNA5 are involved in stationary-phase autolysis (Halfmann et al., 2007). In strain D39, the only expression of csRNA1 is induced by competence stimulatory peptide (Tsui et al., 2010). An insertion/deletion mutant exerts strong cis-acting effects on the transcription of the adjacent gene, indicating that this sRNA region is co-transcribed in an operon (Tsui et al., 2010). Apart from 5 pneumococcal csRNA1-5, other 61 csRNA candidates (51 to 202 nt) were revealed from CiaR-activated promoters, being classified into 40 different types. Since CiaRH is highly conserved among streptococci, csRNAs have also been detected in other streptococcal genomes including *S. suis*, *S. mitis*, *S. orais*, *S. mutans*, *S. sanguinis* and *S. thermophiles*, further suggesting that csRNAs are part of the regulon of CiaRH in streptococci (Marx et al., 2010). Interestingly, *S. thermophilus* strain ST0 possesses a csRNA gene on a plasmid. It is the only one of the identified csRNA genes that is not located in the genome and represents a new type with only one stem-loop structure (Marx et al., 2010). Nine new sRNA were later detected in *S. pneumoniae* serotype 2 strain D39, among which Spd-sr17 and Spd-sr37 were functionally analyzed. Spd-sr17 was characterized as a 144 nt sequence that decreased in stationary phase, likely folding into a structure that contains single-stranded regions between hairpin structure, while the 80 nt Spd-sr37 showed strong expression in all growth phases, forming a base-paired structure (Tsui et al., 2010). However, deletion or overexpression of Spd-sr17 and Spd-sr37 failed to cause drastic changes in bacterial growth, specific stress response, global transcription, or virulence in this strain (Tsui et al., 2010). Kumar et al further identified a total of 50 sRNA in *S. pneumoniae* TIGR4 by using a tiling array approach. However, 36 of them had no predicted function (Kumar et al., 2010). Remarkably, no asRNAs or CRISPR RNAs have been identified so far in *S. pneumoniae* (Wilton et al., 2015).

2.3 sRNAs in Other Streptococci Species

Although sRNAs can play an important role in virulence, the current knowledge on sRNAs encoded by the GBS chromosome remains limited. Some in silico studies have been conducted on the sequenced genome of GBS NEM316. Four putative csRNAs (csRNA10, csRNA11, csRNA12, and csRNA13) have been identified by searching for a CiaR-activated promoter and the transcriptional terminator sequences within the intergenic regions (Marx et al., 2010). The expression of csRNA10, csRNA11, and csRNA12 have been confirmed in a further study, being up-regulated in acidic conditions (Rosinski-Chupin et al., 2015). In another in silico study, 197 sRNAs have been detected, among which 10 sRNAs were validated by RT-PCR and northern blot hybridization. SQ18 asRNA and SQ893 sRNA act as negative regulators to their target mRNAs *gbs0031* and *gbs1263* respectively, while SQ485 asRNA positively regulates the expression of *gbs1588/gbs1589* (Pichon et al., 2012). In *S. suis* serotype 2 isolates 05ZYH33 and 98HAH33 that caused severe human infections in China, 6 sRNA candidates related to bacterial virulence have been discovered, including cspA and rli38, which are specific to Chinese isolates(Zhang et al., 2014).

Being specific for particular species, a small number of sRNAs have been discovered in other streptococci. A novel member of the treRNA category, designated as "miRNA-size small RNAs" or "msRNAs" has been described in *S. mutans* (Lee and Hong, 2012). Before that, there was no publication regarding msRNAs in bacteria with comparable size to miRNAs, which exist in eukaryotes with 19-23 nt length transcript. A deep sequencing approach has revealed putative msRNAs by detecting reads corresponding to 15 and 26 nt (Lee and Hong, 2012). These msRNAs may result from processed RNAs and have been confirmed by northern blot as well as qRT-PCR. In a later study, three putative msRNAs targeting vicRKX have been predicted in *S. mutans* by using deep sequencing and bioinformatics analysis, and have been indicated being related to the biofilm forimg(Mao et al., 2016). In *Mycobacteria Marinum*, an msRNA that has been discovered being derived from an RNA stem-loop with the characteristics expected for a pre-microRNA(Furuse et al., 2014). Furthermore, msRNAs (20nt long) molecule named 24B_1 has been identified in *Escherichia coli* after induction of Shiga toxin-converting bacteriophage Φ24B, and it has been predicted impairing expression of the *d_ant* gene coding for an anti-repressor(Nejman-Falenczyk et al., 2015).

2.4 Highly conserved sRNAs in Streptococci

Genome-wide searches in streptococcal genome have identified several highly conserved sRNAs with housekeeping or stress-related functions, among which contain sRNA component of RNase P, tmRNA (transfer-messenger RNA), 4.5S RNA (as mentioned above), riboswitches and T-box. RNases P is the endoribonuclease that generates mature 5'-ends of tRNA by removal of the 5'-leader elements of precursor-tRNA (Frank and Pace, 1998; Silvaggi et al., 2006). A single catalytic RNA component constitutes RNase P with a protein subunit, while the RNA subunit is catalytically active in vitro in the absence of the protein subunit (Frank and Pace, 1998). tmRNA, also designated as ssrA RNA, harbors properties of both a tRNA and an mRNA (Keiler, 2008). tmRNA together with the small protein SmpB, mediates peptide tagging of the nascent protein and release of the stalled ribosome, ensuring the protein synthesis with high fidelity in trans-translation (Keiler, 2007, 2008). In most bacteria, the inactivation of tmRNA is related to a decreased ability to response to or recover from stress conditions (Brito et al., 2016). Riboswitches are structures located in 5' region of mRNAs that have the ability to regulate the transcription and/or translation of the downstream mRNA, via directly binding intracellular metabolite (Vitreschak et al., 2004). The T-box family commonly control the expression of genes involved in tRNA aminoacylation, amino acid transport and biosynthesis in gram-positive bacteria (Gutierrez-Preciado et al., 2009).

2.5 Protein co-factors in sRNA-mediated Activities

Although protein interaction with streptococcal sRNAs is rarely reported, protein co-factors are demonstrated playing roles in sRNA activities. It is widely recognized that the RNA chaperone protein Hfq is required in sRNA targeting mRNA for most Gram-negative bacteria, however *hfq* homologues have also been found in the genome of some Gram-positive bacteria (Storz et al., 2011; Vogel and Luisi, 2011). Thus, the presence and function of Hfq have been questioned in streptococci (Chao and Vogel, 2010; Storz et al., 2011). Substituting for Hfq, proteins such as RNA-binding protein with chaperone or helicase may associate with the streptococcal traRNAs. Furthermore, interactions of streptococcal sRNAs do not require protein co-factor or require another type of co-factors (Le Rhun and Charpentier, 2012). Ribonucleases (RNase) exert critical effects on the maturation of sRNAs. Seeming to be universal, endogenous RNase III has been demonstrated playing roles in tracrRNA-directed crRNA maturation(Deltcheva et al., 2011). In addition, RNase is also involved in the interaction between sRNAs and mRNAs. The 5'-3' exo-ribonucleases J1 and J2 was predicted to be the candidate RNase in ska mRNA decay process, and FasX enhanced the stability of ska mRNA and protects target mRNA from degradation via base-pairing to ska mRNA (Ramirez-Pena et al., 2010). Cas proteins of CRISPR/Cas system were discovered functioning in RNA binding, RNA nuclease and crRNA-guided DNA cleavage (Deveau et al., 2010; Makarova et al., 2011).

3. sRNAs regulate the physiological phenotypes of Streptococci

sRNAs are commonly involved in the stress response and drug resistance of streptococci. In GBS, 10 sRNAs have been identified differentially expressed in response to acidic stress (Pichon et al., 2012). An acidic condition of pH5.2 (natural pH of the genitourinary tract) induced the expression of Srn015 (csRNA10), Srn024 (csRNA11), Srn070 (csRNA12), Srn085 (csRNA13), Srn071 and Srn082, but downregulated the expression of Srn046, Srn056, Srn057 and Srn073, underlining the potential of these sRNAs controlled by CiaRH in acidic stress adaptation during bacterial colonization to the urogenital tract (Rosinski-Chupin et al., 2015). The ability to survive at low pH is intimately links to the carbohydrate metabolism and tooth surface colonization of *S. mutans*. L10-leader sRNA, which belongs to treRNA, has been identified as a pH-dependent sRNA and exhibits an elevated expression in the *S. mutans* strain possessing a stronger adherence capacity and acidogenicity, suggesting that L10-Leader may regulate gene expression of *S. mutans* related to cell adhesion and acid production (Xia et al., 2012)..

In *S. pneumoniae*, the phenotype of enhanced β-lactan resistance was demonstrated being dependent on the csRNAs (Schnorpfeil et al., 2013). One of the controlling csRNAs, spr2043, was confirmed negatively regulating the comC gene that codes the precursor of the competence stimulating pheromone (CSP), indicating that a hyperactive CiaRH system prevents competence development by csRNA-dependent post-transcriptional repression of CSP production (Schnorpfeil et al., 2013). Furthermore, tmRNA also plays roles in drug resistance of *S. pneumoniae*. Fluoroquinolones, such as levofloxacin and moxifloxacin, are currently used for the treatment of pneumococcal pneumonia. This cell-killing process requires ongoing protein synthesis and is contributed by reactive oxygen species (ROS). The tmRNA deletion mutant was found significantly more resistant to fluoroquinolones compared to its wild type strain. A decreased accumulation of intracellular ROS was found in the deletion mutant, and the tmRNA deficiency was shown to prevent the levofloxacin-induced chromosome fragmentation, mainly by inhibiting protein synthesis (Brito et al., 2016).

Some sRNAs have also been identified to regulate other physiological traits of streptococci. In *S. pneumoniae*, competence in S. pneumoniae is controlled by 5 csRNAs via post-transcriptional regulation of comC, encoding the precursor of the competence stimulating peptide, which is essential to initiate the regulatory cascade leading to competence. It has been demonstrated that combinations of three csRNAs, csRNA1,2,3, or csRNA1,2,4 were sufficient to stop competence gene expression while two csRNAs were not enough to prevent competence(Laux et al., 2015). Sip is named for surface immunogenic protein that elicits protective immunity against GBS. A translational fusion system revealed that the asRNA SQ18 influences the expression of *sip* gene by acting as a negative post-transcriptional regulator (Pichon et al., 2012). In *S. pneumoniae*, stationary-phase autolysis was affected by csRNA4 and csRNA5 (Halfmann et al., 2007). In additional, the competence-inducing (XIP) peptide is predicted controling bacteriocin production via sRNAs in *S. mutans.* It has been discovered that five XIP-responsive intergenic regions express putative sRNAs. Of thses 5 regions, SMU.153-154 region is located at 3′ of bacteriocin synthetic genes, and is believed being involved in bacteriocin induction (Wenderska et al., 2017)

4 sRNAs regulate the pathogenicity of Streptococci

4.1 sRNAs Regulate Bacterial Invasion

4.1.1 Colonization, Adherence and Internalizations

In GAS serotype M49, a significantly decreased adherence and internalization efficiency in fasX deletion mutant was observed in a comparison of HEp-2 cell infected with its wild type strain, indicating the importance of FasX in the interaction with epithelial host cells and tissue colonization (Kreikemeyer et al., 2001). Furthermore, FasX was shown to positively control the cytokine gene transcription, host cell apoptosis and cytotoxicity (Klenk et al., 2005). 4.5S RNA also exerts its effect on modulating the bacterial virulence. Using growth in human saliva as an ex vivo model of upper respiratory tract infection, a 10-fold reduction in colony-forming units were identified in 4.5S RNA mutation of GAS, consistent with the fact that 4.5S RNA contributes to GAS growth and persistence during upper respiratory tract infections (Trevino et al., 2010). In *S. pneumoniae*, by conducting a random transposon insertion, sRNA F20 was demonstrated contributing to adhesion and invasion of host nasopharyngeal (Mann et al., 2012). In addition, the srn157 and tmRNA deletion mutants of *S. pneumoniae* showed decreased adhesion/invasion to nasopharyngeal cells and endothelial cells respectively, along with compromised fitness and competitive index in the nasopharynx and lungs (Wilton et al., 2015).

Pili and capsule are recognized to play pivotal roles in adhesion of streptococci to host tissue. By base-pairing to the mRNA of the pilus biosynthesis operon, which results in the mRNA destabilization, FasX exerts negative effects on GAS pilus (Liu et al., 2012). In addition, FasX has been found inhibiting the translation of *cpa* mRNA that coding a minor pilin protein (Liu et al., 2012). Afimbrial adhesins can also be negatively modulated by FasX, with the evidence in two matrix protein-binding adhesins (fibronectin-binding protein FBP54 and fibrinogen-binding protein MRP) in GAS (Klenk et al., 2005). In *S. sanguinis*, csRNA1-1 has been found negatively regulated the *pILT*, a constituent of the type IV pilus operon, via binding directly to *pilT* mRNA(Ota et al., 2018). In addition, csRNA1-1 and csRNA1-2 were predicted negatively regulated *S. sanguinis* biofilm formation(Ota et al., 2018). sRNA rss04 of *S. suis*, an important pathogen for pigs, has been observed facilitating *S. suis* invasion of mouse brain and lung in vivo, by repressing capsular polysaccharide (Xiao et al., 2017). In *S. mutans*, the ability of exopolysaccharides synthesizing can affect the biofilm formation which involved in dental caries. It has been reported that *rnc* gene positively regulated exopolysaccharide synthesis significantly, meanwhile repress the downstream *vic* locus. By using deep sequencing and bioinformatics analysis, three msRNAs (msRNAs 1701, 3405 and 1657) that negatively correlated with *vicRKX* but positively correlated with *rnc* have been discovered. It has been indicated that *rnc* repressed *vicRKX* expression at the post-transcriptional level via msRNAs (Mao et al., 2016).

4.1.2 Host Defense Evasion

Appearing as hair-like projections on the cell surface, M protein is considered as the major virulence factor of GAS (Swanson et al., 1969) and protects the bacteria from phagocytosis via conferring resistance to complement-mediated killing by polymorphonuclear leukocytes and macrophages (Lancefield, 1962). In addition, M protein is required for attachment of GAS to keratinocytes, playing an important role in skin surface infection. In previous research, an insertion mutant in the *sagA/pel* promoter region led to a decreased transcription of M protein-coding gene *emm* in the GAS M49 serotype, indicating that the locus is needed for the production of M protein (Li et al., 1999). Enhanced *sagA/pel* expression leads to an increased expression for *emm* gene in a wild-type organism (Li et al., 1999). In a subsequent study, although not affecting the transcription of *emm* gene, insertion-deletion mutation in the *sagA/pel* gene of M6 serotype (IRS470) resulted in a truncated M protein, which influenced the anchorage of M protein into the cell wall meanwhile caused a dramatic reduction in cell-surface associated M protein (Biswas et al., 2001). The production of M protein remained the same in the *sagA/pel* transposon insertion mutation of M1 and M18 serotype compared to the wild-type strain, however, the virulence ability was found impaired in a mouse model of subcutaneous infection (Betschel et al., 1998). To sum up, these observations suggest that the Pel regulator act differently on similar genes in different strains. In addition, RivX is also involved in the production of M protein, for the transcript level of the *emm* gene increased in the GAS strain that overexpresses *rivR/X* (Roberts and Scott, 2007).

Encoded by the *scpA* gene, C5a peptidase (ScpA) is a widely conserved enzyme among Streptococcal strains. The substrate for ScpA is the human anaphylatoxin C5a, which plays a critical role in neutrophil activation and recruitment to the infection site (Ji et al., 1997). In GAS, expression of *scpA* gene was lower in the absence of *rivR/X*, while overexpression of either *rivR* or *rivX* alone increased the *scpA* expression, suggesting a positive effect exerted by RivX on the production of C5a peptide (Roberts and Scott, 2007).

The complement system is a major aim of the innate immune system that functions as the first line of defense against invading pathogens by initiating a proteolytic cascade that results in the opsonization of bacteria with C3b for phagocytosis, generation of anaphylatoxins C3a and C5a, and formation of the cytolytic membrane attack complex (MAC) or C5b-9 for direct bacterial killing (Agarwal et al., 2015). In M1 serotype d, Pel was reported to stimulate the expression of the *sic* gene that encodes the streptococcal inhibitor of complement (Mangold et al., 2004).

4.1.3 Pathogen Invasiveness

Streptokinase(SK), originally designated as streptococcal fibrinolysin (Christensen, 1945), binding plasminogen with high affinity, can hydrolytically activate plasminogen molecules to plasmin, a fibrin-degrading protease (Wang et al., 1998). Though being used in treating blood-clotting disorders, streptokinase is regarded as a virulence factor for the association with absence of fibrin in spreading streptococcal disease, contributing to the pathogen invasiveness (Malke et al., 1994). The presence of streptokinase coding gene is a stable characteristic in GAS and human isolates of group C and G streptococci (GCS and GGS) (Steiner and Malke, 2002). In GAS, the production of SKA is mostly stimulated by FasX by trans-acting to *ska* mRNA. It was observed that FasX exerted a positive influencing on SKA production in GAS M49 serotype (CS101) (Klenk et al., 2005; Kreikemeyer et al., 2001) and M1 serotype (MGAS2221) (Ramirez-Pena et al., 2010). In the strain MGAS2221, by base-pairing a short stretch of 9 nt with the very 5'end of *ska* mRNA, FasX enhanced the stability of *ska* transcript, resulting in a 10-fold increase in SKA activity compare to an isogenic FasX mutant (Ramirez-Pena et al., 2010). Further experiment demonstrated that a double-stranded structure at the 5'end of *ska* mRNA was formed by annealing of FasX to *ska* mRNA, and the structure prevented single-stranded protrusion meanwhile the mimicking of a hairpin helped to stabilize the *ska* mRNA (Ramirez-Pena et al., 2010). What's more, binding less than 30 nt upstream of the start codon AUG, which is relatively close to the RBS (ribosome binding site), the FasX positively affects the steady state level of SKA expression via protecting *ska* mRNA from degradation mediated by ribonucleases (Ramirez-Pena et al., 2010). The FasX-*ska* mRNA base-pairing also promotes a structural change within the 5'UTR. The alteration of 5'UTR structural facilitated the ribosome binding to mRNA and subsequently activated translation initiation (Ramirez-Pena et al., 2010). Combined with aforementioned negative effect on pili, FasX seems exerting an intricate effect on GAS virulence. Other than FasX, an insertion mutation to *pel* region in M49 serotype was found decreasing the expression of SKA, suggesting a positive modulatory function exerted by Pel (Li et al., 1999). Known as the strongest SK producer yet, GCS strain H46A, whose *fasCBAX* operon is preserved at high level of primary structure identity between the GAS and GCS gene, FasX regulated SKC expression via affecting *skc* transcript levels (Steiner and Malke, 2002).

4.2 sRNAs Regulate Cytotoxin Secretion

4.2.1 Secretion of Streptolysin

Streptolysin S (SLS) can be synthesized continuously by stationary-phase cell in the presence of a minimal energy source, and primarily exists in a cell-bound form linking to the streptococcal surface by lipoteichoic acid (Nizet, 2002). As mentioned above, the chromosomal locus for SLS production is mapped to a unique ORF encoding 53 amino acid that is named *sagA* for 'streptolysin-associated gene A' (Datta et al., 2005), and is known being identical with pel locus Downstream of *sagA/pel* locus lies eight additional contiguous genes (*sagB-sagI*) that are required for effective SLS production as well (Mangold et al., 2004; Nizet et al., 2000). The relationship between SLS production and the *sagA/pel* locus has been investigated in several different serotypes of GAS, and the role of this locus in modulating the SLS function is demonstrated at the RNA level. The activity of SLS has been observed totally inhibited when Tn917 insertion disrupted the transcription of *pel* region in GAS M49 serotype (Li et al., 1999). Moreover, a reduction SLO secretion was observed in a 4.5S RNA isogenic mutant strain of GAS, which is consistent with the reduced level of transcripts for the Streptolysin O (SLO) gene, compared to the parent strain MGAS2221 (Trevino et al., 2010).

4.2.2 Secretion of Cysteine Protease (SpeB)

Cysteine protease SpeB is one of the most important virulence factors that GAS harbors. It is secreted as a 42-kDa zymogen and auto-catalyzes into an active 28-kDa cysteine protease (Chen et al., 2003). Apart from promoting epithelial cell apoptosis (Tsai et al., 1999), SpeB cleaves or degrades host serum proteins such as human extracellular matrix, immunoglobulins, complement components, and even GAS surface and secreted proteins (Chiang-Ni and Wu, 2008). In an M49 serotype, the loss of *pel* region transcript resulted in the decreased transcription of *speB* gene (Li et al., 1999). In an M1 serotype, Pel exerted its negative effect on processing and maturation of SpeB at post-transcriptional level(Mangold et al., 2004). In the *pel*-deficient mutant, the transcription level of *speB* was not affected, however the onset of SpeB activity was found clearly delayed compared to the wild type strain. The zymogen of SpeB was effeciently secreted but its processing to the active mature form appeared to be delayed (Mangold et al., 2004). Besides, the transcript of *speB* gene dramatically decreased in the *rivRX* deleted mutant compared with parent strain, while the overexpression of *rivR/X*, *rivR* or *rivX* alone led to the overproduction of *speB* transcript (Roberts and Scott, 2007). Furthermore, a 4.5S RNA isogenic mutant strain of GAS also exhibited the reduced secretion of the SpeB protease (Trevino et al., 2010).

5 Concluding Remarks and Future Prospects

Genetic regulation is critical for bacteria adaptation to environmental changes, and strategies to control genetic networks in response to extracellular stimuli has been developed in bacteria. Accredited to the advancement in genome-wide computational and experimental techniques, the sRNAs-mediated regulatory pathways in streptococci have been reported recently. Studies on streptococcal sRNAs are mainly focused on GAS and *S. pneumoniae,* however only a small number of sRNAs have been predicted bioinformatically in other streptococci like GBS, *S. mutans*, *S. thermophiles* and *S.suis*. Most investigated sRNAs are involved in bacterial physiology and pathogenesis, regulating the biological phenotypes and infection process. Particularly, the integration of FasX, Pel, RivX and csRNAs into regulatory networks have been delineated. However, the understanding of sRNAs in streptococci remains limited. What's more, knowledges on the roles of sRNAs during streptococcal infection are still limited. More efforts are still needed to bridge the gap between sRNAs identification and functional validation, and ultimately translate the critical regulatory mechanisms of sRNAs to the better management of streptococcal infections.

Acknowledgements

This work was supported by the National Natural Science Foundation of China (81800989, 81670978, 81771099), a research fund from the Science & Technology Department of Sichuan Province (2018SZ0121), and a special research fund to the Clinical Research Center for Oral Diseases, Sichuan Province.

Reference

Agarwal, V., Talens, S., Grandits, A.M., and Blom, A.M. (2015). A Novel Interaction between Complement Inhibitor C4b-binding Protein and Plasminogen That Enhances Plasminogen Activation. The Journal of biological chemistry 290, 18333-18342. https://dx.doi.org/10.1074/jbc.M114.619494.

Almengor, A.C., Kinkel, T.L., Day, S.J., and McIver, K.S. (2007). The catabolite control protein CcpA binds to Pmga and influences expression of the virulence regulator Mga in the Group A streptococcus. Journal of bacteriology 189, 8405-8416. https://dx.doi.org/10.1128/jb.01038-07.

Bardill, J.P., and Hammer, B.K. (2012). Non-coding sRNAs regulate virulence in the bacterial pathogen Vibrio cholerae. RNA biology 9, 392-401. https://dx.doi.org/10.4161/rna.19975.

Barrangou, R., Fremaux, C., Deveau, H., Richards, M., Boyaval, P., Moineau, S., Romero, D.A., and Horvath, P. (2007). CRISPR provides acquired resistance against viruses in prokaryotes. Science (New York, NY) 315, 1709-1712. https://dx.doi.org/10.1126/science.1138140.

Bassler, B.L. (1999). How bacteria talk to each other: regulation of gene expression by quorum sensing. Current opinion in microbiology 2, 582-587.

Betschel, S.D., Borgia, S.M., Barg, N.L., Low, D.E., and De Azavedo, J.C. (1998). Reduced virulence of group A streptococcal Tn916 mutants that do not produce streptolysin S. Infection and immunity 66, 1671-1679.

Biswas, I., Germon, P., McDade, K., and Scott, J.R. (2001). Generation and surface localization of intact M protein in Streptococcus pyogenes are dependent on sagA. Infection and immunity 69, 7029-7038. https://dx.doi.org/10.1128/iai.69.11.7029-7038.2001.

Blehert, D.S., Palmer, R.J., Jr., Xavier, J.B., Almeida, J.S., and Kolenbrander, P.E. (2003). Autoinducer 2 production by Streptococcus gordonii DL1 and the biofilm phenotype of a luxS mutant are influenced by nutritional conditions. Journal of bacteriology 185, 4851-4860.

Blom, A.M., Bergmann, S., Fulde, M., Riesbeck, K., and Agarwal, V. (2014). Streptococcus pneumoniae phosphoglycerate kinase is a novel complement inhibitor affecting the membrane attack complex formation. The Journal of biological chemistry 289, 32499-32511. https://dx.doi.org/10.1074/jbc.M114.610212.

Brito, L., Wilton, J., Ferrandiz, M.J., Gomez-Sanz, A., de la Campa, A.G., and Amblar, M. (2016). Absence of tmRNA Has a Protective Effect against Fluoroquinolones in Streptococcus pneumoniae. Frontiers in microbiology 7,

2164. https://dx.doi.org/10.3389/fmicb.2016.02164.

Caparon, M.G., and Scott, J.R. (1987). Identification of a gene that regulates expression of M protein, the major virulence determinant of group A streptococci. Proceedings of the National Academy of Sciences of the United States of America 84, 8677-8681.

Carapetis, J.R., Steer, A.C., Mulholland, E.K., and Weber, M. (2005). The global burden of group A streptococcal diseases. The Lancet Infectious diseases 5, 685-694. https://dx.doi.org/10.1016/s1473-3099(05)70267-x.

Chao, Y., and Vogel, J. (2010). The role of Hfq in bacterial pathogens. Current opinion in microbiology 13, 24-33. https://dx.doi.org/10.1016/j.mib.2010.01.001.

Chen, C.Y., Luo, S.C., Kuo, C.F., Lin, Y.S., Wu, J.J., Lin, M.T., Liu, C.C., Jeng, W.Y., and Chuang, W.J. (2003). Maturation processing and characterization of streptopain. The Journal of biological chemistry 278, 17336-17343. https://dx.doi.org/10.1074/jbc.M209038200.

Chiang-Ni, C., and Wu, J.J. (2008). Effects of streptococcal pyrogenic exotoxin B on pathogenesis of Streptococcus pyogenes. Journal of the Formosan Medical Association = Taiwan yi zhi 107, 677-685. https://dx.doi.org/10.1016/s0929-6646(08)60112-6.

Christensen, L.R. (1945). STREPTOCOCCAL FIBRINOLYSIS: A PROTEOLYTIC REACTION DUE TO A SERUM ENZYME ACTIVATED BY STREPTOCOCCAL FIBRINOLYSIN. The Journal of general physiology 28, 363-383.

Costerton, J.W., Stewart, P.S., and Greenberg, E.P. (1999). Bacterial biofilms: a common cause of persistent infections. Science (New York, NY) 284, 1318-1322.

Cunningham, M.W. (2000). Pathogenesis of group A streptococcal infections. Clinical microbiology reviews 13, 470-511.

Dalton, T.L., and Scott, J.R. (2004). CovS inactivates CovR and is required for growth under conditions of general stress in Streptococcus pyogenes. Journal of bacteriology 186, 3928-3937. https://dx.doi.org/10.1128/jb.186.12.3928-3937.2004.

Datta, V., Myskowski, S.M., Kwinn, L.A., Chiem, D.N., Varki, N., Kansal, R.G., Kotb, M., and Nizet, V. (2005). Mutational analysis of the group A streptococcal operon encoding streptolysin S and its virulence role in invasive infection. Molecular microbiology 56, 681-695. https://dx.doi.org/10.1111/j.1365-2958.2005.04583.x.

Deltcheva, E., Chylinski, K., Sharma, C.M., Gonzales, K., Chao, Y., Pirzada, Z.A., Eckert, M.R., Vogel, J., and Charpentier, E. (2011). CRISPR RNA maturation by trans-encoded small RNA and host factor RNase III. Nature 471, 602-607. https://dx.doi.org/10.1038/nature09886.

Deveau, H., Barrangou, R., Garneau, J.E., Labonte, J., Fremaux, C., Boyaval, P., Romero, D.A., Horvath, P., and Moineau, S. (2008). Phage response to CRISPR-encoded resistance in Streptococcus thermophilus. Journal of bacteriology 190, 1390-1400. https://dx.doi.org/10.1128/jb.01412-07.

Deveau, H., Garneau, J.E., and Moineau, S. (2010). CRISPR/Cas system and its role in phage-bacteria interactions. Annual review of microbiology 64, 475-493. https://dx.doi.org/10.1146/annurev.micro.112408.134123.

Echenique, J., Kadioglu, A., Romao, S., Andrew, P.W., and Trombe, M.C. (2004). Protein serine/threonine kinase StkP positively controls virulence and competence in Streptococcus pneumoniae. Infection and immunity 72, 2434-

2437.

Edwards, M.S., Rench, M.A., Haffar, A.A., Murphy, M.A., Desmond, M.M., and Baker, C.J. (1985). Long-term sequelae of group B streptococcal meningitis in infants. The Journal of pediatrics 106, 717-722.

Federle, M.J., and Bassler, B.L. (2003). Interspecies communication in bacteria. The Journal of clinical investigation 112, 1291-1299. https://dx.doi.org/10.1172/jci20195.

Frank, D.N., and Pace, N.R. (1998). Ribonuclease P: unity and diversity in a tRNA processing ribozyme. Annual review of biochemistry 67, 153-180. https://dx.doi.org/10.1146/annurev.biochem.67.1.153.

Furuse, Y., Finethy, R., Saka, H.A., Xet-Mull, A.M., Sisk, D.M., Smith, K.L., Lee, S., Coers, J., Valdivia, R.H., Tobin, D.M., *et al.* (2014). Search for microRNAs expressed by intracellular bacterial pathogens in infected mammalian cells. PloS one 9, e106434. https://dx.doi.org/10.1371/journal.pone.0106434.

Galante, J., Ho, A.C., Tingey, S., and Charalambous, B.M. (2015). Quorum sensing and biofilms in the pathogen, Streptococcus pneumoniae. Current pharmaceutical design 21, 25-30.

Garneau, J.E., Dupuis, M.E., Villion, M., Romero, D.A., Barrangou, R., Boyaval, P., Fremaux, C., Horvath, P., Magadan, A.H., and Moineau, S. (2010). The CRISPR/Cas bacterial immune system cleaves bacteriophage and plasmid DNA. Nature 468, 67-71. https://dx.doi.org/10.1038/nature09523.

Garst, A.D., Edwards, A.L., and Batey, R.T. (2011). Riboswitches: structures and mechanisms. Cold Spring Harbor perspectives in biology 3. https://dx.doi.org/10.1101/cshperspect.a003533.

Giammarinaro, P., and Paton, J.C. (2002). Role of RegM, a homologue of the catabolite repressor protein CcpA, in the virulence of Streptococcus pneumoniae. Infection and immunity 70, 5454-5461.

Gottesman, S. (2004). The small RNA regulators of Escherichia coli: roles and mechanisms*. Annual review of microbiology 58, 303-328. https://dx.doi.org/10.1146/annurev.micro.58.030603.123841.

Gottesman, S., and Storz, G. (2011). Bacterial small RNA regulators: versatile roles and rapidly evolving variations. Cold Spring Harbor perspectives in biology 3. https://dx.doi.org/10.1101/cshperspect.a003798.

Guedon, E., Sperandio, B., Pons, N., Ehrlich, S.D., and Renault, P. (2005). Overall control of nitrogen metabolism in Lactococcus lactis by CodY, and possible models for CodY regulation in Firmicutes. Microbiology (Reading, England) 151, 3895-3909. https://dx.doi.org/10.1099/mic.0.28186-0.

Guenzi, E., Gasc, A.M., Sicard, M.A., and Hakenbeck, R. (1994). A two-component signal-transducing system is involved in competence and penicillin susceptibility in laboratory mutants of Streptococcus pneumoniae. Molecular microbiology 12, 505-515.

Gutierrez-Preciado, A., Henkin, T.M., Grundy, F.J., Yanofsky, C., and Merino, E. (2009). Biochemical features and functional implications of the RNA-based T-box regulatory mechanism. Microbiology and molecular biology reviews : MMBR 73, 36-61. https://dx.doi.org/10.1128/mmbr.00026-08.

Halfmann, A., Kovacs, M., Hakenbeck, R., and Bruckner, R. (2007). Identification of the genes directly controlled by the response regulator CiaR in Streptococcus pneumoniae: five out of 15 promoters drive expression of small non-coding RNAs. Molecular microbiology 66, 110-126. https://dx.doi.org/10.1111/j.1365-2958.2007.05900.x.

Han, H., Liu, C., Wang, Q., Xuan, C., Zheng, B., Tang, J., Yan, J., Zhang, J., Li, M., Cheng, H., *et al.* (2012). The two-component system Ihk/Irr contributes to the virulence of Streptococcus suis serotype 2 strain 05ZYH33 through alteration of the bacterial cell metabolism. Microbiology (Reading, England) 158, 1852-1866. https://dx.doi.org/10.1099/mic.0.057448-0.

Hanks, S.K., Quinn, A.M., and Hunter, T. (1988). The protein kinase family: conserved features and deduced phylogeny of the catalytic domains. Science (New York, NY) 241, 42-52.

Heath, A., DiRita, V.J., Barg, N.L., and Engleberg, N.C. (1999). A two-component regulatory system, CsrR-CsrS, represses expression of three Streptococcus pyogenes virulence factors, hyaluronic acid capsule, streptolysin S, and pyrogenic exotoxin B. Infection and immunity 67, 5298-5305.

Hemsley, C., Joyce, E., Hava, D.L., Kawale, A., and Camilli, A. (2003). MgrA, an orthologue of Mga, Acts as a transcriptional repressor of the genes within the rlrA pathogenicity islet in Streptococcus pneumoniae. Journal of bacteriology 185, 6640-6647.

Hendriksen, W.T., Bootsma, H.J., Estevao, S., Hoogenboezem, T., de Jong, A., de Groot, R., Kuipers, O.P., and Hermans, P.W. (2008). CodY of Streptococcus pneumoniae: link between nutritional gene regulation and colonization. Journal of bacteriology 190, 590-601. https://dx.doi.org/10.1128/jb.00917-07.

Hertzen, E., Johansson, L., Kansal, R., Hecht, A., Dahesh, S., Janos, M., Nizet, V., Kotb, M., and Norrby-Teglund, A. (2012). Intracellular Streptococcus pyogenes in human macrophages display an altered gene expression profile. PloS one 7, e35218. https://dx.doi.org/10.1371/journal.pone.0035218.

Hindley, J. (1967). Fractionation of 32P-labelled ribonucleic acids on polyacrylamide gels and their characterization by fingerprinting. Journal of molecular biology 30, 125-136.

Hondorp, E.R., and McIver, K.S. (2007). The Mga virulence regulon: infection where the grass is greener. Molecular microbiology 66, 1056-1065. https://dx.doi.org/10.1111/j.1365-2958.2007.06006.x.

Ishino, Y., Shinagawa, H., Makino, K., Amemura, M., and Nakata, A. (1987). Nucleotide sequence of the iap gene, responsible for alkaline phosphatase isozyme conversion in Escherichia coli, and identification of the gene product. Journal of bacteriology 169, 5429-5433.

Iyer, R., Baliga, N.S., and Camilli, A. (2005). Catabolite control protein A (CcpA) contributes to virulence and regulation of sugar metabolism in Streptococcus pneumoniae. Journal of bacteriology 187, 8340-8349. https://dx.doi.org/10.1128/jb.187.24.8340-8349.2005.

Ji, Y., Carlson, B., Kondagunta, A., and Cleary, P.P. (1997). Intranasal immunization with C5a peptidase prevents nasopharyngeal colonization of mice by the group A Streptococcus. Infection and immunity 65, 2080-2087.

Jimenez, J.C., and Federle, M.J. (2014). Quorum sensing in group A Streptococcus. Frontiers in cellular and infection microbiology 4, 127. https://dx.doi.org/10.3389/fcimb.2014.00127.

Jin, H., and Pancholi, V. (2006). Identification and biochemical characterization of a eukaryotic-type serine/threonine kinase and its cognate phosphatase in Streptococcus pyogenes: their biological functions and substrate identification. Journal of molecular biology 357, 1351-1372. https://dx.doi.org/10.1016/j.jmb.2006.01.020.

Johansson, J., Mandin, P., Renzoni, A., Chiaruttini, C., Springer, M., and

Cossart, P. (2002). An RNA thermosensor controls expression of virulence genes in Listeria monocytogenes. Cell 110, 551-561.

Joyce, E.A., Kawale, A., Censini, S., Kim, C.C., Covacci, A., and Falkow, S. (2004). LuxS is required for persistent pneumococcal carriage and expression of virulence and biosynthesis genes. Infection and immunity 72, 2964-2975.

Karvelis, T., Gasiunas, G., Miksys, A., Barrangou, R., Horvath, P., and Siksnys, V. (2013). crRNA and tracrRNA guide Cas9-mediated DNA interference in Streptococcus thermophilus. RNA biology 10, 841-851. https://dx.doi.org/10.4161/rna.24203.

Keiler, K.C. (2007). Physiology of tmRNA: what gets tagged and why? Current opinion in microbiology 10, 169-175. https://dx.doi.org/10.1016/j.mib.2007.03.014.

Keiler, K.C. (2008). Biology of trans-translation. Annual review of microbiology 62, 133-151. https://dx.doi.org/10.1146/annurev.micro.62.081307.162948.

Kihlberg, B.M., Cooney, J., Caparon, M.G., Olsen, A., and Bjorck, L. (1995). Biological properties of a Streptococcus pyogenes mutant generated by Tn916 insertion in mga. Microbial pathogenesis 19, 299-315.

Klenk, M., Koczan, D., Guthke, R., Nakata, M., Thiesen, H.J., Podbielski, A., and Kreikemeyer, B. (2005). Global epithelial cell transcriptional responses reveal Streptococcus pyogenes Fas regulator activity association with bacterial aggressiveness. Cellular microbiology 7, 1237-1250. https://dx.doi.org/10.1111/j.1462-5822.2005.00548.x.

Kreikemeyer, B., Boyle, M.D., Buttaro, B.A., Heinemann, M., and Podbielski, A. (2001). Group A streptococcal growth phase-associated virulence factor regulation by a novel operon (Fas) with homologies to two-component-type regulators requires a small RNA molecule. Molecular microbiology 39, 392-406.

Kreikemeyer, B., McIver, K.S., and Podbielski, A. (2003). Virulence factor regulation and regulatory networks in Streptococcus pyogenes and their impact on pathogen-host interactions. Trends in microbiology 11, 224-232.

Kreth, J., Chen, Z., Ferretti, J., and Malke, H. (2011). Counteractive balancing of transcriptome expression involving CodY and CovRS in Streptococcus pyogenes. Journal of bacteriology 193, 4153-4165. https://dx.doi.org/10.1128/jb.00061-11.

Krzysciak, W., Jurczak, A., Koscielniak, D., Bystrowska, B., and Skalniak, A. (2014). The virulence of Streptococcus mutans and the ability to form biofilms. European journal of clinical microbiology & infectious diseases : official publication of the European Society of Clinical Microbiology 33, 499-515. https://dx.doi.org/10.1007/s10096-013-1993-7.

Kumar, R., Shah, P., Swiatlo, E., Burgess, S.C., Lawrence, M.L., and Nanduri, B. (2010). Identification of novel non-coding small RNAs from Streptococcus pneumoniae TIGR4 using high-resolution genome tiling arrays. BMC genomics 11, 350. https://dx.doi.org/10.1186/1471-2164-11-350.

Lamy, M.C., Zouine, M., Fert, J., Vergassola, M., Couve, E., Pellegrini, E., Glaser, P., Kunst, F., Msadek, T., Trieu-Cuot, P., *et al.* (2004). CovS/CovR of group B streptococcus: a two-component global regulatory system involved in virulence. Molecular microbiology 54, 1250-1268. https://dx.doi.org/10.1111/j.1365-2958.2004.04365.x.

Lancefield, R.C. (1933). A SEROLOGICAL DIFFERENTIATION OF HUMAN AND OTHER GROUPS OF HEMOLYTIC STREPTOCOCCI. The Journal of experimental medicine 57, 571-595.

Lancefield, R.C. (1962). Current knowledge of type-specific M antigens of group A streptococci. Journal of immunology (Baltimore, Md : 1950) 89, 307-313.

Laux, A., Sexauer, A., Sivaselvarajah, D., Kaysen, A., and Bruckner, R. (2015). Control of competence by related non-coding csRNAs in Streptococcus pneumoniae R6. Frontiers in genetics 6, 246. https://dx.doi.org/10.3389/fgene.2015.00246.

Le Rhun, A., Beer, Y.Y., Reimegard, J., Chylinski, K., and Charpentier, E. (2016). RNA sequencing uncovers antisense RNAs and novel small RNAs in Streptococcus pyogenes. RNA biology 13, 177-195. https://dx.doi.org/10.1080/15476286.2015.1110674.

Le Rhun, A., and Charpentier, E. (2012). Small RNAs in streptococci. RNA biology 9, 414-426. https://dx.doi.org/10.4161/rna.20104.

Lee, H.J., and Hong, S.H. (2012). Analysis of microRNA-size, small RNAs in Streptococcus mutans by deep sequencing. FEMS microbiology letters 326, 131-136. https://dx.doi.org/10.1111/j.1574-6968.2011.02441.x.

Levesque, C.M., Mair, R.W., Perry, J.A., Lau, P.C., Li, Y.H., and Cvitkovitch, D.G. (2007). Systemic inactivation and phenotypic characterization of two-component systems in expression of Streptococcus mutans virulence properties. Letters in applied microbiology 45, 398-404. https://dx.doi.org/10.1111/j.1472-765X.2007.02203.x.

Levin, J.C., and Wessels, M.R. (1998). Identification of csrR/csrS, a genetic locus that regulates hyaluronic acid capsule synthesis in group A Streptococcus. Molecular microbiology 30, 209-219.

Li, W., Liu, L., Qiu, D., Chen, H., and Zhou, R. (2010). Identification of Streptococcus suis serotype 2 genes preferentially expressed in the natural host. International journal of medical microbiology : IJMM 300, 482-488. https://dx.doi.org/10.1016/j.ijmm.2010.04.018.

Li, Z., Sledjeski, D.D., Kreikemeyer, B., Podbielski, A., and Boyle, M.D. (1999). Identification of pel, a Streptococcus pyogenes locus that affects both surface and secreted proteins. Journal of bacteriology 181, 6019-6027.

Liu, M., Hanks, T.S., Zhang, J., McClure, M.J., Siemsen, D.W., Elser, J.L., Quinn, M.T., and Lei, B. (2006). Defects in ex vivo and in vivo growth and sensitivity to osmotic stress of group A Streptococcus caused by interruption of response regulator gene vicR. Microbiology (Reading, England) 152, 967-978. https://dx.doi.org/10.1099/mic.0.28706-0.

Liu, Y., and Burne, R.A. (2009). Multiple two-component systems of Streptococcus mutans regulate agmatine deiminase gene expression and stress tolerance. Journal of bacteriology 191, 7363-7366. https://dx.doi.org/10.1128/jb.01054-09.

Liu, Z., Trevino, J., Ramirez-Pena, E., and Sumby, P. (2012). The small regulatory RNA FasX controls pilus expression and adherence in the human bacterial pathogen group A Streptococcus. Molecular microbiology 86, 140-154. https://dx.doi.org/10.1111/j.1365-2958.2012.08178.x.

Lynskey, N.N., Reglinski, M., Calay, D., Siggins, M.K., Mason, J.C., Botto, M., and Sriskandan, S. (2017). Multi-functional mechanisms of immune evasion by the streptococcal complement inhibitor C5a peptidase. PLoS pathogens 13, e1006493. https://dx.doi.org/10.1371/journal.ppat.1006493.

Lyon, W.R., Madden, J.C., Levin, J.C., Stein, J.L., and Caparon, M.G. (2001). Mutation of luxS affects growth and virulence factor expression in Streptococcus pyogenes. Molecular microbiology 42, 145-157.

Makarova, K.S., Haft, D.H., Barrangou, R., Brouns, S.J., Charpentier, E., Horvath, P., Moineau, S., Mojica, F.J., Wolf, Y.I., Yakunin, A.F., *et al.* (2011). Evolution and classification of the CRISPR-Cas systems. Nature reviews Microbiology 9, 467-477. https://dx.doi.org/10.1038/nrmicro2577.

Malke, H., and Ferretti, J.J. (2007). CodY-affected transcriptional gene expression of Streptococcus pyogenes during growth in human blood. Journal of medical microbiology 56, 707-714. https://dx.doi.org/10.1099/jmm.0.46984-0.

Malke, H., Mechold, U., Gase, K., and Gerlach, D. (1994). Inactivation of the streptokinase gene prevents Streptococcus equisimilis H46A from acquiring cell-associated plasmin activity in the presence of plasminogen. FEMS microbiology letters 116, 107-112.

Malke, H., Steiner, K., McShan, W.M., and Ferretti, J.J. (2006). Linking the nutritional status of Streptococcus pyogenes to alteration of transcriptional gene expression: the action of CodY and RelA. International journal of medical microbiology : IJMM 296, 259-275. https://dx.doi.org/10.1016/j.ijmm.2005.11.008.

Mangold, M., Siller, M., Roppenser, B., Vlaminckx, B.J., Penfound, T.A., Klein, R., Novak, R., Novick, R.P., and Charpentier, E. (2004). Synthesis of group A streptococcal virulence factors is controlled by a regulatory RNA molecule. Molecular microbiology 53, 1515-1527. https://dx.doi.org/10.1111/j.1365-2958.2004.04222.x.

Mann, B., van Opijnen, T., Wang, J., Obert, C., Wang, Y.D., Carter, R., McGoldrick, D.J., Ridout, G., Camilli, A., Tuomanen, E.I., *et al.* (2012). Control of virulence by small RNAs in Streptococcus pneumoniae. PLoS pathogens 8, e1002788. https://dx.doi.org/10.1371/journal.ppat.1002788.

Mao, M.Y., Yang, Y.M., Li, K.Z., Lei, L., Li, M., Yang, Y., Tao, X., Yin, J.X., Zhang, R., Ma, X.R., *et al.* (2016). The rnc Gene Promotes Exopolysaccharide Synthesis and Represses the vicRKX Gene Expressions via MicroRNA-Size Small RNAs in Streptococcus mutans. Frontiers in microbiology 7, 687. https://dx.doi.org/10.3389/fmicb.2016.00687.

Marouni, M.J., and Sela, S. (2003). The luxS gene of Streptococcus pyogenes regulates expression of genes that affect internalization by epithelial cells. Infection and immunity 71, 5633-5639.

Marraffini, L.A., and Sontheimer, E.J. (2010). CRISPR interference: RNA-directed adaptive immunity in bacteria and archaea. Nature reviews Genetics 11, 181-190. https://dx.doi.org/10.1038/nrg2749.

Marx, P., Nuhn, M., Kovacs, M., Hakenbeck, R., and Bruckner, R. (2010). Identification of genes for small non-coding RNAs that belong to the regulon of the two-component regulatory system CiaRH in Streptococcus. BMC genomics 11, 661. https://dx.doi.org/10.1186/1471-2164-11-661.

Mellin, J.R., and Cossart, P. (2012). The non-coding RNA world of the bacterial pathogen Listeria monocytogenes. RNA biology 9, 372-378. https://dx.doi.org/10.4161/rna.19235.

Mizuno, T., Chou, M.Y., and Inouye, M. (1984). A unique mechanism regulating gene expression: translational inhibition by a complementary RNA transcript (micRNA). Proceedings of the National Academy of Sciences of the United States of America 81, 1966-1970.

Mraheil, M.A., Billion, A., Kuenne, C., Pischimarov, J., Kreikemeyer, B., Engelmann, S., Hartke, A., Giard, J.C., Rupnik, M., Vorwerk, S., *et al.* (2010). Comparative genome-wide analysis of small RNAs of major Gram-positive pathogens: from identification to application. Microbial biotechnology 3, 658-676. https://dx.doi.org/10.1111/j.1751-7915.2010.00171.x.

Nejman-Falenczyk, B., Bloch, S., Licznerska, K., Dydecka, A., Felczykowska, A., Topka, G., Wegrzyn, A., and Wegrzyn, G. (2015). A small, microRNA-size, ribonucleic acid regulating gene expression and development of Shiga toxin-converting bacteriophage Phi24Beta. Scientific reports 5, 10080. https://dx.doi.org/10.1038/srep10080.

Nielsen, J.S., Lei, L.K., Ebersbach, T., Olsen, A.S., Klitgaard, J.K., Valentin-Hansen, P., and Kallipolitis, B.H. (2010). Defining a role for Hfq in Gram-positive bacteria: evidence for Hfq-dependent antisense regulation in Listeria monocytogenes. Nucleic acids research 38, 907-919. https://dx.doi.org/10.1093/nar/gkp1081.

Nizet, V. (2002). Streptococcal beta-hemolysins: genetics and role in disease pathogenesis. Trends in microbiology 10, 575-580.

Nizet, V., Beall, B., Bast, D.J., Datta, V., Kilburn, L., Low, D.E., and De Azavedo, J.C. (2000). Genetic locus for streptolysin S production by group A streptococcus. Infection and immunity 68, 4245-4254.

Novakova, L., Saskova, L., Pallova, P., Janecek, J., Novotna, J., Ulrych, A., Echenique, J., Trombe, M.C., and Branny, P. (2005). Characterization of a eukaryotic type serine/threonine protein kinase and protein phosphatase of Streptococcus pneumoniae and identification of kinase substrates. The FEBS journal 272, 1243-1254. https://dx.doi.org/10.1111/j.1742-4658.2005.04560.x.

Ota, C., Morisaki, H., Nakata, M., Arimoto, T., Fukamachi, H., Kataoka, H., Masuda, Y., Suzuki, N., Miyazaki, T., Okahashi, N., *et al.* (2018). Streptococcus sanguinis Noncoding cia-Dependent Small RNAs Negatively Regulate

Expression of Type IV Pilus Retraction ATPase PilT and Biofilm Formation. Infection and immunity 86. https://dx.doi.org/10.1128/iai.00894-17.

Pan, X., Ge, J., Li, M., Wu, B., Wang, C., Wang, J., Feng, Y., Yin, Z., Zheng, F., Cheng, G., *et al.* (2009). The orphan response regulator CovR: a globally negative modulator of virulence in Streptococcus suis serotype 2. Journal of bacteriology 191, 2601-2612. https://dx.doi.org/10.1128/jb.01309-08.

Park, S.E., Jiang, S., and Wessels, M.R. (2012). CsrRS and environmental pH regulate group B streptococcus adherence to human epithelial cells and extracellular matrix. Infection and immunity 80, 3975-3984. https://dx.doi.org/10.1128/iai.00699-12.

Patenge, N., Fiedler, T., and Kreikemeyer, B. (2013). Common regulators of virulence in streptococci. Current topics in microbiology and immunology 368, 111-153. https://dx.doi.org/10.1007/82_2012_295.

Perez, N., Trevino, J., Liu, Z., Ho, S.C., Babitzke, P., and Sumby, P. (2009). A genome-wide analysis of small regulatory RNAs in the human pathogen group A Streptococcus. PloS one 4, e7668. https://dx.doi.org/10.1371/journal.pone.0007668.

Petersen, F.C., Ahmed, N.A., Naemi, A., and Scheie, A.A. (2006). LuxS-mediated signalling in Streptococcus anginosus and its role in biofilm formation. Antonie van Leeuwenhoek 90, 109-121. https://dx.doi.org/10.1007/s10482-006-9065-y.

Phares, C.R., Lynfield, R., Farley, M.M., Mohle-Boetani, J., Harrison, L.H., Petit, S., Craig, A.S., Schaffner, W., Zansky, S.M., Gershman, K., *et al.* (2008). Epidemiology of invasive group B streptococcal disease in the United States, 1999-2005. Jama 299, 2056-2065. https://dx.doi.org/10.1001/jama.299.17.2056.

Picazo, J.J. (2009). Management of antibiotic-resistant Streptococcus pneumoniae infections and the use of pneumococcal conjugate vaccines. Clinical microbiology and infection : the official publication of the European Society of Clinical Microbiology and Infectious Diseases 15 Suppl 3, 4-6. https://dx.doi.org/10.1111/j.1469-0691.2009.02723.x.

Pichon, C., du Merle, L., Caliot, M.E., Trieu-Cuot, P., and Le Bouguenec, C. (2012). An in silico model for identification of small RNAs in whole bacterial genomes: characterization of antisense RNAs in pathogenic Escherichia coli and Streptococcus agalactiae strains. Nucleic acids research 40, 2846-2861. https://dx.doi.org/10.1093/nar/gkr1141.

Qi, F., Merritt, J., Lux, R., and Shi, W. (2004). Inactivation of the ciaH Gene in Streptococcus mutans diminishes mutacin production and competence development, alters sucrose-dependent biofilm formation, and reduces stress tolerance. Infection and immunity 72, 4895-4899. https://dx.doi.org/10.1128/iai.72.8.4895-4899.2004.

Quach, D., van Sorge, N.M., Kristian, S.A., Bryan, J.D., Shelver, D.W., and Doran, K.S. (2009). The CiaR response regulator in group B Streptococcus promotes intracellular survival and resistance to innate immune defenses. Journal of bacteriology 191, 2023-2032. https://dx.doi.org/10.1128/jb.01216-08.

Quereda, J.J., and Cossart, P. (2017). Regulating Bacterial Virulence with RNA. Annual review of microbiology 71, 263-280. https://dx.doi.org/10.1146/annurev-micro-030117-020335.

Rajagopal, L., Clancy, A., and Rubens, C.E. (2003). A eukaryotic type serine/threonine kinase and phosphatase in Streptococcus agalactiae reversibly phosphorylate an inorganic pyrophosphatase and affect growth, cell segregation, and virulence. The Journal of biological chemistry 278, 14429-14441. https://dx.doi.org/10.1074/jbc.M212747200.

Ramirez-Pena, E., Trevino, J., Liu, Z., Perez, N., and Sumby, P. (2010). The group A Streptococcus small regulatory RNA FasX enhances streptokinase activity by increasing the stability of the ska mRNA transcript. Molecular microbiology 78, 1332-1347. https://dx.doi.org/10.1111/j.1365-2958.2010.07427.x.

Ribardo, D.A., and McIver, K.S. (2006). Defining the Mga regulon: Comparative transcriptome analysis reveals both direct and indirect regulation by Mga in the group A streptococcus. Molecular microbiology 62, 491-508. https://dx.doi.org/10.1111/j.1365-2958.2006.05381.x.

Roberts, S.A., Churchward, G.G., and Scott, J.R. (2007). Unraveling the regulatory network in Streptococcus pyogenes: the global response regulator CovR represses rivR directly. Journal of bacteriology 189, 1459-1463. https://dx.doi.org/10.1128/jb.01026-06.

Roberts, S.A., and Scott, J.R. (2007). RivR and the small RNA RivX: the missing links between the CovR regulatory cascade and the Mga regulon. Molecular microbiology 66, 1506-1522. https://dx.doi.org/10.1111/j.1365-2958.2007.06015.x.

Romby, P., and Charpentier, E. (2010). An overview of RNAs with regulatory functions in gram-positive bacteria. Cellular and molecular life sciences : CMLS 67, 217-237. https://dx.doi.org/10.1007/s00018-009-0162-8.

Romilly, C., Caldelari, I., Parmentier, D., Lioliou, E., Romby, P., and Fechter, P. (2012). Current knowledge on regulatory RNAs and their machineries in Staphylococcus aureus. RNA biology 9, 402-413.

https://dx.doi.org/10.4161/rna.20103.

Rosini, R., and Margarit, I. (2015). Biofilm formation by Streptococcus agalactiae: influence of environmental conditions and implicated virulence factors. Frontiers in cellular and infection microbiology 5, 6. https://dx.doi.org/10.3389/fcimb.2015.00006.

Rosinski-Chupin, I., Sauvage, E., Sismeiro, O., Villain, A., Da Cunha, V., Caliot, M.E., Dillies, M.A., Trieu-Cuot, P., Bouloc, P., Lartigue, M.F., *et al.* (2015). Single nucleotide resolution RNA-seq uncovers new regulatory mechanisms in the opportunistic pathogen Streptococcus agalactiae. BMC genomics 16, 419. https://dx.doi.org/10.1186/s12864-015-1583-4.

Saier, M.H., Jr., Chauvaux, S., Cook, G.M., Deutscher, J., Paulsen, I.T., Reizer, J., and Ye, J.J. (1996). Catabolite repression and inducer control in Gram-positive bacteria. Microbiology (Reading, England) 142 (Pt 2), 217-230. https://dx.doi.org/10.1099/13500872-142-2-217.

Sapranauskas, R., Gasiunas, G., Fremaux, C., Barrangou, R., Horvath, P., and Siksnys, V. (2011). The Streptococcus thermophilus CRISPR/Cas system provides immunity in Escherichia coli. Nucleic acids research 39, 9275-9282. https://dx.doi.org/10.1093/nar/gkr606.

Schauder, S., Shokat, K., Surette, M.G., and Bassler, B.L. (2001). The LuxS family of bacterial autoinducers: biosynthesis of a novel quorum-sensing signal molecule. Molecular microbiology 41, 463-476.

Schnorpfeil, A., Kranz, M., Kovacs, M., Kirsch, C., Gartmann, J., Brunner, I., Bittmann, S., and Bruckner, R. (2013). Target evaluation of the non-coding csRNAs reveals a link of the two-component regulatory system CiaRH to competence control in Streptococcus pneumoniae R6. Molecular microbiology

89, 334-349. https://dx.doi.org/10.1111/mmi.12277.

Shanker, E., and Federle, M.J. (2017). Quorum Sensing Regulation of Competence and Bacteriocins in Streptococcus pneumoniae and mutans. Genes 8. https://dx.doi.org/10.3390/genes8010015.

Sharma, C., and Heidrich, N. (2012). Small RNAs and virulence in bacterial pathogens. RNA biology 9, 361-363. https://dx.doi.org/10.4161/rna.20517.

Shelburne, S.A., 3rd, Keith, D., Horstmann, N., Sumby, P., Davenport, M.T., Graviss, E.A., Brennan, R.G., and Musser, J.M. (2008). A direct link between carbohydrate utilization and virulence in the major human pathogen group A Streptococcus. Proceedings of the National Academy of Sciences of the United States of America 105, 1698-1703. https://dx.doi.org/10.1073/pnas.0711767105.

Shelburne, S.A., 3rd, Sumby, P., Sitkiewicz, I., Granville, C., DeLeo, F.R., and Musser, J.M. (2005). Central role of a bacterial two-component gene regulatory system of previously unknown function in pathogen persistence in human saliva. Proceedings of the National Academy of Sciences of the United States of America 102, 16037-16042. https://dx.doi.org/10.1073/pnas.0505839102.

Siller, M., Janapatla, R.P., Pirzada, Z.A., Hassler, C., Zinkl, D., and Charpentier, E. (2008). Functional analysis of the group A streptococcal luxS/AI-2 system in metabolism, adaptation to stress and interaction with host cells. BMC microbiology 8, 188. https://dx.doi.org/10.1186/1471-2180-8-188.

Silvaggi, J.M., Perkins, J.B., and Losick, R. (2006). Genes for small, noncoding RNAs under sporulation control in Bacillus subtilis. Journal of bacteriology 188, 532-541. https://dx.doi.org/10.1128/jb.188.2.532-541.2006.

Solano-Collado, V., Espinosa, M., and Bravo, A. (2012). Activator role of the pneumococcal Mga-like virulence transcriptional regulator. Journal of bacteriology 194, 4197-4207. https://dx.doi.org/10.1128/jb.00536-12.

Sonenshein, A.L. (2005). CodY, a global regulator of stationary phase and virulence in Gram-positive bacteria. Current opinion in microbiology 8, 203-207. https://dx.doi.org/10.1016/j.mib.2005.01.001.

Sonnleitner, E., Romeo, A., and Blasi, U. (2012). Small regulatory RNAs in Pseudomonas aeruginosa. RNA biology 9, 364-371. https://dx.doi.org/10.4161/rna.19231.

Steiner, K., and Malke, H. (2001). relA-Independent amino acid starvation response network of Streptococcus pyogenes. Journal of bacteriology 183, 7354-7364. https://dx.doi.org/10.1128/jb.183.24.7354-7364.2001.

Steiner, K., and Malke, H. (2002). Dual control of streptokinase and streptolysin S production by the covRS and fasCAX two-component regulators in Streptococcus dysgalactiae subsp. equisimilis. Infection and immunity 70, 3627-3636.

Storz, G., Vogel, J., and Wassarman, K.M. (2011). Regulation by small RNAs in bacteria: expanding frontiers. Molecular cell 43, 880-891. https://dx.doi.org/10.1016/j.molcel.2011.08.022.

Swanson, J., Hsu, K.C., and Gotschlich, E.C. (1969). Electron microscopic studies on streptococci. I. M antigen. The Journal of experimental medicine 130, 1063-1091.

Thomason, M.K., and Storz, G. (2010). Bacterial antisense RNAs: how many are there, and what are they doing? Annual review of genetics 44, 167-188. https://dx.doi.org/10.1146/annurev-genet-102209-163523.

Trevino, J., Perez, N., Ramirez-Pena, E., Liu, Z., Shelburne, S.A., 3rd, Musser, J.M., and Sumby, P. (2009). CovS simultaneously activates and inhibits the

CovR-mediated repression of distinct subsets of group A Streptococcus virulence factor-encoding genes. Infection and immunity 77, 3141-3149. https://dx.doi.org/10.1128/iai.01560-08.

Trevino, J., Perez, N., and Sumby, P. (2010). The 4.5S RNA component of the signal recognition particle is required for group A Streptococcus virulence. Microbiology (Reading, England) 156, 1342-1350. https://dx.doi.org/10.1099/mic.0.036558-0.

Tsai, P.J., Lin, Y.S., Kuo, C.F., Lei, H.Y., and Wu, J.J. (1999). Group A Streptococcus induces apoptosis in human epithelial cells. Infection and immunity 67, 4334-4339.

Tsui, H.C., Mukherjee, D., Ray, V.A., Sham, L.T., Feig, A.L., and Winkler, M.E. (2010). Identification and characterization of noncoding small RNAs in Streptococcus pneumoniae serotype 2 strain D39. Journal of bacteriology 192, 264-279. https://dx.doi.org/10.1128/JB.01204-09.

Vahling, C.M., and McIver, K.S. (2006). Domains required for transcriptional activation show conservation in the mga family of virulence gene regulators. Journal of bacteriology 188, 863-873. https://dx.doi.org/10.1128/jb.188.3.863-873.2006.

Vendeville, A., Winzer, K., Heurlier, K., Tang, C.M., and Hardie, K.R. (2005). Making 'sense' of metabolism: autoinducer-2, LuxS and pathogenic bacteria. Nature reviews Microbiology 3, 383-396. https://dx.doi.org/10.1038/nrmicro1146.

Vidal, J.E., Ludewick, H.P., Kunkel, R.M., Zahner, D., and Klugman, K.P. (2011). The LuxS-dependent quorum-sensing system regulates early biofilm formation by Streptococcus pneumoniae strain D39. Infection and immunity 79, 4050-4060. https://dx.doi.org/10.1128/iai.05186-11.

Vitreschak, A.G., Rodionov, D.A., Mironov, A.A., and Gelfand, M.S. (2004). Riboswitches: the oldest mechanism for the regulation of gene expression? Trends in genetics : TIG 20, 44-50. https://dx.doi.org/10.1016/j.tig.2003.11.008.

Vogel, J., and Luisi, B.F. (2011). Hfq and its constellation of RNA. Nature reviews Microbiology 9, 578-589. https://dx.doi.org/10.1038/nrmicro2615.

Voyich, J.M., Sturdevant, D.E., Braughton, K.R., Kobayashi, S.D., Lei, B., Virtaneva, K., Dorward, D.W., Musser, J.M., and DeLeo, F.R. (2003). Genome-wide protective response used by group A Streptococcus to evade destruction by human polymorphonuclear leukocytes. Proceedings of the National Academy of Sciences of the United States of America 100, 1996-2001. https://dx.doi.org/10.1073/pnas.0337370100.

Wahlestedt, C. (2013). Targeting long non-coding RNA to therapeutically upregulate gene expression. Nature reviews Drug discovery 12, 433-446. https://dx.doi.org/10.1038/nrd4018.

Wang, X., Lin, X., Loy, J.A., Tang, J., and Zhang, X.C. (1998). Crystal structure of the catalytic domain of human plasmin complexed with streptokinase. Science (New York, NY) 281, 1662-1665.

Wang, Y., Zhang, W., Wu, Z., Zhu, X., and Lu, C. (2011). Functional analysis of luxS in Streptococcus suis reveals a key role in biofilm formation and virulence. Veterinary microbiology 152, 151-160. https://dx.doi.org/10.1016/j.vetmic.2011.04.029.

Waters, L.S., and Storz, G. (2009). Regulatory RNAs in bacteria. Cell 136, 615-628. https://dx.doi.org/10.1016/j.cell.2009.01.043.

Wenderska, I.B., Latos, A., Pruitt, B., Palmer, S., Spatafora, G., Senadheera, D.B., and Cvitkovitch, D.G. (2017). Transcriptional Profiling of the Oral

Pathogen Streptococcus mutans in Response to Competence Signaling Peptide XIP. mSystems 2. https://dx.doi.org/10.1128/mSystems.00102-16.

Wessels, M.R., Moses, A.E., Goldberg, J.B., and DiCesare, T.J. (1991). Hyaluronic acid capsule is a virulence factor for mucoid group A streptococci. Proceedings of the National Academy of Sciences of the United States of America 88, 8317-8321.

Willenborg, J., Fulde, M., de Greeff, A., Rohde, M., Smith, H.E., Valentin-Weigand, P., and Goethe, R. (2011). Role of glucose and CcpA in capsule expression and virulence of Streptococcus suis. Microbiology (Reading, England) 157, 1823-1833. https://dx.doi.org/10.1099/mic.0.046417-0.

Wilton, J., Acebo, P., Herranz, C., Gomez, A., and Amblar, M. (2015). Small regulatory RNAs in Streptococcus pneumoniae: discovery and biological functions. Frontiers in genetics 6, 126. https://dx.doi.org/10.3389/fgene.2015.00126.

Winkler, M.E., and Hoch, J.A. (2008). Essentiality, bypass, and targeting of the YycFG (VicRK) two-component regulatory system in gram-positive bacteria. Journal of bacteriology 190, 2645-2648. https://dx.doi.org/10.1128/jb.01682-07.

Xia, L., Xia, W., Li, S., Li, W., Liu, J., Ding, H., Li, J., Li, H., Chen, Y., Su, X., *et al.* (2012). Identification and expression of small non-coding RNA, L10-Leader, in different growth phases of Streptococcus mutans. Nucleic acid therapeutics 22, 177-186. https://dx.doi.org/10.1089/nat.2011.0339.

Xiao, G., Tang, H., Zhang, S., Ren, H., Dai, J., Lai, L., Lu, C., Yao, H., Fan, H., and Wu, Z. (2017). Streptococcus suis small RNA rss04 contributes to the induction of meningitis by regulating capsule synthesis and by inducing biofilm formation in a mouse infection model. Veterinary microbiology 199, 111-119.

https://dx.doi.org/10.1016/j.vetmic.2016.12.034.

Yoshida, A., Ansai, T., Takehara, T., and Kuramitsu, H.K. (2005). LuxS-based signaling affects Streptococcus mutans biofilm formation. Applied and environmental microbiology 71, 2372-2380. https://dx.doi.org/10.1128/aem.71.5.2372-2380.2005.

Zhang, D., Du, N., Ma, S., Hu, Q., Lu, G., Chen, W., and Zeng, C. (2014). In vitro transcriptome analysis of two Chinese isolates of Streptococcus suis serotype 2. Genomics, proteomics & bioinformatics 12, 266-275. https://dx.doi.org/10.1016/j.gpb.2014.11.001.

Zheng, F., Zook, C., Campo, L., Henault, M., Watson, H., Wang, Q.M., and Peng, S.B. (2002). Identification and characterization of Streptococcus pneumoniae Ffh, a homologue of SRP54 subunit of mammalian signal recognition particle. Biochemical and biophysical research communications 292, 601-608. https://dx.doi.org/10.1006/bbrc.2002.6694.

Zorgani, M.A., Quentin, R., and Lartigue, M.F. (2016). Regulatory RNAs in the Less Studied Streptococcal Species: From Nomenclature to Identification. Frontiers in microbiology 7, 1161. https://dx.doi.org/10.3389/fmicb.2016.01161.

https://doi.org/10.21775/9781912530229.03

Signal Transduction of Streptococci by Cyclic Dinucleotide Second Messengers

Zhengyi Li[1#], Xueqin Zhang[1#], Lei Cheng[1,2], Xin Xu[1,2], Xuedong Zhou[1,2], Hui Wu[3], Xian Peng[1*]

[1] State Key Laboratory of Oral Diseases, National Clinical Research Center for Oral Diseases, West China Hospital of Stomatology, Sichuan University, Chengdu, China

[2] State Key Laboratory of Oral Diseases, National Clinical Research Center for Oral Diseases, Department of Cariology and Endodontics, West China Hospital of Stomatology, Sichuan University, Chengdu, China

[3] Department of Pediatric Dentistry, University of Alabama at Birmingham, School of Dentistry, Birmingham, AL, USA.

*Correspondence: pengx@scu.edu.cn

Abstract

Since the discovery of cyclic dimeric guanosine 3',5'-monophosphate (c-di-GMP) in 1987, the role of cyclic dinucleotides in signal pathways has been extensively studied. Many receptors and effectors of cyclic dinucleotides have been identified which play important roles in cellular processes. Example of such effectors include cyclic dimeric adenosine 3',5'-monophosphate (c-di-AMP)-binding proteins and endoplasmic reticulum membrane adaptor. Accumulating evidence indicate that cyclic dinucleotides act as second messengers that not only regulate the bacterial physiological processes but also affect host immune responses during infections. Streptococci species, which produce cyclic dinucleotides, are responsible for many human diseases. Numerous studies suggest that the cyclic dinucleotides are vital in signal transduction pathways as second messengers and influence the progression of infectious diseases. Here, we provide an overview of the molecular principles of cyclic dinucleotides synthesis and degradation and discuss recent progress on streptococcal signal transduction pathways by cyclic dinucleotide second messengers and their role in regulating host immune reaction. This review will provide a better understanding of the molecular mechanisms of streptococcal cyclic dinucleotide second messengers thereby revealing novel targets for preventing infections.

Introduction

In natural environments, bacteria are exposed to rapidly changing nutrient availability and physical conditions. To adapt to their environments, bacteria use efficient signal transduction systems to detect changes and to transduce the information (Galperin 2018). Therefore, signaling molecules are the most crucial part of these systems. Over the past few decades, nucleotide messengers, including linear and cyclic nucleotides have been shown to control important biological signaling cascades in prokaryotic cells (Gomelsky 2011, Kalia et al. 2013).

Linear nucleotides include guanosine 3',5'-bispyrophosphate (ppGpp) and guanosine 3'-diphosphate, 5'-triphosphate (pppGpp) (collectively known as (p)ppGpp) are important second messengers which are produced by bacteria in response to nutrient starvation, which reduces the growth rate of cells. (p)ppGpp affects several vital biochemical processes in the bacteria including transcription, translation and DNA replication, thus regulating the virulence factors and biofilm formation (Kalia et al. 2013, Liu et al. 2015). Studies on *Streptococcus mutans* revealed the role of (p)ppGpp in stress tolerance, peptide signaling and competence development (Kaspar et al. 2016). Moreover, (p)ppGpp and its synthetases are involved in the pathogenesis of zoonotic *Streptococcus suis* (Zhu et al. 2016).

The cyclic mononucleotides, cyclic guanosine 3',5'-monophosphate (cAMP) and cyclic adenosine 3',5'-monophosphate (cGMP) are extensively used as second messengers in bacteria. cAMP is involved in transcription regulation by binding to cAMP-receptor protein (Crp), a transcription factor that regulates the transcription of a large number of promoters (Hengge et al. 2016). A study on type III CRISPR-Cas systems of *Streptococcus thermophilus* demonstrated that the cAMP synthesized by Cas10 binds the ribonuclease Csm6 thereby activating its nonspecific RNA degradation (Kazlauskiene et al. 2017). The role of cGMP in bacterial signaling become clear when it was detected in *Rhodospirillum centenum* and acted as a regulator of developmental process via cGMP-dependent transcription factor (Marden et al. 2011). Although numerous studies have investigated cAMP signaling in streptococci, the biology of cGMP in streptococci is not fully understood.

Recently, the cyclic dinucleotides (CDNs) including c-di-GMP, c-di-AMP and cyclic GMP-AMP (cGAMP) have emerged as key components in cellular signal transduction pathways (Drexler et al. 2017). They are also released or secreted by the bacteria and are recognized by the host innate immune system

(Danilchanka et al. 2013).

C-di-GMP was first described to mediate signals of allosteric activators of cellulose synthase in *Gluconacetobacter xylinus* (Ross et al. 1987). Since then, c-di-GMP and its metabolizing enzymes have been identified in many Gram-negative bacteria such as *Caulobacter crescentus, Pseudomonas aeruginosa, Vibrio cholera* and *Escherichia coli* (Kalia, Merey et al. 2013) and a few Gram-positive bacteria such as *Bacillus subtilis* and *Listeria monocytogenes* (Yan et al. 2010, Kalia, Merey et al. 2013). C-di-GMP mainly regulates the motility-to-sessility transition, and it is also associated with biofilm formation and virulence factor production (Commichau et al. 2018)

C-di-AMP was added to the list of nucleotide second messengers in 2009. Since then, it has been identified in several Gram-positive bacteria including *Bacillus subtilis* (Gándara and Alonso 2015), *Listeria monocytogenes* (Woodward et al. 2010), *Staphylococcus aureus* (Corrigan et al. 2011), and *Streptococcus* species such as *S. mutans* (Peng et al. 2016), *S. pneumoniae* (Bai et al. 2014), *S. pyogenes* (Kamegaya et al. 2011) *S. agalactiae* (the Group B *Streptococcus* [GBS]) (Andrade et al. 2016), and in a few Gram-negative bacteria including *Chlamydia trachomatis* (Barker et al. 2013) and *Borrelia burgdorferi* (Ye et al. 2014). The functions of c-di-AMP have been extensively studied in *streptococcus* such as biofilm formation, osmolyte transport, bacterial growth, and virulence factor expression (Corrigan and Grundling 2013, Commichau et al. 2018, Hall et al. 2018).

cGAMP is the most recently identified cyclic dinucleotide second messenger.

There are two types of cGAMP: 3',3'-cGAMP and 2'3'-cGAMP. 3',3'-cGAMP was first identified as a second messenger in *V. cholerae and* was required for chemotaxis and intestinal colonization (Davies et al. 2012). 2'3'-cGAMP was uncovered as a key factor that regulates innate immune signaling and antiviral response in eukaryotes (Krasteva et al. 2017).

The metabolism of cyclic dinucleotide second messengers

c-di-GMP

C-di-GMP is synthesized from two GTP molecules by diguanylate cyclases (DGCs) and is degraded into pGpG by phosphodiesterases (PDEs). DGC was first purified and tested in *G. xylinus* by Benziman and colleagues (Ross et al. 1987). Further biochemical and structural studies on DGCs provided the mechanistic understanding on c-di-GMP synthesis. In 1995, the gene for the pleiotropic protein D (PleD, one of the DGCs) was characterized and a novel C-terminal GGDEF (Gly-Gly-Asp-Glu-Phe) domain was identified (Hecht and Newton 1995). Further studies revealed the crystal structure of *C. crescent* PleD and showed that PleD is composed of two response regulator domains, receiver (REC) and GGDEF (Chan et al. 2004, Wassmann et al. 2007). The GGDEF (contains GGDEF and GGEEF, collectively called GGDEF) motif is conserved in most DGCs (Tal et al. 1998, Simm et al. 2004, De et al. 2009). Two Mg^{2+} or Mn^{2+} are required for the formation of phosphoester bond and the first two Gly residues of the GGDEF motif are responsible for GTP binding whereas the forth Glu residue is involved in metal ion coordination (Pei et al. 2001). The third Asp/Glu is responsible for catalysis and plays a role in metal coordination.

Hydrolysis of c-di-GMP is triggered by proteins carrying EAL (Glu-Ala-Leu) and

HD-GYP (His-Asp and Gly-Tyr-Pro) domains. EAL domain-containing PDE cleaves c-di-GMP into pGpG (5'-phosphoguanylul-guanosine) and further into two GMPs by oligoribonuclease (Orr et al. 2015). The activity of EAL-containing PDEs requires either Mg^{2+} or Mn^{2+} while its activity is strongly inhibited by Ca^{2+} and Zn^{2+} (Rao et al. 2008, Tchigvintsev et al. 2010). The activity of EAL domain-containing PDEs depends on the structure of a two-metal cation cluster which is required for the binding of divalent cations and c-di-GMP, and for catalysis (Salter et al. 2007). The EAL-containing PDEs that only have a single cation are catalytically inactive. Additional mutagenesis and structural studies classified the EAL domains into three categories: enzymatically active, potentially enzymatically active, and enzymatically inactive (Rao et al. 2009). HD-GYP domain containing proteins have resisted high-resolution crystallography, however, the crystal structure of a few HD-GYP containing PDEs has been reported (Kalia et al. 2013). One prototype of HD-GYP domain-containing protein is RpFg, which transforms c-di-GMP to GMP via 5'-pGpG as an intermediate (Nickels et al. 2011). Some hybrid proteins that contain both GGDEF and EAL or HD-GYP domains have also been reported. Although the majority of such proteins contain one intact active site, a few proteins have dual functions of DGC and PDE (Ferreira et al. 2008, Seshasayee et al. 2010).

c-di-AMP

C-di-AMP is synthesized by diadenylyl cyclase (DAC) enzymes from two molecules of ATP and is broken down to pApA by PDE enzymes. C-di-AMP is synthesized through the reaction of two molecules of ATP carried out by the nucleophilic attack at each 3'-OH group on the α-phosphate of the other ATP molecule, which leads to the ring closure accompanied by the release of two molecules of pyrophosphate (Commichau et al. 2018). The enzymes responsible for producing c-di-AMP share a common catalytically active domain known as DAC domain. C-di-AMP has not been detected in an organism which does not encode an enzyme containing a DAC domain. In these enzymes, the DAC domains are often accompanied by other domains that regulate the enzymatic activity (Commichau et al. 2018). So far, five classes of cyclases with different regulatory domains have been identified: DisA (DNA integrity scanning protein A), CdaA, CdaS, CdaM, and CdaZ (Commichau et al. 2015, Kellenberger et al. 2015, Blötz et al. 2017, Commichau et al. 2018). Although most organisms possess one type of diadenylate cylase, some contain multiple enzymes such as *Clostridium* spp., which contains both CdaA and DisA. *B. subtilis* has three types of DACs, DisA, CdaA, and CdaS (Fahmi et al. 2017). The most common motifs that reside in DAC domain proteins are DGA (Asp-Gly-Ala) and RHR (Arg-His-Arg) which possess cyclase activity required for the synthesis of c-di-AMP (Gundlach et al. 2015, Rosenberg et al. 2015). Structural analysis of DACs has revealed that DisA, CdaS and CdaA have a conserved DAC domain of about 150 amino acids. In addition, a structure-based sequence alignment of these three proteins revealed sequence identities of 28% to 49% among their DAC domains (Huynh and Woodward 2016).

The c-di-AMP hydrolyzing enzyme PDE was first identified in *B. subtilis* and since then it has been found in a variety of bacteria including *S. aureus*, *L. monocytogenes*, and *Streptococcus* spp. (Rao et al. 2010, Huynh et al. 2015). PDEs degrade C-di-AMP into two molecules of AMP with an intermediate linear form of phosphoadenyl adenosine. Four classes of PDEs that specifically degrade c-di-AMP have been identified: GdpP, DhhP, PgpH and CdnP (cyclic dinucleotide phosphodiesterase) (Huynh et al. 2016). The PDEs of PgpH and GdpP families cleave c-di-AMP to a linear dinucleotide pApA whereas Dhhp-type PDEs cleave pApA to the final product AMP. The cell wall-anchored PDE CdnP is responsible for degrading extracellular c-di-AMP to AMP (Commichau et al. 2018). The expression of PDEs varies in different bacterial species. For example, *L. monocytogenes* encodes GdpP and PgpH, whereas *Streptococcus* and *Staphylococcus* species express GdpP and Dhhp. The PDEs appear to be stimulated by both intracellular and extracellular stimuli but the specific mechanisms remain unknown. The GdpP-type PDE was first described in *L. lactis*, and later, gdpP orthologs were found in Firmicutes such as *E. aecalis, E. faecium, S. aureus, S. pneumoniae, S. mutans,* and *S. pyogenes* (Commichau et al. 2018). The GdpP-type PDEs contain two N-terminally located transmembrane helices which are accompanied by a PAS motif, a degenerated GGDEF motif, DHH and DHHA1 motifs (Huynh et al. 2016). DHH/DHHA1 domain family proteins are named due to the presence of a conserved Asp-His-His motif in their active sites which is essential for cleaving c-di-AMP to 5'-pApA (Rao et al. 2010, Corrigan et al. 2011). Dhhp also contains a DHH/DHHA1 domain which has been identified in several bacteria including *S. pneumoniae* (Cron et al. 2011)*, S. aureus* (Bowman et al. 2016)*, Mycobacterium spp*. (Tang et al. 2015), and *Borrelia burgdoferi* (Ye et al. 2014). The DhhP-type PDEs degrade c-di-AMP into AMP in a two-step process involving the formation of an intermediate 5'-pApA (Manikandan et al. 2014). PgpH-type PDEs are widespread among bacteria especially in Firmicutes. PgpH contains a histidine-aspartate (HD) domain in its C-terminus that is responsible for specific binding and degradation of c-di-AMP into 5'-pApA. In addition, PgpH possesses a 7TMR-HDEDD domain and seven transmembrane helices (Huynh et al. 2015). Cdnp is a recently discovered extracellular cell-wall-anchored c-di-AMP hydrolase in GBS (Andrade et al. 2016), consisting of a 5'-nucleotidase domain and a metallophosphoesterase domain essential for enzyme activity.

cGAMP

3',3'-cGAMP and its synthase DncV were discovered in 2012 by Mekalano and colleagues while studying the contribution of VSP-1 (*Vibrio* seventh pandemic island-1) to the pathogenesis (Davies et al. 2012). DncV can also synthesize c-di-AMP and c-di-GMP in vitro, but cGAMP is the major product of DncV. Subsequently, three cGAMP-specific PDEs named V-cGAP1/2/3 in *V. cholerae* were reported, all of which contained the HD-GYP domains. Although all the three PDEs hydrolyze 3',3'-cGAMP to a linear 5'-pApG, only V-cGAP1 harbors 5'-nuleostidase activity which enables cleavage of 5'-pApG to 5'-ApG (Gao et al. 2015). In 2015, a study on GGDEF domain-containing enzymes isolated from *G. sulfurreducens* identified dubbed hybrid promiscuous (Hypr) GGDEF enzymes that produce either c-di-AMP or cGAMP, except c-di-GMP, depending on ATP-to-GTP ratios (Hallberg et al. 2016).

2',3'-cGAMP was first identified as a cyclic dinucleotide second messenger in mammals (Ablasser et al. 2013). One phosphodiester bond of 2',3'-cGAMP has a unique linkage between 2'-OH of GMP and 5'-phosphate of AMP, distinguishing the mammalian 2'3'-cGAMP from other bacterial cyclic dinucleotides (Zhang et al. 2013). 2'3'-GAMP is synthesized by a cytoplasmic nucleotidyl transferase cyclic GMP-AMP synthase (cGAS). The synthesis of cGAMP is a two-step reaction process with an intermediate product ppGpp(2'-5')A (Gao et al. 2013). Interestingly, the double stranded DNA in the cytoplasm can stimulate the enzymatic activity of cGAS to produce cGAMP thereby inducing host immune reaction (Gao et al. 2013). In 2014, a glycoprotein named ENPP1 was reported on the plasma membrane and endoplasmic reticulum which can degrade 2',3'-cGAMP into AMP and GMP (Li et al. 2014).

The regulation of cyclic di-nucleotides in *Streptococci*

C-di-GMP can be sensed by RNAs, proteins, and is involved in many biological processes including switching between motile and sessile biofilm states, virulence, biofilm matrix formation and quorum sensing (Romling et al. 2005, Shanahan et al. 2012). C-di-GMP binding RNAs contain the GEMM (Genes for the Environment, Membranes and Motility) motif (Weinberg et al. 2007), and these RNAs are upstream of genes that encode diguanylate cyclases, phosphodiesterases and some other c-di-GMP-associated proteins. C-di-GMP binding proteins carry the GGDEF domains and the EAL domains containing the diguanylate cyclase and phosphodiesterase activity, respectively (Ramesh 2015). The discovery of c-di-GMP in *Streptococci* has not been reported, but the impact of extracellular c-di-GMP on *S. mutans* is not fully understood. It was shown that c-di-GMP reduced biofilm formation in *S. mutans* UA159 and inhibited the adherence of *S. mutans* to tooth surfaces suggesting that c-di-GMP may be used as a novel antimicrobial agent (Yan et al. 2010). Indeed, c-di-GMP can inhibit the occurrence and development of dental caries in rats (Yan et al.), but has no effect on the number of bacterial plagues. Moreover, C-di-GMP may indirectly affect *S. mutans* via other bacteria or mechanisms. However, further studies are needed to validate the protective effect of c-di-GMP on dental caries to boost the exploitation of its therapeutic effects.

Effectors that bind to c-di-AMP in *Streptococcus* include CabPA, CabPB, KtrA and TrkH potassium transporter subunits, OpuCA transporter subunit, as well as BusR transcriptional factor which are closely associated with bacterial biofilm formation and osmotic homeostasis (Peng et al. 2016, Devaux et al. 2018). In addition, c-di-AMP may influence *Streptococcus* cell phenotypes such as growth, fitness, and virulence. However, the signal transduction pathways downstream of c-di-AMP remain elusive. Apart from binding to protein adaptors, c-di-AMP can also bind some bacterial signal-sensing riboswitches such as *ydaO* riboswitch with high affinity and specificity. In this way, they regulate the cell wall metabolism, sporulation, and germination in eubacteria (Nelson et al. 2013).

The role of c-di-AMP in *S. mutans* has been extensively studied and the CabPA/VicR/GtfB signaling pathway regulated by c-di-AMP during *S. mutans* biofilm formation has been recently reported. C-di-AMP interacts with VicR through its binding protein (CabPA), and up-regulates *gtfB* expression. The *gtfB* gene encodes a glucosyltransferase which regulates the synthesis of water-insoluble glucans. Therefore, the effect of C-di-AMP on *gtfB* expression augmented EPS (extracellular polysaccharides) production and biofilm formation in *S. mutans*. Deletion of *pdeA* gene, which encodes PDE, increased c-di-AMP levels in *S. mutans* and upregulated the expression of *gtfB*. Moreover, CabPA, a c-di-AMP binding protein containing TrkA_C and TrkA_N domains with a molecular mass of ~25 kDa, can specifically interact with VicR, a response regulator of a two-component regulatory system in *S. mutans*. This provides the basic stimulus-response coupling mechanism allowing organisms to perceive and respond to changes in environmental conditions. It was demonstrated that increased c-di-AMP level in *S. mutans* due to *pdeA* deficiency promoted bacterial colonization in Drosophila (Peng, Zhang et al. 2016). Besides, Cheng et al. (Cheng et al. 2016) showed that in-frame deletion of *cdaA* gene in *S. mutans* decreased c-di-AMP level and increased not only the production of EPS but also the sensitivity to hydrogen peroxide. The global analyses of gene expression revealed that genes with significantly altered expression were clustered in bacterial polysaccharide biosynthesis and oxidoreductase process. The deletion of *cdaA* which encodes CdaA in *S. mutans* resulted in a striking decrease in c-di-AMP levels and bacterial growth rate but increase in cell lysis. The *cdaA* deficiency altered the cell shape and size in *S. mutans* compared to the wild type bacteria. Furthermore, the Δ*cdaA* mutant displayed decreased competitiveness against *Streptococcus sanguinis* which increased the sensitivity of *S. mutans* Δ*cdaA* mutant to H_2O_2 produced by *S. sanguinis* and the production of EPS. A markedly enhanced glucosyltransferases (GTFs) activity was also observed in Δ*cdaA* mutant. Based on whole-genome microarrays, transcriptional changes in a set of functional genes were identified in the Δ*cdaA* mutant. The up-regulated genes included genes associated with cell envelope biogenesis, ribosomal structure and biogenesis, coenzyme metabolism, protein turnover, chaperones, post-translational modification, and translation. The down-regulated genes comprised of genes involved in carbohydrate transport and metabolism, cell secretion, and DNA replication, recombination and repair (Cheng et al. 2016). However, Peng et al. (Peng et al. 2016) reported opposite findings that DAC deficiency in *S. mutans* decreased the production of EPS and that the response of DAC mutant to oxidative stress is independent of gtfB. They observed that DAC deficiency significantly reduced the expression of gtfB, resulting in decreased biofilm formation in the mutant. This inhibited bacterial colonization by the mutant and reduced the cariogenicity of SmDAC deficiency on buccal and sulcal surfaces of teeth. The authors also showed that the attenuated response to hydrogen peroxide stress in SmDAC deficiency contributes to low virulence. Moreover, Peng et al. provided corroborating evidence that *smDAC* mutant exhibits the typical pattern of Gtfs, but complementation of the mutant in a study by Cheng et al. failed to restore its EPS production ability compared to wild type. In contrast to the conclusion reached by Cheng et al. that EPS is required for bacterial resistance to hydrogen peroxide challenge, Peng et al. found that *SmDAC* mutant decreased GtfB and EPS levels and that deletion of *gtfB* did not affect the EPS levels of the *smDAC gtfB* double mutant (Cheng et al. 2016). These contradictory findings point to the complexity of the c-di- AMP signaling network in *S. mutans* and warrant further investigation.

In *S. pneumoniae*, deletion of *pde1* and *pde2* genes encoding PDE markedly increased c-di-AMP levels. Homeostasis of c-di-AMP is essential for pneumococcal growth and its associated pathogenesis which includes adherence to human epithelial cells, colonization and infection. Various mice models have shown that Pde1 and Pde2 are essential for bacterial biology and bacterial-induced diseases (Bai et al. 2013). In a subsequent study, CabP which directly interacts with a potassium transporter (SPD_0076) was identified in *S. pneumoniae*. C-di-AMP regulates potassium uptake by *S. pneumoniae* via its specific binding protein CabP which interacts with a potassium transporter. On the other hand, c-di-AMP reduces the interaction between CabP and SPD_0076 and impairs the ability of potassium uptake by CabP-SPD_0076 complex in *S. pneumonia* (Bai et al. 2014). It was found that high c-di-AMP levels enhanced the susceptibility of *S. pneumoniae* phosphodiesterase mutants to stress conditions such as heat shock, acidic environment, and osmotic stress. Thus, it is likely that accumulation of c-di-AMP is toxic under specific incubation conditions and inhibits adaptation to the stresses in *S. pneumoniae*. Given that the deletion of c-di-AMP effector protein CabP reduced c-di-AMP levels in *S. pneumoniae*, it can be inferred that c-di-AMP and its binding protein mutually affect each other through a feedback loop and this relationship may is dependent on the pneumococcal potassium uptake process. In *S. pneumonia,* it was also demonstrated that mutations in *cdaA* reduced its enzymatic activity and the levels of c-di-AMP, leading to increased resistance to environmental stress in the Δ*pde1* Δ*pde2* mutant. Although CdaA plays a major role in restoring stress resistance, the effect of TrkH in reducing c-di-AMP levels in isolated suppressor mutant is minimal (Zarrella et al. 2018). The c-di-AMP/CabP/SPD_0076 pathway in *S. pneumoniae* regulates potassium uptake, but the mechanism of c-di-AMP in pneumococcal stress resistance is unknown.

In *S. pyogene* (Cho et al. 2013), c-di-AMP phosphodiesterase GdpP influences the maturation of SpeB, a major cysteine protease which can destroy human matrices such as fibronectin and vitronectin thus damaging the host tissues. The absence of GdpP decreases the ability to produce active SpeB. GdpP also influences the virulence of *S. pyogene* by decreasing the sensitivity to β-lactam antibiotic ampicillin in the GdpP null mutant. Re-introduction of GdpP enhanced SpeB processing in the ΔGdpP mutant, although the underlying mechanisms of this effect remain elusive. Moreover, the virulence of ΔGdpP declined in a murine model of soft tissue infection. All these results suggest that increased c-di-AMP levels resulting from GdpP deletion promote SpeB processing, pathogenicity of *S. pyogene* and its response to ampicillin. In S. pyogenes, SpyDAC was found to catalyze the in vivo biosynthesis of c-di-AMP from ATP.

C-di-AMP synthesis is essential for GBS in standard growth conditions but becomes dispensable afterwards due to the accumulation of other compensatory mutations, especially mutations in the genes encoding BusAB transporter responsible for osmolyte import in GBS. Previous experiments demonstrated that c-di-AMP binding transcriptional factor BusR is essential for osmotic homeostasis and that BusR negatively regulates *busAB* transcription by directly binding to the *busAB* promoter. Deletion of *gdpP* which encodes c-di-AMP phosphodiesterase increased c-di-AMP levels in GBS and led to hyperosmotic susceptibility by influencing the BusR transcriptional factor. C-di-AMP is not required for growth in a media without osmolyte which is attributed to the combination of dysregulated homeostasis of potassium and osmolyte uptake. However, high c-di-AMP levels avoids inhibitory effect in the presence of the osmolytes. Adaptation to the absence of c-di-AMP synthesis in Δ*dacA* mutants involves intertwined mutations including necessary mutations in the two *oppC* and *busB* genes as well as other compensatory mutations which alleviates BusB toxicity. In conclusion, c-di-AMP is a central regulator of osmotic homeostasis in GBS and acts by binding to KtrA and TrkH potassium transporter subunits, OpuCA osmolyte transporter subunit, as well as BusR transcriptional factor which represses the expression of BusAB osmolyte transporter. Furthermore, binding of full-length protein Eric containing chloride channel protein to c-di-AMP was not found in this study (Devaux et al. 2018). This intertwined network involves more than four kinds of c-di-AMP binding proteins responsible for potassium uptake or osmolyte import, and both of these processes are essential for osmotic homeostasis.

SSU98_1483 (ssDacA) is a diadenylate cyclase found in *S. suis serotype* 2 (SS2)

(Du et al. 2015). GdpP is a c-di-AMP specific phosphodiesterase that degrades c-di-AMP and the *gdpP* mutant markedly increased the c-di-AMP levels in SS2. This suggests that c-di-AMP is a biological substrate of GdpP and that GdpP contributes to the homeostasis of c-di-AMP in SS2. Studies showed that the *gdpP* mutant strain exhibited lower growth rate than the wild type and complementation of *gdpP* mutant partly restored the growth rate. The deletion of *gdpP* also increased biofilm formation in SS2 and the mutant exhibited a significant reduction in hemolytic activity, invasion and adherence to HEp-2 cells compared to the parental strain which indicates that GdpP contributes to the virulence of SS2. GdpP or c-di-AMP influences the expression of virulence-related genes *cps2*, *sly, fpbs*, *mrp*, *ef* and *gdh.* These genes are significantly decreased in *gdpP* mutant based on qRT-PCR analysis. In murine infection models, the pathogenicity of *gdpP* mutant strain was attenuated compared with the wild-type strain, indicating that inactivation of GdpP weakens the virulence of SS2. Also, in vivo colonization experiments demonstrated that more wild-type strains were recovered from brain, spleen, lung and liver than the *gdpP* mutant strains after intraperitoneal inoculation with *gdp*P mutant or the wild type strain. This result reveals the significance of c-di-AMP homeostasis during SS2 infection. Deletion of *gdp*P weakened the growth and virulence but improved the biofilm formation ability of SS2 (Du et al. 2014).

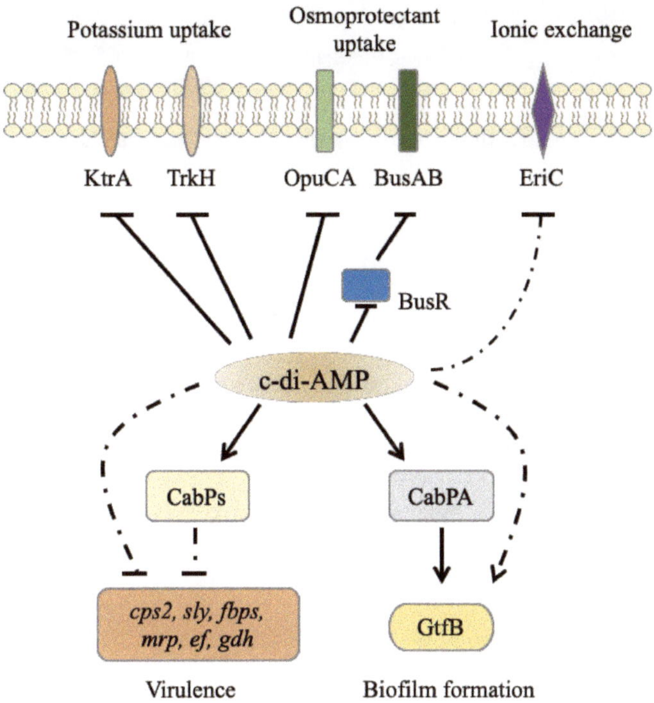

Figure 1. c-di-AMP signaling in *streptococcus*. The KtrA and TrkH potassium transporter subunits and OpuCA osmolyte transporter subunits are conserved c-di-AMP binding proteins. C-di-AMP can bind to BusR transcriptional factor, which is a repressor of osmolyte transporter BusAB. C-di-AMP might also regulate EriC, a chloride channel protein with 11 transmembrane domains. High level of c-di-AMP binds to CabPA, which regulates the expression of *gtfB* intermediated by histidine two-component system response regulator VicR. c-di-AMP also affect streptococcal virulence by influencing the putative expression of virulence-related genes *cps2*, *sly*, *fpbs*, *mrp*, *ef* and *gdh.*

A recent study showed that (Severin et al. 2018) cGAMP can activate a phospholipase Vibrio (CapV) which degrades phospholipids in El Tor biotype of *Vibrio cholerae* cell membrane. However, the presence of this phospholipase in *Streptococcus* has not been reported given that it has recently been identified as a second messenger in bacteria.

The interaction between host cells and *Streptococcus* are regulated by CDNs

With the development of new techniques, novel host cell receptors sensing c-di-AMP or c-di-GMP have been reported. C-di-AMP and c-di-GMP trigger type I interferons via STING including MITA and MPYS which further activate TBK1 kinase and IRF3 transcription factor (Burdette et al. 2011, Chen et al. 2016, Ruiz-Moreno et al. 2018). Besides, helicase DEAD box polypeptide 41 (DDX41) was also found to be one of a pattern recognition receptor (PRR) which senses c-di-AMP or c-di-GMP, and interacts and co-localizes with the STING adaptor (Parvatiyar et al. 2012).The oxidoreductase RECON is another CDN host receptor which antagonizes STING activation and downregulates NF-kB activation. And CDN inhibition or gene knockout of RECON leads to increasing NF-kB activation and a decline in bacterial survival (McFarland et al. 2017). Recently, ERAdP that acts as a direct sensor of c-di-AMP in *L. monocytogenes* infection has been reported. C-di-AMP binding to a C-terminal domain of ERAdP results in dimerization of ERAdP leading to the activation of the kinase TAK1 which activates NF-κB to induce the production of cytokines in host innate immune cells (Moretti et al. 2017).

c-di-GMP is a unique bacterial intracellular second messenger which promotes protective innate and adaptive immunity against various invasive bacterial infections including, but not limited to *S. pneumoniae and S. pyogenes* (Ogunniyi et al. 2008, Yan et al. 2009). Intranasal pretreatment of mice with c-di-GMP led to a significant decrease in pneumococcal bacterial load in lungs and blood after serotypes 2 and 3 challenge, and also reduced lung titers after serotype 4 challenge. Coincidentally, intraperitoneal co-administration of c-di-GMP significantly increased antigen-specific antibody titers as well as the survival of mice. It was also observed that c-di-GMP has no significant direct short-term influence on *S. pneumoniae* growth either *in vitro* or *in vivo* (Ogunniyi et al. 2008). Intranasal c-di-GMP administration induces pulmonary dendritic cell recruitment and proinflammatory cytokine/chemokine responses in mice. Similarly, co-administration of c-di-GMP with pneumococcal surface adhesion A (PsaA) led to a strong antigen-specific serum immunoglobulin G (IgG) and secretory IgA antibody response at multiple mucosal surfaces. Importantly, the immunized mice showed a significantly reduced nasopharyngeal *S. pneumoniae* colonization (Yan et al. 2009). Moreover, c-di-GMP not only induced a higher Ab production including IgG1 and IgA but also conferred a better protection against *S. pneumoniae* infection than the mammalian 2'3'-cGAMP. However, c-di-GMP did not cause strong inflammatory responses or lung injury but enhanced type II (IFNγ) and III IFN (IFN λ) production and generated of T_H1, T_H2, and T_H17 responses through the production of polarizing cytokines including T_H1 polarizing cytokines IL-12p70 and IFNγ, T_H2 polarizing cytokines IL-5 and IL-13, and T_H17 polarizing cytokines IL-23, IL-6 and TGF-β1 by activating pinocytosis-efficient-DCs. C-di-GMP also enhanced Ag uptake dependent on MPYS (STING, MITA) expression in $CD11C^+$ cells (Blaauboer et al. 2015).

In addition, c-di-AMP influences the interaction between host cells and *Streptococcus* by signaling through factors that regulate this process. In GBS, IFN-β production is mainly regulated through cGAS-STING-dependent recognition of bacterial DNA. GBS expresses an ectonucleotidase CdnP which hydrolyzes extracellular bacterial c-di-AMP and reduces its levels. Inactivation of CdnP leads to c-di-AMP accumulation outside the bacteria and an increase in IFN-β levels *in vivo* which promotes GBS killing by the host and reduces GBS virulence. This process is mainly due to enhanced DNA sensing by cGAS and STING-dependent sensing of c-di-AMP. Therefore, GBS expresses CdnP at its surface to promote its virulence (Andrade et al. 2016). Furthermore, a degraded RNA sensor in monocytes and macrophages of human Toll-Like Receptor 8 (TLR8) and IFN regulatory factor 5 (IRF5) form a signaling pathway that regulates GBS sensing and cytokine activation in host cells (Ehrnstrom et al. 2017).

Studies have shown that cGAMP is an important node in host innate immune responses to bacterial DNAs and is a strong activator of STING which induces interferon-β production through activation of IRF3 in mammalian cells transfected with either the DNA or infected with DNA virus (Wu et al. 2013). It was reported that cGAMP activates B cells through STING pathway when cytoplasmic DNA is sensed *in vivo.* Human B cells can be activated by cGAMP *in vitro* thus increasing IgM production (Chen et al. 2016). cGAS senses and binds to cytosolic DNA and activates the synthesis of cGAMP and subsequent production of type I interferon (Cai et al. 2014). During the defense against *Neisseria gonorrhoeae* (GC), both cGAMP and TLR4, known as a host pattern recognition receptor, contribute to IFN-β activation in the immune cells. TLRs and cGAS functions synergistically when host innate immunity is confronted with GC invasion (Andrade et al. 2016).

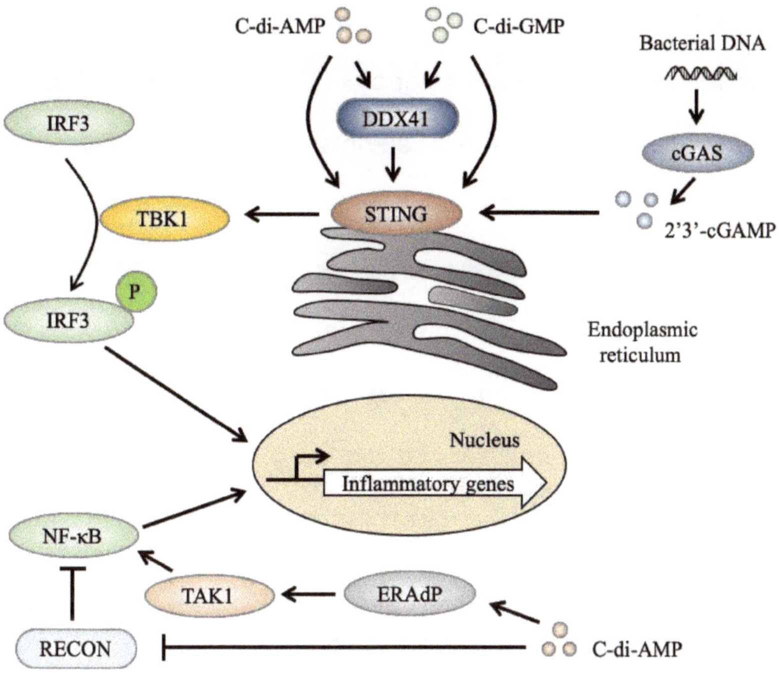

Fig 2. Induction of host immune responses by CDNs. Bacterial c-di-AMP and c-di-GMP bind to eukaryotic helicase DDX41, which in turn interacts and activates the co-receptor STING. Some evidence suggest that these two dinucleotides interact directly with STING. Activated STING can then activate the kinase TBK1, leading to phosphorylation of transcription factor IRF3 triggering its translocating to the nucleus, further promoting the expression of IFNβ and activation of type I interferon response. In addition, bacterial double-strand DNA may be sensed by eukaryotic cGAMP synthase cGAS. High levels of 2'3'-cGAMP interacts and activates STING, leading to activation of type I interferon response through TBK1/IRF3 pathway. The oxidoreductase RECON is another cyclic dinucleotide host receptor which prevents the activation of transcription factor NF-κB. Recently, it was found that the ER membrane adaptor ERAdP acts as a direct sensor for c-di-AMP, which causes dimerization of ERAdP. This in turn interacts and activates the kinase TAK1.Activated TAK1 initiates activation and translocation of NF-κB, leading to the production of cytokines.

Conclusions and prospect

In this review, the regulatory networks of c-di-GMP, c-di-AMP, and cGAMP cyclic nucleotide second messengers are summarized. Cyclic nucleotide second messengers play important roles in both prokaryotes and eukaryotes. They mediate bacterial growth, fitness, virulence, and regulate innate immune responses in host cells via intricate signaling networks. Cyclic nucleotide signaling molecules are rapidly internalized by cellular diffusion and act in highly specific manner forming a multilayered signaling scaffold (Jenal et al. 2017). However, our understanding on the role of cyclic dinucleotide signaling is far from being complete. In *Streptococcus*, a diverse group of more than 100 recognized species which cause invasive diseases (Andam et al. 2015), it is important to explore the specific mechanisms of signal transduction based on nucleotide signaling molecules to develop effective strategies for controlling infections caused by these pathogens. It is possible that there are other unidentified cyclic dinucleotides which may be discovered in future research. In addition, future studies on cyclic dinucleotide should focus on identifying more cyclic di-nucleotides receptor and effector proteins to fill the gaps in our understanding of the cyclic di-nucleotide messengers-based network. *yadO* riboswitch is the only identified c-di-AMP binding RNA and majority of bacteria that produce c-di-AMP do not contain this riboswitch (Fahmi et al. 2017). Thus, identification of potential c-di-AMP binding RNA should also be a focus of further research. Moreover, the specific environmental input signals that influence cyclic di-nucleotide second messenger pathways remain to be identified.

It has been proposed that c-di-GMP can be administered as an effective broad spectrum immunomodulator and vaccine adjuvant to prevent infectious diseases.

This also implies that c-di-GMP has the potential to be a mucosal vaccine adjuvant. Moreover, a combination of LCP based peptide vaccine with c-di-AMP targeted against *S. pyogenes* infection reduced antigen dose and maintained vaccine efficacy providing a strong evidence that c-di-GMP can be used as a vaccine adjuvant (Schulze et al. 2017). In a study on mice model, it was found that c-di-GMP exhibited better mucosal adjuvant response than cGAMP.

Compared to c-di-GMP, the c-di-AMP signaling network appears to be far simpler since most organisms contain only a single DAC enzyme. However, in view of the various phenotypes associated with the levels of c-di-AMP, it is possible that there are many c-di-AMP effectors. This might be the case for other second messenger systems which function through many effectors and regulate many physiological processes. cGAMP has been identified in bacteria and mammals, however, the presence of cGAMP in bacteria is limited to an only a few species. Given that the enzymes responsible for cGAMP synthesis also synthesize c-di-GMP or c-di-AMP, we postulate that cGAMP signaling network may be interrelated with other cyclic di-nucleotide signal networks. Moreover, there are no reports on cGAMP producing *Streptococcus* or cGAMP signal pathway related physiological processes in *Streptococcus* and warrant further investigation.

Given the indispensable roles of cyclic dinucleotide second messengers in bacteria and in host cells, there have been efforts to interfere with cyclic dinucleotide signaling (Opoku-Temeng et al. 2016). However, only a handful of inhibitors of cyclic dinucleotide signaling have been identified. We believe that with the development of high-throughput screening methods, other small molecule regulators of cyclic signal pathways will be identified. Considering the importance of cyclic dinucleotide in bacteria cellular process and host immune reactions, the development of novel cyclic nucleotide inhibitors may have a tremendous impact on clinical practice.

Acknowledgements

This work was supported by the National Natural Science Foundation of China (81700963) and Sichuan Science and Technology Program (2018JY0561).

References

Ablasser, A., et al. (2013). cGAS produces a 2 '-5 '-linked cyclic dinucleotide second messenger that activates STING. Nature 498(7454): 380-384. doi: 10.1038/nature12306

Andam, C. P. and, W. P. (2015). Mechanisms of genome evolution of Streptococcus. Infect Genet Evol 33: 334-342. doi: 10.1016/j.meegid.2014.11.007

Andrade, W. A., et al. (2016). Type I Interferon Induction by Neisseria gonorrhoeae: Dual Requirement of Cyclic GMP-AMP Synthase and Toll-like Receptor 4. Cell Rep 15(11): 2438-2448. doi: 10.1016/j.celrep.2016.05.030

Andrade, Warrison A., et al. (2016). Group B Streptococcus Degrades Cyclic-di-AMP to Modulate STING-Dependent Type I Interferon Production. Cell Host Microbe 20(1): 49-59. doi: 10.1016/j.chom.2016.06.003

Bai, Y., et al. (2013). Two DHH subfamily 1 proteins in Streptococcus pneumoniae possess cyclic di-AMP phosphodiesterase activity and affect bacterial growth and virulence. J Bacteriol 195(22): 5123-5132. doi: 10.1128/JB.00769-13

Bai, Y., et al. (2014). Cyclic Di-AMP Impairs Potassium Uptake Mediated by a Cyclic Di-AMP Binding Protein in Streptococcus pneumoniae. J Bacteriol 196(3): 614. doi: 10.1128/JB.01041-13

Barker, J. R., et al. (2013). STING-dependent recognition of cyclic di-AMP mediates type I interferon responses during Chlamydia trachomatis infection. MBio 4(3): e00018-00013. doi: 10.1128/mBio.00018-13

Blaauboer, S. M., et al. (2015). The mucosal adjuvant cyclic di-GMP enhances antigen uptake and selectively activates pinocytosis-efficient cells in vivo. Elife 4: e06670. doi: 10.7554/eLife.06670

Blötz, C., et al. (2017). Identification of the components involved in cyclic di-AMP signaling in Mycoplasma pneumoniae. Front Microbiol 8: 1328. doi: 10.3389/fmicb.2017.01328

Bowman, L., et al. (2016). New insights into the cyclic di-adenosine monophosphate (c-di-AMP) degradation pathway and the requirement of the cyclic-dinucleotide for acid stress resistance in Staphylococcus aureus. J Biol Chem 291(53): 26970-26986. doi: 10.1074/jbc.M116.747709

Burdette, D. L., et al. (2011). STING is a direct innate immune sensor of cyclic di-GMP. Nature 478(7370): 515-518. doi: 10.1038/nature10429

Cai, X., et al. (2014). The cGAS-cGAMP-STING pathway of cytosolic DNA sensing and signaling. Mol Cell 54(2): 289-296. doi: 10.1016/j.molcel.2014.03.040

Chan, C., et al. (2004). Structural basis of activity and allosteric control of diguanylate cyclase. Proc Natl Acad Sci U S A 101(49): 17084. doi:

10.1073/pnas.0406134101

Chen, Q., et al. (2016). Regulation and function of the cGAS-STING pathway of cytosolic DNA sensing. Nat Immunol 17(10): 1142-1149. doi: 10.1038/ni.3558

Cheng, X., et al. (2016). Regulation of oxidative response and extracellular polysaccharide synthesis by a diadenylate cyclase in Streptococcus mutans. Environ Microbiol 18(3): 904-922. doi: 10.1111/1462-2920.13123

Cho, K. H. and S. O. Kang (2013). Streptococcus pyogenes c-di-AMP Phosphodiesterase, GdpP, Influences SpeB Processing and Virulence. PLoS One 8(7): e69425. doi: 10.1371/journal.pone.0069425

Commichau, F. M., et al. (2015). A jack of all trades: the multiple roles of the unique essential second messenger cyclic di-AMP. Mol Microbiol 97(2): 189-204. doi: 10.1111/mmi.13026

Commichau, F. M., et al. (2018). A Delicate Connection: c-di-AMP Affects Cell Integrity by Controlling Osmolyte Transport. Trends Microbiol 26(3): 175-185. doi: 10.1016/j.tim.2017.09.003

Commichau, F. M., et al. (2018). Making and breaking of an essential poison: the cyclases and phosphodiesterases that produce and degrade the essential second messenger cyclic di-AMP in bacteria. J Bacteriol 201(1). pii: e00462-18. doi: 10.1128/JB.00462-18

Corrigan, R. M., et al. (2011). c-di-AMP is a new second messenger in Staphylococcus aureus with a role in controlling cell size and envelope stress. PLoS Pathog 7(9): e1002217. doi: 10.1371/journal.ppat.1002217

Corrigan, R. M. and A. Grundling (2013). Cyclic di-AMP: another second messenger enters the fray. Nat Rev Microbiol 11(8): 513-524. doi: 10.1038/nrmicro3069

Cron, L., et al. (2011). Two DHH subfamily 1 proteins contribute to pneumococcal virulence and confer protection against pneumococcal disease. Infect Immun 79(9):3697-710. doi: 10.1128/IAI.01383-10

Danilchanka, O. and J. J. Mekalanos (2013). Cyclic dinucleotides and the innate immune response. Cell 154(5): 962-970. doi: 10.1016/j.cell.2013.08.014

Davies, B. W., et al. (2012). Coordinated regulation of accessory genetic elements produces cyclic di-nucleotides for V. cholerae virulence. Cell 149(2): 358-370. doi: 10.1016/j.cell.2012.01.053

Davies, Bryan W., et al. (2012). Coordinated Regulation of Accessory Genetic Elements Produces Cyclic Di-Nucleotides for V. cholerae Virulence. Cell 149(2): 358-370. doi: 10.1016/j.cell.2012.01.053

De, N., et al. (2009). Determinants for the activation and autoinhibition of the diguanylate cyclase response regulator WspR. J Mol Biol 393(3): 619-633. doi: 10.1016/j.jmb.2009.08.030

Devaux, L., et al. (2018). Cyclic di-AMP regulation of osmotic homeostasis is essential in Group B Streptococcus. PLoS Genet 14(4): e1007342. doi: 10.1371/journal.pgen.1007342

Drexler, D. J., et al. (2017). Structural and Biophysical Analysis of the Soluble DHH/DHHA1-Type Phosphodiesterase TM1595 from Thermotoga maritima. Structure 25(12): 1887-1897 e1884. doi: 10.1016/j.str.2017.10.001

Du, B., et al. (2014). Functional analysis of c-di-AMP phosphodiesterase, GdpP, in Streptococcus suis serotype 2. Microbiol Res 169(9-10): 749-758. doi: 10.1016/j.micres.2014.01.002

Du, B. and J. H. Sun (2015). Diadenylate cyclase evaluation of ssDacA (SSU98_1483) in Streptococcus suis serotype 2. Genet Mol Res 14(2): 6917-6924. doi: 10.4238/2015.June.18.34

Ehrnstrom, B., et al. (2017). Toll-Like Receptor 8 Is a Major Sensor of Group B

Streptococcus But Not Escherichia coli in Human Primary Monocytes and Macrophages. Front Immunol 8: 1243. doi: 10.3389/fimmu.2017.01243

Fahmi, T., et al. (2017). c-di-AMP: An Essential Molecule in the Signaling Pathways that Regulate the Viability and Virulence of Gram-Positive Bacteria. Genes (Basel) 8(8). doi: 10.3390/genes8080197

Ferreira, R. B., et al. (2008). Vibrio parahaemolyticus ScrC modulates cyclic dimeric GMP regulation of gene expression relevant to growth on surfaces. J Bacteriol 190(3): 851-860. doi: 10.1128/JB.01462-07

Galperin, M. Y. (2018). What bacteria want. Environ Microbiol. doi: 10.1111/1462-2920.14398

Gándara, C. and J. C. Alonso (2015). DisA and c-di-AMP act at the intersection between DNA-damage response and stress homeostasis in exponentially growing Bacillus subtilis cells. DNA Repair 27: 1-8. doi: 10.1016/j.dnarep.2014.12.007

Gao, J., et al. (2015). Identification and characterization of phosphodiesterases that specifically degrade 3′3′-cyclic GMP-AMP. Cell Research 25: 539. doi: 10.1038/cr.2015.40

Gao, P., et al. (2013). Cyclic [G(2′,5′)pA(3′,5′)p] Is the Metazoan Second Messenger Produced by DNA-Activated Cyclic GMP-AMP Synthase. Cell 153(5): 1094-1107. doi: 10.1016/j.cell.2013.04.046

Gomelsky, M. (2011). cAMP, c-di-GMP, c-di-AMP and now cGMP: bacteria use them all! Mol Microbiol 79(3): 562-565. doi: 10.1111/j.1365-2958.2010.07514.x

Gundlach, J., et al. (2015). An essential poison: synthesis and degradation of cyclic di-AMP in Bacillus subtilis. J Bacteriol 197(20):3265-74. doi: 10.1128/JB.00564-15

Hall, C. L. and V. T. Lee (2018). Cyclic-di-GMP regulation of virulence in bacterial pathogens. Wiley Interdiscip Rev RNA 9(1). doi: 10.1002/wrna.1454

Hallberg, Z. F., et al. (2016). Hybrid promiscuous (Hypr) GGDEF enzymes produce cyclic AMP-GMP (3', 3'-cGAMP). Proc Natl Acad Sci U S A 113(7): 1790-1795. doi: 10.1073/pnas.1515287113

Hecht, G. B. and A. Newton (1995). Identification of a novel response regulator required for the swarmer-to-stalked-cell transition in Caulobacter crescentus. J Bacteriol 177(21): 6223-6229. doi: 10.1128/jb.177.21.6223-6229.1995

Hengge, R., et al. (2016). Bacterial Signal Transduction by Cyclic Di-GMP and Other Nucleotide Second Messengers. J Bacteriol 198(1): 15-26. doi: 10.1128/jb.00331-15

Huynh, T. N., et al. (2015). An HD-domain phosphodiesterase mediates cooperative hydrolysis of c-di-AMP to affect bacterial growth and virulence. Proc Natl Acad Sci U S A 112(7): E747-E756. doi: 10.1073/pnas.1416485112

Huynh, T. N. and J. J. Woodward (2016). Too much of a good thing: regulated depletion of c-di-AMP in the bacterial cytoplasm. Curr Opin Microbiol 30: 22-29. doi: 10.1016/j.mib.2015.12.007

Jenal, U., et al. (2017). Cyclic di-GMP: second messenger extraordinaire. Nat Rev Microbiol 15(5): 271-284. doi: 10.1038/nrmicro.2016.190

Kalia, D., et al. (2013). Nucleotide, c-di-GMP, c-di-AMP, cGMP, cAMP, (p)ppGpp signaling in bacteria and implications in pathogenesis. Chem Soc Rev 42(1): 305-341. doi: 10.1039/c2cs35206k

Kamegaya, T., et al. (2011). Identification of a Streptococcus pyogenes SF370 gene involved in production of c-di-AMP. Nagoya J Med Sci 73(1-2): 49-57. doi: 10.18999/nagjms.73.1-2.49

Kaspar, J., et al. (2016). An Essential Role for (p)ppGpp in the Integration of Stress Tolerance, Peptide Signaling, and Competence Development in

Streptococcus mutans. Front Microbiol 7:1162. doi: 10.3389/fmicb.2016.01162

Kazlauskiene, M., et al. (2017). A cyclic oligonucleotide signaling pathway in type III CRISPR-Cas systems. Science 357(6351): 605-609. doi: 10.1126/science.aao0100

Kellenberger, C. A., et al. (2015). RNA-based fluorescent biosensors for live cell imaging of second messenger cyclic di-AMP. J Am Chem Soc 137(20): 6432-6435. doi: 10.1021/jacs.5b00275

Krasteva, P. V. and H. Sondermann (2017). Versatile modes of cellular regulation via cyclic dinucleotides. Nat Chem Biol 13(4): 350-359. doi: 10.1038/nchembio.2337

Li, L., et al. (2014). Hydrolysis of 2′3′-cGAMP by ENPP1 and design of nonhydrolyzable analogs. Nat Chem Biol 10: 1043. doi: 10.1038/nchembio.1661

Liu, K., et al. (2015). Diversity in (p)ppGpp metabolism and effectors. Curr Opin Microbiol 24: 72-79. doi: 10.1016/j.mib.2015.01.012

Manikandan, K., et al. (2014). Two-step synthesis and hydrolysis of cyclic di-AMP in Mycobacterium tuberculosis. PLoS One 9(1): e86096. doi: 10.1371/journal.pone.0086096

Marden, J. N., et al. (2011). Cyclic GMP controls Rhodospirillum centenum cyst development. Mol Microbiol 79(3): 600-615. doi: 10.1111/j.1365-2958.2010.07513.x

McFarland, A. P., et al. (2017). Sensing of Bacterial Cyclic Dinucleotides by the Oxidoreductase RECON Promotes NF-kappaB Activation and Shapes a Proinflammatory Antibacterial State. Immunity 46(3): 433-445. doi: 10.1016/j.immuni.2017.02.014

Moretti, J., et al. (2017). STING Senses Microbial Viability to Orchestrate Stress-Mediated Autophagy of the Endoplasmic Reticulum. Cell 171(4): 809-823 e813. doi: 10.1016/j.cell.2017.09.034

Nelson, J. W., et al. (2013). Riboswitches in eubacteria sense the second messenger c-di-AMP. Nat Chem Biol 9(12): 834-839. doi: 10.1038/nchembio.1363

Nickels, B. E. and S. L. Dove (2011). NanoRNAs: a class of small RNAs that can prime transcription initiation in bacteria. J Mol Biol 412(5): 772-781. doi: 10.1016/j.jmb.2011.06.015

Ogunniyi, A. D., et al. (2008). c-di-GMP is an effective immunomodulator and vaccine adjuvant against pneumococcal infection. Vaccine 26(36): 4676-4685. doi: 10.1016/j.vaccine.2008.06.099

Opoku-Temeng, C., et al. (2016). Cyclic dinucleotide (c-di-GMP, c-di-AMP, and cGAMP) signalings have come of age to be inhibited by small molecules. Chem Commun (Camb) 52(60): 9327-9342. doi: 10.1039/c6cc03439j

Orr, M. W., et al. (2015). Oligoribonuclease is the primary degradative enzyme for pGpG in Pseudomonas aeruginosa that is required for cyclic-di-GMP turnover. Proc Natl Acad Sci U S A 112(36):E5048-57. doi: 10.1073/pnas.1507245112

Parvatiyar, K., et al. (2012). The helicase DDX41 recognizes the bacterial secondary messengers cyclic di-GMP and cyclic di-AMP to activate a type I interferon immune response. Nat Immunol 13(12): 1155-1161. doi: 10.1038/ni.2460

Pei, J. and N. V. Grishin (2001). GGDEF domain is homologous to adenylyl cyclase. Proteins. 2001 Feb 1;42(2):210-6. doi: 10.1002/1097-0134(20010201)42:2<210::AID-PROT80>3.0.CO;2-8

Peng, X., et al. (2016). Effects of diadenylate cyclase deficiency on synthesis of extracellular polysaccharide matrix of Streptococcus mutans revisit. Environ

Microbiol 18(11): 3612-3619. doi: 10.1111/1462-2920.13440

Peng, X., et al. (2016). Cyclic di-AMP mediates biofilm formation. Mol Microbiol 99(5): 945-959. doi: 10.1111/mmi.13277

Ramesh, A. (2015). Second messenger - Sensing riboswitches in bacteria. Semin Cell Dev Biol 47-48: 3-8. doi: 10.1016/j.semcdb.2015.10.019

Rao, F., et al. (2009). The functional role of a conserved loop in EAL domain-based cyclic di-GMP-specific phosphodiesterase. J Bacteriol 191(15): 4722-4731. doi: 10.1128/JB.00327-09

Rao, F., et al. (2010). YybT is a signaling protein that contains a cyclic dinucleotide phosphodiesterase domain and a GGDEF domain with ATPase activity. J Biol Chem 1;285(1):473-82. doi: 10.1074/jbc.M109.040238

Rao, F., et al. (2008). Catalytic mechanism of cyclic di-GMP-specific phosphodiesterase: a study of the EAL domain-containing RocR from Pseudomonas aeruginosa. J Bacteriol 190(10): 3622-3631. doi: 10.1128/JB.00165-08

Romling, U., et al. (2005). C-di-GMP: the dawning of a novel bacterial signalling system. Mol Microbiol 57(3): 629-639. doi: 10.1111/j.1365-2958.2005.04697.x

Rosenberg, J., et al. (2015). Structural and biochemical analysis of the essential diadenylate cyclase CdaA from Listeria monocytogenes. J Bacteriol 190(10):3622-31. doi: 10.1128/JB.00165-08

Ross, P., et al. (1987). Regulation of cellulose synthesis in Acetobacter xylinum by cyclic diguanylic acid. Nature 325: 279. doi: 10.1038/325279a0

Ruiz-Moreno, J. S., et al. (2018). The cGAS/STING Pathway Detects Streptococcus pneumoniae but Appears Dispensable for Antipneumococcal Defense in Mice and Humans. Infect Immun 86(3). doi: 10.1128/IAI.00849-17

Salter, E. A. and A. Wierzbicki (2007). The mechanism of cyclic nucleotide hydrolysis in the phosphodiesterase catalytic site. J Phys Chem B 111(17): 4547-4552. doi: 10.1021/jp066582+

Schulze, K., et al. (2017). Bivalent mucosal peptide vaccines administered using the LCP carrier system stimulate protective immune responses against Streptococcus pyogenes infection. Nanomedicine 13(8): 2463-2474. doi: 10.1016/j.nano.2017.08.015

Seshasayee, A. S., et al. (2010). Comparative genomics of cyclic-di-GMP signalling in bacteria: post-translational regulation and catalytic activity. Nucleic Acids Res 38(18):5970-81. doi: 10.1093/nar/gkq382

Severin, G. B., et al. (2018). Direct activation of a phospholipase by cyclic GMP-AMP in El Tor Vibrio cholerae. Proc Natl Acad Sci U S A 115(26): E6048-E6055. doi: 10.1073/pnas.1801233115

Shanahan, C. A. and S. A. Strobel (2012). The bacterial second messenger c-di-GMP: probing interactions with protein and RNA binding partners using cyclic dinucleotide analogs. Org Biomol Chem 10(46): 9113-9129. doi: 10.1039/c2ob26724a

Simm, R., et al. (2004). GGDEF and EAL domains inversely regulate cyclic di-GMP levels and transition from sessility to motility. Mol Microbiol 53(4): 1123-1134. doi: 10.1111/j.1365-2958.2004.04206.x

Tal, R., et al. (1998). Three cdg operons control cellular turnover of cyclic di-GMP in Acetobacter xylinum: genetic organization and occurrence of conserved domains in isoenzymes. J Bacteriol 180(17): 4416-4425. https://jb.asm.org/content/180/17/4416.short

Tang, Q., et al. (2015). Functional analysis of a c-di-AMP-specific phosphodiesterase MsPDE from Mycobacterium smegmatis. Int J Biol Sci 11(7):813-24. doi: 10.7150/ijbs.11797

Tchigvintsev, A., et al. (2010). Structural insight into the mechanism of c-di-GMP hydrolysis by EAL domain phosphodiesterases. J Mol Biol 402(3):524-38. doi: 10.1016/j.jmb.2010.07.050

Wassmann, P., et al. (2007). Structure of BeF3−-modified response regulator PleD: implications for diguanylate cyclase activation, catalysis, and feedback inhibition. Structure 15(8): 915-927. doi: 10.1016/j.str.2007.06.016

Weinberg, Z., et al. (2007). Identification of 22 candidate structured RNAs in bacteria using the CMfinder comparative genomics pipeline. Nucleic Acids Res 35(14): 4809-4819. doi: 10.1093/nar/gkm487

Woodward, J. J., et al. (2010). c-di-AMP secreted by intracellular Listeria monocytogenes activates a host type I interferon response. Science 328(5986): 1703-1705. doi: 10.1126/science.1189801

Wu, J., et al. (2013). Cyclic GMP-AMP is an endogenous second messenger in innate immune signaling by cytosolic DNA. Science 339(6121): 826-830. doi: 10.1126/science.1229963

Yan, H., et al. (2009). 3',5'-Cyclic diguanylic acid elicits mucosal immunity against bacterial infection. Biochem Biophys Res Commun 387(3): 581-584. doi: 10.1016/j.bbrc.2009.07.061

Yan, W., et al. (2010). The effect of c-di-GMP (3'-5'-cyclic diguanylic acid) on the biofilm formation and adherence of Streptococcus mutans. Microbiol Res 165(2): 87-96. doi: 10.1016/j.micres.2008.10.001

Yan, W., et al. (2012). [Exogenous 3', 5'-cyclic diguanylic acid prevents caries formation in rats]. Nan Fang Yi Ke Da Xue Xue Bao 32(5):639-42.

Ye, M., et al. (2014). DhhP, a c-di-AMP phosphodiesterase of Borrelia burgdorferi, is essential for cell growth and virulence. Infect Immun 82(5):1840-9. doi: 10.1128/IAI.00030-14

Zarrella, T. M., et al. (2018). Stress Suppressor Screening Leads to Detection of Regulation of Cyclic di-AMP Homeostasis by a Trk Family Effector Protein in Streptococcus pneumoniae. J Bacteriol 200(12). doi: 10.1128/JB.00045-18

Zhang, X., et al. (2013). Cyclic GMP-AMP Containing Mixed Phosphodiester Linkages Is An Endogenous High-Affinity Ligand for STING. Molecular Cell 51(2): 226-235. doi: 10.1016/j.molcel.2013.05.022

Zhu, J., et al. (2016). (p)ppGpp synthetases regulate the pathogenesis of zoonotic Streptococcus suis. Microbiol Res 191: 1-11. doi: 10.1016/j.micres.2016.05.007

 https://doi.org/10.21775/9781912530229.04

Mobile Genetic Elements in Streptococci

Miao Lu[#], Tao Gong[#], Anqi Zhang, Boyu Tang, Jiamin Chen, Zhong Zhang, Yuqing Li*, Xuedong Zhou*

State Key Laboratory of Oral Diseases, National Clinical Research Center for Oral Diseases, West China Hospital of Stomatology, Sichuan University, Chengdu, PR China.

[#]Miao Lu and Tao Gong contributed equally to this work.

*Address correspondence to: liyuqing@scu.edu.cn, zhouxd@scu.edu.cn

Abstract

Streptococci are a group of Gram-positive bacteria belonging to the family Streptococcaceae, which are responsible of multiple diseases. Some of these species can cause invasive infection that may result in life-threatening illness. Moreover, antibiotic-resistant bacteria are considerably increasing, thus imposing a global consideration. One of the main causes of this resistance is the horizontal gene transfer (HGT), associated to gene transfer agents including transposons, integrons, plasmids and bacteriophages. These agents, which are called mobile genetic elements (MGEs), encode proteins able to mediate DNA movements. This review briefly describes MGEs in streptococci, focusing on their structure and properties related to HGT and antibiotic resistance.

Introduction

Streptococci are a group of Gram-positive bacteria widely distributed across human and animals. Unlike the *Staphylococcus* species, streptococci are catalase negative and are subclassified into the three subspecies alpha, beta and gamma according to the partial, complete or absent hemolysis induced, respectively. The beta hemolytic streptococci species are further classified by the cell wall carbohydrate composition (Lancefield, 1933) and according to human diseases in Lancefield groups A, B, C and G. group C and group G antigen are relatively rare compared with group A and group B antigen (Korona-Glowniak et al., 2015). Streptococci cause numerous diseases in human and animals, and many of them can cause serious and invasive infections damaging public health.

Group A streptococcus (GAS), also known as *Streptococcus pyogenes,* causes serious infections such as streptococcal toxic shock syndrome necrotizing fasciitis, and causes mild human infections such as impetigo and pharyngitis. Repeated GAS infections can cause autoimmune diseases, such as acute rheumatic fever, rheumatic heart disease and acute post-streptococcal glomerulonephritis (Parks et al., 2015). Group B streptococcus (GBS), also known as *Streptococcus agalactiae*, is included in the normal flora of the rectum and female urogenital tract, which under normal circumstances causes no harm, but it may trigger infection when the body defense system is weak. The infection can be transferred to a neonate from the mother colonized with GBS, causing meningitis and sepsis in newborns (Verani et al., 2010). The most represented specie belonging to the groups C and G streptococci is the *Streptococcus equisimilis,* a subspecies of *Streptococcus dysgalactiae*, which is usually present in human upper airways

(Jensen and Kilian, 2012). *S. dysgalactiae* is able to colonize other sites like skin, causing superficial and invasive infections such as septic arthritis, skin and tissue infections (Brandt and Spellerberg, 2009). *Streptococcus pneumoniae* is part of the normal flora of the human upper respiratory tract. The primary site of pneumococcal colonization in human body is the nasopharynx. It can cause a wide spectrum of infections, including invasive and non-invasive diseases. Pneumococcal infection can occur at multiple sites including the meninges and lung parenchyma, and is the primary cause of community acquired pneumonia (CAP) and a main cause of bacteremia, especially in infants and older adults (Remington and Sligl, 2014). The viridans streptococci, belonging to a heterogeneous group, are commensal flora of the oral cavity, upper respiratory, gastrointestinal and female genital tracts. These species can be divided into five groups named *Anginosus*, *Mutans*, *Mitis*, *Salivarius* and *Sanguinis* (Facklam, 2002). The viridians streptococci can cause severe infections including infective endocarditis, dental caries and deep seated abscesses (Menon, 2016). The *Streptococcus anginosus* group, previously known as *Streptococcus milleri*, includes three distinct species: *Streptococcus intermedius*, *Streptococcus anginosus* and *Streptococcus constellatus* (Claridge et al., 2001). These species, constituting the normal flora of the gastrointestinal tract and oropharynx, can form abscesses, which may alter the spectrum of related clinical disease. The most common infections caused by the *S. anginosus* group are empyema, infective endocarditis, cellulitis, and abscesses in the head, neck, lung and abdomen (Kowlessar et al., 2006; Noguchi et al., 2015).

Many diseases are caused by or associated with streptococci, but the treatment options are limited. Currently, one of the biggest problems is the bacterial antibiotic resistance. It has been reported that some GAS isolates are resistant to macrolides, tetracycline and fluoroquinolones (Walker et al., 2014). Among GBS, several isolates resulted resistant to clindamycin and macrolides, and the percentage may be as high as 30% in some areas (Compain et al., 2014). The predominant antibiotic resistance of *S. dysgalactiae* subsp. *equisimilis* isolates are erythromycin, clindamycin and tetracycline (Lu et al., 2016).

Some bacterial strains have intrinsic resistance, while others gain antibiotic resistance by mutation, horizontal gene acquisition and recombination with foreign DNA. In most instances, these mechanisms occur together. Mutations can be considered the workhorse of bacteria evolution, while the horizontal gene transfer (HGT) can immediately transfer a resistant gene (Perry et al., 2014). HGT is the sharing of genetic material among organisms other than by the DNA transmission from parent to offspring, which enable organisms to acquire new genes and phenotypes.

Microbe employs different mechanisms to transfer genes horizontally, including transformation, transduction and conjugation. In natural transformation, cells uptake naked DNA from the environment. It was reported that there exist about 80 species capable of transformation, and they are evenly divided between Gram-positive and Gram-negative bacteria (Johnston et al., 2014). Transduction is a process in which foreign DNA is transferred into a cell with the help of a virus or viral vector, transferring virulence genes between bacteria. Physical contact between the cell providing the DNA and the cell receiving the DNA is not required. Conjugation is the transfer of DNA between cells by a bridge-like connection or by direct cell-to-cell contact. During conjugation the cell donates a conjugative or mobilizable genetic element that is often a transposon or a plasmid. The result is frequently beneficial to the recipient, gaining xenobiotic tolerance, antibiotic resistance or new metabolic capability. Agents responsible of DNA movement in HGT are called mobile genetic elements (MGEs). Here, we reviewed principal MGEs in streptococci species, including transposable elements, integrons, plasmids and bacteriophages.

Transposable elements

A transposable element, is a segment of DNA that can be integrated in different sites along a chromosome, sometimes creating or reversing mutations and altering the cell genetic identity and genome size. Transposable elements were discovered by the American geneticist McClintock in 1951 (Korona-Glowniak et al., 2015), and were widely recognized in 1967 when Shapiro found transposable elements in *Escherichia coli*. Initially transposable elements were considered unusual (Fedoroff, 2012a), but then they were regarded as 'selfish DNA' being able to replicate autonomously. More evidences showed that transposons play a crucial role in genome evolution (Fedoroff, 2012b). Transposable elements are commonly found in chromosomes, integrative conjugal elements (ICEs), bacteriophages, and plasmids. In prokaryotes, transposons can mainly fall into two categories, the insertion sequence (IS) and transposon (Tn).

Insertion sequences

An insertion sequence, also known as insertion sequence element, or IS, is a short DNA sequence that acts as a simple transposable element. It is defined as a small (2.5-kb), phenotypically cryptic segments of DNA with a simple genetic organization, and capable of inserting at multiple sites in a target molecule (Mahillon and Chandler, 1998). ISs present two major characteristics: they are small compared with other transposable elements and they only encode proteins relevant for transposition activity. These proteins usually catalyze enzymatic reactions, help the ISs movements, and can also stimulate or inhibit the transposition activity. However, in many cases, the exact role and activity of the protein are still unclear (Siguier et al., 2014).

Many studies have been conducted about the relationship between ISs and the antibiotic resistance in *Streptococcus*. Beigverdi et al. isolated forty-one *S. agalactiae* from pregnant women at 35-37 weeks of gestation, and found that 97.6%, 24.4%, and 14.6% of them were resistant to tetracycline, erythromycin and clindamycin, respectively. The insertion sequence IS1548 was found in 63.4% of isolates, which flanking the antibiotic gene (Beigverdi et al., 2014). Another study showed that the presence of rib, IS861 and GBSi1 group II intron within the C5a peptidase gene might accentuate the group B *Streptococcus* virulence potential and ecological success (Florindo et al., 2014). Tn*Ss1* flanked by direct repeats of an IS6-family element may be related to the chloramphenicol resistance of *Streptococcus suis* (Takamatsu et al., 2003). In foodborne strains of *S. bovis*, there was a subregion of about 4 kb containing a novel mosaic tetracycline resistance gene *tet*(S/M) flanked by two copies of the IS1216 mobile element (Barile et al., 2012).

Transposons

Transposons can be described as either composite transposons or complex transposons (Bennett, 2004). Composite transposons, for example Tn*5*, Tn*9* and Tn*10*, are flanked by two separate identical or different IS elements. A composite transposon contains protein coding sequences encoding resistance to one or more antibiotics. A complex transposon has a more complicated structure than a composite transposon, being usually flanked by short IRs (Mahillon and Chandler, 1998). Between IRs there are genes related to transposition and antibiotic resistance (table.1).

Table 1. Details of Transposons discussed in this review

Name	Approximate size (kb)	Antibiotic resistance gene
Tn*916*	16.4	*tet*(M)
Tn*917*	5.3	*tet*(M)and *erm*(B)
Tn*3872*	24	*tet*(M)and *erm*(B)
Tn*5801*	25.8	*tet*(M)
Tn*6002*	20.9	*erm*(B) and *tet*(M)
Tn*6003*	25.1	*tet*(M) and *erm*(B)
Tn*1545*	25.3	*erm*(AM), *aph*A-3 and *tet*(M)
Tn*2010*	26.4	*mega*, *tet* (M) and *erm* (B)
Tn*2009*	23.5	*mega*, *tet*(M)
Tn*1207.1*	7.2	*mef*(A)
Tn*1207.3*	52.5	*mef*(A)
Tn*2017*	28.5	*mega*, *tet*(M), *erm*(B)
Tn*5251*	18	*tet*(M)
Tn*5253*	64.5	*cat* and *tet*(M)
Tn*5252*	47.5	*cat* and *tet*(M)

Tn*916*

Tn*916* is a 16.4 kb transposon containing a tetracycline resistance determinant. Tn*916* can transfer through conjugation at frequencies of 10^{-8} to 10^{-5} per donor in the absence of plasmid DNA, and it is also capable of insertion at different locations of the recipient chromosome (Johnson and Grossman, 2015). Some of the trans-conjugants possess more than two non-tandem copies (Gawron-Burke and Clewell, 1982), and have been named "conjugative transposon", or "integrative conjugative element". The recombination module of all Tn*916*-related elements is located at one end of the element with imperfect repeat sequences. Tn*916* contains four functional modules: regulation, conjugation, recombination and accessory genes not involved in conjugative transposition (e.g. tetracycline resistance). The proposed regulatory region of Tn*916* includes *orf12*, *orf9*, *orf7* and *orf8*. *Orf6* and *orf10* are also in this region, but their functions are not clear yet (Figure.1) (Celli and Trieu-Cuot, 1998). Homologous segments have been found in the transposon junction regions but there was no direct duplication of the target sequence (Clewell et al., 1988). Tn*916* replicates autonomously *via* a rolling-circle mechanism, which depends on the relaxase encoded by *orf20* (Wright and Grossman, 2016). It has been reported that Tn*916* transposes by an excision-insertion mechanism (Caparon and Scott, 1989), in which it is firstly excised by a cut from donor cells and transformed into a circular form, then it is transferred by cell-cell contact, and finally it is inserted into receiving cells through recombinase (Roberts and Mullany, 2011). Two transposon-encoded proteins, Xis-Tn (an accessory protein) and Int-Tn (a tyrosine recombinase), are needed for excision. Int-Tn alone is enough for integration. The conjugative transposition of Tn*916* needs a functional integrase in both recipient and donor strain (Storrs et al., 1991). The DNA sequences of the target sites of many Tn*916* insertions showed that Tn*916* prefers to use A-T rich targets in different hosts (Wang et al., 2000). The most important regulatory event in Tn*916* for conjugative transposition is the interaction between ribosomes and tetracycline, which makes the transcriptional terminator destroyed at the end of *orf12* and continues transcribing the *tet*(M) and the downstream genes. Tetracycline can inactivate most of the ribosomes, and a few ribosomes can remain active by the protection of the basal expression of *tet*(M) (Roberts and Mullany, 2009).

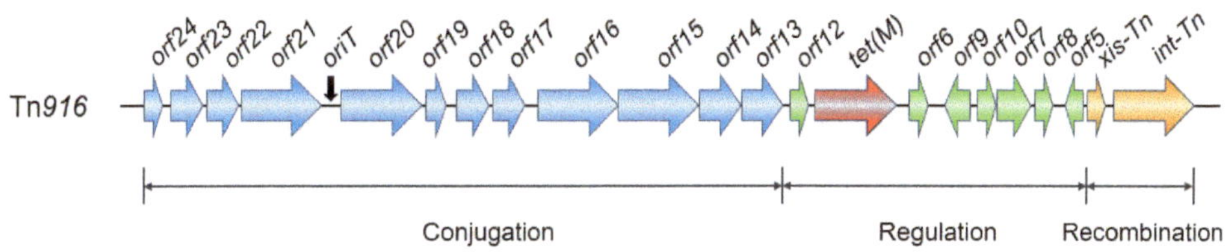

Figure 1. Schematic representation of Tn*916*.

The Tn*916* family is widespread among numerous commensal and pathogenic bacteria. Tn*916* has an extensive host range, both *in vitro* and *in vitro* (Scott, 2002). In *S. suis* serotype 2, the presence of Tn*916* with *tet*(M) has favored the pathogenicity and dissemination of this pathogen (Zhang et al., 2015). In 138 antibiotic resistant pneumococcal isolated strains from nasopharynx of healthy young children from Lublin of Poland, Tn*916* was detected in 16 strains (Korona-Glowniak et al., 2015). Moreover, among 62 tetracycline-resistant *tet*(M)-positive *S. agalactiae* isolate strains, 70% of them harboring Tn*916* (Mingoia et al., 2013).

Tn*917*

Tn*917* was identified on pAD2, a nonconjugative plasmid of *Enterococcus faecalis* strain DS16 (Tomich et al., 1980). Erythromycin resistance associated with Tn*917* is inducible, and the exposition of cells to low concentrations of erythromycin can induce transposition from pAD2 to pAD1. Tn*917* is 5,257 bp in length and has six open reading frames. *Orf2* is *erm*(B), which is responsible for erythromycin resistance. In 263 viridans group streptococci strains, the long-established transposon Tn*917* has been found in seven isolates by PCR mapping, which including one *Streptococcus mitis* strain, three *Streptococcus oralis* strains, two *S. sanguinis* strains and one *S. anginosus* strain (Brenciani et al., 2014).

Tn*3872*

Tn*3872* is found by DNA-DNA hybridization, and nucleotide sequence analyses revealed that they were made up by the insertion of Tn*917* into *orf9* of Tn*916* (McDougal et al., 1998). With the *tet*(M) gene of Tn*916* linked to the *erm*(B) gene of Tn*917*, Tn*3872* can lead to the erythromycin and tetracycline resistance. *Tet*(M) on the Tn*3872*-like transposon in the *S. agalactiae* isolates has been confirmed by PCR with the primers *tet*(M)-forward and *erm*(B) (Sadowy et al., 2010). Thirteen strains had Tn*3872* among twenty-one clinical *Streptococcus salivarius*, and two of them carried a Tn*3872* with a silent *tet*(M) gene (Chaffanel et al., 2015). In pneumococcus, *tet*(M) can evolve through homologous recombination after Tn*3872* integration into the host chromosome (Chaffanel et al., 2015). It has been reported that Tn*916* can transfer from an enterococcal donor to a pneumococcal recipient, but the composite transposon Tn*3872* cannot transfer from a pneumococcal to another pneumococcal or from an enterococcal to a pneumococcal. The low transfer frequency of Tn*3872* or the inactivation of mobility of Tn*916* due to the insertion of Tn*917* may be the reason (Swartley et al., 1993). However, in *Streptococcus defectives* strains, the frequency of conjugative transposition of Tn*3872* was ranging from $4x10^{-6}$ to $2.8x10^{-8}$ (Poyart et al., 2000). Tn*3872*-related elements were distributed in many serotypes of pneumococcus, i.e. 6B, 23F, and 14 (Quintero et al., 2011; Ramos et al., 2014). In some strains, Tn*3872* and Tn*3872*-related elements were lacking *tet*(M) gene (Cochetti et al., 2007; Quintero et al., 2011). Genetic elements relevant to drug resistance can insert in Tn*3872*, and the strain will display resistance to other antibiotics. For example, *mega*, a genetic element carrying *mef*(E), has been found transferred by conjugation to *S. pneumoniae* and then linked with Tn*3872*-like transposon by recombinases (Figure.2) (Del Grosso et al., 2006; Santagati et al., 2009). In some strains, Tn*3872* carries transposase genes *tnpA* and *tnpR* (Korona-Glowniak et al., 2016; Korona-Glowniak et al., 2015).

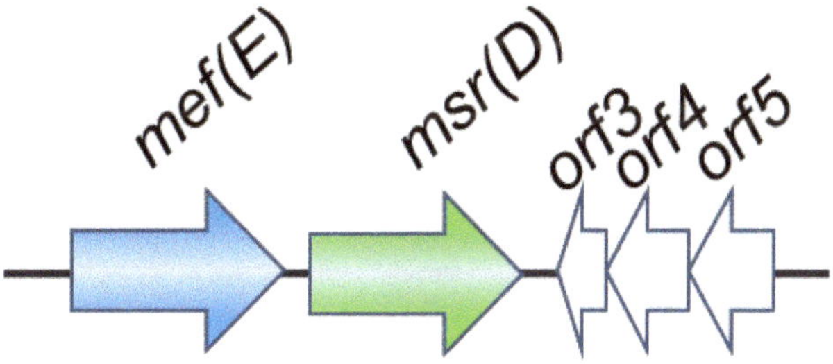

Figure 2. *Mega* of *S. pneumoniae*.

Tn*5801*

Tn*5801*, 25.8 kb in size, was originally detected in *Staphylococcus aureus* Mu50 (Kuroda et al., 2001). It is a Tn*916* family element, and the *int* and *xis* genes of Tn*916* are replaced by a unique gene (*int5801*) encoding different types of tyrosine recombinases. The *tet*(M) genes of Tn*5801* and Tn*916* are similar, with about 97.7% DNA identity (de Vries et al., 2016). Tn*5801*-like transposons have been detected in some streptococci, including *S. mitis* B6 (Denapaite et al., 2010), S. *agalactiae* (Da Cunha et al., 2014; Mingoia et al., 2013) and *S. oralis* (Brenciani et al., 2014). In *tet*(M)-positive tetracycline resistant *S. agalactiae*, 30% harbors Tn*5801*-like element (Mingoia et al., 2013).

Tn*6002*

Tn*6002*, a 20.9 kb element, is formed by the insertion of an *erm*(B)-carrying DNA fragment into Tn*916*, which is located between *orf20* and *orf19* (Brenciani et al., 2007). It was originally discovered in *Streptococcus cristatus* and designated as Tn916Erm, whose complete sequence is available in GenBank since 2005 (accession no. AY898750) (Varaldo et al., 2009). It has also been found in *S. pyogenes* (Ardanuy et al., 2010; Brenciani et al., 2007), *S. salivarius* (Chaffanel et al., 2015) and S. *pneumoniae* (Calatayud et al., 2007; Korona-Glowniak et al., 2016; Quintero et al., 2011). In *erm*(B)-positive pneumococci, Tn*6002* was the most common *erm*(B)-carrying Tn*916*-related element (Cochetti et al., 2008; Korona-Glowniak et al., 2015). The spread of Tn*6002* from the ST386/ST4310/ST4825 lineage to the serotype 6C of the pneumococci isolates lead to the increase of the antimicrobial non-susceptibility of serotype 6C in the late-PCV7 period (Rolo et al., 2011). In viridans group streptococci (VGS), Tn*6002* were also detected into *S. mitis* and *S. oralis* (Brenciani et al., 2014). In oral streptococci isolated from healthy human, Tn*6002* was detected from *Streptococcus infantis*, *S. mitis*, *S. sanguinis*, *Streptococcus australis*, *S. oralis* and *Streptococcus parasanguini.*

Tn*6003*

Tn*6003*, a new Tn*6002*-related composite element, found in 2007 (Cochetti et al., 2007), contains *tet*(M) and *erm*(B) gene, as well as the *aadE-sat4-aphA*-3 gene cluster that forms a 4.2 kb fragment called MAS (macrolideaminoglycoside-streptothricin). Tn*6003* is 25.1 kb (accession number: AM410044). The transfer frequency of Tn*6003* from *S. pneumoniae* strain Ar4 to *E. faecalis* JH2-2 is 1.7×10^{-7} (Cochetti et al., 2007). Tn*6003* is a common Tn*916*-related element in *erm*(B)-positive pneumococci (Quintero et al., 2011).

Tn*1545*

Tn*1545*, a 25.3 kb element, was originally detected in the chromosome of *S. pneumoniae* BM4200 (an antibiotics-resistant clinical isolates) (Courvalin and Carlier, 1987). It confers resistance to erythromycin (*erm*AM), kanamycin (*aphA*-3) and tetracycline (*tet*M) and was the first genetic element found carrying the *erm*(B) gene in *S. pneumoniae*. Compared with Tn*916*, between *orf20* and *orf19*, Tn*1545* showed the insertion of two *erm*(B) and MAS elements; and between *orf13* and *tet*(M), Tn*1545* had the insertion of IS*1239* (Quintero et al., 2011). Tn*1545* is self-transferable to various streptococci, including *Streptococcus cremoris*, *S. sanguinis*, *Streptococcus lactis*, *Streptococcus diacetylactis, and E. faecalis*. It is a conjugative shuttle transposon, which can transpose from the hybrid plasmid to various new sites of the new host chromosome and then re-transpose from the chromosome to the plasmid at a frequency of 5×10^{-9}. Though the absence of recombinase genes, the MAS element of Tn*1545* is able to exploit long flanking repeats for excision in a circular form (Palmieri et al., 2012).

Tn*2010*

Tn*2010*, 26.4 kb in length, is a composite element carrying both *mega*, *tet*(M) and *erm*(B) genes (Del Grosso et al., 2007; Talebi et al., 2016). Tn*2010* is similar to Tn*6002*, including sequence and structure, and it has the insertion of the *mega* element into *orf6* of the Tn*6002,* as well as 70 nucleotide mutations (Li et al., 2011). It has been indicated that Tn*2010* could not transfer macrolide resistance through conjugation (Del Grosso et al., 2006). Zhou et al. proved that Tn*2010* is more likely to be transferred by transformation other than conjugation in *S. pneumoniae*, and the transformation frequency is $3x10^{-7}$ transformants per cfu (Zhou et al., 2014).

Tn*2010* is common in *S. pneumoniae* (Bowers et al., 2012; Korona-Glowniak et al., 2015), and recently it has been found in *S. mitis* (Brenciani et al., 2014). In *S. pneumoniae*, most strains owning Tn*2010* belonging to serotype 19A or 19F (Korona-Glowniak et al., 2016). The wide spreading of the macrolide resistance genes is highly relevant to Tn*2010*. *S. pneumoniae* A026, isolated from cerebrospinal fluid, contains Tn*2010* and is resistant to penicillin, trimethoprim, erythromycin, sulfonamides, and tetracycline but is susceptible to levofloxacin (Sui et al., 2013).

Tn*2009*

Tn*2009* is a composite element of approximately 23.5 kb, originally found in the chromosomes of the two *S. pneumonia* strains (Del Grosso et al., 2004). Subsequently, it was detected in *S. mitis* (Brenciani et al., 2014), *S. pyogenes* (Del Grosso et al., 2011) and *S. salivarius* (Chaffanel et al., 2015). Tn*2009* has the insertion of the *mega* element into the *orf6* of a Tn*916*-like transposon containing *tet*(M). In many Tn*2009* element of *S. pyogenes* strains, *mef*(E) is included in a typical *mega* element, while in some strains, *mef(E)* is physically linked with *tet*(M) (Del Grosso et al., 2011). Tn*2009* contains determinants for tetracycline, macrolide and erythromycin resistance. In *S. pneumoniae* and *S. pyogenes* the conjugal transfer of Tn*2009* has not been detected, unlike the transfer by transformation (Del Grosso et al., 2016; Del Grosso et al., 2004). It has been reported that the conjugative Tn*2009* in a Gram-negative commensal bacteria *Acinetobacter junii 326* was indistinguishable from the *S. pneumoniae* Tn*2009* element (Ojo et al., 2006).

Tn*1207.1*

Tn*1207.1*, 7,244 bp in size, is the first described *mef*-carrying genetic element. It has been found in both *S. pneumoniae* and *S. pyogenes* (Santagati et al., 2000). Tn*1207.1* consists in first 5 ORFs with the same direction of transcription and the last 3 ORFs in the opposite direction (Pozzi et al., 2004). The *orf2* encodes a site-specific recombinase, while *orf5*, starting from the downstream of the *mef* (A), is homologous to *vga*(A) and *msr*(SA) (Pozzi et al., 2004). The *mrs*(SA) and *mef*(A) are two important parts of the *mega* element. Tn*1207.1* is not a conjugative element (Santagati et al., 2000).

Tn*1207.3*

Tn*1207.3*, homologous to *Tn1207.1*, is resistant to macrolides and carries the *mef*(A) gene. Tn*1207.3* was found in *S. pyogenes*, *S. pneumoniae*, S. *dysgalactiae* subsp. *Equisimilis* (Figueiredo et al., 2006; Marimon et al., 2005; Rato et al., 2010), *S. gordonii* (Santagati et al., 2003) and *S. agalactiae* (Figueiredo et al., 2006; Marimon et al., 2005). Its nucleotide sequence is highly identical to that of Tn*1207.1,* and only the right end of the Tn*1207.3* has a new sequence (Pozzi et al., 2004; Santagati et al., 2003). Tn*1207.3* can transfer from *S. pyogenes* to *S. pneumoniae* by conjugation, and from *S. pneumoniae* trans-conjugant MF2 to *S. pyogenes* and *S. gordonii* (Santagati et al., 2003). The transfer frequency of Tn*1207.3* to various *S. pyogenes* is significantly different, and the mean frequencies range from 7.20×10^{-8} to 1.13×10^{-6} conjugant per recipient (Hadjirin et al., 2014). The integration of Tn*1207.3* did not cause deletions, while integration of Tn*1207.1* caused a 1947 bp deletion (Pozzi et al., 2004). The DNA sequencing of Tn*1207.3* indicated that it is actually a prophage, the same with the right end of ϕ10394.4 (Iannelli et al., 2014a).

Tn*2017*

Tn*2017*, a Tn*916*-like element, was found in *S. pneumoniae* and beta-hemolytic *Streptococcus* (Korona-Glowniak et al., 2015). Tn*2017* resembles Tn*3872*, which is the insertion of *mega* element into the *orf6* of Tn*3872*, thus being resistant to erythromycin, tetracycline and macrolide. Tn*2017* also contains the *tnpA* and *tnpR* genes. In erythromycin and tetracycline resistant beta-hemolytic

Streptococcus isolates, Tn*2017* is the most prevalent transposon, while in *S. pneumonia* isolates it is the second most common transposon (Tantivitayakul et al., 2016).

Tn*5251*

Tn*5251*, 18 kb in length, is like Tn*916* in structure and size. It contains 22 ORFs, of which 20 have the same direction and transcription, and half are annotated (Santoro et al., 2010). Tn*5251* carries *tet*(M), *int* and *six* genes. Tn*5251* cannot transfer independently from Tn*5253* through conjugation (Kilic et al., 1994), however recently it has been proposed that Tn*5251* can transfer autonomously in *S. pneumoniae* and *E. faecalis*, and the autonomous copies can moved into *S. pneumoniae*, *S. gordonii*, *S. agalactiae*, *S. pyogenes*, *E. faecalis* and *Bacillus subtilis* (Santoro et al., 2010).

Tn*5253*

Tn*5253*, originally found in the chromosome of *S. pneumoniae* BM6001 (Dang-Van et al., 1978), was formerly called the Ω(*cat-tet*) element, carrying the *cat* and *tet*(M) genes (Vijayakumar et al., 1986). Tn*5253* is 64.5 kb in size, and contains 79 ORFs in total, 38 of which have annotations (Iannelli et al., 2014b). Tn*5251* and Tn*5252* were found integrated into the Tn*5253*, demonstrating that Tn*5253* was a composite element (Ayoubi et al., 1991). The highest conjugation frequency of Tn*5253* was 6.7×10^{-3} trans-conjugants/donor when S*. pyogenes* was the donor (Iannelli et al., 2014b). Tn*5253* is common in the drug resistance of *S. pneumoniae* strains, especially in the serotype 6B strains, but the *cat* gene may not be detected for the modifications of the transposon (Quintero et al., 2011). Among clinical isolates of *S. pneumoniae*, Tn*5253*-related elements are extremely different (Henderson-Begg et al., 2009; Mingoia et al., 2011). In streptococci, *cat*(Q) and *mef*(I) are linked in a genetic module called IQ module that has been detected in *S. pneumoniae*, *S. pyogenes* and viridans group streptococci. The IQ module in *S. pyogenes* belongs to the Tn*5253* family (Mingoia et al., 2014a; Mingoia et al., 2014b; Mingoia et al., 2007).

Tn*5252*

Tn*5252*, 47.5 kb in size, is a streptococcal conjugative transposon (Kilic et al., 1994). It was originally found in the chromosome of *S. pneumoniae* BM6001 as a part of Tn*5253* (Shoemaker et al., 1979). Until now, Tn*5252* has only been partially sequenced, showing the presence of an integrase gene, *tet*(M), and *cat*. While Tn*916* class of element can insert at several spots, Tn*5252* prefer to integrate at a specific site, which is drove by *int* gene. *Cat* gene can encode chloramphenicol acetyltransferases (CATs) that lead to the enzymatic inactivation of the drug, and Tn*5252* can be detected in the clinical chloramphenicol resistant strains (Korona-Glowniak et al., 2015). Since the discovery of Tn*5252*, its homologues have been detected in many streptococcal species, such as *S. suis*, *S. pneumoniae* and GBS (Kayansamruaj et al., 2015; Wyres et al., 2013).

Integron

Integrons are genetic elements that can help bacteria to enhance their ability of adaptation and evolution through the acquisition, stockpiling, excision and differential expression of new genes (Escudero et al., 2015). The genes are usually contained in a gene structure called cassette, a circular element carrying one promoterless gene together with a recombination site (*attC*) (Figure.3). Another structure included in integrons is a stable platform. The platform has several parts, and the first is *intI*, a gene encoding the integrase, which is a member of tyrosine recombinase family (Messier and Roy, 2001). The second part is the *attI* site, an integron-associated recombination site, which is in the upstream of the *intI* gene (Partridge et al., 2000). The third part is *Pc*, an integron-associated promoter, located between the *intI* and the *attI* site or within the *intI* gene (Collis and Hall, 1995). Cassettes are non-replicative elements, and they play their roles by integrating into the platform and expression from the *Pc* promoter. With the constant integration taking place, an array of cassettes will be assembled, and the cassettes' excised and integrated into the first position have the maximal expression (Escudero et al., 2015). There are two key advantages owning by the integron system for the genomic studies. First, the new cassette is integrated into the bacterial genome at *attI* site, which will not disturb the existing genes. Second, the newly integrated cassettes are expressed by the promoter *Pc* that it is ready for the natural selection (Gillings, 2014).

At present, there are five classes of mobile integrons associated with antibiotic resistance: classes 1, 2 and 3, obtained from clinical contexts (Partridge, 2011); class 4, detected in *Vibrio cholerae*; and class 5, found on the plasmid of *Alivibrio salmonicida* (Gillings, 2014). All five classes are linked to mobile elements, such as conjugative plasmids, insertion sequences and transposons, which can work as vehicles for the gene transmission (Mazel, 2006).

Integrons play an important role in the spread of antibiotic resistance, especially in Gram-negative pathogens. Nonetheless, recent studies focused on the role of integrons in antibiotic resistance in Gram-positive microorganisms (Deng et al., 2015; Li et al., 2017), such as *Staphylococcus, Enterococcus, Corynebacterium* and *Streptococcus*. In streptococci, *S. pneumoniae* was proved to carry class 1 integron (Shi et al., 2006).

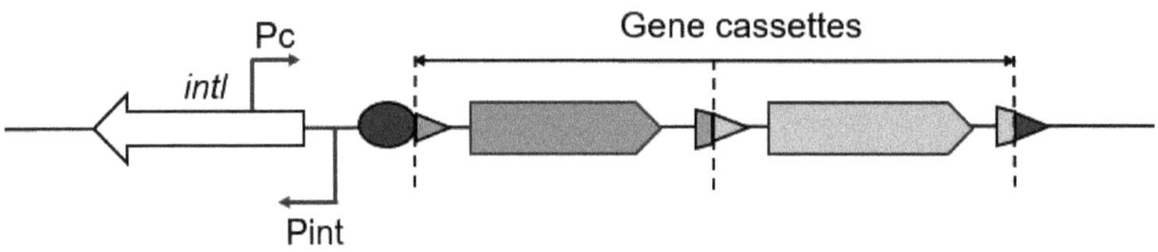

Figure 3. General structure of class 1 integrons.

Plasmids

Plasmids are a group of extrachromosomal DNA molecules that can self-replicate, are usually circular and does not contain genes required for essential cellular functions. The typical plasmid is a circular, closed double-stranded DNA molecule; nonetheless, linear double-stranded plasmids have been found (Stewart et al., 2005). In general, a plasmid includes essential genes related to replicative functions and accessory genes encoding different functions (Fernandez-Lopez et al., 2006) including drug resistance, fitness and virulence factors (Gyles and Boerlin, 2014). Plasmids are inherited from parent to daughter cells at cell division by partitioning. For some plasmids carrying accessory genes, they are therefore large and always present in low copy numbers in their host to ensure that the metabolic burden of maintaining and duplicating their genomes are not excessive. And the low copy number would increase their chance of being faithfully transmitted to daughter cells during division. Therefore, in order to decrease random losses, the plasmid should be inherited stably at a short time (Sengupta and Austin, 2011).

Some plasmids carry genes allowing them to transfer by conjugation, while others are lacking such genes and count on the transferable function of other plasmids (Smillie et al., 2010). The conjugative genes trigger DNA transfer from the donor cell to the recipient cell, and other genes guarantee the DNA survival in the new host. The transfer genes also encode proteins able to 'pump' the DNA into recipient cells, and these proteins all belong to the TraG-like family of ATPases (Schroder et al., 2002). Relaxase are also important in many conjugative systems that can nick DNA to give a single-stranded substrate suitable for transfer, and the nicking process takes place in a specific site. In Gram-negative bacteria the pilus is an important section of conjugative systems. Pilus assembly plays the role of a type IV secretion system (T4SS) (Lybarger and Sandkvist, 2004), which help bacteria in two aspects: the genetic exchange and the transmission of effector molecules to target cells (Cascales and Christie, 2003).

In Gram-positive bacteria conjugative mechanisms include the pheromone-independent system that have a cell surface protein able to trigger mating-pair formation and pheromone-dependent mating systems in enterococci (Frost et al., 2005). In streptococci, the transfer of chromosomal DNA may be affected by the plasmid transfer in horizontal gene transfer, due to the chromosome mobilization ability of plasmids and the high frequency of recombination mode of the F plasmids (Lawley et al., 2003). Plasmids have been grouped according to their replicons, and the streptococcal plasmid pMV158 belongs to the Rep_2 family (PF01719). There are several parts related to the mobilization of plasmid pMV158: prototype of the MOBV relaxase family, the product of the *mobM* gene, and its cognate origin of transfer, *oriT*. The *mobM* gene encodes the relaxase MobM, the relaxase of pMV158. MobM forms a dimer through its C-terminal domain, anchoring the protein to the cell membrane and participating in T4SS protein-protein interactions (Fernandez-Lopez et al., 2013). Plasmid pLS1 is a deleted derivative of pMV158, which has lost the ability of conjugative mobilization, thus decreasing the host cell growth by 8-9% in *S. pneumonia* (Hernandez-Arriaga et al., 2012). Many toxins and drug resistance genes in streptococci are encoded by plasmid. The β plasmid from *E. faecalis* can transfer the erythromycin and lincomycin resistance genes by conjugation (LeBlanc et al., 1978). The diplococci-producing ability can be transferred from the plasmid of *Streptococcus cremoris* 346 to the plasmid-free S. *cremoris* by conjugation with high frequency (Davey, 1984). The minocycline resistance in an oral S*treptococcus infantis* isolate may be due to the *tet(S)* gene which is carried by a novel low copy number plasmid (Ciric et al., 2014).

Bacteriophages

Certain drug resistance in bacteria are found to be relevant to phages (Brussow et al., 2004; Lekunberri et al., 2017), and there are many antibiotic resistant genes carried by the bacteriophages, such as *tetA*, *tetW*, *intI1*, *intI2*, *intI3*, *blaOXA-2* and *blaTEM* (Anand et al., 2016). Phage genomes may contain DNA or RNA, and the DNA can be single- or double-stranded. The essential genes of phages include the replicase genes, genes encoding for proteins involved in the DNA packaging, and genes encoding phage components used to disturb the host cell replicative machinery (Frost et al., 2005).

Phage-mediated horizontal gene transfer occurs via lysogenic conversion or transduction. Moreover, some phages can play a significant role in the emergence of pathogens, including *S. aureus* (food poisoning), *S. pyogenes* (scarlet fever), *E. coli* (Shiga toxin) and *Clostridium botulinum* (botulism). In *S. mitis*, phage tail genes *pblA* and *pblB* encode adhesion proteins for bacterial host attachment. The process of phage DNA packaging is highly accurate, but the host genome DNA fragment may be packaged at a finite frequency compared with the phage DNA, resulting in phage particles that can attach and deliver the DNA into suitable bacteria, with the injected exogenous bacterial DNA integrated into the genome. The observation of toxin genes in *S. pyogenes* and *Corynebacterium* located next to the phage attachment site raised the hypothesis that the prophage genes actually stand for bacterial genes obtained by a false excision from a previous bacterial host (Brussow et al., 2004).

There are likely many new variants of MGEs, which can contain other resistant genes able to transfer to broad host range. Besides, MGEs may evolve complicated mechanisms of transcriptional regulation which likely minimize the biological cost to the host. Furthermore, many of the current studies are focus on laboratory strains, which may not reflect the variability of natural isolates. Therefore, it is urgent to clarify the structure and function of the DNA integrated in a new host, the rate of the gene transfer in both laboratory and natural bacteria, as well as the effects of MGEs on the transcriptional regulation of the gene expression.

Conclusions

Here, we have reviewed principal MGEs in streptococci, including transposable elements, integrons, plasmids, and bacteriophages. MGEs have a strong environmental and clinical impact due to their role in the spread of antibiotic resistance in different bacteria. Transposable elements, mainly containing IS and transposons, are strictly related to the resistance to erythromycin, tetracycline, kanamycin and macrolide. The relationship between antibiotic resistance and integrons is well studied. Recently class 1 integron has been found in *S. pneumoniae*. Many toxins and the spread of drug resistance in *Streptococcus* are plasmid encoded, including the resistance to erythromycin, minocycline, and lincomycin. Finally, phages can mediate horizontal gene transfer *via* lysogenic conversion or transduction.

Acknowledgements

This work was supported by the National Natural Science Foundation of China (31200985 and 31870065). The authors declare no potential conflicts of interest regarding the authorship and/or publication of this article.

References

Anand, T., Bera, B.C., Vaid, R.K., Barua, S., Riyesh, T., Virmani, N., Hussain, M., Singh, R.K., and Tripathi, B.N. (2016). Abundance of antibiotic resistance genes in environmental bacteriophages. J Gen Virol 97, 3458-3466 DOI: 10.1099/jgv.0.000639.

Ardanuy, C., Domenech, A., Rolo, D., Calatayud, L., Tubau, F., Ayats, J., Martin, R., and Linares, J. (2010). Molecular characterization of macrolide- and multidrug-resistant *Streptococcus pyogenes* isolated from adult patients in Barcelona, Spain (1993-2008). J Antimicrob Chemother 65, 634-643 DOI: 10.1093/jac/dkq006.

Ayoubi, P., Kilic, A.O., and Vijayakumar, M.N. (1991). Tn*5253*, the pneumococcal omega (*cat tet*) BM6001 element, is a composite structure of two conjugative transposons, Tn*5251* and Tn*5252*. J Bacteriol 173, 1617-1622

Barile, S., Devirgiliis, C., and Perozzi, G. (2012). Molecular characterization of a novel mosaic *tet*(S/M) gene encoding tetracycline resistance in foodborne strains of *Streptococcus bovis*. Microbiology 158, 2353-2362 DOI: 10.1099/mic.0.058206-0.

Beigverdi, R., Jabalameli, F., Mirsalehian, A., Hantoushzadeh, S., Boroumandi, S., Taherikalani, M., and Emaneini, M. (2014). Virulence factors, antimicrobial susceptibility and molecular characterization of *Streptococcus agalactiae* isolated from pregnant women. Acta Microbiol Immunol Hung 61, 425-434 DOI: 10.1556/AMicr.61.2014.4.4.

Bennett, P.M. (2004). Genome plasticity: insertion sequence elements, transposons and integrons, and DNA rearrangement. Methods Mol Biol 266, 71-113 DOI: 10.1385/1-59259-763-7:071.

Bowers, J.R., Driebe, E.M., Nibecker, J.L., Wojack, B.R., Sarovich, D.S., Wong, A.H., Brzoska, P.M., Hubert, N., Knadler, A., Watson, L.M., et al. (2012). Dominance of multidrug resistant CC271 clones in macrolide-resistant *streptococcus pneumoniae* in Arizona. BMC Microbiol 12, 12 DOI: 10.1186/1471-2180-12-12.

Brandt, C.M., and Spellerberg, B. (2009). Human infections due to *Streptococcus dysgalactiae* subspecies *equisimilis*. Clin Infect Dis 49, 766-772 DOI: 10.1086/605085.

Brenciani, A., Bacciaglia, A., Vecchi, M., Vitali, L.A., Varaldo, P.E., and Giovanetti, E. (2007). Genetic elements carrying *erm*(B) in *Streptococcus pyogenes* and association with *tet*(M) tetracycline resistance gene. Antimicrob Agents Chemother 51, 1209-1216 DOI: 10.1128/AAC.01484-06.

Brenciani, A., Tiberi, E., Tili, E., Mingoia, M., Palmieri, C., Varaldo, P.E., and Giovanetti, E. (2014). Genetic determinants and elements associated with antibiotic resistance in viridans group streptococci. J Antimicrob Chemother 69, 1197-1204 DOI: 10.1093/jac/dkt495.

Brussow, H., Canchaya, C., and Hardt, W.D. (2004). Phages and the evolution of bacterial pathogens: from genomic rearrangements to lysogenic conversion. Microbiol Mol Biol Rev 68, 560-602, table of contents DOI: 10.1128/MMBR.68.3.560-602.2004.

Calatayud, L., Ardanuy, C., Cercenado, E., Fenoll, A., Bouza, E., Pallares, R., Martin, R., and Linares, J. (2007). Serotypes, Clones, and Mechanisms of Resistance of Erythromycin-Resistant *Streptococcus pneumoniae* Isolates Collected in Spain. Antimicrob Agents Chemother 51, 3240-3246 DOI:

10.1128/AAC.00157-07.

Caparon, M.G., and Scott, J.R. (1989). Excision and insertion of the conjugative transposon Tn*916* involves a novel recombination mechanism. Cell 59, 1027-1034

Cascales, E., and Christie, P.J. (2003). The versatile bacterial type IV secretion systems. Nat Rev Microbiol 1, 137-149 DOI: 10.1038/nrmicro753.

Celli, J., and Trieu-Cuot, P. (1998). Circularization of Tn*916* is required for expression of the transposon-encoded transfer functions: characterization of long tetracycline-inducible transcripts reading through the attachment site. Mol Microbiol 28, 103-117

Chaffanel, F., Charron-Bourgoin, F., Libante, V., Leblond-Bourget, N., and Payot, S. (2015). Resistance genes and genetic elements associated with antibiotic resistance in clinical and commensal isolates of *Streptococcus salivarius*. Appl Environ Microbiol 81, 4155-4163 DOI: 10.1128/AEM.00415-15.

Ciric, L., Brouwer, M.S., Mullany, P., and Roberts, A.P. (2014). Minocycline resistance in an oral *Streptococcus infantis* isolate is encoded by *tet*(S) on a novel small, low copy number plasmid. FEMS Microbiol Lett 353, 106-115 DOI: 10.1111/1574-6968.12410.

Claridge, J.E., 3rd, Attorri, S., Musher, D.M., Hebert, J., and Dunbar, S. (2001). Streptococcus intermedius, *Streptococcus constellatus*, and *Streptococcus anginosus* ("Streptococcus milleri group") are of different clinical importance and are not equally associated with abscess. Clin Infect Dis 32, 1511-1515 DOI: 10.1086/320163.

Clewell, D.B., Flannagan, S.E., Ike, Y., Jones, J.M., and Gawron-Burke, C. (1988). Sequence analysis of termini of conjugative transposon Tn*916*. J Bacteriol 170, 3046-3052

Cochetti, I., Tili, E., Mingoia, M., Varaldo, P.E., and Montanari, M.P. (2008). erm(B)-carrying elements in tetracycline-resistant pneumococci and correspondence between Tn*1545* and Tn*6003*. Antimicrob Agents Chemother 52, 1285-1290 DOI: 10.1128/AAC.01457-07.

Cochetti, I., Tili, E., Vecchi, M., Manzin, A., Mingoia, M., Varaldo, P.E., and Montanari, M.P. (2007). New Tn*916*-related elements causing *erm*(B)-mediated erythromycin resistance in tetracycline-susceptible pneumococci. J Antimicrob Chemother 60, 127-131 DOI: 10.1093/jac/dkm120.

Collis, C.M., and Hall, R.M. (1995). Expression of antibiotic resistance genes in the integrated cassettes of integrons. Antimicrob Agents Chemother 39, 155-162

Compain, F., Hays, C., Touak, G., Dmytruk, N., Trieu-Cuot, P., Joubrel, C., and Poyart, C. (2014). Molecular characterization of *Streptococcus agalactiae* isolates harboring small *erm*(T)-carrying plasmids. Antimicrob Agents Chemother 58, 6928-6930 DOI: 10.1128/AAC.03855-14.

Courvalin, P., and Carlier, C. (1987). Tn*1545*: a conjugative shuttle transposon. Mol Gen Genet 206, 259-264

Da Cunha, V., Davies, M.R., Douarre, P.E., Rosinski-Chupin, I., Margarit, I., Spinali, S., Perkins, T., Lechat, P., Dmytruk, N., Sauvage, E., et al. (2014). *Streptococcus agalactiae* clones infecting humans were selected and fixed through the extensive use of tetracycline. Nat Commun 5, 4544 DOI: 10.1038/ncomms5544.

Dang-Van, A., Tiraby, G., Acar, J.F., Shaw, W.V., and Bouanchaud, D.H. (1978). Chloramphenicol resistance in *Streptococcus pneumoniae*: enzymatic acetylation and possible plasmid linkage. Antimicrob Agents Chemother 13, 577-583

Davey, G.P. (1984). Plasmid associated with diplococcin production in *Streptococcus cremoris*. Appl Environ Microbiol 48, 895-896

de Vries, L.E., Hasman, H., Jurado Rabadan, S., and Agerso, Y. (2016). Sequence-Based Characterization of Tn*5801*-Like Genomic Islands in Tetracycline-Resistant *Staphylococcus pseudintermedius* and Other Gram-positive Bacteria from Humans and Animals. Front Microbiol 7, 576 DOI: 10.3389/fmicb.2016.00576.

Del Grosso, M., Camilli, R., Barbabella, G., Blackman Northwood, J., Farrell, D.J., and Pantosti, A. (2011). Genetic resistance elements carrying *mef* subclasses other than *mef*(A) in *Streptococcus pyogenes*. Antimicrob Agents Chemother 55, 3226-3230 DOI: 10.1128/AAC.01713-10.

Del Grosso, M., Camilli, R., Iannelli, F., Pozzi, G., and Pantosti, A. (2006). The *mef*(E)-carrying genetic element (mega) of *Streptococcus pneumoniae*: insertion sites and association with other genetic elements. Antimicrob Agents Chemother 50, 3361-3366 DOI: 10.1128/AAC.00277-06.

Del Grosso, M., Camilli, R., Rizzi, E., Pietrelli, A., De Bellis, G., and Pantosti, A. (2016). ICESpy009, a Conjugative Genetic Element Carrying *mef*(E) in *Streptococcus pyogenes*. Antimicrob Agents Chemother 60, 3906-3912 DOI: 10.1128/AAC.03082-15.

Del Grosso, M., Northwood, J.G., Farrell, D.J., and Pantosti, A. (2007). The macrolide resistance genes *erm*(B) and *mef*(E) are carried by Tn*2010* in dual-gene *Streptococcus pneumoniae* isolates belonging to clonal complex CC271. Antimicrob Agents Chemother 51, 4184-4186 DOI: 10.1128/AAC.00598-07.

Del Grosso, M., Scotto d'Abusco, A., Iannelli, F., Pozzi, G., and Pantosti, A. (2004). Tn*2009*, a Tn*916*-like element containing *mef*(E) in *Streptococcus pneumoniae.* Antimicrob Agents Chemother 48, 2037-2042 DOI: 10.1128/AAC.48.6.2037-2042.2004.

Denapaite, D., Bruckner, R., Nuhn, M., Reichmann, P., Henrich, B., Maurer, P., Schahle, Y., Selbmann, P., Zimmermann, W., Wambutt, R., et al. (2010). The genome of *Streptococcus mitis* B6--what is a commensal? PLoS One 5, e9426 DOI: 10.1371/journal.pone.0009426.

Deng, Y., Bao, X., Ji, L., Chen, L., Liu, J., Miao, J., Chen, D., Bian, H., Li, Y., and Yu, G. (2015). Resistance integrons: class 1, 2 and 3 integrons. Ann Clin Microbiol Antimicrob 14, 45 DOI: 10.1186/s12941-015-0100-6.

Escudero, J.A., Loot, C., Nivina, A., and Mazel, D. (2015). The integron: adaptation on demand. Microbiol Spectr 3, MDNA3-0019-2014 DOI:10.1128/microbiolspec.MDNA3-0019-2014.

Facklam, R. (2002). What happened to the streptococci: overview of taxonomic and nomenclature changes. Clin Microbiol Rev 15, 613-630

Fedoroff, N.V. (2012a). McClintock's challenge in the 21st century. Proc Natl Acad Sci U S A 109, 20200-20203 DOI: 10.1073/pnas.1215482109.

Fedoroff, N.V. (2012b). Presidential address. Transposable elements, epigenetics, and genome evolution. Science 338, 758-767

Fernandez-Lopez, C., Pluta, R., Perez-Luque, R., Rodriguez-Gonzalez, L.,

Espinosa, M., Coll, M., Lorenzo-Diaz, F., and Boer, D.R. (2013). Functional properties and structural requirements of the plasmid pMV158-encoded MobM relaxase domain. J Bacteriol 195, 3000-3008 DOI: 10.1128/JB.02264-12.

Fernandez-Lopez, R., Garcillan-Barcia, M.P., Revilla, C., Lazaro, M., Vielva, L., and de la Cruz, F. (2006). Dynamics of the IncW genetic backbone imply general trends in conjugative plasmid evolution. FEMS Microbiol Rev 30, 942-966 DOI: 10.1111/j.1574-6976.2006.00042.x.

Figueiredo, T.A., Aguiar, S.I., Melo-Cristino, J., and Ramirez, M. (2006). DNA methylase activity as a marker for the presence of a family of phage-like elements conferring efflux-mediated macrolide resistance in streptococci. Antimicrob Agents Chemother 50, 3689-3694 DOI: 10.1128/AAC.00782-06.

Florindo, C., Damiao, V., Silvestre, I., Farinha, C., Rodrigues, F., Nogueira, F., Martins-Pereira, F., Castro, R., Borrego, M.J., Santos-Sanches, I., et al. (2014). Epidemiological surveillance of colonising group B *Streptococcus* epidemiology in the Lisbon and Tagus Valley regions, Portugal (2005 to 2012): emergence of a new epidemic type IV/clonal complex 17 clone. Euro Surveill 19

Frost, L.S., Leplae, R., Summers, A.O., and Toussaint, A. (2005). Mobile genetic elements: the agents of open source evolution. Nat Rev Microbiol 3, 722-732 DOI: 10.1038/nrmicro1235.

Gawron-Burke, C., and Clewell, D.B. (1982). A transposon in *Enterococcus faecalis* with fertility properties. Nature 300, 281-284

Gillings, M.R. (2014). Integrons: past, present, and future. Microbiol Mol Biol Rev 78, 257-277 DOI: 10.1128/MMBR.00056-13.

Gyles, C., and Boerlin, P. (2014). Horizontally transferred genetic elements and their role in pathogenesis of bacterial disease. Vet Pathol 51, 328-340 DOI: 10.1177/0300985813511131.

Hadjirin, N.F., Harrison, E.M., Holmes, M.A., and Paterson, G.K. (2014). Conjugative transfer frequencies of *mef*(A)-containing Tn*1207.3* to macrolide-susceptible *Streptococcus pyogenes* belonging to different *emm* types. Lett Appl Microbiol 58, 299-302 DOI: 10.1111/lam.12213.

Henderson-Begg, S.K., Roberts, A.P., and Hall, L.M. (2009). Diversity of putative Tn*5253*-like elements in *Streptococcus pneumoniae*. Int J Antimicrob Agents 33, 364-367 DOI: 10.1016/j.ijantimicag.2008.10.002.

Hernandez-Arriaga, A.M., Espinosa, M., and del Solar, G. (2012). Fitness of the pMV158 replicon in *Streptococcus pneumoniae*. Plasmid 67, 162-166 DOI: 10.1016/j.plasmid.2012.01.005.

Iannelli, F., Santagati, M., Santoro, F., Oggioni, M.R., Stefani, S., and Pozzi, G. (2014a). Nucleotide sequence of conjugative prophage Phi1207.3 (formerly Tn1207.3) carrying the *mef*(A)/*msr*(D) genes for e fflux resistance to macrolides in *Streptococcus pyogenes*. Front Microbiol 5, 687 DOI: 10.3389/fmicb.2014.00687.

Iannelli, F., Santoro, F., Oggioni, M.R., and Pozzi, G. (2014b). Nucleotide sequence analysis of integrative conjugative element Tn*5253* of *Streptococcus pneumoniae*. Antimicrob Agents Chemother 58, 1235-1239 DOI: 10.1128/AAC.01764-13.

Jensen, A., and Kilian, M. (2012). Delineation of *Streptococcus dysgalactiae*, its subspecies, and its clinical and phylogenetic relationship to *Streptococcus pyogenes*. J Clin Microbiol 50, 113-126 DOI: 10.1128/JCM.05900-11.

Johnson, C.M., and Grossman, A.D. (2015). Integrative and conjugative elements (ICEs): what they do and how they work. Annu Rev Genet 49, 577-601 DOI: 10.1146/annurev-genet-112414-055018.

Johnston, C., Martin, B., Fichant, G., Polard, P., and Claverys, J.P. (2014). Bacterial transformation: distribution, shared mechanisms and divergent control. Nat Rev Microbiol 12, 181-196 DOI: 10.1038/nrmicro3199.

Kayansamruaj, P., Pirarat, N., Kondo, H., Hirono, I., and Rodkhum, C. (2015). Genomic comparison between pathogenic *Streptococcus agalactiae* isolated from Nile tilapia in Thailand and fish-derived ST7 strains. Infect Genet Evol 36, 307-314 DOI: 10.1016/j.meegid.2015.10.009.

Kilic, A.O., Vijayakumar, M.N., and al-Khaldi, S.F. (1994). Identification and nucleotide sequence analysis of a transfer-related region in the streptococcal conjugative transposon Tn*5252*. J Bacteriol 176, 5145-5150

Korona-Glowniak, I., Maj, M., Siwiec, R., Niedzielski, A., and Malm, A. (2016). Molecular epidemiology of *Streptococcus pneumoniae* Isolates from children with recurrent upper respiratory tract infections. PLoS One 11, e0158909 DOI:10.1371/journal.pone.0158909.

Korona-Glowniak, I., Siwiec, R., and Malm, A. (2015). Resistance determinants and their association with different transposons in the antibiotic-resistant *Streptococcus pneumoniae*. Biomed Res Int 2015, 836496 DOI: 10.1155/2015/836496.

Kowlessar, P.I., O'Connell, N.H., Mitchell, R.D., Elliott, S., and Elliott, T.S. (2006).

Management of patients with *Streptococcus milleri* brain abscesses. J Infect 52, 443-450 DOI: 10.1016/j.jinf.2005.08.028.

Kuroda, M., Ohta, T., Uchiyama, I., Baba, T., Yuzawa, H., Kobayashi, I., Cui, L., Oguchi, A., Aoki, K., Nagai, Y., et al. (2001). Whole genome sequencing of meticillin-resistant *Staphylococcus aureus*. Lancet 357, 1225-1240

Lancefield, R.C. (1933). A serological differentiation of human and other groups of hemolytic Streptococci. J Exp Med 57, 571-595

Lawley, T.D., Klimke, W.A., Gubbins, M.J., and Frost, L.S. (2003). F factor conjugation is a true type IV secretion system. FEMS Microbiol Lett 224, 1-15 DOI: 10.1016/S0378-1097(03)00430-0.

LeBlanc, D.J., Hawley, R.J., Lee, L.N., and St Martin, E.J. (1978). "Conjugal" transfer of plasmid DNA among oral Streptococci. Proc Natl Acad Sci USA 75, 3484-3487

Lekunberri, I., Subirats, J., Borrego, C.M., and Balcazar, J.L. (2017). Exploring the contribution of bacteriophages to antibiotic resistance. Environ Pollut 220, 981-984 DOI: 10.1016/j.envpol.2016.11.059.

Li, Y., Tomita, H., Lv, Y., Liu, J., Xue, F., Zheng, B., and Ike, Y. (2011). Molecular characterization of *erm*(B)- and *mef*(E)-mediated erythromycin-resistant *Streptococcus pneumoniae* in China and complete DNA sequence of Tn*2010*. J Appl Microbiol 110, 254-265 DOI: 10.1111/j.1365-2672.2010.04875.x.

Li, Y., Yang, L., Fu, J., Yan, M., Chen, D., and Zhang, L. (2017). Microbial pathogenicity and virulence mediated by integrons on Gram-positive microorganisms. Microb Pathog 111, 481-486 DOI:

10.1016/j.micpath.2017.09.035.

Lu, B., Fang, Y., Huang, L., Diao, B., Du, X., Kan, B., Cui, Y., Zhu, F., Li, D., and Wang, D. (2016). Molecular characterization and antibiotic resistance of clinical *Streptococcus dysgalactiae* subsp. *equisimilis* in Beijing, China. Infect Genet Evol 40, 119-125 DOI: 10.1016/j.meegid.2016.01.030.

Lybarger, S.R., and Sandkvist, M. (2004). Microbiology. A hitchhiker's guide to type IV secretion. Science 304, 1122-1123 DOI: 10.1126/science.1098806.

Mahillon, J., and Chandler, M. (1998). Insertion sequences. Microbiol Mol Biol Rev 62, 725-774

Marimon, J.M., Valiente, A., Ercibengoa, M., Garcia-Arenzana, J.M., and Perez-Trallero, E. (2005). Erythromycin resistance and genetic elements carrying macrolide efflux genes in *Streptococcus agalactiae*. Antimicrob Agents Chemother 49, 5069-5074 DOI: 10.1128/AAC.49.12.5069-5074.2005.

Mazel, D. (2006). Integrons: agents of bacterial evolution. Nat Rev Microbiol 4, 608-620 DOI: 10.1038/nrmicro1462.

McDougal, L.K., Tenover, F.C., Lee, L.N., Rasheed, J.K., Patterson, J.E., Jorgensen, J.H., and LeBlanc, D.J. (1998). Detection of Tn*917*-like sequences within a Tn*916*-like conjugative transposon (Tn*3872*) in erythromycin-resistant isolates of *Streptococcus pneumoniae*. Antimicrob Agents Chemother 42, 2312-2318

Menon, T. (2016). Understanding the viridians group Streptococci: Are we there yet? Indian J Med Microbiol 34, 421-426 DOI: 10.4103/0255-0857.195371.

Messier, N., and Roy, P.H. (2001). Integron integrases possess a unique additional domain necessary for activity. J Bacteriol 183, 6699-6706 DOI: 10.1128/JB.183.22.6699-6706.2001.

Mingoia, M., Morici, E., Brenciani, A., Giovanetti, E., and Varaldo, P.E. (2014a). Genetic basis of the association of resistance genes *mef*(I) (macrolides) and *catQ* (chloramphenicol) in Streptococci. Front Microbiol 5, 747 DOI: 10.3389/fmicb.2014.00747.

Mingoia, M., Morici, E., Morroni, G., Giovanetti, E., Del Grosso, M., Pantosti, A., and Varaldo, P.E. (2014b). Tn*5253* family integrative and conjugative elements carrying *mef*(I) and *catQ* determinants in *Streptococcus pneumoniae* and *Streptococcus pyogenes*. Antimicrob Agents Chemother 58, 5886-5893 DOI: 10.1128/AAC.03638-14.

Mingoia, M., Morici, E., Tili, E., Giovanetti, E., Montanari, M.P., and Varaldo, P.E. (2013). Characterization of Tn*5801.Sag*, a variant of *Staphylococcus aureus* Tn*916* family transposon Tn*5801* that is widespread in clinical isolates of *Streptococcus agalactiae*. Antimicrob Agents Chemother 57, 4570-4574 DOI: 10.1128/AAC.00521-13.

Mingoia, M., Tili, E., Manso, E., Varaldo, P.E., and Montanari, M.P. (2011). Heterogeneity of Tn*5253*-like composite elements in clinical *Streptococcus pneumoniae* isolates. Antimicrob Agents Chemother 55, 1453-1459 DOI: 10.1128/AAC.01087-10.

Mingoia, M., Vecchi, M., Cochetti, I., Tili, E., Vitali, L.A., Manzin, A., Varaldo, P.E., and Montanari, M.P. (2007). Composite structure of *Streptococcus pneumoniae* containing the erythromycin efflux resistance gene *mef*(I) and the chloramphenicol resistance gene *catQ*. Antimicrob Agents Chemother 51, 3983-

3987 DOI: 10.1128/AAC.00790-07.

Noguchi, S., Yatera, K., Kawanami, T., Yamasaki, K., Naito, K., Akata, K., Shimabukuro, I., Ishimoto, H., Yoshii, C., and Mukae, H. (2015). The clinical features of respiratory infections caused by the *Streptococcus anginosus* group. BMC Pulm Med 15, 133 DOI: 10.1186/s12890-015-0128-6.

Ojo, K.K., Ruehlen, N.L., Close, N.S., Luis, H., Bernardo, M., Leitao, J., and Roberts, M.C. (2006). The presence of a conjugative Gram-positive Tn*2009* in Gram-negative commensal bacteria. J Antimicrob Chemother 57, 1065-1069 DOI: 10.1093/jac/dkl094.

Palmieri, C., Mingoia, M., Massidda, O., Giovanetti, E., and Varaldo, P.E. (2012). *Streptococcus pneumoniae* transposon Tn*1545*/Tn*6003* changes to Tn*6002* due to spontaneous excision in circular form of the *erm*(B)- and aphA3-containing macrolide-aminoglycoside-streptothricin (MAS) element. Antimicrob Agents Chemother 56, 5994-5997 DOI: 10.1128/AAC.01487-12.

Parks, T., Barrett, L., and Jones, N. (2015). Invasive streptococcal disease: a review for clinicians. Br Med Bull 115, 77-89 DOI: 10.1093/bmb/ldv027.

Partridge, S.R. (2011). Analysis of antibiotic resistance regions in Gram-negative bacteria. FEMS Microbiol Rev 35, 820-855 DOI: 10.1111/j.1574-6976.2011.00277.x.

Partridge, S.R., Recchia, G.D., Scaramuzzi, C., Collis, C.M., Stokes, H.W., and Hall, R.M. (2000). Definition of the *attI1* site of class 1 integrons. Microbiology 146, 2855-2864 DOI: 10.1099/00221287-146-11-2855.

Perry, J.A., Westman, E.L., and Wright, G.D. (2014). The antibiotic resistome:

what's new? Curr Opin Microbiol 21, 45-50 DOI: 10.1016/j.mib.2014.09.002.

Poyart, C., Quesne, G., Acar, P., Berche, P., and Trieu-Cuot, P. (2000). Characterization of the Tn*916*-like transposon Tn*3872* in a strain of *Abiotrophia defectiva* (*Streptococcus defectivus*) causing sequential episodes of endocarditis in a child. Antimicrob Agents Chemother 44, 790-793

Pozzi, G., Iannelli, F., Oggioni, M.R., Santagati, M., and Stefani, S. (2004). Genetic elements carrying macrolide efflux genes in Streptococci. Curr Drug Targets Infect Disord 4, 203-206

Quintero, B., Araque, M., van der Gaast-de Jongh, C., and Hermans, P.W. (2011). Genetic diversity of Tn*916*-related transposons among drug-resistant *Streptococcus pneumoniae* isolates colonizing healthy children in Venezuela. Antimicrob Agents Chemother 55, 4930-4932 DOI: 10.1128/AAC.00242-11.

Ramos, V., Duarte, C., Diaz, A., and Moreno, J. (2014). Mobile genetic elements associated with erythromycin-resistant isolates of *Streptococcus pneumoniae* in Colombia. Biomedica 34 Suppl 1, 209-216 DOI: 10.1590/S0120-41572014000500023.

Rato, M.G., Bexiga, R., Nunes, S.F., Vilela, C.L., and Santos-Sanches, I. (2010). Human group A Streptococci virulence genes in bovine group C Streptococci. Emerg Infect Dis 16, 116-119 DOI: 10.3201/eid1601.090632.

Remington, L.T., and Sligl, W.I. (2014). Community-acquired pneumonia. Curr Opin Pulm Med 20, 215-224 DOI: 10.1097/MCP.0000000000000052.

Roberts, A.P., and Mullany, P. (2009). A modular master on the move: the Tn*916* family of mobile genetic elements. Trends Microbiol 17, 251-258 DOI:

10.1016/j.tim.2009.03.002.

Roberts, A.P., and Mullany, P. (2011). Tn*916*-like genetic elements: a diverse group of modular mobile elements conferring antibiotic resistance. FEMS Microbiol Rev 35, 856-871 DOI: 10.1111/j.1574-6976.2011.00283.x.

Rolo, D., Fenoll, A., Ardanuy, C., Calatayud, L., Cubero, M., de la Campa, A.G., and Linares, J. (2011). Trends of invasive serotype 6C pneumococci in Spain: emergence of a new lineage. J Antimicrob Chemother 66, 1712-1718 DOI: 10.1093/jac/dkr193.

Sadowy, E., Matynia, B., and Hryniewicz, W. (2010). Population structure, virulence factors and resistance determinants of invasive, non-invasive and colonizing *Streptococcus agalactiae* in Poland. J Antimicrob Chemother 65, 1907-1914 DOI: 10.1093/jac/dkq230.

Santagati, M., Iannelli, F., Cascone, C., Campanile, F., Oggioni, M.R., Stefani, S., and Pozzi, G. (2003). The novel conjugative transposon Tn*1207.3* carries the macrolide efflux gene *mef*(A) in *Streptococcus pyogenes*. Microb Drug Resist 9, 243-247 DOI: 10.1089/107662903322286445.

Santagati, M., Iannelli, F., Oggioni, M.R., Stefani, S., and Pozzi, G. (2000). Characterization of a genetic element carrying the macrolide efflux gene *mef*(A) in *Streptococcus pneumoniae*. Antimicrob Agents Chemother 44, 2585-2587

Santagati, M., Lupo, A., Scillato, M., Di Martino, A., and Stefani, S. (2009). Conjugal mobilization of the mega element carrying *mef*(E) from *Streptococcus salivarius* to *Streptococcus pneumoniae*. FEMS Microbiol Lett 290, 79-84 DOI: 10.1111/j.1574-6968.2008.01408.x.

Santoro, F., Oggioni, M.R., Pozzi, G., and Iannelli, F. (2010). Nucleotide sequence and functional analysis of the *tet*(M)-carrying conjugative transposon Tn*5251* of *Streptococcus pneumoniae*. FEMS Microbiol Lett 308, 150-158 DOI: 10.1111/j.1574-6968.2010.02002.x.

Schroder, G., Krause, S., Zechner, E.L., Traxler, B., Yeo, H.J., Lurz, R., Waksman, G., and Lanka, E. (2002). TraG-like proteins of DNA transfer systems and of the *Helicobacter pylori* type IV secretion system: inner membrane gate for exported substrates? J Bacteriol 184, 2767-2779

Scott, K.P. (2002). The role of conjugative transposons in spreading antibiotic resistance between bacteria that inhabit the gastrointestinal tract. Cell Mol Life Sci 59, 2071-2082

Sengupta, M., and Austin, S. (2011). Prevalence and significance of plasmid maintenance functions in the virulence plasmids of pathogenic bacteria. Infect Immun 79, 2502-2509 DOI: 10.1128/IAI.00127-11.

Shi, L., Zheng, M., Xiao, Z., Asakura, M., Su, J., Li, L., and Yamasaki, S. (2006). Unnoticed spread of class 1 integrons in Gram-positive clinical strains isolated in Guangzhou, China. Microbiol Immunol 50, 463-467

Shoemaker, N.B., Smith, M.D., and Guild, W.R. (1979). Organization and transfer of heterologous chloramphenicol and tetracycline resistance genes in pneumococcus. J Bacteriol 139, 432-441

Siguier, P., Gourbeyre, E., and Chandler, M. (2014). Bacterial insertion sequences: their genomic impact and diversity. FEMS Microbiol Rev 38, 865-891 DOI: 10.1111/1574-6976.12067.

Smillie, C., Garcillan-Barcia, M.P., Francia, M.V., Rocha, E.P., and de la Cruz, F. (2010). Mobility of plasmids. Microbiol Mol Biol Rev 74, 434-452 DOI: 10.1128/MMBR.00020-10.

Stewart, P.E., Byram, R., Grimm, D., Tilly, K., and Rosa, P.A. (2005). The plasmids of *Borrelia burgdorferi*: essential genetic elements of a pathogen. Plasmid 53, 1-13 DOI: 10.1016/j.plasmid.2004.10.006.

Storrs, M.J., Poyart-Salmeron, C., Trieu-Cuot, P., and Courvalin, P. (1991). Conjugative transposition of Tn*916* requires the excisive and integrative activities of the transposon-encoded integrase. J Bacteriol 173, 4347-4352

Sui, Z., Zhou, W., Yao, K., Liu, L., Zhang, G., Yang, Y., and Feng, J. (2013). Complete genome sequence of *Streptococcus pneumoniae* strain A026, a clinical multidrug-resistant isolate carrying Tn*2010*. Genome Announc 1 DOI: 10.1128/genomeA.01034-13.

Swartley, J.S., McAllister, C.F., Hajjeh, R.A., Heinrich, D.W., and Stephens, D.S. (1993). Deletions of Tn*916*-like transposons are implicated in *tetM*-mediated resistance in pathogenic Neisseria. Mol Microbiol 10, 299-310 DOI: 10.1111/j.1365-2958.1993.tb01956.x.

Takamatsu, D., Osaki, M., and Sekizaki, T. (2003). Chloramphenicol resistance transposable element Tn*Ss1* of *Streptococcus suis*, a transposon flanked by IS*6*-family elements. Plasmid 49, 143-151

Talebi, M., Azadegan, A., Sadeghi, J., Ahmadi, A., Ghanei, M., Katouli, M., Owlia, P., and Pourshafie, M.R. (2016). Determination of characteristics of erythromycin resistant *Streptococcus pneumoniae* with preferred PCV usage in Iran. PLoS One 11, e0167803 DOI: 10.1371/journal.pone.0167803.

Tantivitayakul, P., Lapirattanakul, J., Vichayanrat, T., and Muadchiengka, T. (2016). Antibiotic resistance patterns and related mobile genetic elements of Pneumococci and beta-Hemolytic Streptococci in Thai Healthy Children. Indian J Microbiol 56, 417-425 DOI: 10.1007/s12088-016-0607-3.

Tomich, P.K., An, F.Y., and Clewell, D.B. (1980). Properties of erythromycin-inducible transposon Tn*917* in *Streptococcus faecalis*. J Bacteriol 141, 1366-1374

Varaldo, P.E., Montanari, M.P., and Giovanetti, E. (2009). Genetic elements responsible for erythromycin resistance in Streptococci. Antimicrob Agents Chemother 53, 343-353 DOI: 10.1128/AAC.00781-08.

Verani, J.R., McGee, L., Schrag, S.J. (2010). Prevention of perinatal group B Streptococcal disease-revised guidelines from CDC, 2010. MMWR Recomm Rep 59, 1-36

Vijayakumar, M.N., Priebe, S.D., and Guild, W.R. (1986). Structure of a conjugative element in *Streptococcus pneumoniae*. J Bacteriol 166, 978-984

Walker, M.J., Barnett, T.C., McArthur, J.D., Cole, J.N., Gillen, C.M., Henningham, A., Sriprakash, K.S., Sanderson-Smith, M.L., and Nizet, V. (2014). Disease manifestations and pathogenic mechanisms of group A Streptococcus. Clin Microbiol Rev 27, 264-301 DOI: 10.1128/CMR.00101-13.

Wang, H., Roberts, A.P., and Mullany, P. (2000). DNA sequence of the insertional hot spot of Tn*916* in the *Clostridium difficile* genome and discovery of a Tn*916*-like element in an environmental isolate integrated in the same hot spot. FEMS Microbiol Lett 192, 15-20 DOI: 10.1111/j.1574-6968.2000.tb09352.x.

Wright, L.D., and Grossman, A.D. (2016). Autonomous replication of the conjugative transposon Tn*916*. J Bacteriol 198, 3355-3366 DOI: 10.1128/JB.00639-16.

Wyres, K.L., van Tonder, A., Lambertsen, L.M., Hakenbeck, R., Parkhill, J., Bentley, S.D., and Brueggemann, A.B. (2013). Evidence of antimicrobial resistance-conferring genetic elements among pneumococci isolated prior to 1974. BMC Genomics 14, 500 DOI: 10.1186/1471-2164-14-500.

Zhang, C., Zhang, Z., Song, L., Fan, X., Wen, F., Xu, S., and Ning, Y. (2015). Antimicrobial resistance profile and genotypic characteristics of *Streptococcus suis* capsular type 2 isolated from clinical carrier sows and diseased pigs in China. Biomed Res Int 2015, 284303 DOI: 10.1155/2015/284303.

Zhou, W., Yao, K., Zhang, G., Yang, Y., Li, Y., Lv, Y., and Feng, J. (2014). Mechanism for transfer of transposon Tn*2010* carrying macrolide resistance genes in *Streptococcus pneumoniae* and its effects on genome evolution. J Antimicrob Chemother 69, 1470-1473 DOI: 10.1093/jac/dku019.

 https://doi.org/10.21775/9781912530229.05

The VicRK Two-Component System Regulates *Streptococcus mutans* Virulence

Lei Lei[1,3#], Li Long[2#], Xin Yang[2], Yang Qiu[2], Yanglin Zeng[2], Tao Hu[1], Shida Wang[2*], Yuqing Li[2*]

[1] State Key Laboratory of Oral Diseases, National Clinical Research Center for Oral Diseases, Department of Preventive Dentistry, West China Hospital of Stomatology, Sichuan University, Chengdu, China

[2] State Key Laboratory of Oral Diseases, National Clinical Research Center for Oral Diseases, West China Hospital of Stomatology, Sichuan University, Chengdu, China

[3] Department of Microbiology, The Forsyth Institute, Cambridge, Massachusetts, USA

[#] Co-first author

* Corresponding authors: liyuqing@scu.edu.cn, 61454699@qq.com

Abstract:

Streptococcus mutans is considered the predominant etiological agent of dental caries with the ability to form biofilm on the tooth surface. And, its abilities to obtain nutrients and metabolize fermentable dietary carbohydrates to produce acids contribute to its pathogenicity. The responses of *S. mutans* to environmental stresses are essential for its survival and role in cariogenesis. The VicRK system is one of the 13 putative TCS of *S. mutans*. The conserved functions of the VicRK signal transduction system is the key regulator of bacterial oxidative stress responses, acidification, cell wall metabolism, and biofilm formation. In this paper, it was discussed how the VicRK system regulates *S. mutans* virulence including bacterial physiological function, operon structure, signal transduction, and even post-transcriptional control in its regulon. Thus, this emerging subspecialty of the VicRK regulatory networks in *S. mutans* may strengthen our understandings aimed at providing a basis for the prevention of dental caries.

Introduction

Dental plaque is a multi-species microbial biofilm that form on the tooth surfaces. Dental caries is promoted by environmental changes (e.g., changes in pH) that cause ecological shifts to plaque residents that favor the proliferation of aciduric bacteria (Kazor *et al.*, 2003; Paster *et al.*, 2001; Paster *et al.*, 2002). Among the hundreds of bacterial species that colonize and persist in the oral cavity, *Streptococcus mutans* is among the few species that have been consistently linked with caries formation (Loesche *et al.*, 1986). In addition, the ability of *S. mutans* to synthesize extracellular polysaccharides that promote formation of the plaque biofilm also contributes to its pathogenicity.

Two-component signal transduction systems (TCS) are among the regulatory networks that are essential for bacterial adaptation, survival, and virulence in response to changes in the external environment. Typically, signal transduction is accomplished via two regulatory elements comprising a membrane-associated histidine kinase (HK) and a cytoplasmic response regulator (RR). Upon exposure to an environmental change such as pH, osmolarity, or oxidation-reduction potential, the HK becomes autophosphorylated at a conserved histidine residue. Following the transfer of this phosphate group to a RR, the regulator can control transcription of target genes by binding to their promoter regions (Stein *et al.*, 2002; Davies *et al.*, 1998; Mascher *et al.*, 2006). The most widely distributed essential TCS, the WalK/WalR system, was originally described in *Bacillus subtilis* (Fukuchi *et al.*, 2000). This system has since been reported as essential for cell viability in several closely related pathogens (*Staphylococcus aureus*, *Streptococcus pneumoniae*, *S. mutans*, and *Streptococcus pyogenes*), and referred to under various designations (YycG/YycF, VicK/VicR, MicA/MicB) (Martin *et al.*, 1999; Lange *et al.*, 1999; Wagner *et al.*, 2002). And, attempts to inactivate the *walRK* locus in *Listeria monocytogenes*, *Enterococcus faecalis, Staphylococcus epidermidis* and *Str. pneumoniae* were unsuccessful, indicating that it is likely to be essential in these bacteria (Lange *et al.*, 1999, Kallipolitis *et al.*, 2001; Hancock *et al.*, 2004; Dhiman *et al.*, 2014).

The VicRK TCS is one of 13 such systems found in *S. mutans* UA159 (Senadheera *et al.*, 2005). Based on sequence homology, the VicRK genes encode a surface-associated HK (VicK) and an intracellular RR (VicR). The *vic* operon comprises three genes encoding the following: VicR, a RR of the OmpR family; VicK, its cognate HK; and VicX, a putative protein sharing 55% identity to YycJ in *B. subtilis* (Fukuchi *et al.*, 2000). A putative transcriptional start site for the *vic* operon was mapped 16 bp upstream of the ATG codon of *vicR*. VicK, an atypical PAS domain-containing HK, can be autophosphorylated *in vitro*, and

VicR functions as a phospho-acceptor protein *in vitro* (Ajdic *et al.*, 2002; Senadheera *et al.*, 2007; Durso *et al.*, 2014). YycFG (VicRK) TCS has been the topic of many recent studies among these quorum-sensing systems. In streptococci, the YycFG (VicRK) system regulates processes involved in cell wall metabolism, nutrient uptake, osmotic protection, cell division (Barendt *et al.*, 2009), cell wall biogenesis (Moraes *et al.*, 2014), antibiotic-resistance (Li *et al.*, 2009) and resistance to complement immunity(Alves *et al.*, 2017). Furthermore, *vicRK* genes in 242 *S. mutans* clinical isolates were sequenced, locus 470 showed higher missense mutation in high-severity caries group than control group, indicating The locus 470 missense mutation of the vicK gene may be related to caries in children with *S. mutans* (Zhuang *et al.*, 2018). The VicRK system has attracted considerable interests in *S. mutans*, yielding insights into its physiological function, its potential activating signal and virulence (Liu *et al.*, 2006). In this review, we summariz how the VicRK system regulates *S. mutans* virulence including bacterial physiological function, operon structure, signal transduction, and even post-transcriptional control in its regulon. The current knowledge on the system of the VicRK regulatory networks in *S. mutans* may strengthen our understandings of bacterial virulence and provided new aspects for the management of dental caries.

VicRK-dependent phenotype

The WalK/WalR system has been reported as essential for cell viability in several closely related pathogens (*Sta. aureus*, *Str. pneumoniae*, *S. mutans*, and *Str. pyogenes*) and referred to under various designations (YycG/YycF,

VicK/VicR, MicA/MicB). The WalK/WalR (YycFG) TCS is highly conserved and specific to low G+C Gram-positive bacteria. *B. subtilis* and *L. monocytogenes* have a 6-gene operon (*walR*, *walK*, *yycH*, *yycI*, *yycJ*, *yycK*), *E. faecalis* lacks the last gene, while only three are present in streptococci (*walR*, *walK*, *yycJ*; Figure. 1).

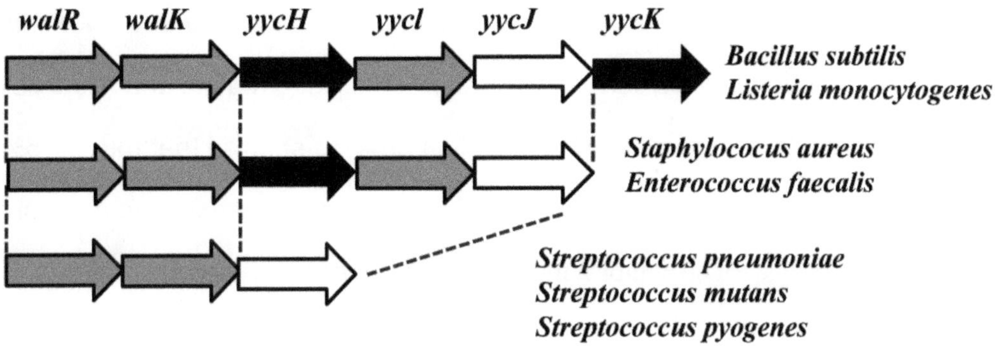

Figure 1. Genetic map of the *walRK* (*vicRK*) operon.

Oxidative stress response

After the initial colonization stage, *S. mutans* undergoes continuous dynamic challenges to which it must respond and adapt. There is much evidence suggesting a relation between oxidative stress and biofilm formation (Sampathkumar *et al.*, 2006; Loo *et al.*, 2004) because oxidative stress response genes were up-regulated during biofilm growth. D.M. Deng *et al* (Deng *et al.*, 2004) constructed a strain which expressed of a green fluorescent reporter-protein (GFP) under control of the autoregulated *vicRK* promoter, showing increased fluorescence intensity under oxidative stress. After a clean knockout of the *vicK* gene, the mutant proved to be more sensitive to H_2O_2 than the wild-type parent strain, leading to the hypothesis that there was a link between oxidative/redox stress and the VicRK TCS of *S. mutans*.

Additionally, oxygen is required by several oral bacteria for respiration and energy generation. Although oral streptococci do not possess a full electron transport chain and cannot carry out oxidative phosphorylation, these organisms maintain a high capacity to metabolize oxygen, primarily through NADH oxidase enzymes (Higuchi *et al.*, 1993; Marquis *et al.*, 1995). It has been shown that exposure of *S. mutans* to oxygen strongly inhibits biofilm formation and alters cell surface biogenesis. The results of transcription profiling showed that roughly 5% of the genes of the organism are differentially expressed in response to aeration. Consistent with this observation, the ability of *S. mutans* to form biofilms is severely impaired by exposure to oxygen. Real-time reverse-transcription polymerase chain-reaction revealed that transcription of *gtfB*, but not *gtfC*, was responsive to oxygen and that aeration causes major changes in the amount and degree of cell association of the Gtf enzymes. Moreover, inactivation of the VicK sensor kinase affected the expression and localization the GtfB and GtfC enzymes. These studies provide insights into the complex transcriptional and posttranscriptional regulatory networks in *S. mutans* that modulate virulence gene expression in response to changes in oxygen availability (Ahn *et al.*, 2007).

Acid production and acid survival

S. mutans can rapidly metabolize dietary sucrose, producing lactic acid as a metabolic end product, which can lower the plaque pH and erode the tooth enamel, thereby initiating dental decay (cariogenesis). Studies have shown that if *S. mutans* cultures are pre-exposed to an adaptive acidic environment (approximately pH 5.0 to 5.5), the survival rates of the bacteria are dramatically enhanced at approximately 3.5 of pH value (Hamilton *et al.*, 1998; Hamilton *et al.*, 1991; Quivey *et al.*, 2000). The regulation of acid production and the tolerance to low pH of *S. mutans* have gained considerable attention since both properties contribute substantially to the virulence of this organism. Senadheera *et al.* (Senadheera *et al.*, 2009) reported that the VicK sensor kinase of *S. mutans* was involved in both the acidogenicity and the aciduricity. When cultures were supplemented with glucose, the glycolytic rate of a *vicK* null mutant was significantly decreased, compared with the wild type, suggesting the impaired acid production for the mutant. When *S. mutans* wild-type UA159 and *vicK* deletion mutant strains grown at neutral and low pH values, global transcriptional analysis using DNA microarray revealed that loss of VicK significantly affected expression of 89 transcripts at pH 5.5. Using a cut-off of two-fold change, 38 genes were significantly up-regulated while 51 genes were suppressed. The results provide insights into the acid-inducible regulon of *S. mutans*, and the affected transcripts included genes with putative functions in transport and maintenance of cell membrane integrity, implying a novel role for VicK in regulating intracellular pH homeostasis in *S. mutans* (Senadheera *et al.*, 2009).

Exopolysacharrides synthesis

In part, the virulence of the dental caries pathogen *S. mutans* relies on the sucrose-dependent synthesis of and interaction with glucan, a major component of the extracellular matrix of tooth biofilms. Glucan, including water-insoluble glucan (WIG) and water-soluble glucan (WSG), undergo structural modifications resulting from the effects of glucosyltransferases (GtfB/C/D) and fructosyltransferase (Ftf) during glucan synthesis (Rolla *et al.*, 1983). A *vicK* deletion mutant was isolated by Senadheera *et al* (Senadheera *et al.*, 2009), while a *vicR* (putative RR) null mutation was apparently lethal. Compared with the wild-type UA159, the *vicK* mutant biofilm formed in sucrose-supplemented medium was easily detachable. The rate of total dextran formation by *vicK* mutant was remarkably reduced compared to the wild type. Furthermore, a recombinant VicR fusion protein was shown to bind the promoter regions of the *gtfB*, *gtfC*, and *ftf* genes. Study also showed that addition of high-fructose corn syrup-5, a substitute for sucrose, up-regulated the expression of *vicR* and *gtfB* and reduced the microhardness of teeth (Sun. *et al.*, 2014). Animal studies conducted using *vicK* mutant in a specific-pathogen free (SPF) rat model resulted in a significant reduction in *vicK* mutant CFU counts compared to that of the wild-type UA159.

Additionally, it has been suggested that other surface proteins, such as glucan-binding proteins (Gbps), with affinity for glucan contribute to *S. mutans* biofilm growth by mediating bacterial interaction with extracellular glucan (Banas *et al.*, 2003. A conditional knockdown mutant that expressed *gbpB* antisense RNA under the control of a tetracycline-inducible promoter was constructed in wild type strain UA159 (Duque *et al.*, 2011). It was established that GbpB depletion impaired initial phases of sucrose-dependent biofilm formation. Additionally, GbpB was directly regulated by VicR, and exogenous native GbpB partially restored the biofilm phenotype. Several cellular phenotypes were significantly affected by GbpB depletion, including altered bacterial cell shape, decreased autolysis, increased cell hydrophobicity, and sensitivity to antibiotics and osmotic and oxidative stresses.

Cell wall biosynthesis

Defining the specific functions of genes involved in cell wall biogenesis is important for clarifying the transition mechanism of *S. mutans* from planktonic to biofilm growth in the presence of sucrose. Using transcriptomic analysis comparing the wild-type UA159 strain with a *vicK* deletion mutant together with electrophoretic motility shift assays, Stipp *et al.* (Stipp *et al.*, 2013) identified genes directly regulated by both VicR and GcrR (CovR) with putative functions in cell wall biogenesis, including *gbpB*, *wapE*, *smaA*, *SMU.2146c*, and *lysM*. Deletion mutants of genes (*wapE*, *lysM*, *smaA*) promote significant alterations in biofilm formation, including increased fragility, defects in microcolony aggregation, and atypical cell morphology. Additionally, VicR, CovR and VicRK/CovR targets (*gbpB*, *wapE*, *smaA*, *SMU.2146c*, *lysM*, *epsC*) are up-regulated in UA159 during biofilm initiation, in a sucrose-dependent manner. These results support a model in which VicRK and CovR coordinate cell division and surface biogenesis with the extracellular polysaccharides synthesis for formation of structurally stable biofilm.

Cell division

In *S. mutans*, *vicK*-defective mutant also forms long chains. However, GbpB, homologues of PcsB, seems to work as a glucan-binding protein in *S. mutans*. Abnormal cell morphology but no long chain was observed after bringing down expression of GbpB. Cristiane Duque *et al.* (Duque *et al.*, 2011) constructed an antisense RNA to disturb the expression of *gbpB* and this strain formed chains of shorter cocci but not long chains, which is different to a *vicK*-defective mutant. Previous study demonstrated that a 60-kilodalton immunodominant glycoprotein contributed to cell wall integrity and maintains cell shape in *S. mutans* (Chia *et al.*, 2001).

Bacteriocin production and autolysin

Bacteriocins are peptides that have antimicrobial activity against other strains or related species. Although bacteriocin producers are resistant to their own bacteriocins through the action of immunity proteins, production of these bacteriolytic or bacteriostatic molecules can be costly due to constitutively expressed plasmid carriage, and possible lethality of production. *S. mutans* is a primary pathogen associated with dental caries and its bacteriocin (mutacin) production ability is thought to play an important role in maintaining competitiveness in the multispecies oral biofilm. The signaling networks that regulate bacteriocin production and transport as well as immunity to bacteriocins involve peptide pheromone sensing pathways which is very similar to those involved in genetic competence (Kleerebezem *et al.*, 2001). Currently, two classes of mutacins have been identified: Lantibiotics are ribosomal synthesized and undergo extensive posttranslational modifications that create the unusual amino acids lanthionine or b-methyllanthionine, whereas non-lantibiotics consist of either one or two small unmodified peptides (Qi *et al.*, 2001). It has been demonstrated that a putative inducible repressor, *irvA*, seems to be involved in the gene regulation pathway of the LuxS-mediated lantibiotic, mutacin I(Tsang *et al.*, 2006). In this study, it was found that all these genetic loci (*vicK, pttB hk03, hrcA, adhE, Smu1281,* and *ciaH*) influenced mutacin I production at the transcriptional level. It has been demonstrated that these multiple inputs can be divided into two pathways: *irvA*-dependent and *irvA*-independent. Mutations in genes such as *luxS*, *hk03*, *vicK*, and *pttB* activate the expression of a common regulator, *irvA*. Thus, current studies suggest that VicRK signals may modulate the transcription of the mutacin operon.

Senadheera *et al.* (Senadheera *et al.*, 2012) showed that the VicRK modulates bacteriocin production and cell viability, in part by direct modulation of competence-stimulating peptide (CSP) production in *S. mutans*. VicR binding to the *comC* coding region has been confirmed since a well-conserved VicR binding site was identified within the *comC* open reading frame. Global transcriptome and real-time transcriptional analysis of the VicK-deficient strain revealed significant modulation of several bacteriocin-related genes, including *nlmAB*, *nlmC*, and *nlmD*, suggesting a role for the VicRK in producing mutacins IV, V, and VI (Qi *et al.*, 2001; Qi *et al.*, 1999; Qi *et al.*, 1999). The results may reveal a novel regulatory link between the VicRK and ComDE systems to modulate bacteriocin production and autolysis of *S. mutans*.

Also, the *atlA* gene in *S. mutans* encodes an autolysin required for biofilm maturation and biogenesis of a normal cell surface (Ahn *et al.*, 2006). The formation of long chains, a characteristic of *atlA* mutant, was less evident in cells grown with aeration. Deletion of the *vicK* gene also led to inhibition of processing of *atlA*, and the mutant was more resistant to autolysis (Qi *et al.*, 1999; Ahn *et al.*, 2007). When grown under aerobic conditions, the *vicK* mutant also showed significantly increased biofilm formation compared to wild-type UA159 strain.

Taken together, these investigations illustrate the crucial role of AtlA and VicK in orchestrating growth on surfaces and envelope biogenesis in response to autolysis.

Carbohydrate substrates metabolism

Although other microorganisms may also be involved in dental plaque, *S. mutans* is a key contributor in the formation of biofilms associated with sucrose ingestion (Beighton *et al.*, 2005; Kanasi *et al.*, 2010). *S. mutans* produces several extracellular sucrose-metabolizing enzymes, such as glucosyltransferases and fructosyltransferase. The cooperative action of these enzymes is essential for sucrose-dependent cellular adhesion and biofilm formation. A global RR (VicR) plays important roles in *ftf* and *gtf* expression in response to environmental changes. Real-time reverse-transcription polymerase chain-reaction were used to quantify the relative levels of *ftf*, *gtfB*, *gtfC*, *gtfD* and *vicR* transcription of *S. mutans* in the presence of various dietary carbohydrates: sucrose, D-glucose, D-fructose, D-glucitol (D-sorbitol), D-mannitol and xylitol. The expression of *vicR* was induced only at the presence of xylitol at late exponential phase and declined at early exponential phase. Taken together, these findings show that dietary carbohydrates have a major influence on the transcription of *ftf*, *gtfB*, *gtfC* and *gtfD*, but less on *vicR* in *S. mutans* (Shemesh *et al.,* 2006).

To avoid the role of sucrose on the production of virulence factors by *S. mutans*, sugar substitutes are commonly consumed because of their potential to lower or block the production of acids and interfere with biofilm formation (Rozen *et al.*, 2001; Rozen *et al.*, 2004). The presence of sugars or sugar substitutes profoundly affected the expression of *spaP, gtfB, gtfC, gbpB, ftf, vicR* and *vicX* in either biofilm or planktonic cells. The substitution of sucrose induced a down-regulation of *vicR* and *vicX* in sucrose-dependent colonization in biofilm cells. However, sucralose but not sorbitol fulfilled the purpose of reducing the cariogenic potential, since it induced the biofilm formation with the lowest biomass, did not change the pH value of the medium (Durso et al., 2014).

Biochemistry of VicRK operon structures and gene function

Structural features of the VicK kinase in S. mutans

Prokaryotic organisms commonly regulate gene expression in response to environment changes using TCS (Parkinson *et al.*, 1993; Zhulin *et al.*, 1997). Such systems typically comprise a membrane associated sensory kinase and a cytoplasmic RR. Wang *et al* (Wang *et al.*, 2013) reported the crystal structure of a complete cytoplasmic portion of sensory kinase VicK from *S. mutans*. The overall structure of VicK is a long-rod dimer that anchors four connected segment domains: HAMP, Per-ARNT-SIM (PAS), DHp, and catalytic and ATP binding domain. The HAMP, a signal transducer, and the PAS domain, major sensor, adopt canonical folds with dyad symmetry. In contrast, the dimer of the DHp and ATP binding domains is asymmetric because of different helical bends in the DHp domain and spatial positions of the ATP binding domains. Moreover, a conserved proline, is adjacent to the phosphoryl acceptor histidine. This structure contributes to helical bending, which is essential for the autokinase and phosphatase activities. Together, the elegant architecture of VicK with a signal transducer and sensor domain suggests a model where DHp helical bending and an ATP binding domain swing movement are likely coordinated for autokinase activation (Wang *et al.*, 2013).

DNA binding sites of VicR

The binding sequence for the VicR homolog in *B. subtilis* and *Sta. aureus*, WalR, has been well characterized. The consensus sequence consists of two hexameric half sites separated by a non-conserved 5 base pair (Howell *et al.*, 2003). The upstream regions of virulence genes (*gtfB, gtfC* and *ftf*) in *S. mutans* were analyzed previously and were found to be match to the WalR consensus in the promoter regions (Senadheera *et al.*, 2005). Ayala *et al.* (Ayala *et al.*, 2014) investigated the biochemical characteristics of the *S. mutans* protein VicR. They dissected the DNA binding requirements of the recognition sequences for VicR in its unphosphorylated and phosphorylated forms. In *S. mutans*, CovR (RR of GbpC) is an orphan RR that is involved in biofilm formation, sucrose mediated adhesion and the acid tolerance response (Dunning *et al.*, 2008). In addition, CovR regulates virulence traits that overlap with the VicRK TCS and it has been identified a perfect match to the WalR consensus binding site (TGTTATagaacTGTAAT) 94 bp upstream of the start codon on the forward strand. The VicR consensus sequence of *nlmC, bmsH, gtfC, gbpB, gcrR* and *plsX* are located on the coding strand of each gene. In contrast, the consensuses for *gtfB, wapA, atlaA, relP, glnQ,* and *copY* are found on the non-coding strand (Ayala *et al.*, 2014).

The role of vicX in S. mutans virulence

Despite our knowledge of the various physiological properties that are subject to control by VicRK, a putative role for the third gene (*vicX*) in the tricistronic operon is not well known. Senadheera *et al.* (Senadheera *et al.*, 2007) constructed a nonpolar deletion mutation in the *vicX* coding region in *S. mutans* UA159. The growth kinetics of the *vicX* mutant (SmuvicX) showed that there was considerable sensitivity to paraquat-induced oxidative stress. Interestingly, *vicX*-deficient cells grown in a glucose-supplemented medium exhibited significantly increased *gtfB/C* expression compared with expression in the wild type. In the absence of *vicX*, the ability of *S. mutans* to grow under oxidative stress conditions and its ability to take up foreign DNA are drastically compromised (Senadheera et al., 2007). What's more, a decrease of WIG synthesis and WIG/ WSG were observed in *vicX*-deficient cells (Lei *et al.*, 2015). However, the information regarding the structural features and biochemistry characteristics of the VicX is still limited.

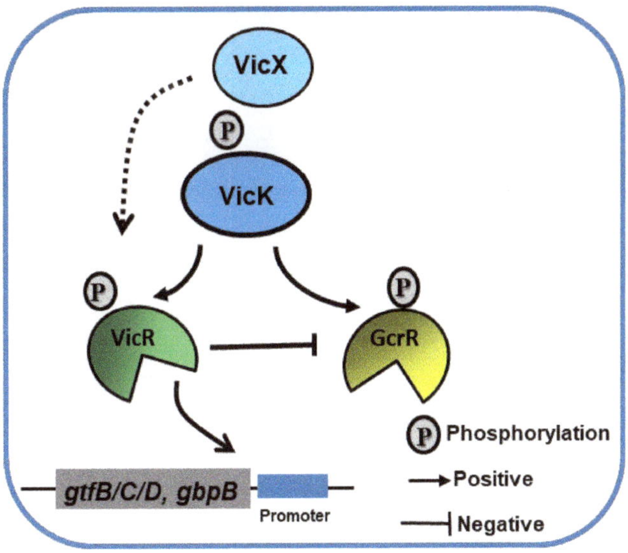

Figure 2. The interaction of VicRK TCS in *S. mutans*.

VicRK-mediated signal transduction in sensing and co-regulating cell physiology

VicK and CovR (GcrR)

Previously, we discussed the role of the *S. mutans* VicRK TCS in modulating biofilm formation, oxidative stress and acid tolerance responses. Using *in vitro* phosphorylation assays, Downey *et al* (Downey *et al.*, 2014) demonstrated that activating its cognate RR protein, the sensor kinase, VicK, can transphosphorylate a non-cognate stress regulatory RR, CovR, in the presence of manganese. Overlapped DNA binding specificities for VicR and CovR were observed in native promoters, consistent with these proteins being part of the same transcriptional regulon. Their findings supported the regulatory complexities observed with the *S. mutans* manganese-dependent response, which involves cross-talk between non-cognate signal transduction systems (VicRK and CovR) to modulate stress response pathways (Figure 2).

VicRK and ADS

The oral commensal *Streptococcus gordonii* must adapt to constantly fluctuating and often hostile environmental conditions to persist in oral biofilms through the hydrolysis of arginine in saliva (Burne *et al.*, 2000). The arginine deiminase system (ADS) of *S. gordonii* enables bacterial cells to produce, ornithine, ammonia, CO_2, and ATP from arginine hydrolysis, augmenting the acid tolerance of the organism. It has been demonstrated that the CiaRH and ComDE TCS are required for low-pH-dependent expression of ADS genes in *S. gordonii*. Further, the VicRK TCS is required for optimal ADS gene expression under anaerobic conditions and enhances the sensitivity of the operon to oxygen (Liu *et al.*, 2009). The authors observed mutants of *S. gordonii* lacking components of the CiaRH,

ComDE, or VicRK grew more slowly in acidified media and were more sensitive to killing at lethal pH values and to agents that induce oxidative stress. Simultaneously, the VicRK, ComDE, and CiaRH two-component systems affect AgDS gene expression in response to acidic stresses in *S. mutans*(Liu *et al.*, 2009). These studies reveal some notable differences in the contribution of CiaRH, ComDE, and VicRK to stress tolerance between the *S. gordonii* and *S. mutans*.

VicRK and CRISPR-Cas Systems

CRISPR-Cas (clustered regularly interspaced short palindromic repeats) systems provide adaptive microbial immunity against invading viruses, plasmids, and transposable elements (Barrangou *et al.*, 2008; Brouns *et al.*, 2008). *S. mutans* UA159 has two CRISPR-Cas systems: CRISPR1 (type II-A) and CRISPR2 (type I-C). Using plasmid transformation experiments, Serbanescu *et al.* (Serbanescu *et al.*, 2015) demonstrated that the CRISPR1-Cas system inhibits transformation of *S. mutans* by plasmids. Functional analysis of *cas* deletion mutants revealed that in addition to a role in plasmid targeting, both CRISPR systems also contribute to the regulation of bacterial physiology in *S. mutans*. Compared to wild-type cells, the *cas* deletion mutant strain displayed diminished growth under oxidative stress, enhanced growth under low pH, and reduced survival under heat shock conditions. Transcriptional analysis revealed that VicRK differentially modulates expression of *cas* genes within CRISPR-Cas systems, suggesting that VicRK might coordinate the expression of two CRISPR-Cas systems.

VicRK and c-di-AMP signaling pathway

Cyclic di-AMP (c-di-AMP) is an emerging second messenger in bacteria (Oppenheimer-Shaanan *et al.*, 2011). The level of c-di-AMP is modulated by activity of diadenylate cyclase (CdaA) that produces c-di-AMP and phosphodiesterase (PDE) that degrades c-di-AMP (Romling *et al.*, 2008). Cheng *et al.* (Cheng *et al.*, 2015) observed genes with increased or decreased expression in the *cdaA* mutant were clustered in cellular polysaccharide biosynthetic or oxidoreductase activity, respectively. Previous studies have also found that VicRK is involved in cell wall and a *vicK* knockout mutant in *S. mutans* would likely result in enhanced cell death, increased lysis (Senadheera *et al.*, 2009) and increased sensitivity to oxidative stress (Deng *et al.*, 2004). This study suggests a probable interplay between c-di-AMP signaling and TCS in this bacterium. Interestingly, it has been revealed that only CabPA deficiency inhibited both the increased biofilm formation and the up-regulated expression of GtfB observed in the *pdeA* mutant (Peng *et al.*, 2005). In addition, CabPA but not CabPB interacted with VicR, a known transcriptional factor that regulates expression of *gtfB*, suggesting that a signaling link between CabPA and GtfB through VicR. They performed a pull-down assay using recombinant CabPA, CabPB and VicR, and confirmed that VicR directly interacted with CabPA not CabPB *in vitro*. Take together, these studies revealed a new role of c-di-AMP in mediating biofilm formation through a CabPA/VicR/GtfB signaling network in which c-di-AMP mediates GtfB expression and biofilm formation through the interaction between CabPA and VicR in *S. mutans*.

Interaction with LiaRS in S. mutans

Previous study shows that expression of *vicR* is pH-dependent and LiaRS signal system in *S. mutans* controls acid adaption and envelope stress response. These evidences suggest a model which LiaS senses external stimulus and autophosphorylation, then passing a phosphate group to LiaR and phosphorylated LiaR regulates expression of *vicR* (Tremblay *et al.*, 2009).

Response of VicRK system to natural products treatment

Carolacton is a newly identified secondary metabolite with high inhibitory activity by inducing damage of *S. mutans* biofilms. Carolacton was also found to induce dose-dependent damage of *S. mutans* biofilms over a wide concentration range (Kunze *et al.*, 2010). The sub-networks modulated by the VicR could putatively be related to the physiological effect of carolacton, and the microarray data show that expression of *vicRK*, one of the earliest responding TCSs to carolacton treatment, was down-regulated. The network approach revealed *vicRK* regulators and interactions as part of the response mechanisms of *S. mutans* biofilm cells to carolacton.

Although several antiplaque agents have been used, some studies reported that several natural products derived from herbs, such as *Mentha longifolia* L., *Aralia continentalis* and *Curcuma longa L*, showed inhibitory effects on dental biofilms. Jeong *et al.* (Jeong *et al.*, 2013) isolated a single chemical compound from *A. continentalis* for investigation of anticariogenic properties. Inhibitory effects of the extracted chemical compound on growth, acid production, biofilm formation, and adherence of *S. mutans* were evaluated. Real-time PCR analysis was performed and showed that the expression of *gtfB, gtfC, gbpB, spaP, brpA, relA*, and *vicR* were significantly decreased in *S. mutans* after treatment with the compound.

Another natural product extracted from *Chrysanthemum boreale* was also evaluated for antibacterial activity against *S. mutans*. This essential oil exhibited significant inhibition of growth, adherence capacity, and acid production. Furthermore, real-time PCR analysis showed that gene expression of several virulence factors such as *gtfB*, *gtfC*, *gtfD*, *gbpB*, *spaP*, *brpA*, *relA*, and *vicR* was significantly decreased in a dose-dependent manner. These results suggest that natural products may inhibit growth, adhesion, acid tolerance, biofilm formation, and related virulence factors of *S. mutans* through the VicRK regulatory mechanism.

Small regulatory noncoding RNA control of *vic* operon

Recently, we identified and confirmed a noncoding antisense RNA (ASvicR) with the potential to attenuate the functions of an essential RR, VicR, that covered the *vicR* coding sequence, including the DNA binding and dimerization domains. Furthermore, it has previously been shown that the overexpression of ASvicR resulted in a reduction in biofilm formation and acid production (Lei *et al.*, 2018). A novel category of miRNA-size small RNAs (msRNAs) targeting *vicR*, which reduced dental caries by interfering with bacterial exopolysaccharide metabolism, has been described in *S. mutans*. Notably, a putative RNase III-encoding gene (*rnc*) that may play a critical role in the specific cleavage of dsRNA structures located downstream from the *vicRKX* operon (Mao *et al.*, 2016). Thus, this emerging subspecialty of bacterial regulatory RNAs may reshape our understanding of *S. mutans* gene regulation from its transcriptional level.

Anti-TCS drugs

TCSs play a role in the virulence of *S. mutans*, making them attractive targets for drug development. Anti-TCS drugs that repress virulence have mostly targeted the sensory domains of their target HKs, and since such domains are unique to each HK, anti-TCS drugs should potentially be effective only to a specific TCS (Gotoh *et al.*, 2010). One of the targeted TCSs for HK inhibitors is the WalK/WalR (formerly YycG/YycF) system, which plays a key role in cell wall metabolism (Watanabe *et al.*, 2003; Watanabe *et al.*, 2008)]. The effect of Walkmycin C (WKM C), a HK inhibitor, against *S. mutans* has been evaluated. This compound inhibited the *in vitro* autophosphorylation activity of three purified *S. mutans* HKs: VicK, CiaH, and LiaS. This inhibitory effect suggests that blocking the autophosphorylation of multiple HKs may inhibit phosphotransfer to VicR from VicK and other HKs. In the presence of WKM C, *S. mutans* formed abnormal biofilms and showed a defect in competence. When pretreated with the inhibitor, cells showed an increase in acid sensitivity. These results indicated the possibility of developing HK inhibitors into anti-virulence drugs for *S. mutans*.

Concluding remarks and future directions

S. mutans has few alternative sigma factors encoding genes in its genome (Ajdic *et al.*, 2002), therefore TCSs are considered to play a central role in stress tolerance (Lemos *et al.*, 2008). It will be important to further perform structure-function studies on the sensory domains of VicR which binds to the promoters of *gtfBCD* and positively regulates their expression. Both the unknown structural features of the VicK HK and its interaction with RRs are worthy of further study. Furthermore, little is known on the structural features and biochemistry characteristics of the VicX hydrolase. With the increasing number of bacteria that are rapidly becoming resistant to antibiotics, the VicRK system provide promising targets for a new class of antimicrobials and effective dental caries therapy.

Acknowledgements

We greatly appreciate and thank Prof. Margaret J. Duncan from The Forsyth Institute for her insightful comments and revising regarding this manuscript. This work was supported by the National Natural Science Foundation of China (81800964 and 31870065).

Disclosure statement

The authors do not have financial or commercial competing interests.

References

Ahn SJ, Burne RA. (2007). Effects of oxygen on biofilm formation and the AtlA autolysin of *Streptococcus mutans*. J Bacterio. 189(17), 6293-6302 DOI: 10.1128/JB.00546-07

Ahn SJ, Burne RA. (2006). The *atlA* operon of *Streptococcus mutans*: role in autolysin maturation and cell surface biogenesis. J Bacteriol. 188(19), 6877-6888 DOI: 10.1128/JB.00536-06

Ahn SJ, Wen ZT, Burne RA. (2007). Effects of oxygen on virulence traits of

Streptococcus mutans. J Bacteriol. 189(23), 8519-8527 DOI: 10.1128/JB.01180-07

Ajdic D, McShan WM, McLaughlin RE, Savic G, Chang J, Carson MB, Primeaux C, Tian R, Kenton S, Jia H, Lin S, Qian Y, Li S, Zhu H, Najar F, Lai H, White J, Roe BA, Ferretti JJ. (2002). Genome sequence of *Streptococcus mutans* UA159, a cariogenic dental pathogen. Proc Natl Acad Sci U S A. 99(22), 14434-14439 DOI: 10.1073/pnas.172501299

Alves, L. A., et al. (2017). The two-component system VicRK regulates functions associated with *Streptococcus mutans* resistance to complement immunity. Mol Oral Microbiol. 32(5): 419-431 DOI: 10.1111/omi.12183

Ayala E, Downey JS, Mashburn-Warren L, Senadheera DB, Cvitkovitch DG, Goodman SD. (2014). A biochemical characterization of the DNA binding activity of the response regulator VicR from *Streptococcus mutans*. PLoS One. 9(9), e108027 DOI: 10.1371/journal.pone.0108027

Banas JA, Vickerman MM. (2003). Glucan-binding proteins of the oral streptococci. Crit Rev Oral Biol Med. 14(2), 89-99 DOI : 10.1177/154411130301400203

Barendt SM, Land AD, Sham LT, Ng WL, Tsui HC, Arnold RJ, Winkler ME. (2009). Influences of capsule on cell shape and chain formation of wild-type and *pcsB* mutants of serotype 2 *Streptococcus pneumoniae*. J Bacteriol. 191(9), 3024-3040 DOI: 10.1128/JB.01505-08

Barrangou R, Fremaux C, Deveau H, Richards M, Boyaval P, Moineau S, Romero DA, Horvath P. (2007). CRISPR provides acquired resistance against viruses in prokaryotes. Science. 315(5819), 1709-1712 DOI: 10.1126/science.1138140

Beighton D. (2005). The complex oral microflora of high-risk individuals and groups and its role in the caries process. Community Dent Oral Epidemiol. 33(4), 248-255 DOI: 10.1111/j.1600-0528.2005.00232.x

Brouns SJ, Jore MM, Lundgren M, Westra ER, Slijkhuis RJ, Snijders AP, Dickman MJ, Makarova KS, Koonin EV, van der Oost J. (2008). Small CRISPR RNAs guide antiviral defense in prokaryotes. Science. 321(5891), 960-964 DOI: 10.1126/science.1159689

Burne RA, Marquis RE. (2000). Alkali production by oral bacteria and protection against dental caries. FEMS Microbiol Lett. 193(1), 1-6 DOI: 10.1111/j.1574-6968.2000.tb09393.x

Cheng X, Zheng X, Zhou X, Zeng J, Ren Z, Xu X, Cheng L, Li M, Li J, Li Y. (2015). Regulation of oxidative response and extracellular polysaccharide synthesis by a diadenylate cyclase in *Streptococcus mutans*. Environ Microbiol DOI: 10.1111/1462-2920.13123.

Chia JS, Chang LY, Shun CT, Chang YY, Tsay YG, Chen JY. (2001). A 60-Kilodalton Immunodominant Glycoprotein Is Essential for Cell Wall Integrity and the Maintenance of Cell Shape in *Streptococcus mutans*. Infect. Immun. 69(11): 696987–6998 DOI: 10.1128/IAI.69.11.6987-6998.2001

Davies DG, Parsek MR, Pearson JP, Iglewski BH, Costerton JW, Greenberg EP. (1998). The involvement of cell-to-cell signals in the development of a bacterial biofilm. Science. 280(5361), 295-298 DOI: 10.1126/science.280.5361.295

Deng DM, Liu MJ, Ten Cate JM, Crielaard W. (2007). The VicRK system of *Streptococcus mutans* responds to oxidative stress. J Dent Res. 86(7), 606-610 DOI: 10.1177/154405910708600705

Dhiman A, Bhatnagar S, Kulshreshtha P, Bhatnagar R. (2014). Functional characterization of WalRK: A two-component signal transduction system from *Bacillus anthracis*. FEBS Open Bio. 4, 65-76 DOI: 10.1016/j.fob.2013.12.005

Downey JS, Mashburn-Warren L, Ayala EA, Senadheera DB, Hendrickson WK, McCall LW,Sweet JG, Cvitkovitch DG, Spatafora GA, Goodman SD. (2014). *In vitro* manganese-dependent cross-talk between *Streptococcus mutans* VicK and GcrR: implications for overlapping stress response pathways. PLoS One. 9(12), e115975 DOI: 10.1371/journal.pone.0115975

Dunning DW, McCall LW, Powell WF Jr, Arscott WT, McConocha EM, McClurg CJ, Goodman SD, Spatafora GA. (2008). SloR modulation of the *Streptococcus mutans* acid tolerance response involves the GcrR response regulator as an essential intermediary. Microbiology. 154(Pt 4), 1132-1143 DOI: 10.1099/mic.0.2007/012492-0

Duque C, Stipp RN, Wang B, Smith DJ, Hofling JF, Kuramitsu HK, Duncan MJ, Mattos-Graner RO. (2011). Downregulation of GbpB, a component of the VicRK regulon, affects biofilm formation and cell surface characteristics of *Streptococcus mutans*. Infect Immun. 79(2), 786-796 DOI: 10.1128/IAI.00725-10

Durso SC, Vieira LM, Cruz JN, Azevedo CS, Rodrigues PH, Simionato MR. (2014). Sucrose substitutes affect the cariogenic potential of *Streptococcus mutans* biofilms. Caries Res. 48(3), 214-222 DOI: 10.1159/000354410

Fukuchi K, Kasahara Y, Asai K, Kobayashi , Moriya S, Ogasawara N. (2000). The essential two-component regulatory system encoded by *yycF* and *yycG* modulates expression of the *ftsAZ* operon in *Bacillus subtilis*. Microbiology. 146 (Pt 7), 1573-1583 DOI: 10.1099/00221287-146-7-1573

Gotoh Y, Eguchi Y, Watanabe T, Okamoto S, Doi A, Utsumi R. (2010). Two-component signal transduction as potential drug targets in pathogenic bacteria. Curr Opin Microbiol. 13(2), 232-239 DOI: 10.1016/j.mib.2010.01.008

Hamilton IR, Buckley ND. (1991). Adaptation by *Streptococcus mutans* to acid tolerance. Oral Microbiol Immunol. 6(2), 65-71 DOI: 10.1111/j.1399-302X.1991.tb00453.x

Hamilton IR, Svensater G. (1998). Acid-regulated proteins induced by *Streptococcus mutans* and other oral bacteria during acid shock. Oral Microbiol Immunol. 13(5), 292-300 https://doi.org/10.1111/j.1399-302X.1998.tb00710.x

Hancock LE, Perego M. (2004). Systematic inactivation and phenotypic characterization of two-component signal transduction systems of *Enterococcus faecalis* V583. J Bacteriol. 186(23), 7951-7958 DOI: 10.1128/JB.186.23.7951-7958.2004

Higuchi M, Shimada M, Yamamoto Y, Hayashi T, Koga T, Kamio Y. (1993). Identification of two distinct NADH oxidases corresponding to H_2O_2-forming oxidase and H_2O-forming oxidase induced in *Streptococcus mutans*. J Gen Microbiol. 139(10), 2343-2351 DOI: 10.1099/00221287-139-10-2343

Howell A, Dubrac S, Andersen KK, Noone D, Fert J, Msadek T, Devine K. (2003). Genes controlled by the essential YycG/YycF two-component system of *Bacillus subtilis* revealed through a novel hybrid regulator approach. Mol Microbiol. 49(6), 1639-1655 DOI: 10.1046/j.1365-2958.2003.03661.x

Jeong SI, Kim BS, Keum KS, Lee KH, Kang SY, Park BI, Lee YR, You YO. (2013). Kaurenoic Acid from Aralia continentalis Inhibits Biofilm Formation of *Streptococcus mutans*. Evid Based Complement Alternat Med. 160592 DOI: 10.1155/2013/160592

Kallipolitis BH, Ingmer H. (2001). *Listeria monocytogenes* response regulators important for stress tolerance and pathogenesis. FEMS Microbiol Lett. 204(1),

111-115 DOI: 10.1111/j.1574-6968.2001.tb10872.x

Kanasi E, Johansson I, Lu SC, Kressin NR, Nunn ME, Kent R Jr, Tanner AC. (2010). Microbial risk markers for childhood caries in pediatricians' offices. J Dent Res. 89(4), 378-383 DOI: 10.1177/0022034509360010

Kazor CE, Mitchell PM, Lee AM, Stokes LN, Loesche WJ, Dewhirst FE, Paster BJ. (2003). Diversity of bacterial populations on the tongue dorsa of patients with halitosis and healthy patients. J Clin Microbiol. 41(2), 558-563 DOI: 10.1128/JCM.41.2.558-563.2003

Kleerebezem M, Quadri LE. (2001). Peptide pheromone-dependent regulation of antimicrobial peptide production in Gram-positive bacteria: a case of multicellular behavior. Peptides. 22(10), 1579-1596 DOI: 10.1016/S0196-9781(01)00493-4

Kunze B, Reck M, Dotsch A, Lemme A, Schummer D, Irschik H, Steinmetz H, Wagner-Dobler I. (2010). Damage of *Streptococcus mutans* biofilms by carolacton, a secondary metabolite from the myxobacterium *Sorangium cellulosum*. BMC Microbiol. 10, 199 DOI: 10.1186/1471-2180-10-199

Lange R, Wagner C, de Saizieu A, Flint N, Molnos J, Stieger M, Caspers P, Kamber M, Keck W, Amrein KE. (1999). Domain organization and molecular characterization of 13 two-component systems identified by genome sequencing of *Streptococcus pneumoniae*. Gene. 237(1), 223-234 DOI: 10.1016/S0378-1119(99)00266-8

Lei L, Stipp RN, Chen T, Wu SZ, Hu T, Duncan MJ. (2018). Activity of *Streptococcus mutans* VicR Is Modulated by Antisense RNA. J Dent Res. 97(13):1477-1484 DOI: 10.1177/0022034518781765

Lei, L., et al. (2015). Modulation of Biofilm Exopolysaccharides by the *Streptococcus mutans vicX* Gene. Front Microbiol. 6:1432 DOI:

10.3389/fmicb.2015.01432

Lemos JA, Burne RA. (2008). A model of efficiency: stress tolerance by *Streptococcus mutans*. Microbiology. 154(Pt 11), 3247-3255 DOI: 10.1099/mic.0.2008/023770-0

Li N, Wang F, Niu S, Cao J, Wu K, Li Y, Yin N, Zhang X, Zhu W, Yin Y. (2009). Discovery of novel inhibitors of *Streptococcus pneumoniae* based on the virtual screening with the homology-modeled structure of histidine kinase (VicK). BMC Microbiol. 9, 129 DOI: 10.1186/1471-2180-9-129

Liu M, Hanks TS, Zhang J, McClure MJ, Siemsen DW, Elser JL, Quinn MT, Lei B. (2006). Defects *in ex vivo* and *in vivo* growth and sensitivity to osmotic stress of group A *Streptococcus* caused by interruption of response regulator gene *vicR*. Microbiology. 152(Pt 4), 967-978 DOI: 10.1099/mic.0.28706-0

Liu Y, Burne RA. (2009). Multiple two-component systems modulate alkali generation in *Streptococcus gordonii* in response to environmental stresses. J Bacteriol. 191(23), 7353-7362 DOI: 10.1128/JB.01053-09

Liu, Y. and R. A. Burne (2009). Multiple two-component systems of *Streptococcus mutans* regulate agmatine deiminase gene expression and stress tolerance. J Bacteriol. 191(23): 7363-7366 DOI: 10.1128/JB.01054-09

Loesche WJ. (1986). Role of *Streptococcus mutans* in human dental decay. Microbiol Rev. 50(4), 353-380 DOI: 10.1016/0882-4010(86)90044-6

Loo CY, Mitrakul K, Jaafar S, Gyurko C, Hughes CV, Ganeshkumar N. (2004). Role of a *nosX* homolog in *Streptococcus gordonii* in aerobic growth and biofilm formation. J Bacteriol. 186(24), 8193-8206 DOI: 10.1128/JB.186.24.8193-8206.2004

Mao MY, Yang YM, Li KZ, Lei L, Li M, Yang Y, Tao X, Yin JX, Zhang R, Ma XR,

Hu T. (2016). The *rnc* gene regulates exopolysaccharides and affects the VicRK signal transduction system via microRNA- size small RNAs in *Streptococcus mutans*. Front Microbiol. 7:687 DOI: 10.3389/fmicb.2016.00687

Marquis RE. (1995). Oxygen metabolism, oxidative stress and acid-base physiology of dental plaque biofilms. J Ind Microbiol. 15(3), 198-207 DOI: 10.1007/BF01569826

Martin PK, Li T, Sun D, Biek DP, Schmid MB. (1999). Role in cell permeability of an essential two-component system in *Staphylococcus aureus*. J Bacteriol. 181(12), 3666-3673 DOI: http://dx.doi.org/

Mascher T, Helmann JD, Unden G. (2006). Stimulus perception in bacterial signal-transducing histidine kinases. Microbiol Mol Biol Rev. 70(4), 910-938 DOI: 10.1128/MMBR.00020-06

Moraes JJ, Stipp RN, Harth-Chu EN, Camargo TM, Hofling JF, Mattos-Graner RO. (2014). Two-component system VicRK regulates functions associated with establishment of *Streptococcus sanguinis* in biofilms. Infect Immun. 82(12), 4941-4951 DOI: 10.1128/IAI.01850-14

Oppenheimer-Shaanan Y, Wexselblatt E, Katzhendler J, Yavin E, Ben-Yehuda S. (2011). c-di-AMP reports DNA integrity during sporulation in *Bacillus subtilis*. EMBO Rep. 12(6), 594-601 DOI: 10.1038/embor.2011.77

Parkinson JS. (1993). Signal transduction schemes of bacteria. Cell. 73(5), 857-871 DOI: 10.1016/0092-8674(93)90267-T

Paster BJ, Boches SK, Galvin JL, Ericson RE, Lau CN, Levanos VA, Sahasrabudhe A, Dewhirst FE. (2001). Bacterial diversity in human subgingival plaque. J Bacteriol.183(12), 3770-3783 DOI: 10.1128/JB.183.12.3770-3783.2001

Paster BJ, Falkler Jr WA Jr, Enwonwu CO, Idigbe EO, Savage KO, Levanos VA, Tamer MA, Ericson RL, Lau CN, Dewhirst FE. (2002). Prevalent bacterial species and novel phylotypes in advanced noma lesions. J Clin Microbiol. 40(6), 2187-2191 DOI: 10.1128/JCM.40.6.2187-2191.2002

Peng X, Zhang Y, Bai G, Zhou X, Wu H. (2015). Cyclic di-AMP mediates biofilm formation. Mol Microbiol. 99(5):945-59 DOI: 10.1111/mmi.13277.

Qi F, Chen P, Caufield PW. (1999). Functional analyses of the promoters in the lantibiotic mutacin II biosynthetic locus in *Streptococcus mutans*. Appl Environ Microbiol. 65(2), 652-658 DOI:10.1089/oli.1.1999.9.105

Qi F, Chen P, Caufield PW. (1999). Purification of mutacin III from group III *Streptococcus mutans* UA787 and genetic analyses of mutacin III biosynthesis genes. Appl Environ Microbiol. 65(9), 3880-3887

Qi F, Chen P, Caufield PW. (2001). The group I strain of *Streptococcus mutans*, UA140, produces both the lantibiotic mutacin I and a nonlantibiotic bacteriocin, mutacin IV. Appl Environ Microbiol. 67(1), 15-21 DOI: 10.1128/AEM.67.1.15-21.2001

Quivey RG Jr, Kuhnert WL, Hahn K. (2000). Adaptation of oral streptococci to low pH. Adv Microb Physiol. 42, 239-274 DOI: 10.1016/S0065-2911(00)42004-7

Rolla G, Ciardi JE, Schultz SA. (1983). Adsorption of glucosyltransferase to saliva coated hydroxyapatite. Possible mechanism for sucrose dependent bacterial colonization of teeth. Scand J Dent Res. 91(2), 112-117 DOI: 10.1111/j.1600-0722.1983.tb00786.x

Romling U. (2008) Great times for small molecules: c-di-AMP, a second messenger candidate in Bacteria and Archaea. Sci Signa. 1(33), pe39 DOI: 10.1126/scisignal.133pe39

Rozen R, Bachrach G, Bronshteyn M, Gedalia I, Steinberg D. (2001). The role of fructans on dental biofilm formation by *Streptococcus sobrinus*, *Streptococcus mutans*, *Streptococcus gordonii* and *Actinomyces viscosus*. FEMS Microbiol Lett. 195(2), 205-210 DOI: 10.1111/j.1574-6968.2001.tb10522.x

Rozen R, Bachrach G, Steinberg D. (2004). Effect of carbohydrates on fructosyltransferase expression and distribution in *Streptococcus mutans* GS-5 biofilms. Carbohydr Res. 339(18), 2883-2888 DOI:10.1016/j.carres.2004.07.028

Sampathkumar B, Napper S, Carrillo CD, Willson P, Taboada E, Nash JH, Potter AA, Babiuk LA, Allan BJ. (2006). Transcriptional and translational expression patterns associated with immobilized growth of Campylobacter jejuni. Microbiology. 152(Pt 2), 567-577 DOI: 10.1099/mic.0.28405-0

Senadheera D, Krastel K, Mair R, Persadmehr A, Abranches J, Burne RA, Cvitkovitch DG. (2009). Inactivation of VicK affects acid production and acid survival of *Streptococcus mutans*. J Bacterio. 191(20), 6415-6424 DOI: 10.1128/JB.00793-09

Senadheera DB, Cordova M, Ayala EA, Chavez De Paz LE, Singh K, Downey JS, Svensater G, Goodman SD, Cvitkovitch DG. (2012). Regulation of bacteriocin production and cell death by the VicRK signaling system in *Streptococcus mutans*. J Bacteriol. 194(6), 1307-1316 DOI: 10.1128/JB.06071-11

Senadheera MD, Guggenheim B, Spatafora GA, Huang YC, Choi J, Hung DC, Treglown JS,Goodman SD, Ellen RP, Cvitkovitch DG. (2005). A VicRK signal transduction system in *Streptococcus mutans* affects gtfBCD, gbpB, and ftf expression, biofilm formation, and genetic competence development. J Bacteriol. 187(12), 4064-4076 DOI: 10.1128/JB.187.12.4064-4076.2005

Senadheera MD, Lee AW, Hung DC, Spatafora GA, Goodman SD, Cvitkovitch

DG. (2007). The *Streptococcus mutans* vicX gene product modulates gtfB/C expression, biofilm formation, genetic competence, and oxidative stress tolerance. J Bacteriol. 189(4), 1451-1458 DOI: 10.1128/JB.01161-06

Serbanescu MA, Cordova M, Krastel K, Flick R, Beloglazova N, Latos A, Yakunin AF, Senadheera DB, Cvitkovitch DG. (2015). Role of the *Streptococcus mutans* CRISPR-Cas systems in immunity and cell physiology. J Bacteriol. 197(4), 749-761 DOI: 10.1128/JB.02333-14.

Shemesh M, Tam A, Feldman M, Steinberg D. (2006). Differential expression profiles of *Streptococcus mutans ftf*, *gtf* and *vicR* genes in the presence of dietary carbohydrates at early and late exponential growth phases. Carbohydr Res. 341(12), 2090-2097 DOI: 10.1016/j.carres.2006.05.010

Stein T, Borchert S, Kiesau P, Heinzmann S, Kloss S, Klein C, Helfrich M, Entian KD. (2002). Dual control of subtilin biosynthesis and immunity in *Bacillus subtilis*. Mol Microbiol. 44(2), 403-416 DOI: 10.1046/j.1365-2958.2002.02869.x

Stipp RN, Boisvert H, Smith DJ, Hofling JF, Duncan MJ, Mattos-Graner RO. (2013). CovR and VicRK regulate cell surface biogenesis genes required for biofilm formation in *Streptococcus mutans*. PLoS One. 8(3), e58271 DOI: 10.1371/journal.pone.0058271

Sun, M., et al. (2014). "Effect of high-fructose corn syrup on *Streptococcus mutans* virulence gene expression and on tooth demineralization." Eur J Oral Sci. 122(3): 216-222 DOI: 10.1111/eos.12132

Tremblay YD, Lo H, Li YH, Halperin SA, Lee SF. (2009). Expression of the *Streptococcus mutans* essential two-component regulatory system VicRK is pH and growth-phase dependent and controlled by the LiaFSR three-component regulatory system. Microbiology. 155(Pt 9):2856-2865 DOI:

10.1099/mic.0.028456-0

Tsang P, Merritt J, Shi W, Qi F. (2006). IrvA-dependent and IrvA-independent pathways for mutacin gene regulation in *Streptococcus mutans*. FEMS Microbiol Lett. 261(2), 231-234 DOI: 10.1111/j.1574-6968.2006.00351.x

Wagner C, Saizieu Ad Ad, Schonfeld HJ, Kamber M, Lange R, Thompson CJ, Page MG (2002). Genetic analysis and functional characterization of the *Streptococcus pneumoniae vic* operon. Infect Immun. 70(11), 6121-6128 DOI: 10.1128/IAI.70.11.6121-6128.2002

Wang C, Sang J, Wang J, Su M, Downey JS, Wu Q, Wang S, Cai Y, Xu X, Wu J, Senadheera DB, Cvitkovitch DG, Chen L, Goodman SD, Han A. (2013). Mechanistic insights revealed by the crystal structure of a histidine kinase with signal transducer and sensor domains. PLoS Biol. 11(2), e1001493 DOI: 10.1371/journal.pbio.1001493

Watanabe T, Hashimoto Y, Yamamoto K, Hirao K, Ishihama A, Hino M, Utsumi R. (2003). Isolation and characterization of inhibitors of the essential histidine kinase, YycG in Bacillus subtilis and *Staphylococcus aureus*. J Antibiot (Tokyo). 56(12), 1045-1052 DOI：10.7164/antibiotics.56.1045

Watanabe T, Okada A, Gotoh Y, Utsumi R. (2008). Inhibitors targeting two-component signal transduction. Adv Exp Med Biol. 631, 229-236 DOI: 10.1007/978-0-387-78885-2_16

Zhuang, P. L., et al. (2018). Relationship between the genetic polymorphisms of vicR and vicK *Streptococcus mutans* genes and early childhood caries in two-year-old children. BMC Oral Health. 18(1): 39 DOI: 10.1186/s12903-018-0501-y

Zhulin IB, Taylor BL, Dixon R. (1997). PAS domain S-boxes in Archaea, Bacteria and sensors for oxygen and redox. Trends Biochem Sci. 22(9), 331-333 DOI: 10.1016/S0968-0004(97)01110-9

 https://doi.org/10.21775/9781912530229.06

ComDE Two-Component Signal Transduction Systems in Oral Streptococci: Structure and Function

Yue Zu[#], Wei Li[#], Qi Wang, Jiahao Chen, Qiang Guo*

State Key Laboratory of Oral Diseases, National Clinical Research Center for Oral Diseases, West China Hospital of Stomatology, Sichuan University, Chengdu, China.

[#] Yue Zu and Wei Li contributed equally to this review

*Corresponding author: guoqiang2014@scu.edu.cn

Abstract

Oral streptococci depend on two-component signal transduction systems (TCSs), the most widespread regulatory systems in bacteria, to detect and respond to diverse stresses in oral environment. Among the larger panel of TCSs equipped by oral streptococci, ComDE TCS is thought to be one of the most classical TCSs. So far, it has been proved that ComDE TCS could play critical roles in environmental stress responses to acid, antibiotic, oxidative pressures and so on, and modulating multiple virulence traits like biofilm formation, bacteriocin production, competence, autolysis. Here, the well characterized *Streptococcus mutans* and *Streptococcus pneumoniae* are chosen as the representative species to introduce the composition, signaling pathways and regulated phenotypes of

ComDE TCS in oral streptococci. The potential ComDE TCS-targeted antimicrobial applications are also discussed at last.

Introduction

In nature, bacteria are often confronted with complex and fluctuating environmental conditions including nutrient availability, pH, temperature, redox potential, osmotic pressure and toxic chemicals. Not only for adapting to changing habitats, but also for expression of virulence factors, sophisticated signaling pathways are evolved by bacteria to detect and respond to diverse stresses in the environment. It is believed that two-component signal transduction systems (TCSs) are the most widespread regulatory systems in bacteria to cope with various external stimuli (Mitrophanov and Groisman, 2008). A prototypical TCS consists of two conserved components, a specific histidine kinase (HK) and a cognate response regulator (RR) (Figure 1). The HK functions as a transmembrane sensor protein to sense environmental signals and once it is activated, there is a phosphotransfer reaction between its histidine residue and an aspartate residue of corresponding cytosolic RR protein, thus the transmembrane information transmissions occur (Figure 1). The RR undergoes a conformational change, enabling DNA binding and activating or repressing transcription of target genes (Ann M. Stock et al., 2000; Hoch, 2000) (Figure 1).

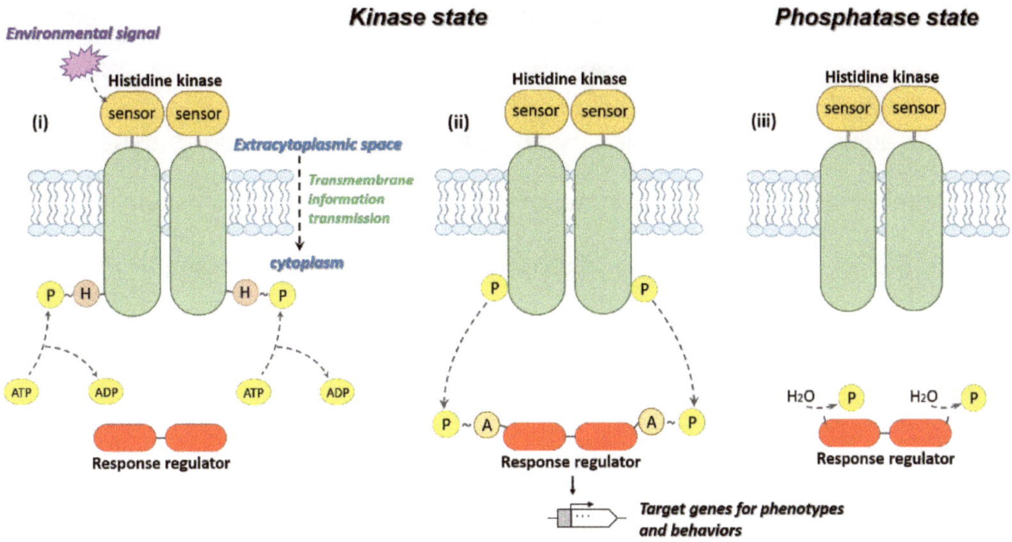

Figure 1. Schematic diagram of the structure and work processes of two-component systems (TCSs). (i) When the external environment changes, the transmembrane histidine kinases (HKs) recognize the signal, rapidly dimerize and catalyze the ATP-dependent autophosphorylation at the specific histidine (H) residue. (ii) Subsequently, HKs signal to its corresponding cytosolic response regulator (RR) protein by transferring the high-energy phosphoryl group to an aspartate (A) residue of RR, which thus underwent a conformational change. The activated RR proteins bind to the target genes to adjust the physiology and behaviors of bacteria. (iii) Phosphate groups are transferred from RR proteins to water molecules in the hydrolysis reaction, or the system returns to the ground state through RR's dephosphorylation.

Although oral cavity provides diverse habitats, rich foods and comfortable temperature for many microorganisms to colonize and survive, it is undeniable that oral microorganisms need to adapt to the frequent changing environment which is caused by human activities like diet, breath, taking antibiotics. As a major part of the normal microbiota in humans, oral streptococci adapt to diverse oral sites, including different mucosal sites and supra-gingival or sub-gingival plaque biofilms. More than 10 TCSs have been identified in many oral streptococci (Mattos-Graner and Duncan, 2017) and depending on which, these commensal bacteria adjust their cell physiology and behaviors in response to dynamic oral environment and even cause infections such as dental caries, when the host-dental biofilm homeostasis is disrupted. Hence, it is necessary to deeply investigate the TCS regulatory circuits and the environmental signals that trigger expression of virulence factors in oral streptococci.

Streptococcus mutans is known as the primary cariogenic dental pathogen of human. A variety of studies have linked the regulation of TCSs with the known virulence traits of this organism, including biofilm formation, acidogenicity, aciduricity, genetic transformation and bacteriocin production (Senadheera and Cvitkovitch, 2008). It is discovered that *S. mutans* strains harbor 12-14 different and complete TCSs in their genomes (Song et al., 2012). In fact, the discovery of part of these TCSs in *S. mutans* was largely dependent on the information of their homologues in the closely related species, *Streptococcus pneumoniae*, whose TCSs have been best characterized and made great contributions to the research of TCSs in other members of the genus *Streptococcus. S. pneumoniae* is an opportunistic pathogen in the normal human nasopharyngeal microflora as well as oral microflora, although the role in oral cavity is still unclear. It is found that *S. pneumoniae* is involved in a myriad of infections and is a major cause of pneumonia, sinusitis, otitis media, bacteremia and meningitis (Musher, 2010). Pneumococcal pathophysiological processes, such as competence, autolysis, bacteriocin production and virulence factors expression, have been demonstrated to be driven by multiple TCSs (Gomez-Mejia et al., 2018). So far, 13 complete TCSs and an orphan RR have been identified in the genome of *S. pneumoniae*

(Lange et al., 1999; Throup et al., 2000). The large number of different types of TCSs expressed by *S. mutans* and *S. pneumoniae* suggest the important roles of TCS in tight regulation of their behaviors and pathogenicity.

Among the TCSs processed by *S. mutans* and *S. pneumoniae*, four highly conserved TCSs are shared by both of them, including ComDE TCS, CiaRH TCS, VicRK TCS and LiaRS TCS. ComDE TCS is one of the classical TCSs in oral streptococci and it is proved that multiple virulence traits are under the tight control of this TCS. The pneumococcal genes encoding the HK protein (ComD) and its cognate RR protein (ComE) belong to the operon *com*, which was first described systematically by Cheng *et al.* in 1997 (Cheng Q et al., 1997). In *S. mutans, comDE* genes are identified by searching *S. mutans* genome database with their homologs in *S. pneumoniae* (Li et al., 2001b).

Based on streptococcal 16S rRNA sequences, numerous species of oral streptococci are classified into five groups: (1) Mutans, (2) Salivarius, (3) Anginosus, (4) Sanguinus, and (5) Mitis (Facklam, 2002). However, only Mutans, Anginosus and Mitis groups are found to have ComDE TCS (Havarstein, 2010). The members of these groups diverge in preferred habitats, physiological and pathogenic properties. As a result, it is unrealistic to describe each ComDE TCS distributed in different genomes of these streptococci within the scope of this review. Instead, the extensively studied *S. mutans* (Mutans group) and *S. pneumoniae* (Mitis group) are chosen as the representative species for ComDE TCS research of oral streptococci and we compare the similarities and differences of the composition, target genes and regulated phenotypes of ComDE TCS in the two streptococci, thus providing insights into the structure and function of this TCS in oral streptococci. Finally, the potential ComDE TCS-targeted antimicrobial applications are also discussed.

The composition of ComDE TCS

ComDE TCS is also known as a quorum sensing (QS) system and CSP (competence-stimulating peptide) is employed as the specific signal peptide to activate the TCS. The precursor of CSP is encoded by *comC* gene and processed and exported by the ComAB secretion apparatus. When extracellular CSP increases to the threshold concentration, it is bound by the transmembrane HK receptor ComD, which in turn activates the intercellular RR ComE through phosphoryl transfer (Boudes et al., 2014). The activation of ComE enhances the expression of a series of genes including *comAB*, *comC*, *comDE* and the gene encoding the alternative sigma factor ComX (σ^X, sometimes also called SigX), which controls the expression of late competence genes that encode the proteins for DNA binding, import and recombination (Martin et al., 2013; Pestova et al., 1996; Pozzi et al., 1996; Ween et al., 1999). Different from *S. pneumoniae*, Phosphorylated ComE does not activate the transcription of *comX* directly in *S. mutans*. In fact, the ComDE TCS in *S. mutans* depends on another regulatory system named ComRS, which is proven to be the proximal regulator to *comX,* to relay signals to activate *comX*, although the mechanism is still unclear (Mashburn-Warren et al., 2010).

CSP

In 1995, Håvarstein *et al.* showed that a 17-residue peptide produced by *S. pneumoniae* was able to induce the development of competence. The peptide was named CSP and was found to be encoded by *comC* gene, then cleaved and secreted to extracellular environment by ComAB transporter, serving as an autoinducer molecule for QS (Håvarstein et al., 1995) (Figure 2A). Actually, complete *comC* expression product should be a 44-residule peptide which contains a recognition site for the ComAB (a specific Gly-Gly site) (Hui and Morrison, 1991; Whatmore et al., 1999). For many streptococci including *S. pneumoniae,* their co*mDE* operons start with *comC*, which is followed by *comD*, and *comE* is immediately after the *comD* (Li et al., 2001a) (Figure 2A). However, the *comC gene* in *S. mutan* was located at complementary strand and in the opposite direction of *comDE* (SMU_1916-1917, UA159) operon in 5' to 3' direction (Li et al., 2001a) (Figure 3A). In addition, *comC* in *S. mutans* encodes a polypeptide which consists of 46 residues. After cleavage, the polypeptide is secreted extracellularly as a 21-residue peptide (Li et al., 2001a), and then it will be processed by protease SepM into the mature 18-residue CSP (Figure 3A), giving it a lower concentration but a higher efficiency (Hossain and Biswas, 2012). Biswas *et al.* found SepM homologs from 4 streptococci, indicating it might play a broader role in interspecies communication (Biswas et al., 2016). The sequence of CSP is highly variable in streptococci, and the recognition of CSP by ComD receptor is species-specific even strain-specific (Petersen et al., 2005). It has been reported that strains of different subtypes can only be induced by their specific CSP (except Pn15, VA1) (Petersen et al., 2005). Surprisingly, CSP molecules produced by *S. mutans* are also proven to affect other microorganisms, for example, inhibiting the germ tubes of *Candida albicans* (Jarosz et al., 2009).

ComD

Two main genes, *comD* and *comE,* were first described in 1996 by Pestova *et al.* in *S. pneumoniae*. ComD, a highly conserved protein encoded by *comD*, functions as a sensory HK and is responsible for activating its cognate RR ComE by phosphorylation (Pestova et al., 1996) (Figure 2A). ComD is a transmembrane protein containing 7 transmembrane alpha-helices. Its N-terminus is the CSP binding domain, and the C-terminus locates inside the cell and contains a kinase domain with a conserved sequence. ComD in *S. mutans* can detect extracellular accumulation of CSP and result in conformational changes and autophosphorylation of the kinase domain in ComD (Figure 3A), just like *S. pneumoniae*. Moreover, Prudhomme *et al.* found that the membrane protein ComD retained the CSP molecules, limiting its free diffusion in the medium (Prudhomme et al., 2016). The peptide ComD of *S. mutans* belongs to the HPK10 subfamily of HK (Hvarstein, 2003), consisting of 441 amino acid residues (Ajdić et al., 2002). The identity between *S. mutans* and *S. pneumoniae* is only 22%, and the similarity is 44% (Hvarstein, 2003). Dong *et al.* found that there were two hydropathically distinct regions in the ComD protein of *S. mutans,* the N-terminal membrane-spanning region (1-220 residues) and the C-terminal hydrophilic region (221-441 residues) inside the cytoplasm. They also suggested that three extracellular loops (loopA, loopB, loopC) of the N-terminal membrane domain of ComD protein might be the receptor for CSP recognition, of which loopB and loopC were essential for CSP recognition while loopA contributed little in CSP detection (Dong et al., 2016).

ComE

In *S. pneumoniae*, ComE is encoded by *comE* and belongs to the AlgR/AgrA/LytR transcription factors subfamily. Activated ComE exhibits following effects: (1) combining with the promotor of *comCDE* and enhancing its expression, which establishes a positive-feedback loop via increasing the level of CSP production, output, and signal transduction capacities (Martin et al., 2013); (2) inducing *comX* and *comW*, both of which contribute to the expression of several late competence genes (Ween et al., 1999) (Figure 2A). In addition, Martin *et al.* found that when the progressive accumulation of ComE outcompeted ComE~P, the expression of early genes would decelerate rapidly (Martin et al., 2013). Mirouze *et al.* suggested DprA interacted with ComE~P, promoting the shut-off of competence (Mirouze et al., 2013) (Figure 2A). ComE in *S. mutans* is structurally similar to that in *S. pneumoniae*, while ComE in *S. mutans* acts as a repressor of *comC* gene expression in the absence of CSP induction (Kreth et al., 2007). There are two 11 bp direct repeats, DRI and DRII, which are required for ComE binding between *nlmC* and *comC* promoter regions. Furthermore, the two repeats are also essential for the expression of *nlmC* and *comC* genes (Kreth et al., 2007). Hung *et al.* demonstrated that ComE did not require two repeats for high-affinity binding. Different from *S. pneumoniae*, there is no ComE binding site found in *S. mutans* *comX* promoter. Instead, the *comX* promoter has a motif that is most probably recognized by ComR, the cytoplasmic Rgg-type transcriptional regulator of ComRS regulatory system (Mashburn-Warren et al., 2010). In addition, the activated ComE regulator in *S. mutans* directly activates the genes encoding a diverse repertoire of bacteriocins (Perry et al., 2009) (Figure 3A). As a result, Martin *et al.* suggested that the *S. mutans* ComDE TCS had orthologs to the *S. pneumoniae* BlpRH TCS that controlled the synthesis of bacteriocin-like peptides (Martin et al., 2006).

ComAB

In 1991, Hui *et al.* determined the complete nucleotide sequence of *comA* and suggested the possible function of ComA in the transport of pneumococcal competence factor (CF). The sequence contained a single large open reading frame (ORF) ORF1 of 2,151 bp and a Shine-Dalgarno sequence (AAAGGAG) which laid immediately upstream of ORF1, as well as another sequence (TTtAat-17bp-TAaAAT) similar to the *Escherichia coli* sigma 70 promoter consensus locating 410 bp upstream of ORF1 (Hui and Morrison, 1991). It is found that the *comAB* genes encode proteins of 77 kDa and 50 kDa, respectively, and belong to the ABC transporter family, which has been demonstrated to have a dual function: (1) removal of the leader peptide from its substrate, and (2) translocation of its substrate across the cytoplasmic membrane. It has been reported that ComA and ComB assist to process and secret CSP (Håvarstein et al., 1996; Hui and Morrison, 1991) (Figure 2A). In 2000, Petersen identified a cysteine-protease ABC transporter/accessory proteins named CslAB (SMU.1897 and SMU.1898) in *S. mutans* by comparing *csl* locus with the *comAB* homologous genes of *S. pneumoniae* and *Streptococcus gordonii* (Petersen and Scheie, 2000). In fact, the CSP peptide is exported by another ABC transporter named NlmTE. The NlmTE was first demonstrated as an ABC transporter in 2005, which has been found to transfer mutacins out of cells(Hale et al., 2005). Hossain MS et al, demonstrated for the first time that NlmTE was the true transporter of CSP. Because of the proteolytic activity of NlmTE, the leader sequence of the 46-residue-long CSP was cleaved off after the GG motif (present at position 25)(Hossain and Biswas, 2012), thus generated the 21-residue-long CSP(CSP-21) which can be recognized and specifically cleaved by SepM (Biswas et al., 2016) (Figure 3A).

ComX

In *S. pneumoniae, comX* is the core regulator of competence development and encodes the competence-specific alternative sigma factor ComX, which composes a holoenzyme with RNA polymerase (RNAP) to enable RNA polymerase to recognize a specific 8 bp sequence (*com* box or *cin* box) and thus to turn on late competence gene transcription needed for DNA uptake and homologous recombination (Figure 2A). The late genes consist of 12 major operons, and encode 28 competence-specific proteins, which facilitate the efficient transport, preservation and recombination of extracellular genomic fragments (Campbell et al., 1998; Luo and Morrison, 2003). In addition to the regulation by ComE~P and ComE, ComX also depends on *comW* which is only found in *S. pneumoniae* and other 10 species of streptococci. It is recognized that ComW can increase both activity and stability of ComX as well as withstand the negative regulation of ComX by the ClpP protease complex (Peterson et al., 2004; Sung and Morrison, 2005) (Figure 2A). Similarly, ComX in *S. mutans* is a sigma factor inducing the late competence genes. In *S. mutans*, the activation of ComE can induce the expression of *comX*, but in an indirect way via ComR (Figure 3A). Aspiras *et al.* showed the positive regulation of ComX in natural transformation and competence development in *S. mutans* (Aspiras et al., 2004). The adaptor protein MecA and ClpCP form the MecA-ClpCP Protease Complex acting as a negative regulator of ComX stability in *S. mutans* (Dong et al., 2014; Wahl et al., 2014).

Target genes and phenotypes under regulation of ComDE TCS

Downstream target genes in S. pneumoniae

In 2004, Peterson *et al.* constructed DNA microarrays and analyzed the temporal expression profiles of unique predicted ORFs presented in the *S. pneumoniae* TIGR4 genome, and divided them into four temporally distinct expression profiles: early, late and delayed gene induction, and gene repression. Moreover, this study increased the number of identified CSP-responsive genes from approximately 40 identified previously to 188 and found that among 124 CSP-inducible genes, 67 were dispensable for transformation, whereas 23 were required for transformation (Peterson et al., 2004). Proteins encoded by late genes which are responsible for DNA uptake and recombination include the protein-processing DprA, recombination-promoting CoiA (Desai and Morrison, 2007), MssA related to homologous recombination, HexA and HexB related to mismatch repair (Prudhomme et al., 1989; Prudhomme et al., 1991) and Cgl assembled into a type 4 pilus-like DNA uptake device. SSB proteins protect ssDNA from degradation (Hedayati et al., 2005), and the endonuclease (EndA) participates in the transport of ssDNA (Puyet et al., 1989). In addition, CibAB (a crucial trigger factor for autolysis) (van Belkum et al., 1991), CbpD (participate in cell killing and release DNA) (Johnsborg et al., 2008) and LytA (cytosolic N-acetylmuramoyl-l-alanine amidase released by bacterial lysis) are essential for autolysis.

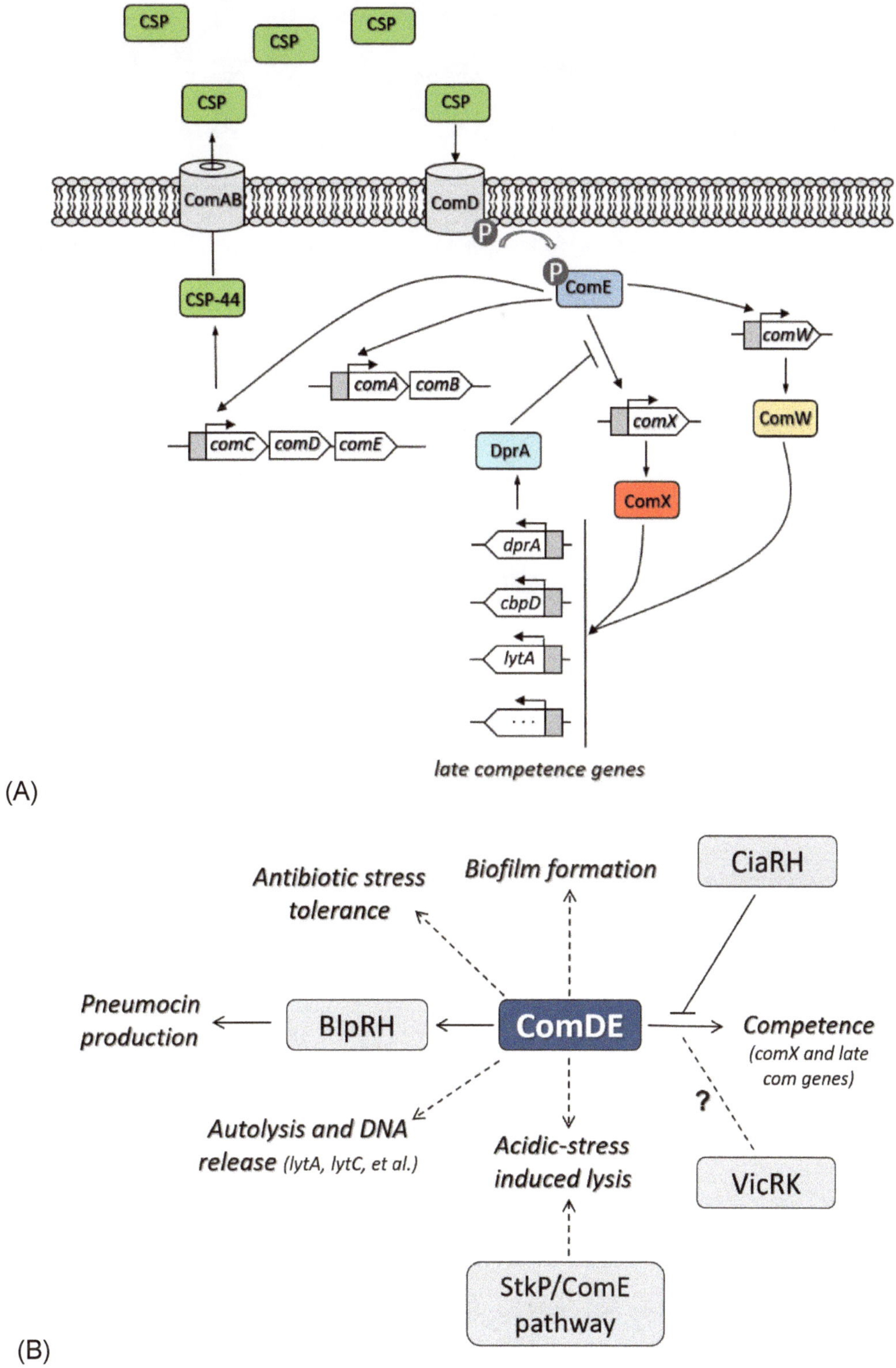

Figure 2. Signaling pathways, regulated phenotypes and interacted regulatory systems of ComDE

TCS in *S. pneumoniae.*

(A) Signaling pathways of ComDE TCS in *S. pneumoniae.* In *S. pneumoniae, comC*, *comD* and *comE* genes of ComDE TCS belong to the *comCDE* operon which starts with *comC,* following immediately by *comD* and *comE.* CSP is encoded by *comC* as a 44-aa peptide that is processed and exported by the ComAB secretion apparatus as a 17-aa peptide, where it serves as the signal peptide for the ComDE TCS. When extracellular CSP increases to the threshold concentration, it is bound by the transmembrane HK receptor ComD, which then activates the intercellular RR ComE via phosphoryl transference. Activated ComE upregulates transcription of *comAB*, *comCDE*, *comW* and *comX* genes. ComX, an alternate sigma factor, as well as the core competence gene, together with RNA polymerase triggers transcription of a group of late competence genes that encode the proteins for DNA binding, import and recombination, such as *dprA*, *cbpD* and so on. ComW could increase both activity and stability of ComX. It is reported that DprA interacts with ComE~P to block ComE-driven transcription, chiefly impacting ComX production, thus shutting off competence. **(B) Regulated phenotypes and interacted regulatory systems of ComDE TCS in *S. pneumoniae.*** In *S. pneumoniae*, ComDE TCS is known to control competence for natural genetic transformation directly and also trigger autolysis and DNA releasing via the induction of competence. In addition, ComDE signaling pathway is involved in biofilm formation and antibiotic stress response. Interestingly, ComDE TCS in *S. pneumoniae* is reported to activate ASIL (acidic-stress induced lysis) and repress acid tolerance response (ATR) thereby facilitating the release of DNA and virulence factors from the bacteria. ComDE TCS have multiple interaction with several regulators in *S. pneumoniae*. For example, CiaRH TCS regulates competence by interfering in CSP accumulation, thus negatively affecting competence development induced by ComDE. Moreover, the interaction between ComDE TCS and BlpRH TCS influences the secretion of Pneumocin (Pneumococcal bacteriocin), which is encoded by *blpMN* genes. It is recently reported that StkP/ComE pathway, independent of the HK-dependent ComDE system, mainly triggers the acidic stress response by inducing ASIL and inhibiting ATR. VicRK TCS may be involved in competence control, however, the interaction with ComDE TCS is still unknown.

Downstream target genes in S. mutans

Through using a high-resolution tiling array, Khan *et al.* found 20 genes in four distinct transcriptionally active regions (TARs) and 83 genes in 29 TARs were upregulated at early or late phases of the CSP response, respectively. The early-phase responding genes are associated with bacteriocin production and immunity, including *nlmA*, *nlmB*, *nlmC*, *nlmD*, *cipB*, *bsmK*, *bsmL*, *immA*, *immB* and so on. They demonstrated the expression of late-phase responding genes was dependent on the expression of *comS* and 4 TARs containing 23 genes were directly regulated by the ComE regulon (Khan et al., 2016).The promoters of *bsmB, bsmC, nlmAB, nlmC*, *immAB, cslAB* and *comC* genes showed high binding affinities with ComE (Hung et al., 2011; Kreth et al., 2007). The ComX regulon regulates the expression of 21 operons (Khan et al., 2016), including late competence genes (*dprA*, *ssbB*, *recA*, *dpnB*, *comEC*, etc) (Eaton and Jacques, 2010; Khan et al., 2016; Quivey and Faustoferri, 1992), attacking or self-killing (*lytF*) (Eaton and Jacques, 2010; Okinaga et al., 2010), pilus formation (*pilC*, *comGG-comGA*) (Dufour et al., 2011; Eaton and Jacques, 2010; Merritt et al., 2007; Okinaga et al., 2010), restriction-modification system (*radC*) (Okinaga et al., 2010), and others like *ackA* (Merritt et al., 2005), *cinA* (Mair et al., 2012).

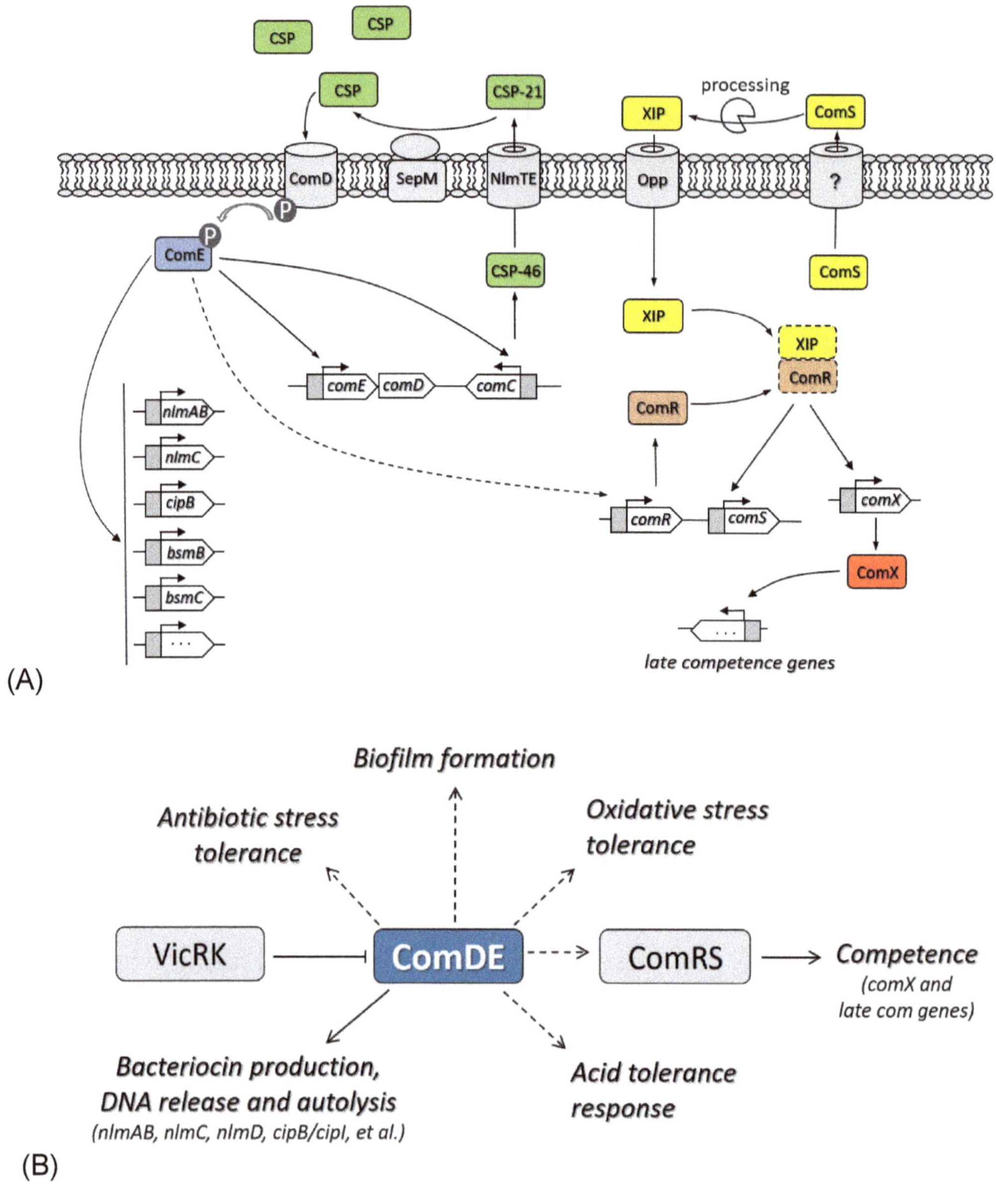

Figure 3. Signaling pathways, regulated phenotypes and interacted regulatory systems of ComDE TCS in *S. mutans.*

(A) Signaling pathways of ComDE TCS in *S. mutans.* The ComDE TCS in *S. mutans* acts as a typical two-component system with CSP as the signaling peptide, encoded by the *comC* gene. Different from *S. pneumoniae*, the *comC gene* in *S. mutan* was located at complementary strand and in the opposite direction of *comDE* operon. The immature CSP translated from *comC* consists of 46 residues (CSP-46). After cleavage and secrete extracellularly by NlmTE transporter, the polypeptide becomes a 21-residue peptide (CSP-21), and then it is further processed by protease SepM into the mature 18-residue CSP. When extracellular CSP increases to the threshold concentration, it is bound by the transmembrane HK receptor ComD, which then activates the ComDE signaling pathway. The intercellular RR ComE is essential for the expression *comC* gene, but it also acts as a repressor of *comC* gene in the absence of CSP induction. Notably, ComE regulator in *S. mutans* directly activates the genes controlling bacteriocins production, autolysis and DNA release, including *nlmAB*, *nlmC*, *cipB* and so on. The phosphorylated ComE also induces the expression of the alternative sigma factor ComX, but in an indirect way via ComRS system, which employs XIP, a 7-residue peptide deriving from ComS peptide, as the signal peptide. ComX is the core regulator of competence development and controls the expression of late competence genes that encode the proteins for DNA binding, import and recombination. **(B) Regulated phenotypes and interacted regulatory systems of ComDE TCS in *S. mutans.*** Different from *S. pneumoniae,* ComDE TCS in *S. mutans* directly control mutacin production, autolysis and DNA release and regulates competence development via signaling to ComRS system. In addition, biofilm formation, stress responses to acid, antibiotic and oxygen are also related to ComDE TCS in *S. mutans*, although the mechanisms are still unclear. It is reported that there is complex interaction between ComDE TCS and VicRK TCS. The RR regulator VicR could bind domain at *comC* coding region and affect the transcription of *comC*, *comDE* and *comX*. Moreover, the transcription of *nlmAB*, *nlmC*, and *nlmD* genes which encode Mutacins IV, V, and VI respectively increased when the HK VicK was inactivated.

The phenotypes regulated by ComDE TCS

In *S. pneumoniae*, ComDE TCS is known to affect competence for natural genetic transformation together with CiaRH, and also trigger autolysis and DNA releasing via the induction of competence. Moreover, the interaction between ComDE TCS and BlpRH TCS is recently proofed to influence the secretion of Pneumocin (Pneumococcal bacteriocin) (Kjos et al., 2016; Wholey et al., 2016). In addition, ComDE is dedicated to biofilm formation, toxin production and responses for environmental changes including acid stress, antibiotic pressure (Figure 2B). Similar with *S. pneumoniae*, ComDE TCS also regulates competence development, biofilm formation, bacteriocin production, autolysis and DNA release, as well as stress responses to acid, antibiotic and oxygen in *S. mutans* (Figure 3B). However, difference in phenotypes regulated by this TCS has been observed in different strains of *S. mutans*. For example, defects in *comDE* genes significantly affect biofilm formation, acid tolerance and competence development in *S. mutans* NG8, but only competence in strain UA159 (Ahn et al., 2006; Li et al., 2002).

Competence development

Bacterial transformation, a programmed mechanism for gene transfer, is initially discovered in *S. pneumoniae*. *S. pneumoniae* could develop competence during growth, a state in which foreign DNA can be integrated into the genome, leading to stably inherited changes in virulence and resistance of the bacteria. It has been reported that two distinct systems were responsible for the control of the development of competence in streptococci: ComDE system and ComRS system. *S. pneumoniae* has only ComDE system, while *S. mutans* has both.

In *S. pneumoniae*

Traditionally, we presumed that when the accumulation of CSP reached the threshold, ComDE TCS would be activated by QS. Recently it has been showed that competence induction can be simultaneously influenced by cell density, external pH, antibiotic-induced stress, and cell history, rather than mere cell density (Claverys et al., 2006; Moreno-Gamez et al., 2017). In addition, data from Prudhomme *et al.* demonstrated that competence shift in the population relied on a self-activated cell fraction via a growth time-dependent mechanism. During competence induction, CSP remains bound to cells, and the rate of competence development relies on the contact between activated and quiescent cells to propagate competence (Prudhomme et al., 2016). The activated ComDE TCS triggers the development of competence and forms a positive feedback loop, thereby amplifying the transcription of other genes in the competent regulator (Figure 2A), including up-regulation of approximately 80 genes expression. Among them, 14 genes encode essential proteins for DNA uptake and recombination, including RecA, LytA, DprA, CbpD and Cgl (Peterson et al., 2004). Karlsson *et al.* suggested that a putative *comX*-dependent repressor shut off the competence via inhibiting the expression of *comDE* and *comX*. However, subsequent studies showed that late-gene DprA secluded or dephosphorylated ComE~P and the shut-off of competence was regulated by direct interference with *comDE* and *comX* expression (Mirouze et al., 2013). DprA is conserved among all transformable species (Johnston et al., 2014), and the *dprA* gene of *S. pneumoniae* has following three functions: (i) load the recombinase RecA onto the internalized ssDNA and then recombine the homologous DNA into the genome; (ii) protect internalized DNA such as transforming ssDNA from degradation; (iii) interact strongly with ComE~P to promote the shut-off of competence (Figure 2A).

In *S. mutans*

ComDE system also functions as a pathway of competence development in *S.*

mutans (Li et al., 2001b). ComX encoded by *comX* is the core compentence regulatory component in *S. mutans* as same as in *S. pneumoniae* (Fontaine et al., 2015), however, the coexistence of ComDE and ComRS systems in *S. mutans* makes it more complicated to regulate competence. As mentioned above, in the regulatory network of competence in *S. mutans*, ComDE system locates upstream of ComRS system and depends on the latter to activate *comX* expression, while ComRS could directly activate the transcription of *comX* even in the absence of ComDE system or CSP signal (Figure 3A, 3B). It is also reported that the composition of the growth medium determines which extracellular signal peptide, CSP or XIP (*sigX*-inducing peptide, the signal peptide of ComRS system), affects *comX* induction and competence development in *S. mutans* (see more details in "Interaction with ComRS system in *S. mutans*" part). Because of the different regulations of competence development between *S. mutans* and other streptococci, only a small subpopulation (between 1% and 50%) of *S. mutans* cells can become competent after stimulated by CSP (Aspiras et al., 2004; Lemme et al., 2011; Son et al., 2012). In *S. mutans*, ComE~P controls the expression of 4 operons, ComR-XIP controls 2, and ComX 21 (Khan et al., 2016). Different from *S. pneumoniae*, genetic competence in *S. mutans* is shut off by adaptor protein MecA. The N-terminal domain of MecA could interact with ComX and the C-terminal domain could interact with protease ClpC, thus forming a ternary ComX-MecA-ClpC complex (Bachtiar et al., 2016) in order to isolate ComX and further degrade it by the protease complex ClpC/ClpP, with little effect on the transcription of *comX* in *S. mutans* (Dong et al., 2014). Interestingly, when *S. mutans* grows in CDM, MecA has little impact on the development of competence, probably because the high level of concentration of ComX leads to the saturation of MecA (Tian et al., 2013).

Biofilm formation

In *S. pneumoniae*

The *S. pneumoniae* microbial biofilm was first directly microscopy analyzed on human mucosa and proved the characteristics of a biofilm-like state (Sanderson et al., 2006). The *S. pneumoniae* microbial biofilm was widely discovered to be regulated by LuxS/autoinducer 2 (AI-2), an interspecies QS system (Trappetti et al., 2011; Vidal et al., 2011). However, Oggioni MR *et al.* found the addition of CSP to planktonic cultures was able to increase early formation of biofilms on plastic surfaces. In 2013, Vidal JE *et al.* demonstrated that two QS systems, ComDE and LuxS/AI-2 both regulated the early *S. pneumoniae* biofilms by using the bioreactor and immobilized human cells. Their data confirmed that, in the first 8 h, the ComDE TCS did not regulate biofilm formation on abiotic surfaces but on immobilized lung and pharyngeal cells. In addition, they found autolysis of wild-type biofilms was triggered 16 h after inoculation while those produced by the *luxS* deletion mutant or the *comC* deletion mutant were still culturable, thus they suggested that both of the two systems may also control biofilm lysis (Vidal et al., 2013).

In *S. mutans*

Li *et al.* used *S. mutans* wild-type strain NG8 and its mutants defective in *comC, comD, comE, comX, comCDE* to carry out a study and found that the *comC* mutant formed abnormal biofilms with altered architecture and the *comD* and *comE* mutants formed biofilms with reduced biomass. The different biofilms between *comC* mutants and others suggested that CSP could activate more signal transduction pathways to form biofilms apart from ComDE TCS (Li et al.,

2002). In the strain UA159, the defective of *comC* was found no effect on biofilm formation (Ahn et al., 2006). CSP is believed to act as a biofilm switch and mediate many processes in a biofilm (Aspiras et al., 2004). Liu's group found that *comC* and *gtfB* genes were downregulated in the hypertonic solution, and in this way biofilm diffusion was initiated (Liu et al., 2013). Bachtiar *et al.* used a specific IgY to act against ComD and found that it inhibited biofilm formation (Bachtiar et al., 2016).

Stress responses

Acid and Antibiotic stress responses in *S. pneumoniae*

In *S. pneumoniae*, ComDE and CiaRH are the principal TCSs responsible for the response to acidic stress and they control survival and suicide in two contrasting mechanisms (Cortes et al., 2015; Pinas et al., 2008). Lethal acid stress results in the activation of the acid tolerance response (ATR) and protection against acidic-stress induced lysis (ASIL) by CiaRH. In 2015, Cortes *et al.* demonstrated that F(0)F(1)-ATPase was involved in the response to acidic stress as it repressed the LytA expression on the pneumococcal surface by expressing protecting proteins on the pneumococcal surface to benefit the intracellular survival of the pneumococcus in macrophages (Balachandran et al., 2001; Cortes et al., 2015; Guiral et al., 2005). On the contrary, ComDE performs the activation of ASIL and represses ATR thereby facilitating the release of DNA and virulence factors from the bacteria (Cortes et al., 2015; Pinas et al., 2008) (Figure 2B). Either cultivated in *vitro* or in pneumocyte cultures, all experimental results suggest that both TCSs and F(0)F(1)-ATPase control the stress response by independent pathways (Cortes et al., 2015).

Conjugation, transduction and natural transformation are three important mechanisms for antibiotic resistant bacteria to spread multidrug resistant clones rapidly. And they all have examples of antibiotics that at subinhibitory concentrations they stimulate DNA transfer, which contrasts with the traditional view that antibiotics act on particular target sites specifically (Nagel et al., 2011; Prudhomme et al., 2006). It has been reported that antibiotics may regulate as much as 20% of the genes in *S. pneumoniae* genome (Rogers et al., 2007). Importantly, subinhibitory concentrations of several antibiotics activate the expression of competence genes in *S. pneumoniae* (Prudhomme et al., 2006; Slager et al., 2014). Sturod *et al.* investigated the effect of 6 antibiotics at subinhibitory concentrations on *comX* expression and found that antibiotics might vary in their effects on competence, ranging from inhibitory to stimulatory effects, and that responses affecting transcription of *comX* weren't always related to kinetic changes in transformability (Sturod et al., 2018).

Acid, Antibiotic and Oxidative stress responses in *S. mutans*

S. mutans lives in oral cavity where pH can shift frequently from 7 to 3. Thus, *S. mutans* has to survive in the variable environment. The adaptation of *S. mutans* to survive at pH 5.5 includes upregulating 169 genes and downregulating 108 genes. Many genes encoding TCS including CiaRH, LevSR, LiaSR, ScnKR, Hk/Rr1037/1038 and ComDE are upregulated for acid adaptation (Gong et al., 2009). Li *et al.* found that mutants with the deficiency of *comC, -D, or -E* genes had a lower log-phase ATR. Moreover, they showed that adding CSP to mutant *comC* could regain ATR (Li et al., 2001a).

ComDE system also takes part in the stress response of *S. mutans* to antibiotics.

Perry *et al.* showed that a subinhibitory concentration of the protein synthesis-inhibitor antibiotic spectinomycin significantly induced the transcription of *comC* gene (8.9±3.8-fold) (Perry et al., 2009). Further deletion of *comC* gene resulted in a significant increase of lag phase when exposed to the spectinomycin stress, while the complemented strain showed better recovery (Perry et al., 2009). There are always some bacteria cells surviving in the situation of high dose of antibiotics. These cells will neither grow nor die in the presence of microbicidal antibiotics called 'persisters' (Keren et al., 2004). Interestingly, unlike those antibiotic-resistant mutants, the daughter cells of persisters act as regular cells when facing the antibiotics again (Wiuff et al., 2005). Leung *et al.* demonstrated the persister cells existed in *S. mutans* and found the inactivation of CipB (Mutacin V, a self-acting bacteriocin) completely abolished the presisters (Leung and Levesque, 2012). In 2015, they showed the importance of an intact CSP-dependent competence pathway to develop persisters and found that the LexA (encoded by SMU.2070) induced by CSP played a significant role in forming multidrug-tolerant persisters (Leung et al., 2015).

It is reported that level of *comD* transcript was upregulated about 2.3-fold in the presence of oxygen (Ahn et al., 2007), suggesting the involvement of ComDE TCS in oxidative stress response in *S. mutans*. The *rcrRPQ* operon of *S. mutans* encodes many products which contribute to oxidative stress response (Seaton et al., 2011) and influences the expression of *comX* and *xrpA*, a novel gene internal to *comX*, to modulate competence development and stress response. The overexpression of *xrpA* impairs the growth in aerobic conditions (Kaspar et al., 2015).

Autolysis and DNA release

In *S. pneumoniae*

In 2004, Moscoso *et al.* proved that DNA release in *S. pneumoniae* was triggered by competence induction and depended on the major autolytic amidase, LytA, and an autolytic lysozyme, LytC (Dagkessamanskaia et al., 2004).When entering competence, *cinA-recA-dinF-lytA* operon is activated as ComX binds to p1 promoter and LytA molecules accumulate on the cell envelope (Fernandez-Tornero et al., 2001; Zhu et al., 2015). When other factors choline binding protein D (CbpD) and CibAB are produced, LytA could cleave the lactyl-amide bond that links the stem peptides and the glycan strands of the peptidoglycan, in other words, the cell expresses its effective cytolytic activity during the competence. In 2011, Zhu L *et al.* prepared a mutated peptide, CSP1-E1A, which inhibited development of competence in DNA transformation, and found it reduced the expression of pneumococcal virulence factors CbpD and LytA in *vitro*. Therefore, competence is important for the induction of autolysis. In addition, some of proteins encoded by late genes such as CbpD, LytA and LytC (Eldholm et al., 2009), constitute a killing mechanism used by competent cells to acquire DNA from non-competent pneumococci. This mechanism was first described by Steinmoen *et al.* in 2003. To be more detailed, competent *S. pneumoniae* cells synthesize both bacteriocins and bacteriolysins and cause the release of chromosomal DNA, Pneumolysin and a cytoplasmic β-galactosidase from non-competent cells (Guiral et al., 2005; Steinmoen et al., 2003; Zhu and Lau, 2011).

In *S. mutans*

In streptococci, CSP is thought to be involved in competence development and autolysis. In 2005, Petersen *et al.* showed the addition of synthetic CSP promoted significant release of DNA during culture growth of *S. mutans* (Petersen et al., 2005). Different from *S. pneumonia*, CSP only induces the autolysis of a fraction of population in *S. mutans* (Perry et al., 2009). Autolysis in the presence of CSP in *S. mutans* is triggered through the intracellular accumulation of a self-acting bacteriocin, CipB (Mutacin V), and the bacteriocin immunity protein CipI confers immunity against CipB. Perry *et al.* confirmed that the *nlmC* gene encoding Mutacin V was regulated directly by ComE and its transcription level was in direct proportion to the concentration of CSP in cultures (Perry et al., 2009). *S. mutans* responds to CSP signal and then is divided into competent and non-competent cells. On the activation of ComX, some cells induce the competence while others commit autolysis in an early stage of competence initiation (Lemme et al., 2011). Competent cells could gain DNA fragments released by lysed cells to acquire new genes or to repair themselves (Martin et al., 2006) or to build biofilm (Perry et al., 2009; Petersen et al., 2005). In addition, production of mutacins also result in enhanced cell death and increased autolysis (Senadheera et al., 2012).

Bacteriocin production

In *S. pneumoniae*

Pneumocin MN is a two-peptide bacteriocin involved in intraspecies competition during the nasopharyngeal colonization of *S. pneumoniae*. Pneumocin MN is encoded by *blpMN* genes, which are parts of a much larger regulon *blp* encoding BlpAB (an ABC transporter), BlpC (a peptide pheromone) and BlpRH (TCS). Recently, the proof that the crosstalk between Com and Blp systems contributes to the secretion of Pneumocin was frequently proposed (Figure 2B). Son *et al.* demonstrated that *S. pneumoniae* with *blpAB* lesions could not secrete Pneumocins and BlpC, but was able to respond to peptides from coexisting strains (Son et al., 2011). Kjos M *et al* pointed out that the induction by coexisting strains could be due to endogenous BlpC as both BlpC and Pneumocins could be exported via ComAB, and activation of ComE~P might also directly induce expression of the *blp* promoters, both suggesting a tightly regulatory connection between Pneumocin and ComDE system (Kjos et al., 2016). Moreover, Wholey WY *et al.* further suggested that ComAB could both cleave and secrete a fraction of BlpC when lacking a functional *blpA* gene after CSP stimulation (Wholey et al., 2016). Importantly, CSP induction only occurs in ComAB sufficient background, rather than in ComAB absent condition, or in the presence of an intact blpA gene. Once under conditions adverse to competence, all BlpC secretion is primarily mediated by BlpAB rather than ComAB.

In *S. mutans*

Mutacin refers to the bacteriocin produced by *S. mutans* in particular. There are two types of mutacins, lantibiotic mutacins and non-lantibiotic mutacins. In general, lantibiotic mutacins have a wide spectrum of killing bacteria, while non-lantibiotic mutacins can only kill a few. Now it is clear that ComDE system in involved in regulation of several mutacins, including one lantibiotic mutacin named Mutacin Smb and three non-lantibiotic mutacins, Mutacin IV, V (CipB) and VI (Figure 3B). CSP is shown to be required for the transcription of *smbA* and *smbB* in *S. mutans* strain GS5, and deletion of *comC* gene attenuated the production of Mutacin Smb, while addition of synthetic CSP to the culture of this mutant restored the production of this bacteriocin (Yonezawa and Kuramitsu, 2005). Mutacin IV is encoded by *nlmAB* genes (Balakrishnan et al., 2000), and acts against many streptococci including *S. mutans*, but not staphylococci and enterococci (Qi et al., 2001). The group of *S. mutans* becomes more competent via producing mutacins and killing other cells. Kreth *et al.* confirmed the temporary coordination between Mutacin IV production and competence development induced by CSP, and showed that two hours after the activation of Mutacin IV genes expression by CSP, the cells of *S. mutans* went into competence (Kreth et al., 2005). The transcription of *nlmD*, which encodes the non-lantibiotic Mutacin VI (SMU.423), also requires the CSP-ComDE signaling pathway, and the addition of CSP substantially induces *nlmD* expression (Premnath et al., 2018). In addition, other TCSs of *S. mutans*, like VicRK and HdrRM, are also found to regulate expression of Mutacin IV, V and VI (Merritt and Qi, 2012).

Interactions with other regulatory systems

Interaction with CiaRH system in S. pneumoniae

Besides ComDE TCS, competence development is also modulated by other TCSs such as CiaRH (Mascher et al., 2003). CiaRH is reported to regulate competence by interfering in CSP accumulation (Cassone et al., 2012; Schnorpfeil et al., 2013; Sebert et al., 2005) (Figure 2B). Two ways have been postulated for CiaRH to interfere with CSP accumulation: (i)HtrA, a serine protease encoded from *htrA* and directly controlled by CiaR, leads to the degradation of CSP, though the protein target HtrA acting on is still unclear (Sebert et al., 2005); (ii)5 small non-coding RNAs, designated as csRNAs, suppress the product of CSP by regulating the transcription of *comC*. These csRNAs are encoded from the *ccnA-E* genes, and they work in the form of combination (Laux et al., 2015).

Interaction with VicRK system in S. pneumoniae

VicRK (WalRK, YycFG, MicAB) TCS in *S. pneumoniae* is also described to participate in the regulation of competence. Wagner C *et al.* observed the reduction of transformation frequency when *vicK* was inactivated or truncated at its C terminus (vicKst) and the reduction was still observed even with prolonged incubation in the presence of competence-stimulating peptide (Wagner et al., 2002). Another study showed that both processes of mutations reducing the stability of VicR and abolishing kinase activity in VicK were related to competence derepression under microaerobiosis, suggesting that VicRK was involved in the control of competence under oxidative stress (Echenique and Trombe, 2001). However, the gene detected to be regulated by VicRK in recent experiment didn't include competence operon. Thus, whether the VicRK system is involved in competence control or able to interact with ComDE is still hard to say (Ng et al., 2005) (Figure 2B).

Interaction with BlpRH system in S. pneumoniae

As mentioned above, BlpRH TCS in *S. pneumoniae* modulates the expression of bacteriocins via the mechanism where accumulated BlpC binds to and activates BlpRH TCS which leads to the upregulation of the genes in the BIR that encode bacteriocins (Pneumocins). Although *com* and *blp* were originally considered to operate independently, two recent studies suggested the crosstalk between *com* and *blp* (Figure 2B). They pointed out that activated ComE protein could bind to and upregulate the *blpABC* promoter which regulates the production of BlpABC (Kjos et al., 2016; Wholey et al., 2016). It is proved that ComAB could process and secrete BlpC while Wang CY *et al.* showed that BlpAB could secrete CSP, which explained the BlpC-induced *com* activation they observed (Mitrophanov and Groisman, 2008; Wang et al., 2018).

Interaction with StkP system in S. pneumoniae

StkP, a eukaryotic-like serine/threonine protein kinase, is required for activation of *comDE* upstream of the autoregulated loop orchestrated by CSP in competence, which is the first description of positive regulation of *comDE* transcription, reported by Echenique *et al.* in 2004 (Echenique et al., 2004). Subsequently in 2007, the results from the study by Sasková *et al.* indicated that functional StkP was essential for the resistance of *S. pneumoniae* to various stress conditions (Saskova et al., 2007). In 2018, Pinas *et al.* clearly demonstrated the StkP/ComE pathway, where StkP activated ComE by phosphorylation of the Thr^{128} residue of the latter, which increased both its DNA-binding affinity and dimerization capacities (Pinas et al., 2018). Under acidic conditions, the StkP/ComE, independent of the HK-dependent ComDE system, regulates 104 genes involved in different cellular processes, such as H_2O_2 production and oxidative stress response. It is newly discovered that StkP/ComE pathway mainly triggers the acidic stress response by inducing ASIL and inhibiting ATR (Figure 2B) and the intracellular survival of *S. pneumoniae* in pneumocytes (Pinas et al., 2018).

Interaction with ComRS system in S. mutans

In *S. mutans*, the ComRS system was first defined by Lauren *et al. (Mashburn-Warren et al., 2010)*. ComRS acts downstream of ComDE system to regulate the expression of *comX* (Mashburn-Warren et al., 2010) (Figure 3A, 3B). ComR is a regulator of the Rgg family which regulates genes of glucosyltransferase (Sulavik et al., 1992) and the expression of *comX* depended on the presence of ComR. *S. mutans* synthesizes ComS peptide and exports its cleaved form, XIP, a 7-residue peptide (Khan et al., 2012), as the diffusible signal to drive *comX* (Khan et al.,

2016). Other cells import environmental XIP via the oligopeptide permease Opp and detect it through the intracellular receptor ComR. The ComR/XIP complex stimulates the expression of *comS*, *comX* and competence genes (Figure 3A). In *S. mutans*, CSP induces *comX* expression and the development of competence only in rich peptide-containing media, whereas XIP is effective only in chemically defined media (CDM) which lack peptides (Son et al., 2012). More interestingly, it is reported that there were a bimodal response of *comX* in complex medium, which depended on an intact copy of *comS* and extracellular CSP signal, and an unimodal response of *comX* in chemically defined medium, which depended on an extracellular XIP signal (Son et al., 2012). Adding exogenous CSP could increase the expression of *comS* (Lemme et al., 2011). It is thought that ComE participates in the regulation of a larger auto-feedback, in which CSP and ComS/XIP play inequivalent roles as CSP stimulates the loop whereas ComS/XIP operates as a subsidiary feedback system inside this larger loop (Hagen and Son, 2017).

Interaction with VicRK system in S. mutans

The VicRK system in *S. mutans* modulates genetic transformation, biofilm formation, stress tolerance, and immunity (Lei et al., 2015). The *vicK* gene encodes the HK protein and *vicR* gene encodes the RR regulator. In addition, the roles of VicRK in mutacin production and cell death were also demonstrated (Tsang et al., 2005; Tsang et al., 2006; van der Ploeg, 2005). Senadheera *et al.* demonstrated VicRK played a role in mutacin production by directly modulating the CSP production in *S. mutans*. They observed a VicR binding domain at *comC* coding region which affected the transcription of *comC*, *comDE*, and *comX*, and the transcription of *nlmAB*, *nlmC*, and *nlmD* genes which encode Mutacins IV, V, and VI respectively increased when the HK VicK was inactivated (Senadheera et al., 2012) (Figure 3B). In addition, a number of studies have connected acid tolerance in *S mutans* with expression of VicRK and ComDE (Liu and Burne, 2009), but the cross-regulation between the two TCSs needs more research.

Potential ComDE TCS-targeted antimicrobial applications

So far, the antimicrobial application targeting ComDE TCS is mainly focused on *S. mutans*. *S. mutans* is known as the main cariogenic bacteria, so controlling its virulence is beneficial to the prevention and treatment of caries. Several approaches based on ComDE TCS to inhibit the pathogenicity of *S. mutans* are summarized as follows:

CSP analogues and derivatives

KBI-3221

There is significant specificity in the interaction of CSP with its HK receptor, making it possible to develop novel targeted drugs using CSP analogues or derivatives. The CSP-ComDE system is related to the ability of *S. mutans* to form biofilm. CSP analogue is synthesized to act as a specific competitive inhibitor of the interaction between CSP and its HK receptor ComD, which hinders the induction of "biofilm phenotype" (Suntharalingam and Cvitkovitch, 2005). The peptide analogue (KBI-3221) of CSP molecule of *S. mutans* is obtained by replacing an Arginine of CSP with Glutamate. It specifically targets the ComDE TCS and inhibits the development of biofilm (LoVetri and Madhyastha, 2010), thus reducing the virulence of *S. mutans*. Therefore, it is of great significance for the prevention and treatment of dental caries.

STAMPs

New drugs can be developed by using the targeting property of CSP, and specifically (or selectively) targeted antimicrobial peptides (STAMPs) (Eckert et al., 2006) is the most prominent example. The CSP sequence of *S. mutans* is employed as the target domain of STAMP to mediate the specific delivery of antimicrobial peptide to *S. mutans* cells, so STAMP can eliminate *S. mutans* from a variety of biofilms without affecting non-caries related streptococci (Eckert et al., 2006). C16G2 is a STAMP with antimicrobial specificity on *S. mutans*, consisting of two peptide sequences as particular functional domains. The first peptide is a *S. mutans* selective 'targeting domain' derived from a *S. mutans* CSP fragment, designated C16 (or CSPC16), and consists of C-terminal 16 amino acids of CSP. The second one is the 'kill domain' derived from the broad-spectrum antimicrobial peptide novispirin G10, designated G2, consisting of 16 amino acids. These two peptide sequences are joined by a short peptide linker (GGG), ensuring that both domains function properly (Eckert et al., 2006; Eckert et al., 2012).

P15-CSP

In the context of porous bone space fillers for oral bone reconstruction, peptides that inhibit microbial growth and promote osteoblast function can be used to enhance the performance of porous bone space fillers. P15-CSP, which contains collagen-mimetic osteogenic peptide P15, is a new fusion peptide with new characteristics that P15 or CSP does not have. P15 can enhance osteogenesis and bone formation of human mesenchymal stem cells (MSCs) in vitro, while the CSP moiety can antagonize CSP of S. *mutans* by competing for the specific ComD receptor. The new fusion peptide P15-CSP exhibits both osteogenic and antimicrobial activities, while single peptide P15 and CSP mediate only one of these two functions. The coating of new fusion peptide P15-CSP with hydrophilic anions represents a promising strategy for promoting MSC adhesion and osteogenic differentiation while inhibiting the formation of biofilm of *S. mutans*. In addition, P15-CSP can be used as dry coating. Easy application and storage on a variety of biocompatible surfaces provide a more cost-effective strategy for optimizing bone regeneration and dental bone repair procedures for biomaterials (Li et al., 2015).

Biological or chemical substances interfering with comDE expression

It is found that certain active substances extracted from natural products, as well as chemically synthesized compounds, could interfere with the initiation and expression of *comDE* TCS genes of *S. mutans*. Under their influence, the cariogenic virulence of *S. mutans* may be impaired, thus some of them becoming new potential drugs for the prevention and treatment of caries.

Cyclo (L-leucyl-L-prolyl)

The chloroform extract of cell-free culture supernatant from mangrove rhizosphere bacterium *Bacillus amyloliquefaciens* (MMS-50), is shown to inhibit or influence the adhesion, acid production, acid tolerance, glucan synthesis, biofilm formation, cell-surface hydrophobic of *S. mutans*, although there is no effect on cell viability. The cyclo (L-leucyl-L-valine) is the major compound responsible for the biological activities of MMS-50 and is capable of reducing the expression of virulence genes of *S. mutans*, including *comDE*, thus it may be a promising anti-caries agent (Gowrishankar et al., 2014).

C-30

C-30, one of the halogenated furanone compounds from *Delisea pulchra*, has been extensively studied because it could inhibit the ComDE TCS-regulated behaviors, especially the production of extracellular virulence factors and the development of biofilms of oral streptococci. The results show that the C-30 can inhibit the biofilm formation of *S. mutans*, with no effect on the growth rate of bacteria. When assessing the effect on bacterial communication, C-30 showed no irritation or genotoxicity. This will provide the possibility of using C-30 to prevent biofilm infection of *S. mutans* (He et al., 2012).

DMTU

The aryl isocyanate was synthesized by a simple one pot reaction of aromatic isocyanate with selective amine. Among all the synthesized compounds, aromatic 1,3-dimethylphenylurea (DMTU) is the most effective one on inhibiting the biofilm activity of *S. mutans*. DMTU treatment of MTCC 497 and SM4 strains resulted in down-regulation of genes involved in the abilities of competence development and bacteriocin production, in particular the ComDE TCS genes. Further test on the effect of DMTU indicated that the effect was not temporary and DMTU had an effect on the late-log phase growth of *S. mutans*. DMTU affects major virulence factors such as biofilm formation and bacteriocin production without causing bacterial death. DMTU and lower fluoride concentration could be used together as potential inhibition measures to reduce the occurrence of dental caries without affecting the characteristics of fluoride mineralization (Kaur et al., 2017).

CaF_2-NPs

Nanoscale-based approaches are being widely used in the development of anti-caries agents and their antibiofilm effectiveness have been proven. Calcium fluoride nanoparticles (CaF_2-NPs) are artificially synthesized by a simple coprecipitation method, with an average particle size of 15–25 nm. A study by Kulshrestha *et al* showed that CaF_2-NPs effectively inhibited biofilm formation of *S. mutans* by suppressing both enzymatic activities associated with glucan synthesis and the genes involved in adhesion, acid production, acid tolerance, and quorum sensing, without affecting bacterial viability. Notably, ComDE TCS was also shown to be suppressed considerably. This may lead to impairment of the virulence factors of *S. mutans*, eventually affecting dental caries. As a result, CaF_2-NPs could be used as a potential therapeutic agent against *S. mutans* to inhibit caries-related problems (Kulshrestha et al., 2016).

Conclusions

As a highly conserved two-component signal transduction system across oral streptococci, ComDE TCS plays vital roles in modulating bacteria's physiological and pathogenic properties. ComDE TCS has been well characterized in *S. mutans* and *S. pneumoniae*, nevertheless, the molecular basis and regulatory pathways in other related oral streptococci may be species-specific and quite different. The differences in regulatory mechanisms and functions, as well as genetic evolutionary relationship, need to be further investigated in oral streptococci sharing ComDE TCS. Moreover, it is noticed that oral streptococci are equipped with a considerably large panel of TCSs and sometimes one phenotype may be simultaneously regulated by multiple TCSs. The complicated interactions among these TCSs provide more possibilities for sophisticated and accurate regulations in response to stimuli of specific growth conditions, but also create difficulties for our understanding of regulatory details, which would be one of the key topics in future research of bacterial two-component regulatory systems.

Acknowledgements

This work was supported by the Youth Science Fund Project of National Natural Science Foundation of China, grant number 81500842, and the Undergraduate Innovation and Training Program of Sichuan University, grant number C2018102347.

References

Ahn, S.J., Wen, Z.T., and Burne, R.A. (2006). Multilevel control of competence development and stress tolerance in Streptococcus mutans UA159. Infect Immun 74, 1631-1642.https://dx.doi.org/10.1128/IAI.74.3.1631-1642.2006

Ahn, S.J., Wen, Z.T., and Burne, R.A. (2007). Effects of oxygen on virulence traits of Streptococcus mutans. J Bacteriol 189, 8519-8527.https://dx.doi.org/10.1128/JB.01180-07

Ajdić, D., McShan, W.M., McLaughlin, R.E., Savić, G., Chang, J., Carson, M.B., Primeaux, C., Tian, R., Kenton, S., Jia, H., et al. (2002). Genome sequence of Streptococcus mutans UA159, a cariogenic dental pathogen. Proc Natl Acad Sci U S A 99, 14434-14439.https://dx.doi.org/10.1073/pnas.172501299

Ann M. Stock, Victoria L. Robinson, a., and Goudreau, P.N. (2000). Two-component signal transduction. Annu Rev Biochem 69, 183-215.https://dx.doi.org/10.1146/annurev.biochem.69.1.183

Aspiras, M.B., Ellen, R.P., and Cvitkovitch, D.G. (2004). ComX activity of Streptococcus mutans growing in biofilms. FEMS Microbiol Lett 238, 167-174.https://dx.doi.org/10.1016/j.femsle.2004.07.032

Bachtiar, E.W., Bachtiar, B.M., Soejoedono, R.D., Wibawan, I.W., and Afdhal, A. (2016). Biological and Immunogenicity Property of IgY Anti S. mutans ComD.

Open Dent J 10, 308-314.https://dx.doi.org/10.2174/1874210601610010308

Balachandran, P., Hollingshead, S.K., Paton, J.C., and Briles, D.E. (2001). The autolytic enzyme LytA of Streptococcus pneumoniae is not responsible for releasing pneumolysin. J Bacteriol 183, 3108-3116.https://dx.doi.org/10.1128/JB.183.10.3108-3116.2001

Balakrishnan, M., Simmonds, R.S., Carne, A., and Tagg, J.R. (2000). Streptococcus mutans strain N produces a novel low molecular mass non-lantibiotic bacteriocin. FEMS Microbiol Lett 183, 165-169.https://dx.doi.org/10.1111/j.1574-6968.2000.tb08952.x

Biswas, S., Cao, L., Kim, A., and Biswas, I. (2016). SepM, a Streptococcal Protease Involved in Quorum Sensing, Displays Strict Substrate Specificity. J Bacteriol 198, 436-447.https://dx.doi.org/10.1128/JB.00708-15

Boudes, M., Sanchez, D., Graille, M., van Tilbeurgh, H., Durand, D., and Quevillon-Cheruel, S. (2014). Structural insights into the dimerization of the response regulator ComE from Streptococcus pneumoniae. Nucleic Acids Res 42, 5302-5313.https://dx.doi.org/10.1093/nar/gku110

Campbell, E.A., Choi, S.Y., and Masure, H.R. (1998). A competence regulon in Streptococcus pneumoniae revealed by genomic analysis. Mol Microbiol 27, 929-939.https://dx.doi.org/10.1046/j.1365-2958.1998.00737.x

Cassone, M., Gagne, A.L., Spruce, L.A., Seeholzer, S.H., and Sebert, M.E. (2012). The HtrA protease from Streptococcus pneumoniae digests both denatured proteins and the competence-stimulating peptide. J Biol Chem 287, 38449-38459.https://dx.doi.org/10.1074/jbc.M112.391482

Cheng Q, Campbell E.A, Naughton A.M, Johnson S, and H.R., M. (1997). The com locus controls genetic transformation in Streptococcus pneumoniae. Mol Microbiol 23, 683-692.https://dx.doi.org/10.1046/j.1365-2958.1997.2481617.x

Claverys, J.P., Prudhomme, M., and Martin, B. (2006). Induction of competence regulons as a general response to stress in gram-positive bacteria. Annu Rev Microbiol 60, 451-475.https://dx.doi.org/10.1146/annurev.micro.60.080805.142139

Cortes, P.R., Pinas, G.E., Cian, M.B., Yandar, N., and Echenique, J. (2015). Stress-triggered signaling affecting survival or suicide of Streptococcus pneumoniae. Int J Med Microbiol 305, 157-169.https://dx.doi.org/10.1016/j.ijmm.2014.12.002

Dagkessamanskaia, A., Moscoso, M., Hénard, V., Guiral, S., Overweg, K., Reuter, M., Martin, B., Wells, J., and Claverys, J.-P. (2004). Interconnection of competence, stress and CiaR regulons in Streptococcus pneumoniae_ competence triggers stationary phase autolysis of ciaR mutant cells. Mol Microbiol 51, 1071-1086.https://dx.doi.org/10.1111/j.1365-2958.2003.03892.x

Desai, B.V., and Morrison, D.A. (2007). Transformation in Streptococcus pneumoniae: formation of eclipse complex in a coiA mutant implicates CoiA in genetic recombination. Mol Microbiol 63, 1107-1117.https://dx.doi.org/10.1111/j.1365-2958.2006.05558.x

Dong, G., Tian, X.-L., Gomez, Z.A., and Li, Y.-H. (2014). Regulated proteolysis of the alternative sigma factor SigX in Streptococcus mutans: implication in the escape from competence. BMC Microbiol 14, 183.https://dx.doi.org/10.1186/1471-2180-14-183

Dong, G., Tian, X.L., Cyr, K., Liu, T., Lin, W., Tziolas, G., and Li, Y.H. (2016). Membrane Topology and Structural Insights into the Peptide Pheromone Receptor ComD, A Quorum-Sensing Histidine Protein Kinase of Streptococcus mutans. Sci Rep 6, 26502.https://dx.doi.org/10.1038/srep26502

Dufour, D., Cordova, M., Cvitkovitch, D.G., and Levesque, C.M. (2011).

Regulation of the competence pathway as a novel role associated with a streptococcal bacteriocin. J Bacteriol 193, 6552-6559.https://dx.doi.org/10.1128/JB.05968-11

Eaton, R.E., and Jacques, N.A. (2010). Deletion of competence-induced genes over-expressed in biofilms caused transformation deficiencies in Streptococcus mutans. Mol Oral Microbiol 25, 406-417.https://dx.doi.org/10.1111/j.2041-1014.2010.00589.x

Echenique, J., Kadioglu, A., Romao, S., Andrew, P.W., and Trombe, M.C. (2004). Protein Serine/Threonine Kinase StkP Positively Controls Virulence and Competence in Streptococcus pneumoniae. Infect Immun 72, 2434-2437.https://dx.doi.org/10.1128/iai.72.4.2434-2437.2004

Echenique, J.R., and Trombe, M.C. (2001). Competence repression under oxygen limitation through the two-component MicAB signal-transducing system in Streptococcus pneumoniae and involvement of the PAS domain of MicB. J Bacteriol 183, 4599-4608.https://dx.doi.org/10.1128/JB.183.15.4599-4608.2001

Eckert, R., He, J., Yarbrough, D.K., Qi, F., Anderson, M.H., and Shi, W. (2006). Targeted killing of Streptococcus mutans by a pheromone-guided "smart" antimicrobial peptide. Antimicrob Agents Chemother 50, 3651-3657.https://dx.doi.org/10.1128/AAC.00622-06

Eckert, R., Sullivan, R., and Shi, W. (2012). Targeted antimicrobial treatment to re-establish a healthy microbial flora for long-term protection. Adv Dent Res 24, 94-97.https://dx.doi.org/10.1177/0022034512453725

Eldholm, V., Johnsborg, O., Haugen, K., Ohnstad, H.S., and Havarstein, L.S. (2009). Fratricide in Streptococcus pneumoniae: contributions and role of the cell wall hydrolases CbpD, LytA and LytC. Microbiology 155, 2223-2234.https://dx.doi.org/10.1099/mic.0.026328-0

Facklam, R. (2002). What Happened to the Streptococci: Overview of Taxonomic and Nomenclature Changes. Clin Microbiol Rev 15, 613-630.https://dx.doi.org/10.1128/cmr.15.4.613-630.2002

Fernandez-Tornero, C., Lopez, R., Garcia, E., Gimenez-Gallego, G., and Romero, A. (2001). A novel solenoid fold in the cell wall anchoring domain of the pneumococcal virulence factor LytA. Nat Struct Biol 8, 1020-1024.https://dx.doi.org/10.1038/nsb724

Fontaine, L., Wahl, A., Flechard, M., Mignolet, J., and Hols, P. (2015). Regulation of competence for natural transformation in streptococci. Infect Genet Evol 33, 343-360.https://dx.doi.org/10.1016/j.meegid.2014.09.010

Gomez-Mejia, A., Gamez, G., and Hammerschmidt, S. (2018). Streptococcus pneumoniae two-component regulatory systems: The interplay of the pneumococcus with its environment. Int J Med Microbiol 308, 722-737.https://dx.doi.org/10.1016/j.ijmm.2017.11.012

Gong, Y., Tian, X.L., Sutherland, T., Sisson, G., Mai, J., Ling, J., and Li, Y.H. (2009). Global transcriptional analysis of acid-inducible genes in Streptococcus mutans: multiple two-component systems involved in acid adaptation. Microbiology 155, 3322-3332.https://dx.doi.org/10.1099/mic.0.031591-0

Gowrishankar, S., Poornima, B., and Pandian, S.K. (2014). Inhibitory efficacy of cyclo(L-leucyl-L-prolyl) from mangrove rhizosphere bacterium-Bacillus amyloliquefaciens (MMS-50) toward cariogenic properties of Streptococcus mutans. Res Microbiol 165, 278-289.https://dx.doi.org/10.1016/j.resmic.2014.03.004

Guiral, S., Mitchell, T.J., Martin, B., and Claverys, J.P. (2005). Competence-programmed predation of noncompetent cells in the human pathogen Streptococcus pneumoniae: Genetic requirements. Proc Natl Acad Sci U S A 102,

8710-8715.https://dx.doi.org/10.1073/pnas.0500879102

Hagen, S.J., and Son, M. (2017). Origins of heterogeneity in Streptococcus mutans competence: interpreting an environment-sensitive signaling pathway. Phys Biol 14, 015001.https://dx.doi.org/10.1088/1478-3975/aa546c

Hale, J.D., Heng, N.C., Jack, R.W., and Tagg, J.R. (2005). Identification of nlmTE, the locus encoding the ABC transport system required for export of nonlantibiotic mutacins in Streptococcus mutans. J Bacteriol 187, 5036-5039.https://dx.doi.org/10.1128/JB.187.14.5036-5039.2005

Havarstein, L.S. (2010). Increasing competence in the genus Streptococcus. Molecular Microbiology 78, 541-544.https://dx.doi.org/10.1111/j.1365-2958.2010.07380.x

Håvarstein, L.S., Coomaraswamy, G., and Morrison, D.A. (1995). An Unmodified Heptadecapeptide Pheromone Induces Competence for Genetic Transformation in Streptococcus pneumoniae. Proc Natl Acad Sci U S A 92, 11140-11144.https://dx.doi.org/10.1073/pnas.92.24.11140

Håvarstein, L.S., Gaustad, P., Nes, I.F., and Morrison, D.A. (1996). Identification of the streptococcal competence-pheromone receptor. Mol Microbiol 21, 863-869.https://dx.doi.org/10.1046/j.1365-2958.1996.521416.x

He, Z., Wang, Q., Hu, Y., Liang, J., Jiang, Y., Ma, R., Tang, Z., and Huang, Z. (2012). Use of the quorum sensing inhibitor furanone C-30 to interfere with biofilm formation by Streptococcus mutans and its luxS mutant strain. Int J Antimicrob Agents 40, 30-35.https://dx.doi.org/10.1016/j.ijantimicag.2012.03.016

Hedayati, M.A., Grove, D.E., Steffen, S.E., and Bryant, F.R. (2005). Expression and purification of the SsbB protein from Streptococcus pneumoniae. Protein Expr Purif 43, 133-139.https://dx.doi.org/10.1016/j.pep.2005.03.034

Hoch, J.A. (2000). Two-component and phosphorelay signal transduction. Curr

Opin Microbiol 3, 165-170.https://dx.doi.org/10.1016/S1369-5274(00)00070-9

Hossain, M.S., and Biswas, I. (2012). An extracelluar protease, SepM, generates functional competence-stimulating peptide in Streptococcus mutans UA159. J Bacteriol 194, 5886-5896.https://dx.doi.org/10.1128/JB.01381-12

Hui, F.M., and Morrison, D.A. (1991). Genetic transformation in Streptococcus pneumoniae_ nucleotide sequence analysis shows comA, a gene required for competence induction, to be a member of the bacterial ATP-dependent transport protein family. J Bacteriol 173, 372-381.https://dx.doi.org/10.1128/jb.173.1.372-381.1991

Hung, D.C., Downey, J.S., Ayala, E.A., Kreth, J., Mair, R., Senadheera, D.B., Qi, F., Cvitkovitch, D.G., Shi, W., and Goodman, S.D. (2011). Characterization of DNA binding sites of the ComE response regulator from Streptococcus mutans. J Bacteriol 193, 3642-3652.https://dx.doi.org/10.1128/JB.00155-11

Hvarstein, L.S. (2003). Chapter 16 – Intercellular Communication in Gram-Positive Bacteria Depends on Peptide Pheromones and Their Histidine Kinase Receptors. In Histidine Kinases in Signal Transduction, M. Inouye, and R. Dutta, eds. (San Diego: Academic Press), pp. 341-363.https://dx.doi.org/10.1016/B978-012372484-7/50017-5

Jarosz, L.M., Deng, D.M., van der Mei, H.C., Crielaard, W., and Krom, B.P. (2009). Streptococcus mutans competence-stimulating peptide inhibits Candida albicans hypha formation. Eukaryot Cell 8, 1658-1664.https://dx.doi.org/10.1128/EC.00070-09

Johnsborg, O., Eldholm, V., Bjornstad, M.L., and Havarstein, L.S. (2008). A predatory mechanism dramatically increases the efficiency of lateral gene transfer in Streptococcus pneumoniae and related commensal species. Mol Microbiol 69, 245-253.https://dx.doi.org/10.1111/j.1365-2958.2008.06288.x

Johnston, C., Martin, B., Fichant, G., Polard, P., and Claverys, J.P. (2014). Bacterial transformation: distribution, shared mechanisms and divergent control. Nat Rev Microbiol 12, 181-196.https://dx.doi.org/10.1038/nrmicro3199

Kaspar, J., Ahn, S.J., Palmer, S.R., Choi, S.C., Stanhope, M.J., and Burne, R.A. (2015). A unique open reading frame within the comX gene of Streptococcus mutans regulates genetic competence and oxidative stress tolerance. Mol Microbiol 96, 463-482.https://dx.doi.org/10.1111/mmi.12948

Kaur, G., Balamurugan, P., and Princy, S.A. (2017). Inhibition of the Quorum Sensing System (ComDE Pathway) by Aromatic 1,3-di-m-tolylurea (DMTU): Cariostatic Effect with Fluoride in Wistar Rats. Front Cell Infect Microbiol 7, 313.https://dx.doi.org/10.3389/fcimb.2017.00313

Keren, I., Kaldalu, N., Spoering, A., Wang, Y., and Lewis, K. (2004). Persister cells and tolerance to antimicrobials. FEMS Microbiol Lett 230, 13-18.https://dx.doi.org/10.1016/s0378-1097(03)00856-5

Khan, R., Rukke, H.V., Hovik, H., Amdal, H.A., Chen, T., Morrison, D.A., and Petersen, F.C. (2016). Comprehensive Transcriptome Profiles of Streptococcus mutans UA159 Map Core Streptococcal Competence Genes. mSystems 1.https://dx.doi.org/10.1128/mSystems.00038-15

Khan, R., Rukke, H.V., Ricomini Filho, A.P., Fimland, G., Arntzen, M.O., Thiede, B., and Petersen, F.C. (2012). Extracellular identification of a processed type II ComR/ComS pheromone of Streptococcus mutans. J Bacteriol 194, 3781-3788.https://dx.doi.org/10.1128/JB.00624-12

Kjos, M., Miller, E., Slager, J., Lake, F.B., Gericke, O., Roberts, I.S., Rozen, D.E., and Veening, J.W. (2016). Expression of Streptococcus pneumoniae Bacteriocins Is Induced by Antibiotics via Regulatory Interplay with the Competence System. PLoS Pathog 12, e1005422.https://dx.doi.org/10.1371/journal.ppat.1005422

Kreth, J., Hung, D.C., Merritt, J., Perry, J., Zhu, L., Goodman, S.D., Cvitkovitch, D.G., Shi, W., and Qi, F. (2007). The response regulator ComE in Streptococcus mutans functions both as a transcription activator of mutacin production and repressor of CSP biosynthesis. Microbiology 153, 1799-1807.https://dx.doi.org/10.1099/mic.0.2007/005975-0

Kreth, J., Merritt, J., Shi, W., and Qi, F. (2005). Co-ordinated bacteriocin production and competence development: a possible mechanism for taking up DNA from neighbouring species. Mol Microbiol 57, 392-404.https://dx.doi.org/10.1111/j.1365-2958.2005.04695.x

Kulshrestha, S., Khan, S., Hasan, S., Khan, M.E., Misba, L., and Khan, A.U. (2016). Calcium fluoride nanoparticles induced suppression of Streptococcus mutans biofilm: an in vitro and in vivo approach. Appl Microbiol Biotechnol 100, 1901-1914.https://dx.doi.org/10.1007/s00253-015-7154-4

Lange, R., Wagner, C., de Saizieu, A., Flint, N., Molnos, J., Stieger, M., Caspers, P., Kamber, M., Keck, W., and Amrein, K. (1999). Domain organization and molecular characterization of 13 two-component systems identified by genome sequencing of Streptococcus pneumoniae. Gene 237, 223-234.https://dx.doi.org/10.1016/S0378-1119(99)00266-8

Laux, A., Sexauer, A., Sivaselvarajah, D., Kaysen, A., and Bruckner, R. (2015). Control of competence by related non-coding csRNAs in Streptococcus pneumoniae R6. Front Genet 6, 246.https://dx.doi.org/10.3389/fgene.2015.00246

Lei, L., Yang, Y., Mao, M., Li, H., Li, M., Yang, Y., Yin, J., and Hu, T. (2015). Modulation of biofilm exopolysaccharides by the Streptococcus mutans vicX Gene. Front Microbiol 6, 1432.https://dx.doi.org/10.3389/fmicb.2015.01432

Lemme, A., Grobe, L., Reck, M., Tomasch, J., and Wagner-Dobler, I. (2011). Subpopulation-specific transcriptome analysis of competence-stimulating-peptide-induced Streptococcus mutans. J Bacteriol 193, 1863-1877.https://dx.doi.org/10.1128/JB.01363-10

Leung, V., Ajdic, D., Koyanagi, S., and Levesque, C.M. (2015). The formation of Streptococcus mutans persisters induced by the quorum-sensing peptide pheromone is affected by the LexA regulator. J Bacteriol 197, 1083-1094.https://dx.doi.org/10.1128/JB.02496-14

Leung, V., and Levesque, C.M. (2012). A stress-inducible quorum-sensing peptide mediates the formation of persister cells with noninherited multidrug tolerance. J Bacteriol 194, 2265-2274.https://dx.doi.org/10.1128/JB.06707-11

Li, X., Contreras-Garcia, A., LoVetri, K., Yakandawala, N., Wertheimer, M.R., De Crescenzo, G., and Hoemann, C.D. (2015). Fusion peptide P15-CSP shows antibiofilm activity and pro-osteogenic activity when deposited as a coating on hydrophilic but not hydrophobic surfaces. J Biomed Mater Res A 103, 3736-3746.https://dx.doi.org/10.1002/jbm.a.35511

Li, Y.H., Hanna, M.N., Svensater, G., Ellen, R.P., and Cvitkovitch, D.G. (2001a). Cell density modulates acid adaptation in Streptococcus mutans: implications for survival in biofilms. J Bacteriol 183, 6875-6884.https://dx.doi.org/10.1128/JB.183.23.6875-6884.2001

Li, Y.H., Lau, P.C., Lee, J.H., Ellen, R.P., and Cvitkovitch, D.G. (2001b). Natural genetic transformation of Streptococcus mutans growing in biofilms. J Bacteriol 183, 897-908.https://dx.doi.org/10.1128/JB.183.3.897-908.2001

Li, Y.H., Lau, P.C.Y., Tang, N., Svensater, G., Ellen, R.P., and Cvitkovitch, D.G. (2002). Novel Two-Component Regulatory System Involved in Biofilm Formation and Acid Resistance in Streptococcus mutans. J Bacteriol 184, 6333-6342.https://dx.doi.org/10.1128/jb.184.22.6333-6342.2002

Liu, C., Niu, Y., Zhou, X., Zhang, K., Cheng, L., Li, M., Li, Y., Wang, R., Yang, Y., and Xu, X. (2013). Hyperosmotic response of streptococcus mutans_ from microscopic physiology to transcriptomic profile. BMC Microbiol 13, 275.https://dx.doi.org/10.1186/1471-2180-13-275

Liu, Y., and Burne, R.A. (2009). Multiple two-component systems of Streptococcus mutans regulate agmatine deiminase gene expression and stress tolerance. J Bacteriol 191, 7363-7366.https://dx.doi.org/10.1128/JB.01054-09

LoVetri, K., and Madhyastha, S. (2010). Antimicrobial and antibiofilm activity of quorum sensing peptides and Peptide analogues against oral biofilm bacteria. Methods Mol Biol 618, 383-392.https://dx.doi.org/10.1007/978-1-60761-594-1_24

Luo, P., and Morrison, D.A. (2003). Transient Association of an Alternative Sigma Factor, ComX, with RNA Polymerase during the Period of Competence for Genetic Transformation in Streptococcus pneumoniae. J Bacteriol 185, 349-358.https://dx.doi.org/10.1128/jb.185.1.349-358.2003

Mair, R.W., Senadheera, D.B., and Cvitkovitch, D.G. (2012). CinA is regulated via ComX to modulate genetic transformation and cell viability in Streptococcus mutans. FEMS Microbiol Lett 331, 44-52.https://dx.doi.org/10.1111/j.1574-6968.2012.02550.x

Martin, B., Quentin, Y., Fichant, G., and Claverys, J.P. (2006). Independent evolution of competence regulatory cascades in streptococci? Trends Microbiol 14, 339-345.https://dx.doi.org/10.1016/j.tim.2006.06.007

Martin, B., Soulet, A.L., Mirouze, N., Prudhomme, M., Mortier-Barriere, I., Granadel, C., Noirot-Gros, M.F., Noirot, P., Polard, P., and Claverys, J.P. (2013). ComE/ComE~P interplay dictates activation or extinction status of pneumococcal X-state (competence). Mol Microbiol 87, 394-411. https://dx.doi.org/10.1111/mmi.12104

Mascher, T., Zahner, D., Merai, M., Balmelle, N., de Saizieu, A.B., and Hakenbeck, R. (2003). The Streptococcus pneumoniae cia Regulon: CiaR Target Sites and Transcription Profile Analysis. J Bacteriol 185, 60-70.https://dx.doi.org/10.1128/jb.185.1.60-70.2003

Mashburn-Warren, L., Morrison, D.A., and Federle, M.J. (2010). A novel double-tryptophan peptide pheromone controls competence in Streptococcus spp. via an Rgg regulator. Mol Microbiol 78, 589-606.https://dx.doi.org/10.1111/j.1365-2958.2010.07361.x

Mattos-Graner, R.O., and Duncan, M.J. (2017). Two-component signal transduction systems in oral bacteria. J Oral Microbiol 9, 1400858.https://dx.doi.org/10.1080/20002297.2017.1400858

Merritt, J., and Qi, F. (2012). The mutacins of Streptococcus mutans: regulation and ecology. Molecular Oral Microbiology 27, 57-69.https://dx.doi.org/10.1111/j.2041-1014.2011.00634.x

Merritt, J., Qi, F., and Shi, W. (2005). A unique nine-gene comY operon in Streptococcus mutans. Microbiology 151, 157-166.https://dx.doi.org/10.1099/mic.0.27554-0

Merritt, J., Zheng, L., Shi, W., and Qi, F. (2007). Genetic characterization of the hdrRM operon: a novel high-cell-density-responsive regulator in Streptococcus mutans. Microbiology 153, 2765-2773.https://dx.doi.org/10.1099/mic.0.2007/007468-0

Mirouze, N., Berge, M.A., Soulet, A.L., Mortier-Barriere, I., Quentin, Y., Fichant, G., Granadel, C., Noirot-Gros, M.F., Noirot, P., Polard, P., et al. (2013). Direct involvement of DprA, the transformation-dedicated RecA loader, in the shut-off of pneumococcal competence. Proc Natl Acad Sci U S A 110, E1035-1044.https://dx.doi.org/10.1073/pnas.1219868110

Mitrophanov, A.Y., and Groisman, E.A. (2008). Signal integration in bacterial two-component regulatory systems. Genes Dev 22, 2601-2611.https://dx.doi.org/10.1101/gad.1700308

Moreno-Gamez, S., Sorg, R.A., Domenech, A., Kjos, M., Weissing, F.J., van Doorn, G.S., and Veening, J.W. (2017). Quorum sensing integrates environmental cues, cell density and cell history to control bacterial competence. Nat Commun 8, 854.https://dx.doi.org/10.1038/s41467-017-00903-y

Musher, D. (2010). Streptococcus pneumoniae. In In Mandell G, Bennett J, Dolin R (ed), Mandell, Douglas, and Bennett's principles and practice of infectious disease, 7th, ed. (Philadelphia, PA: Churchill Livingstone Elsevier), pp. 2623–2642

Nagel, M., Reuter, T., Jansen, A., Szekat, C., and Bierbaum, G. (2011). Influence of ciprofloxacin and vancomycin on mutation rate and transposition of IS256 in Staphylococcus aureus. Int J Med Microbiol 301, 229-236.https://dx.doi.org/10.1016/j.ijmm.2010.08.021

Ng, W.L., Tsui, H.C., and Winkler, M.E. (2005). Regulation of the pspA virulence factor and essential pcsB murein biosynthetic genes by the phosphorylated VicR (YycF) response regulator in Streptococcus pneumoniae. J Bacteriol 187, 7444-7459.https://dx.doi.org/10.1128/JB.187.21.7444-7459.2005

Okinaga, T., Xie, Z., Niu, G., Qi, F., and Merritt, J. (2010). Examination of the hdrRM regulon yields insight into the competence system of Streptococcus mutans. Mol Oral Microbiol 25, 165-177.https://dx.doi.org/10.1111/j.2041-1014.2010.00574.x

Perry, J.A., Jones, M.B., Peterson, S.N., Cvitkovitch, D.G., and Levesque, C.M. (2009). Peptide alarmone signalling triggers an auto-active bacteriocin necessary for genetic competence. Mol Microbiol 72, 905-

917.https://dx.doi.org/10.1111/j.1365-2958.2009.06693.x

Pestova, E.V., Håvarstein, L.S., and Morrison, D.A. (1996). Regulation of competence for genetic transformation in Streptococcus pneumoniae by an auto-induced peptide pheromone and a two - component regulatory system. Mol Microbiol 21, 853-862.https://dx.doi.org/10.1046/j.1365-2958.1996.501417.x

Petersen, F.C., and Scheie, A.A. (2000). Genetic transformation in Streptococcus mutans requires a peptide secretion–like apparatus. Oral Microbiol Immun 15, 329-334.https://dx.doi.org/10.1034/j.1399-302x.2000.150511.x

Petersen, F.C., Tao, L., and Scheie, A.A. (2005). DNA binding-uptake system: a link between cell-to-cell communication and biofilm formation. J Bacteriol 187, 4392-4400.https://dx.doi.org/10.1128/JB.187.13.4392-4400.2005

Peterson, S.N., Sung, C.K., Cline, R., Desai, B.V., Snesrud, E.C., Luo, P., Walling, J., Li, H., Mintz, M., Tsegaye, G., et al. (2004). Identification of competence pheromone responsive genes in Streptococcus pneumoniae by use of DNA microarrays. Mol Microbiol 51, 1051-1070.https://dx.doi.org/10.1046/j.1365-2958.2003.03907.x

Pinas, G.E., Cortes, P.R., Orio, A.G., and Echenique, J. (2008). Acidic stress induces autolysis by a CSP-independent ComE pathway in Streptococcus pneumoniae. Microbiology 154, 1300-1308.https://dx.doi.org/10.1099/mic.0.2007/015925-0

Pinas, G.E., Reinoso-Vizcaino, N.M., Yandar Barahona, N.Y., Cortes, P.R., Duran, R., Badapanda, C., Rathore, A., Bichara, D.R., Cian, M.B., Olivero, N.B., et al. (2018). Crosstalk between the serine/threonine kinase StkP and the response regulator ComE controls the stress response and intracellular survival of Streptococcus pneumoniae. PLoS Pathog 14, e1007118.https://dx.doi.org/10.1371/journal.ppat.1007118

Pozzi, G., Masala, L., Iannelli, F., Manganelli, R., Havarstein, L.S., Piccoli, L., Simon, D., and Morrison, D.A. (1996). Competence for genetic transformation in encapsulated strains of Streptococcus pneumoniae_ two allelic variants of the peptide pheromone. J Bacteriol 178, 6087-6090.https://dx.doi.org/10.1128/jb.178.20.6087-6090.

Premnath, P., Reck, M., Wittstein, K., Stadler, M., and Wagner-Dobler, I. (2018). Screening for inhibitors of mutacin synthesis in Streptococcus mutans using fluorescent reporter strains. BMC Microbiol 18, 24.https://dx.doi.org/10.1186/s12866-018-1170-3

Prudhomme, M., Attaiech, L., Sanchez, G., Martin, B., and Claverys, J.-P. (2006). Antibiotic Stress Induces Genetic Transformability in the Human Pathogen <em>Streptococcus pneumoniae</em&gt. Science 313, 89.https://dx.doi.org/10.1126/science.1127912

Prudhomme, M., Berge, M., Martin, B., and Polard, P. (2016). Pneumococcal competence coordination relies on a cell-contact sensing mechanism. PLoS Genet 12, e1006113.https://dx.doi.org/10.1371/journal.pgen.1006113

Prudhomme, M., Martin, B., Mejean, V., and Claverys, J.P. (1989). Nucleotide sequence of the Streptococcus pneumoniae hexB mismatch repair gene_ homology of HexB to MutL of Salmonella typhimurium and to PMS1 of Saccharomyces cerevisiae. J Bacteriol 171, 5332-5338.https://dx.doi.org/10.1128/jb.171.10.5332-5338.1989

Prudhomme, M., Méjean, V., Martin, B., and Claverys, J.P. (1991). Mismatch repair genes of Streptococcus pneumoniae_ HexA confers a mutator phenotype in Escherichia coli by negative complementation. J Bacteriol 173, 7196-7203.https://dx.doi.org/10.1128/jb.173.22.7196-7203.1991

Puyet, A., Greenberg, B., and Lacks, S.A. (1989). The exoA gene of Streptococcus pneumoniae and its product, a DNA exonuclease with apurinic endonuclease activity. J Bacteriol 171, 2278-2286.https://dx.doi.org/10.1128/jb.171.5.2278-2286.1989

Qi, F., Chen, P., and Caufield, P.W. (2001). The group I strain of Streptococcus mutans, UA140, produces both the lantibiotic mutacin I and a nonlantibiotic bacteriocin, mutacin IV. Appl Environ Microbiol 67, 15-21.https://dx.doi.org/10.1128/AEM.67.1.15-21.2001

Quivey, R.G., and Faustoferri, R.C. (1992). In vivo inactivation of the Streptococcus mutans recA gene mediated by PCR amplification and cloning of a recA DNA fragment. Gene 116, 35-42.https://dx.doi.org/10.1016/0378-1119(92)90626-z

Rogers, P.D., Liu, T.T., Barker, K.S., Hilliard, G.M., English, B.K., Thornton, J., Swiatlo, E., and McDaniel, L.S. (2007). Gene expression profiling of the response of Streptococcus pneumoniae to penicillin. J Antimicrob Chemother 59, 616-626.https://dx.doi.org/10.1093/jac/dkl560

Sanderson, A.R., Leid, J.G., and Hunsaker, D. (2006). Bacterial biofilms on the sinus mucosa of human subjects with chronic rhinosinusitis. Laryngoscope 116, 1121-1126.https://dx.doi.org/10.1097/01.mlg.0000221954.05467.54

Saskova, L., Novakova, L., Basler, M., and Branny, P. (2007). Eukaryotic-type serine/threonine protein kinase StkP is a global regulator of gene expression in Streptococcus pneumoniae. Journal of Bacteriology 189, 4168-4179.https://dx.doi.org/10.1128/Jb.01616-06

Schnorpfeil, A., Kranz, M., Kovacs, M., Kirsch, C., Gartmann, J., Brunner, I., Bittmann, S., and Bruckner, R. (2013). Target evaluation of the non-coding csRNAs reveals a link of the two-component regulatory system CiaRH to competence control in Streptococcus pneumoniae R6. Mol Microbiol 89, 334-349.https://dx.doi.org/10.1111/mmi.12277

Seaton, K., Ahn, S.J., Sagstetter, A.M., and Burne, R.A. (2011). A transcriptional regulator and ABC transporters link stress tolerance, (p)ppGpp, and genetic competence in Streptococcus mutans. J Bacteriol 193, 862-874.https://dx.doi.org/10.1128/JB.01257-10

Sebert, M.E., Patel, K.P., Plotnick, M., and Weiser, J.N. (2005). Pneumococcal HtrA protease mediates inhibition of competence by the CiaRH two-component signaling system. J Bacteriol 187, 3969-3979.https://dx.doi.org/10.1128/JB.187.12.3969-3979.2005

Senadheera, D., and Cvitkovitch, D.G. (2008). Quorum Sensing and Biofilm Formation by Streptococcus mutans. In Bacterial Signal Transduction: Networks and Drug Targets, R. Utsumi, ed. (New York, NY: Springer New York), pp. 178-188.https://dx.doi.org/10.1007/978-0-387-78885-2_12

Senadheera, D.B., Cordova, M., Ayala, E.A., Chavez de Paz, L.E., Singh, K., Downey, J.S., Svensater, G., Goodman, S.D., and Cvitkovitch, D.G. (2012). Regulation of bacteriocin production and cell death by the VicRK signaling system in Streptococcus mutans. J Bacteriol 194, 1307-1316.https://dx.doi.org/10.1128/JB.06071-11

Slager, J., Kjos, M., Attaiech, L., and Veening, J.W. (2014). Antibiotic-induced replication stress triggers bacterial competence by increasing gene dosage near the origin. Cell 157, 395-406.https://dx.doi.org/10.1016/j.cell.2014.01.068

Son, M., Ahn, S.J., Guo, Q., Burne, R.A., and Hagen, S.J. (2012). Microfluidic study of competence regulation in Streptococcus mutans: environmental inputs modulate bimodal and unimodal expression of comX. Mol Microbiol 86, 258-272.https://dx.doi.org/10.1111/j.1365-2958.2012.08187.x

Son, M.R., Shchepetov, M., Adrian, P.V., Madhi, S.A., de Gouveia, L., von Gottberg, A., Klugman, K.P., Weiser, J.N., and Dawid, S. (2011). Conserved mutations in the pneumococcal bacteriocin transporter gene, blpA, result in a complex population consisting of producers and cheaters. MBio 2.https://dx.doi.org/10.1128/mBio.00179-11

Song, L., Sudhakar, P., Wang, W., Conrads, G., Brock, A., Sun, J., Wagner-Döbler, I., and Zeng, A.P. (2012). A genome-wide study of two-component signal transduction systems in eight newly sequenced mutans streptococci strains. BMC Genomics 13, 128.https://dx.doi.org/10.1186/1471-2164-13-128

Steinmoen, H., Teigen, A., and Havarstein, L.S. (2003). Competence-Induced Cells of Streptococcus pneumoniae Lyse Competence-Deficient Cells of the Same Strain during Cocultivation. J Bacteriol 185, 7176-7183.https://dx.doi.org/10.1128/jb.185.24.7176-7183.2003

Sturod, K., Salvadori, G., Junges, R., and Petersen, F.C. (2018). Antibiotics alter the window of competence for natural transformation in streptococci. Mol Oral Microbiol 33, 378-387.https://dx.doi.org/10.1111/omi.12240

Sulavik, M.C., Tardif, G., and Clewell, D.B. (1992). Identification of a gene, rgg, which regulates expression of glucosyltransferase and influences the Spp phenotype of Streptococcus gordonii Challis. J Bacteriol 174, 3577-3586.https://dx.doi.org/10.1128/jb.174.11.3577-3586.1992

Sung, C.K., and Morrison, D.A. (2005). Two distinct functions of ComW in stabilization and activation of the alternative sigma factor ComX in Streptococcus pneumoniae. J Bacteriol 187, 3052-3061.https://dx.doi.org/10.1128/JB.187.9.3052-3061.2005

Suntharalingam, P., and Cvitkovitch, D.G. (2005). Quorum sensing in streptococcal biofilm formation. Trends Microbiol 13, 3-

6.https://dx.doi.org/10.1016/j.tim.2004.11.009

Throup, J.P., Koretke, K.K., Bryant, A.P., Ingraham, K.A., Chalker, A.F., Ge, Y., Marra, A., Wallis, N.G., Brown, J.R., Holmes, D.J., et al. (2000). A genomic analysis of two-component signal transduction in Streptococcus pneumoniae. Mol Microbiol 35, 566-576.https://dx.doi.org/10.1046/j.1365-2958.2000.01725.x

Tian, X.L., Dong, G., Liu, T., Gomez, Z.A., Wahl, A., Hols, P., and Li, Y.H. (2013). MecA protein acts as a negative regulator of genetic competence in Streptococcus mutans. J Bacteriol 195, 5196-5206.https://dx.doi.org/10.1128/JB.00821-13

Trappetti, C., Potter, A.J., Paton, A.W., Oggioni, M.R., and Paton, J.C. (2011). LuxS mediates iron-dependent biofilm formation, competence, and fratricide in Streptococcus pneumoniae. Infect Immun 79, 4550-4558.https://dx.doi.org/10.1128/IAI.05644-11

Tsang, P., Merritt, J., Nguyen, T., Shi, W., and Qi, F. (2005). Identification of genes associated with mutacin I production in Streptococcus mutans using random insertional mutagenesis. Microbiology 151, 3947-3955.https://dx.doi.org/10.1099/mic.0.28221-0

Tsang, P., Merritt, J., Shi, W., and Qi, F. (2006). IrvA-dependent and IrvA-independent pathways for mutacin gene regulation in Streptococcus mutans. FEMS Microbiol Lett 261, 231-234.https://dx.doi.org/10.1111/j.1574-6968.2006.00351.x

van Belkum, M.J., Kok, J., Venema, G., Holo, H., Nes, I.F., Konings, W.N., and Abee, T. (1991). The bacteriocin lactococcin A specifically increases permeability of lactococcal cytoplasmic membranes in a voltage-independent, protein-mediated manner. J Bacteriol 173, 7934-7941.https://dx.doi.org/10.1128/jb.173.24.7934-7941.1991

van der Ploeg, J.R. (2005). Regulation of bacteriocin production in Streptococcus mutans by the quorum-sensing system required for development of genetic competence. J Bacteriol 187, 3980-3989.https://dx.doi.org/10.1128/JB.187.12.3980-3989.2005

Vidal, J.E., Howery, K.E., Ludewick, H.P., Nava, P., and Klugman, K.P. (2013). Quorum-sensing systems LuxS/autoinducer 2 and Com regulate Streptococcus pneumoniae biofilms in a bioreactor with living cultures of human respiratory cells. Infect Immun 81, 1341-1353.https://dx.doi.org/10.1128/IAI.01096-12

Vidal, J.E., Ludewick, H.P., Kunkel, R.M., Zahner, D., and Klugman, K.P. (2011). The LuxS-dependent quorum-sensing system regulates early biofilm formation by Streptococcus pneumoniae strain D39. Infect Immun 79, 4050-4060.https://dx.doi.org/10.1128/IAI.05186-11

Wagner, C., Saizieu, A.d., Schonfeld, H.J., Kamber, M., Lange, R., Thompson, C.J., and Page, M.G. (2002). Genetic Analysis and Functional Characterization of the Streptococcus pneumoniae vic Operon. Infection and Immunity 70, 6121-6128.https://dx.doi.org/10.1128/iai.70.11.6121-6128.2002

Wahl, A., Servais, F., Drucbert, A.S., Foulon, C., Fontaine, L., and Hols, P. (2014). Control of natural transformation in salivarius Streptococci through specific degradation of sigmaX by the MecA-ClpCP protease complex. J Bacteriol 196, 2807-2816.https://dx.doi.org/10.1128/JB.01758-14

Wang, C.Y., Patel, N., Wholey, W.Y., and Dawid, S. (2018). ABC transporter content diversity in Streptococcus pneumoniae impacts competence regulation and bacteriocin production. Proc Natl Acad Sci U S A 115, E5776-E5785.https://dx.doi.org/10.1073/pnas.1804668115

Ween, O., Gaustad, P., and Håvarstein, L.S. (1999). Identification of DNA binding sites for ComE, a key regulator of natural competence in Streptococcus pneumoniae. Mol Microbiol 33, 817-827.https://dx.doi.org/10.1046/j.1365-2958.1999.01528.x

Whatmore, A.M., Barcus, V.A., and Dowson, C.G. (1999). Genetic diversity of the streptococcal competence (com) gene locus. J Bacteriol 181, 3144-3154

Wholey, W.Y., Kochan, T.J., Storck, D.N., and Dawid, S. (2016). Coordinated Bacteriocin Expression and Competence in Streptococcus pneumoniae Contributes to Genetic Adaptation through Neighbor Predation. PLoS Pathog 12, e1005413.https://dx.doi.org/10.1371/journal.ppat.1005413

Wiuff, C., Zappala, R.M., Regoes, R.R., Garner, K.N., Baquero, F., and Levin, B.R. (2005). Phenotypic tolerance: antibiotic enrichment of noninherited resistance in bacterial populations. Antimicrob Agents Chemother 49, 1483-1494.https://dx.doi.org/10.1128/AAC.49.4.1483-1494.2005

Yonezawa, H., and Kuramitsu, H.K. (2005). Genetic analysis of a unique bacteriocin, Smb, produced by Streptococcus mutans GS5. Antimicrobial Agents and Chemotherapy 49, 541-548.https://dx.doi.org/10.1128/Aac.49.2.541-548.2005

Zhu, L., and Lau, G.W. (2011). Inhibition of competence development, horizontal gene transfer and virulence in Streptococcus pneumoniae by a modified competence stimulating peptide. PLoS Pathog 7, e1002241.https://dx.doi.org/10.1371/journal.ppat.1002241

Zhu, L., Lin, J., Kuang, Z., Vidal, J.E., and Lau, G.W. (2015). Deletion analysis of Streptococcus pneumoniae late competence genes distinguishes virulence determinants that are dependent or independent of competence induction. Mol Microbiol 97, 151-165.https://dx.doi.org/10.1111/mmi.13016

https://doi.org/10.21775/9781912530229.07

Regulation of Cell Division in Streptococci: Comparing with the Model Rods

Zhenting Xiang[1], Zongbo Li[1], Jumei Zeng[2], Yuqing Li[1*], Jiyao Li[1*]

[1]State Key Laboratory of Oral Diseases, National Clinical Research Center for Oral Diseases, West China Hospital of Stomatology, Sichuan University, Chengdu, PR China.

[2]Department of Infectious Diseases, Boston Children's Hospital, Harvard Medical School, Boston, Massachusetts, USA.

*Corresponding authors: liyuqing@scu.edu.cn, jiyaoliscu@163.com

Abstract

Streptococcus is a genus of oval-shaped bacteria that act as both commensals and pathogens. Streptococcal infections are relevant to high morbidity and huge socioeconomic costs, with drug resistant strains becoming an increasing threat. Cell division plays an essential role during streptococcal colonization and infection, rendering it an ideal target for antibiotics. Substantial progress has been made to uncover the molecular biology and cellular processes of cell division, favoring the target strategies. This review discusses recent advances in our understanding of streptococcal cell division and its regulatory mechanisms regarding the conserved proteins, by comparing with model rods. Peptidoglycan synthesis that involved in septum formation and the maintenance of the unique oval shape have been spatiotemporally controlled in concert with the pace of division. With newly available tools of genetic and cytological study, streptococci will become an additional model bacterial system for cytokinesis and novel therapeutic agents that target cell division.

Introduction

Most bacteria divide into two identical progeny cells by binary fission, which requires the initial formation of a division septum. Extensive studies have elucidated several aspects of cell division in model rod-shaped bacteria *Escherichia coli* and *Bacillus subtilis* (Adams and Errington, 2009; Haeusser and Margolin, 2016). Such studies have revealed that cytokinesis is coordinated by the divisome: a protein complex assembled at the future site of cell division. Cell division begins with the assembly of the conserved tubulin homologue FtsZ, which performs a GTP-dependent polymerization (Scheffers et al., 2002) to form a contractile ring-like structure (Z ring) at the future site of cytokinesis and recruit downstream components of the macromolecular divisome (Egan and Vollmer,

2013). After maturation, divisome constricts the cell membrane in coordination with new peptidoglycan (PG) synthesis at the division plane, leading to the separation of two daughter cells (Adams and Errington, 2009; Haeusser and Margolin, 2016). Bacterial cell division requires tight regulation and precise spatiotemporal coordination of Z ring assembly to ensure equality of the progenies and genomic integrity (Wu and Errington, 2011). The components involved are of critical importance for the propagation and viability of bacteria, and thus play essential roles for pathogenic species in the occurrence of infectious diseases.

The interest in cell division of pathogens started mainly with the intention to identify and characterize targets to develop novel broad-spectrum antimicrobials. The mounting studies on bacterial cell division not only make significant advances in the biological process, they also help us to identify effective cell division inhibitors, especially for pathogens. However, knowledge on the precise roles of cell division proteins mainly lies in model organisms, it is important to characterize and validate the function of cell division proteins in pathogenic species such as *Streptococcus*.

Streptococcus is a genus of Gram-positive bacteria that consists of 122 recognized diverse species (http://www.bacterio.net/streptococcus.html). Despite the relationship between host and streptococci varying from commensal to pathogenic, they can cause serious diseases, ranging from dental caries and pharyngitis to pneumonia and meningitis (Mitchell, 2003; Shak et al., 2013). In addition to numerous deadly infections, non-lethal streptococcal infections are associated with high morbidity and high socioeconomic costs. Drug resistant *Streptococcus pneumoniae,* a human commensal bacterium of streptococcus, was listed in Antibiotic Resistance Threats Reports-2013 to be a severe threat by the Centers for Disease Control and Prevention (CDC) (https://www.cdc.gov/drugresistance/threat-report-2013/).

Streptococcal cell division plays important roles during colonization and infection stages, but many questions regarding molecular biology and cellular processes of cell division remain to be answered (Engholm et al., 2017). Although some fundamental cytokinetic mechanisms in best studied model rods are similar with streptococci, the regulatory systems of bacterial cell division show general specificities, with many additional proteins conserved in the divisome. Herein, after a summary about recent discoveries regarding dynamics of Z-ring assembly and PG synthesis in model rods, we discuss insights into streptococcal Z-ring and cell wall, comparing similarities and differences with model rods. We intend to review these current studies critically and compare similar or distinct views upon streptococcal cell division to provide clues on further research and novel therapeutic agents.

Cell-division pathway in model rods

Z ring dynamics in rods

FtsZ is a cytoplasmtic protein that contains two termini connected by a central core helix. The highly conserved N-terminal globular domain is responsible for polymerization, while the carboxy terminal domain consists of a structured folded part, a flexible linker of variable sequence and length, and a short alpha helix C-terminal end (the central hub) that forms binding sites for several division proteins (Buske and Levin, 2013; Lowe and Amos, 1998; Ma et al., 1996). GTP promotes the FtsZ assembly into tubulin-like protofilaments by head-to-tail monomeric association, leading to initial formation of a ring-like structure (de Boer et al., 1992; Scheffers et al., 2002). When the cell cycle repeats, GTP hydrolysis is coupled to the disassembly of FtsZ.

Although FtsZ is an essential element for membrane constriction during cytokinesis, it does not attach with direct affinity to the cytoplasmic membrane. In *E.coli,* FtsA and ZipA cooperate to tether FtsZ polymers to the inner surface of cytplasmic membrane (Du and Lutkenhaus, 2017; Hale and de Boer, 1997; Ortiz et al., 2016; Pichoff and Lutkenhaus, 2002).

FtsA, which acts as the second most conserved division protein among bacterial species, is structurally homologous to actin. However, unlike other members of the actin family, FtsA contains the four subdomains 1A, 1C, 2A and 2B, with the absence of 1B subdomain adjacent to the 2B subdomain and the addition of 1C subdomain on the opposite site (Szwedziak et al., 2012). The 2B subdomain binds to the central hub of FtsZ while a C-terminal amphipathic helix, which functions as a membrane tageting sequence (MTS), anchors to the membrane (Pichoff and Lutkenhaus, 2005). ATP binding is important for FtsA to recruit FtsZ on a lipid monolayer and to stimulate FtsZ polymerization dynamics, while on the other hand, FtsA exhibits negative regulation on FtsZ filament network organization. The two antagonistic functions of FtsA facilitates Z-ring diameter adjustment during constriction (Loose and Mitchison, 2014). Furthermore, the increase of FtsA oligomerization or polymerization is essential for the integrity of the Z-ring (Shiomi and Margolin, 2007).

Conversely, the bitopic protein ZipA is conserved in Gammaproteobacteria and comprises three domains: a large cytoplasmic C-terminal domain, a basic arm rich in proline and glutamine residues, and an N-terminal transmembrane end (Mosyak et al., 2000). The ZipA C-terminal globular domain contains a binding site that interacts directly with the C-terminal tail of FtsZ to recruit to the assembling Z-ring (Hale and de Boer, 1997). It is thought that ZipA can recruit FtsZ monomers, producing a stronger but inflexible physical link between FtsZ and the membrane, whereas FtsA can only recruit FtsZ polymers, and is a potentially reversible membrane tether (Haeusser and Margolin, 2016; Loose and Mitchison, 2014). FtsA*, a gain-of-function mutant, can compensate for the loss of ZipA during cell division (Geissler et al., 2003). Moreover, a gain-of-functional FtsZ mutant is capable of both bypassing ZipA requirements during divisome assembly and enhancing FtsZ polymer bundling (Haeusser et al., 2015). These results support the hypothesis that ZipA can self-associate into homodimers through its N-terminal domain and could potentially facilitate FtsZ polymer bundling and stability (Skoog and Daley, 2012). In addition, the interaction between FtsZ and ZipA protects FtsZ from degradation by the housekeeping protease complex ClpXP (Pazos et al., 2013). This activity cannot be performed by FtsA*, suggesting additional roles of ZipA other than a Z-ring tether.

In other Gram-positive species that lack ZipA, such as *B. subtilis*, it is speculated that the FtsZ-interacting protein SepF (previously known as YlmF) (Duman et al., 2013) and the FtsZ assembly inhibitor EzrA (Cleverley et al., 2014), which share the same membrane topology with ZipA, may provide alternatives to the role of ZipA. SepF is located in a broadly conserved gene cluster in most Gram-positive bacteria (Hamoen et al., 2006) and serves as an alternative membrane anchor of FtsZ in *B. subtilis* (Duman et al., 2013). In other cases, SepF was reported as an essential tether for FtsZ in mycobacteria (Gupta et al., 2015). Another potential tether in low-GC-content Gram-positive species is EzrA, whose inactivation in *B. subtilis* results in extra Z rings. Containing a single N-terminal transmembrane domain followed by a cytoplasmic domain, EzrA inhibits Z-ring assembly at cell poles and forms an interface between Z-ring and cell wall biosynthesis through an indirect role in recruitment of PBPs (Cleverley et al., 2014; Land et al., 2014).

Spatial regulation in cell division of rods

The Z-ring must tether to the membrane and position precisely at the future division site, usually at the mid-cell. There are two main negative regulatory mechanisms for this in rods: the nucleoid occlusion (NO) and the Min system (Monahan et al., 2014; Rowlett and Margolin, 2015).

The NO system prevents Z ring assembly proximal to the nucleoid or chromosome (Wu and Errington, 2011). SlmA, a DNA-binding protein that serves as an NO system effector in *E. coli,* binds to specific DNA sequences as an orientated dimer that interacts with the FtsZ C-terminal central hub, preventing FtsZ protofilament propagation and bundling (Du and Lutkenhaus, 2014; Schumacher and Zeng, 2016; Tonthat et al., 2013). However, the FtsZ C-terminal domain binds with SlmA as an extended conformation in a narrow, surface-exposed pocket. This conformation occurs rather than as a helix only in the SlmA DNA-bound state, indicating a wide range of conformations are held by the FtsZ C-terminal domain (Schumacher and Zeng, 2016). The GTPase activity of FtsZ is unaffected by the effect of SlmA on accelerating FtsZ fragmentation (Cabre et al., 2015). However, the molecular mechanism by which SlmA inhibits Z-ring assembly is yet to be fully revealed. In *B. subtilis*, Noc, similarly to SlmA, binds to specific DNA sequences, but their modes of action are different. Noc binds to the cell membrane via its N-teminal amphipathic helix, where it simultaneously recruits DNA sequences to function as a division inhibitor (Adams et al., 2015).

The Min system inhibits FtsZ polymerization, blocking Z ring assembly at the cell poles to produce nonviable anucleated minicells. MinD is an ATPase that forms a dimer upon ATP binding in order to associate with the cytoplasmic membrane. Upon binding to MinD, MinC activates and directly interacts with FtsZ to antagonize the longitudinal interactions of FtsZ subunits within a protofilament, the lateral interactions between FtsZ protofilaments, and also the interactions with FtsA and ZipA (Arumugam et al., 2014).

Specifically, to prevent FtsZ assembly at the cell poles, MinE, a topological factor that spatially restricts the inhibitory MinCD complex is needed. In *E. coli*, MinE harbors an amphipathic helix at its N-terminus that serves as an MTS. This helix can stimulate MinD ATPase when sensing a MinD dimer at the membrane. which leads to MinD monomerization and its removal from the membrane (Bonny et al., 2013; Park et al., 2011). The soluble MinCD complex rebinds to the membrane distal to its most recent complex, which is the opposite pole where MinE is absent, through ADP-ATP exchange. The movement of MinCD results in a pole-to-pole oscillation of the Min system governed by MinE conformational changes (Park et al., 2011). More recently, hydrogen-deuterium exchange coupled to mass spectrometry (HDX-MS) was used to examine the conformational dynamics of MinE, suggesting the MTS renders MinE interaction with the membrane dynamic and reversible (Park et al., 2017). Moreover, proteins recruited during early stages of divisome assembly, including FtsZ, ZipA, ZapA, and ZapB, counter-oscillate with the Min system (Bisicchia et al., 2013).

The min system in *B. subtilis* differs from that of *E. coli* because of the lack of MinE. Instead, *B. subtilis* contains DivIVA, which undergoes a curvature-based localization mechanism to localize MinCD to the poles of the cell via the bridging protein MinJ (Bramkamp et al., 2008; Eswaramoorthy et al., 2011). The Min system putatively formed a static bipolar gradient of MinC highest at the cell poles and lowest at the mid-cell to inhibit cell pole Z-ring assembly. However, evidence shows that Min proteins are recruited to the mid-cell prior to septation and prevent polar FtsZ rings forming adjacent to new cell poles, and also promote cytokinetic ring disassembly to establish a new bipolar gradient in daughter cells (Gregory et al., 2008; van Baarle and Bramkamp, 2010).

Recent data show that NO and Min systems are not alone at play in *E. coli* and *B. subtilis*. Other mechanisms, not well identified yet, likely contribute to determining the mid-cell very early during the cell cycle (Rodrigues and Harry, 2012). In bacteria that lack NO and/or Min systems, negative and positive regulators of Z ring formation have been identified, highlighting a diversity of division site selection mechanisms. In *Caulobacter crescentus*, which lack Min and NO proteins, MipZ forms a bipolar gradient that acts as an FtsZ inhibitor, providing spatiotemporal cues for Z ring formation (Kiekebusch et al., 2012). Positive spatial regulators include SsgAB proteins in *Streptomyces coelicolor,* which localize to future sites of multiple septation (Willemse et al., 2011), and PomZ that recruits FtsZ to the division site in *Myxococcus xanthus* (Treuner-Lange et al., 2013).

Peptidoglycan synthesis regulation in rods

Rod-shaped models require robust mechanisms to maintain and pass on their characteristic shape; as a result, cell growth and cell division are closely linked during the cell cycle (Willis and Huang, 2017). First, dispersed elongation is achieved with the cooperation of MreB, an actin-like rod-shape-determining protein. In the process, PG inserts into multiple sites in the lateral wall of newly divided cells to elongate the cell along the longitudinal axis. Soon after, FtsZ localizes to the mid-cell to guide a 'preseptal' phase of cell elongation, followed by 'constrictive' septum synthesis, which enables cell division and daughter cell separation (Typas et al., 2011).

Cell elongation requires a set of proteins to assemble into elongation machinery called the 'elongasome' (Egan et al., 2017; Errington, 2015). The elongasome is a membrane-spanning complex comprising of MreC, MreD, RodA, RodZ and PG cell-wall biosynthetic enzymes PBP1A and PBP2. MreB (and orthologues Mbl and MreBH in *B. subtilis*) localize to the lateral wall of the bacterial cell and spatiotemporally organize the elongasome (Jones et al., 2001). Its crucial role in directional PG insertion was further verified in *E. coli*, where MreB point mutations can tune cell size over a broad range (Monds et al., 2014). Until recently, a generally accepted model of cell wall synthesis proposed MreB guides class A PBPs, which in turn produce PG strands via transglycosylation domains. Class A PBPs (aPBPs) and class B PBPs (bPBPs) then crosslink these strands into a tight PG mesh, fitting new material into cell wall gaps provided by the cleavage activity of autolysins (Gopalani et al., 2016). MreB interacts closely with RodA and RodZ in *E. coli* (Morgenstein et al., 2015) and the latter mediates indirect interaction between MreB and a bPBP. The rotation of MreB patches around the cell depending on bPBP transpeptidation activity and the presence of RodA. The established "Rod complex" (MreB-RodAZ-bPBP) has been suggested by recent works to mediate cell wall synthesis primarily during cell elongation (Zhao et al., 2017).

The relationships between elongation and division machineries are not well-understood, however, there is evidence that a direct interaction between MreB and FtsZ is required for Z-ring contraction, owing to its role in transfer of cell-wall biosynthetic enzymes (Fenton and Gerdes, 2013).

Cell division pathway in *Streptococcus*

Z-ring dynamics in streptococci

FtsA was found to colocalize with FtsZ during exponential growth in *Streptococcus*, consistent with the findings in model rods. The first FtsA protein isolated from *S. pneumoniae* exhibited *in vitro* activity whereby FtsA filaments assembled into large helical structures in the presence of ATP, but did not hydrolyze ATP or depolymerize (Lara et al., 2005). Polymerization and membrane attachment of FtsA is favored by ATP binding through a molecular mechanism by which binding triggers a conformational switch of the amphipathic helix located at the C-terminus. This conformational change accompanies the membrane attachment and facilitates its polymerization (Krupka et al., 2014; Mura et al., 2016). (Table 1)

The FtsZ inhibitor EzrA was hypothesized to function in division site selection in a previous review (Pinho et al., 2013). EzrA is an early cell division protein with a partially redundant role in *B. subtilis.* In *B. subtilis*, EzrA depletion results in a strong phenotype only when combined with other cell division gene mutations, such as *noc* or *gpsB* (a homologue of *divIVA* that is involved in controlling the cell-elongation division cycle, see below). Interestingly, EzrA is essential in *S. pneumoniae* and certain *S. aureus* strains, suggesting it is not as functionally redundant in cocci as *B. subtilis* (Steele et al., 2011; Thanassi et al., 2002). A conflicting result suggests EzrA is non-essential for *S. aureus* viability, but is required for cell size homeostasis through the coordination of proper FtsZ dynamics at mid-cell. EzrA might prevent Z-ring formation near the cell poles of

Table 1. Proteins that involved in cell division in streptococci.

Protein	Function	Localization	References
FtsZ	Tubulin homologue that forms a contractile ring structure (Z ring) at the future cell division site.	Septa	(van Raaphorst et al., 2017)
FtsA	Actin-structural homologue serve as a membrane anchor for the Z ring.	Septa	(Krupka et al., 2014; Mura et al., 2016)
MapZ	Forms ring structures at mid-cell that move apart as peptidoglycan synthesis elongates the cell to mark the future cell division site.	Septa	(Fleurie et al., 2014a; Li et al., 2018)
ZapA, B	Mediate additional stabilization of the Z-ring.	Not determined	(Maggi et al., 2008)
SepF (YlmF)	Cell division protein that is part of the divisome complex and is recruited early to the Z-ring. Its function overlaps with FtsA.	Septa	(Mura et al., 2016)
EzrA	A negative regulator of FtsZ ring formation that modulates the frequency and position of FtsZ ring formation.	Septa	(Rued et al., 2017)
FtsE, X	Cell division ABC transporter ATP-binding protein	Septa	(Sham et al., 2013)
FtsK(SpoIIIE)	SpoE family protein that coordinates cell division and chromosome segregation.	Septa	(Massidda et al., 2013)
DivIB(FtsQ)	Cell division protein that may be involved in stabilizing or promoting the assembly of the division complex	Septa	(Le Gouellec et al., 2008; Masson et al., 2009; Noirclerc-Savoye et al., 2013)
DivIC(FtsB)	Cell division protein that may be involved in stabilizing or promoting the assembly of the division complex	Septa	(Masson et al., 2009; Noirclerc-Savoye et al., 2013)
FtsL	Cell division protein that may play a role in septal PG synthesis.	Septa	(Masson et al., 2009; Noirclerc-Savoye et al., 2013)
FtsW	Septal PG synthesis	Septa	(Maggi et al., 2008; Morlot et al., 2004)
PBP1a	PG glycosyltransferase/transpeptidase for septal and peripheral PG synthesis	Septa	(Land and Winkler, 2011; Wen et al., 2015)
PBP2a	PG glycosyltransferase/transpeptidase, septal PG synthesis?	Septa	(Fenton et al., 2018; Rued et al., 2017)
PBP2b	PG transpeptidases for peripheral PG synthesis.	Septa	(Sharifzadeh et al., 2017; Straume et al., 2017)
PBP2x(FtsI)	PG transpeptidase for septal PG synthesis and remodeling.	Septa	(Sharifzadeh et al., 2017; Tsui et al., 2014)

Table 1. continued.

Protein	Function	Localization	References
PBP3 (DacA)	A D, D-carboxypeptidase that cleaves the ultimate D-Ala from PG peptides.	Cell surface	(Kocaoglu et al., 2015; Morlot et al., 2004; Morlot et al., 2005)
GpsB	A molecular switche that balances septal and peripheral PG synthesis.	Septa	(Rued et al., 2017)
DivIVA	A protein of divisome that is required to correctly localize the elongasome between the septal and lateral cell wall.	Septa and poles	(Ni et al., 2018; Straume et al., 2017)
PcsB (GbpB in *S. mutans*)	PG hydrolase that is required for septa splitting.	Septa	(Bartual et al., 2014; Duque et al., 2011; Sham et al., 2011)
LytB	PG hydrolase that is required for final separation of the daughter cells.	Septa	(Arrigucci and Pozzi, 2017; Zucchini et al., 2017)
MreC, D	Regulator that directs peripheral PG synthesis and control PBP1a localization or activity.	Septa	(Land and Winkler, 2011; Stamsas et al., 2017)
RodZ	Peripheral PG synthesis.	Not determined	(Stamsas et al., 2017; Straume et al., 2017)
RodA	A lipid II flippase and PG polymerase that related to peripheral PG synthesis.	Septa	(Stamsas et al., 2017)
StkP	Serine/threonine kinase; control of cell division	Septa	(Grangeasse, 2016; Zucchini et al., 2017)
PhpP	Phosphatase; control of cell division	Septa	(Martin et al., 2017)
Pmp23	A cell wall hydrolase that is important for proper localization of the Z-ring and MapZ.	Septa	(Jacq et al., 2018; Zucchini et al., 2017)
MacP	A substrate for the kinase StkP that functions as a membrane-anchored cofactor of PBP2a.	Septa	(Fenton et al., 2018)

ovococci, as has been suggested for *B. subtilis.* The model of EzrA with FtsA and FtsZ trapped inside its arch structure was largely based on a recent crystal structure of a homologous protein from *Bacillus subtilis* solved by Cleverley *et al,* who also suggested FtsA and FtsZ filaments position on the side of the EzrA bridge (Cleverley et al., 2014) (Figure 2).

Spatial regulation in cell division of Streptococci

Genomes of *Streptococcus* show an absence of the Min system and nucleoid occlusion effector homologs (Pinho et al., 2013). The lack of NO is further supported by observations that Z-rings in *S. pneumoniae* frequently form over nucleoids and cell constriction occurs concurrently with the separation of the DNA (Land et al., 2013). It was previously speculated that DivIVA functions with ParB to anchor the chromosome and provide a polar gradient of DNA in *S. pneumoniae* (Fadda et al., 2007), but no precise evidence has shown this. This suggests that the septation site selection in *Streptococcus* might be unique, unlike mechanisms present in the rod-shaped *E. coli* and *B. subtilis,* which are based on inhibition of Z-ring assembly.

Recently, studies in *S. pneumoniae*, have demonstrated that a positive mechanism occurs by which FtsZ becomes attracted towards the septation site (Fleurie et al., 2014a; Holeckova et al., 2014). MapZ, also known as LocZ is a transmembrane protein that acts as a molecular beacon to permanently signal the future division site, and was described as an "all-in-one" positive regulatory system (Garcia et al., 2016). Cells lacking MapZ are viable but severely deformed, indicating MapZ is not absolutely required for Z-ring formation, but is essential for its correct placement. MapZ localizes as rings to new division sites prior to FtsZ and FtsA (Holeckova et al., 2014). The MapZ extracellular domain then recognizes and tethers to the nascent PG synthesized at the mid-cell, and its cytoplasmic domain provides a physical anchor for FtsZ. Once the mid-cell Z ring forms, the MapZ-ring splits into two new rings which gradually move apart as the cell elongates from mid-cell to the cell quarters: the sites of the next division. As the two MapZ-rings move, a third MapZ-ring positions at the constricting division site and possibly regulates the cell constriction until the division is complete (Fleurie et al., 2014a). It was thus proposed that MapZ firmly attached to cell wall and shuttled toward the cell equator by the PG elongating the cell (Figure 1). In support of this, further structure-function analysis was performed for molecular characterization of MapZ (Manuse et al., 2016). The extracellular domain of

MapZ undergoes a new bi-modular structure composed of two subdomains: a C-terminal subdomain with conserved patch of amino acids that play a crucial function in binding PG and positioning MapZ at the cell equator, and an N-terminal subdomain that serves as an indispensable pedestal for the C-terminal subdomain. The two subdomains are separated by a flexible serine-rich linker.

Conserved in the *Streptococcaceae* and most other *Lactobacillales*, MapZ is regulated through phosphorylation by the serine/threonine protein kinase StkP, which is involved in the stabilization of the Z-ring and does not affect FtsZ polymerization or GTPase activity (Fleurie et al., 2014a; Li et al., 2018) (Figure 2). Moreover, PG synthesis colocalized with irregular FtsZ structures in *mapZ* mutant, suggesting that cell wall synthesis in *S. pneumoniae* is constantly associated with the divisome (Bramkamp, 2015).

Chromosome segregation in Streptococci

The pneumococcal ParB protein is a homologue of the Noc protein that mediates nucleoid occlusion in *B. subtilis* (Wu and Errington, 2011). However, pneumococcal *ΔparB* mutants are viable and accumulate only a low number (<

5%) of anucleate cells (Minnen et al., 2011). Likewise, pneumococcal Δ*smc* mutants (which lack a nucleoid condensin homologue) and double Δ*smc* Δ*parB* mutants show mild nucleoid segregation defects (Minnen et al., 2011). Another candidate that likely plays an important role in nucleoid segregation in *S. pneumoniae* is FtsK (SpoIIIE), which is a divisome component belonging to a DNA pump superfamily (Demarre *et al.*, 2013). The pneumococcal *ftsK* (*spoIIIE*) gene (*sp0878*) is essential in serotype 4 strain TIGR4, but the role of FtsK (SpoIIIE) in pneumococcal nucleoid separation remains to be determined (Massidda et al., 2013; Pinho et al., 2013; van Raaphorst et al., 2017).

Peptidoglycan synthesis regulation in Streptococci

Cell division requires the coordination of synthesis and remodeling of the cell wall PG. *Streptococcus* are ovoid cocci (named 'ovococci') (Pinho et al., 2013; Zapun et al., 2008) and models for their cell wall synthesis have been well-reported during recent decades (Higgins and Shockman, 1970; Massidda et al., 2013; Morlot et al., 2003; Philippe et al., 2014; Zapun et al., 2008). The two-site model for streptococcal growth and division is based on early ultrastructure reconstruction (Higgins and Shockman, 1976) where two systems, septal and peripheral PG synthesis machineries, both localize to the septum. Unlike the

A.

B.

C.

D.

E.

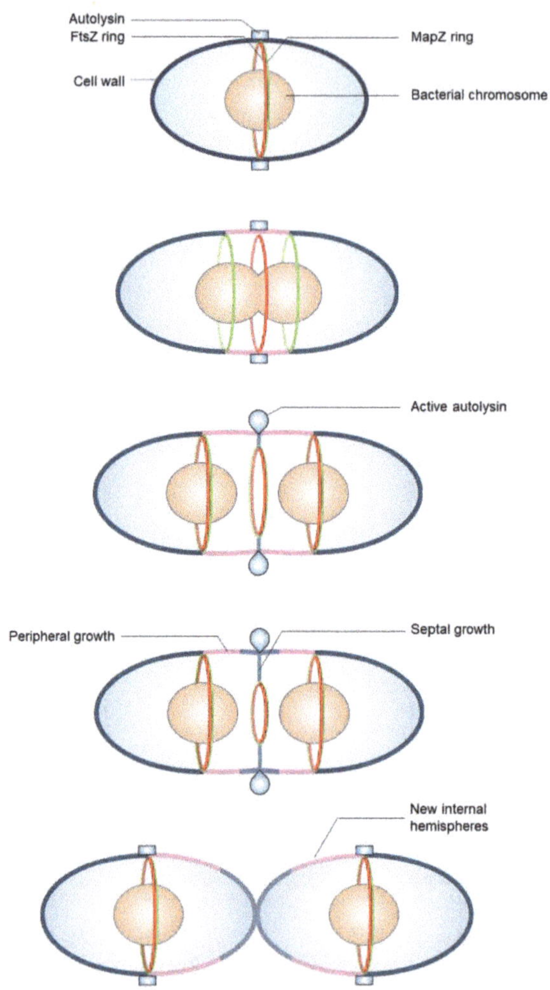

Figure 1. Division site selection and cell wall synthesis during streptococcal cell cycle. (A) Rings formed by FtsZ (red) and MapZ (green) are localized at divisional site. (B) MapZ ring splits into two rings and move apart as cell elongation achieved by PG synthesis. (C) The two MapZ rings migrate to the future division sites of daughter cells while a third MapZ appears at mid-cell. FtsZ rings then relocalize at the beacon of MapZ rings at the future division sites. Peripheral PG synthesis occurs in close proximity to the division site to insert new PG insertion between the present and the future division sites to achieve cell elongation while septal PG synthesis to achieve cell invagination. (D) Rings (FtsZ and MapZ) at mid-cell constrict in concert with new PG synthesis. (E) New internal hemispheres are formed and the resulted daughter cells with FtsZ and MapZ rings localized at mid-cell.

more spherical cocci such as *S. aureus*, streptococcal elongation mode is intimately linked with cell division, whereby cell growth proceeds and overlaps cell division, resulting in the characteristic prolate cell poles.

After an initial inward growth of the cross wall, the ovococcal cell forms an equatorial ring at the mid-cell where cell division begins. Two new equatorial rings are formed on both sides of the original ring, defining the new mid-cell growth zone of the two daughter cells. New PG is inserted in-between the two new equatorial rings, resulting in peripheral growth characterized by simultaneous elongation and constriction. This occurs until the newly formed hemispheres have the same length as the old ones, which is followed by septal PG synthesis to close the septum and PG hydrolyzes to separate the daughter cells (Massidda et al., 2013) (Figure 1). A more recent study confirms this model using super-resolution three-dimensional structured illumination fluorescence microscopy of cells with labeled PG synthesis sites (Wheeler et al., 2011).

The PG synthesis machines produced by *S. pneumoniae* are divided into 3 different types: three class A bifunctional PBPs (PBP1a, PBP1b and PBP2a) that have both transglycosylase (TG) and transpeptidase (TP) activities; two class B monofunctional transpeptidases (TPs) (PBP2x and PBP2b); and one D, D-carboxypeptidase PBP (PBP3 or DacA) (Kocaoglu et al., 2012; Massidda et al., 2013). Both the mid-cell located pneumococcal PBP2x and PBP2b that correspond to PBP3 and PBP2 in *E.coli,* respectively (Egan and Vollmer, 2013), are individually essential, and PBP2x is putatively involved specifically in septum formation, while PBP2b is dedicated to peripheral cell wall synthesis (Berg et al., 2013; Land et al., 2013).

Septal growth

Evidence that PBP2x functions in septal PG synthesis in ovococci was demonstrated by inhibition of PBP2x by the β-lactam antibiotic methicillin, which led to the formation of elongated cells in *Lactococcus lactis* (Sham et al., 2012).

PBP2x inhibition of *S. pneumoniae* with specific antibiotics results in cell elongation, consistent with inhibition of septal PG synthesis (Land et al., 2013; Perez-Nunez et al., 2011). Unexpectedly, depletion of PBP2x in *S. pneumoniae* laboratory strain R6 reportedly led to a mixture of round enlarged, sometimes elongated cells with oddly pointed ends, which was indicative of a complicated division defect (Berg et al., 2013; Peters et al., 2014). This notable difference in appearance suggests that inhibition of PBP2x activity inhibits septal ring closure, whereas the absence of PBP2x may result in a more severe disruption of the division apparatus.

As a key component of divisome, PBP2x employs its extracellular C-terminal PASTA domains for normal localization at the septum (Peters et al., 2014). Fluorescent D-amino acid (FDAA) probes and 3-dimensional Structured Illumination Microscopy (3D-SIM) have been used to demonstrate that PBP2x locates separately from adjacent closing rings including PBP1a, PBP2b, MreC, and StkP to the centers of septa in mid-to-late stages but not in early stages during cell division of pneumococcus D39 cells (Tsui et al., 2014). However, this phenomenon contradicts with the idea discussed by Fleurie *et al* (Fleurie et al., 2014b) of widely separated PG synthesis machines at different mid-cell locations, probably due to limited resolution of approaches. These studies together suggest a localization pattern that, in pre-divisional cells, PBP2x functions for cross-link formation in septal PG synthesis that probably requires PBP2x-StkP interactions (Morlot et al., 2013). At later stages after PBP2x separates to septal centers, adjacent PBP1a and other class A PBPs may continue to synthesize septal PG by glycan strands, indicating that the PG transpeptidase cross-linking activity of

PBP2x may play other roles in PG remodeling rather than septal PG synthesis alone (Tsui et al., 2014).

PcsB is required for septum synthesis and cell separation, although the precise mechanism is yet to be elucidated. PcsB binds to the extracellular part of the division protein complex FtsX and FtsE (Bartual et al., 2014; Sham et al., 2011). FtsEX was placed within the septum membrane in a recent study to illustrate how it bridges the septum width by binding PcsB (Bartual et al., 2014).

PG is not only built, it is also remodeled at the septum by hydrolytic enzymes such as LytB, a process responsible for separating the two daughter cells (Bai et al., 2016). A recent study identified a gene in *Streptococcus gordonii* that putatively encodes LytB (Arrigucci and Pozzi, 2017).

Peripheral growth

In the case of cell wall synthesis, ovococci seem to be similar to rod-shaped bacteria. However, the absence of MreB-like cytoskeletal proteins, which the side-wall PG synthesis of rods are dependent on, indicates peripheral growth in ovococci is mechanistically different from rods (Pinho et al., 2013; Typas et al., 2011).

Studies on *B. subtilis* have revealed that PBP2a, which shares the same subclass with pneumococcal PBP2b, is involved in cell elongation (Sauvage et al., 2008). Further evidence has demonstrated in *Lactococcus lactis*, an ovoid bacterium, that the deletion of PBP2b was not lethal but led cells to the adoption of spherical morphology (Perez-Nunez et al., 2011). A more recent study aimed to uncover the role of PBP2b using a titratable gene depletion method of *pbp2b* in *S. pneumoniae* (Berg et al., 2013). The mutant exhibited a dramatic change in cell morphology whereby individual cells acquired lentil-like appearances in long chains. PBP2b-depleted pneumococci are dependent on a larger proportion of branched muropeptides to fulfil normal growth comparing to wide type (Berg et al., 2013; Straume et al., 2017). In contrast to *S. pneumoniae*, the essentiality of PBP2b homologues in the closely related species *S. sanguinis* and *S. thermophilus* remains to be further investigated (Massidda et al., 2013; Thibessard et al., 2002; Xu et al., 2011).

Depletion of the essential MreC or MreD protein from the *S. pneumoniae* D39 genetic background results in formation of chains of spherical cells remarkably similar to those formed by blocking peripheral PG synthesis by PBP2b depletion (Berg et al., 2013; Land and Winkler, 2011). Taken together, these results suggest that MreCD regulates peripheral PG synthesis in *S. pneumoniae*.

Pneumococcal class A PBPs have been considered functionally redundant. Each of the genes encoding these PBPs can be deleted individually, demonstrating that none are essential for laboratory growth. It is also possible to isolate *pbp1b pbp2a* and *pbp1a pbp1b* double mutants, whereas *pbp1a pbp2a* double mutants are inviable. However, PBP1a and PBP2a are not entirely functionally redundant, because the essentiality of MreCD in *S. pneumoniae* is suppressed in *Δpbp1a* but not *Δpbp2a* mutants from the serotype 2 D39 genetic background (Land and Winkler, 2011). Notably, *Δpbp1a* mutants showed reduced cell length and width compared with their parent strain, an unencapsulated derivative of strain D39 used to reduce masking of cell-shape phenotypes by the capsule (Land and Winkler, 2011). It has been strongly implicated that PBP1a functions in peripheral PG synthesis, although a second role in septal PG synthesis was not ruled out (Land and Winkler, 2011).

Regulation of cell division in *Streptococcus*

DivIVA, GpsB regulatory protein and StkP protein kinase have been proposed as molecular switches that balance septal and peripheral (side-wall like) PG synthesis by the divisome and elongasome in *S. pneumoniae* (Fleurie et al., 2014b).

DivIVA, a switch in peptidoglycan synthesis

In addition to its role in positioning the PBPs, DivIVA may act as a switch to change between septal and peripheral PG synthesis. Although not shown in Figure 2, DivIVA can be connected to the Z ring through EzrA, and it is phosphorylated by the Ser/Thr kinase StkP (Fleurie et al., 2014b).

In bacterial two-hybrid assays *S. pneumoniae* DivIVA interacts with itself, cell division proteins FtsZ, FtsA, ZapA, FtsK and FtsL, PG hydrolase PcsB, and chromosome segregation protein Spo0J. The same study did not detect an interaction of DivIVA with either PBP2X or PBP1a (Le Gouellec et al., 2008). The shorter and rounder appearance of Spn Δ*divIVA* mutant cells in chains suggests a defect in peripheral or in polar PG biosynthesis, as proposed in several papers

(Boersma et al., 2015; Fadda et al., 2007; Fleurie et al., 2014b; Massidda et al., 2013; Straume et al., 2017), but the exact function of DivIVA in Spn remains to be determined.

GpsB, a major regulator of peptidoglycan biosynthesis

GpsB is a cytoplasmic protein conserved in low-GC Gram-positive bacteria. High-resolution crystal structures have further demonstrated that the hexameric GpsB contains two isolated domains, an N-terminal domain (which is highly homologous to DivIVA that interacts with the cytoplasmic face of the cell membrane), and an α-helical C-terminal domain (Cleverley et al., 2016; Rismondo et al., 2016). GpsB is dispensable for growth in *Bacillus subtilis*, where it acts as a guide in PBPs shuttling between septal and lateral sites of PG synthesis. Recent studies have confirmed GpsB plays a role as a switch protein to regulate PBPA1 function and localization in the rod-shaped human pathogen *Listeria monocytogenes* (Rismondo et al., 2016). Being a switch between septal and peripheral PG biosynthesis, however, GpsB in ovococci acts differently to that of rod-shaped bacteria (Figure 2).

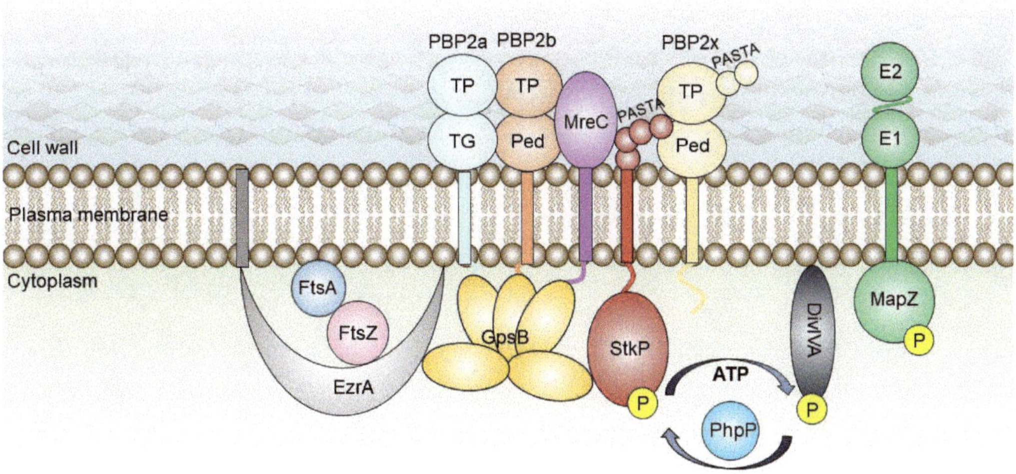

Figure 2. A schematic of the interactions of GpsB and regulatory networks for PG biosynthesis in *Streptococcus* (Rued et al., 2017). GpsB and EzrA form complexes that link the dynamics of divisome to septal and peripheral PG biosynthesis. The level of GpsB can also mediate the phosphorylation cycle of protein including DivIVA and MapZ by StkP kinase and PhpP phosphatase. TP: transpeptidase; TG: transglycosylase; ped: pedestal domains; E1 and E2: extracellular domains of MapZ; P: phosphorylation.

The role of GpsB in cell wall synthesis in *Streptococcus* is crucial, demonstrated by the elongated *S. pneumoniae* D39 growth phenotypes from *gpsB* deletion mutants (Land et al., 2013). The enlarged cell has a defect in cell division septum closure, producing multiple, un-constricted Z-rings along the length of the elongated cell, and is eventually sensitive to lysis (Land et al., 2013). This suggests that GpsB may be a crucial connector in controlling septal ring closure in *S. pneumoniae*. Further studies have investigated the connector mechanisms of

GpsB. PBP2x transpeptidase activity is barely detectable at the centers of division septa upon the depletion of GpsB, suggesting a regulatory role of GpsB in PBP2x function (Rued et al., 2017). Considering the previous hypothesis that extracellular PASTA domains of StkP interact with PBP2x (Morlot et al., 2013), GpsB might positively modulate division ring closure via its interaction with StkP to activate PBP2x in pre-divisional cells. GpsB could also be required continuously during later stages of septal ring closure by associating with other components such as SepF, FtsL and DivIC, but more evidence is needed to support this hypothesis.

In *B. subtilis*, the absence of GpsB causes no change in phenotype under normal conditions, but *gpsB* mutant cells form polar bulges, which is a result of PBP1 delocalization in high-salt media or in combination with the deletion of *ezrA* (Claessen et al., 2008). In *S. pneumoniae*, PBP1a rings are of normal appearance in aseptate cells depleted for GpsB (Land et al., 2013). However, co-immunoprecipitation results indicate a direct interaction between GpsB and proteins involved in peripheral PG synthesis (Rued et al., 2017). The interaction of GpsB directly activates PBP2a (whose exact function in pneumococcal PG synthesis is still unknown) (Tsui et al., 2016), whereas the complexes formed by

GpsB and PBP2b, as well as GpsB and MreC negatively regulate peripheral PG synthesis (Berg et al., 2013; Straume et al., 2017; Tsui et al., 2016). By regulating both septal and peripheral PG synthesis, GpsB is considered to balance the PG synthesis and link PBP regulation to a dynamic divisome through its interaction with EzrA (Fleurie et al., 2014b; Rued et al., 2017).

Regarding localization pattern, GpsB localizes at the lateral walls during cell elongation and moves to the divisional septa before cell constriction in *B. subtilis* (Claessen et al., 2008). Land *et al.* showed that the location patterns of GpsB overlapped with that of FtsZ, but GpsB had a dynamic localization to areas unoccupied by FtsZ in intracellular regions during different stages of cell division (Land et al., 2013).

GpsB is speculated to positively regulate protein phosphorylation levels in several Gram-positive species (Fleurie et al., 2014b; Grangeasse, 2016; Pompeo et al., 2015; Rued et al., 2017). In *B. subtilis,* GpsB is required for PrkC kinase activity, and phosphorylation of GpsB may provide a negative feedback loop by inhibiting PrkC autophosphorylation (Pompeo et al., 2015). In addition, GpsB is phosphorylated *in vivo* by Stp1in *S. agalactiae* (Burnside et al., 2011). However, there is no evidence that GpsB is phosphorylated by StkP, even though GpsB and StkP show overlapping localization during pneumococcal cell division (Rued et al., 2017), in which GpsB is also required for the localization and optimal activity of StkP (Fleurie et al., 2014b). These results may explain epistasis of Δ*divIVA* and Δ*gpsB* in R800 and support the hypothesis that GpsB acts as a regulator through phosphorylation of DivIVA by the StkP kinase to modulate cell elongation (Fleurie et al., 2014b; Grangeasse, 2016; Lewis, 2017; Rued et al., 2017).

It should be noted that, unlike in D39, GpsB was not essential in R800 or R6 background pneumococcal cells (Fleurie et al., 2014b). A similar phenomenon has been found in the specificity of epistasis of *divIVA* mutations to *gpsB* mutations in R800 genetic background cells. Likewise, a requirement for GpsB in localization and kinase activity of StkP toward septal rings is also confined to the R800 strain (Fleurie et al., 2014b; Rued et al., 2017). According to Rued *et al.,* these discrepancies may result from the suppressor accumulation effects on laboratory strains, which was demonstrated in mutant strains using phenotypic and genotypic methods (Lewis, 2017; Rued et al., 2017). This phenomenon suggests an optimal choice for clinical virulent progenitor strains in characterizing cell division mechanisms.

Serine/Threonine kinases and phosphatases

In order to adapt to changes in their environment, bacteria often use phosphorylation/dephosphorylation to transmit cell cycle signals and to respond to external changes, such as nutrients, oxygen, light and osmotic pressure. Two-component signal transduction (TCS) systems are based on transient phosphorylation of a response regulator by a membrane anchored histidine kinase. TCS systems are the most abundant systems in prokaryotes that allow communication between the cell envelope and cytoplasm (Stock et al., 2000). Serine/threonine protein kinases (STPKs) and their cognate Ser/Thr phosphatases represent another major mechanism of transmembrane signaling. Since the early 1990s many eukaryotic-type STPKs have been identified in bacterial genomes. This includes a broad spectrum of pathogens, where STPKs regulate various cellular functions, including cell wall synthesis and cell division through phosphorylating key proteins on Ser/Thr residues to elicit specific downstream effects (Dworkin, 2015; Zhang et al., 2017).

Target cell division-bacterial cell-division inhibitors

With the recent campaign "World Antibiotics Awareness Week" held by the World Health Organization:

(http://www.who.int/campaigns/world-antibiotic-awareness-week/en/), there is little doubt that pathogens resistant to conventional antibiotics are an increasing threat to global health. The emergence of antimicrobial resistance has forced the development of new antimicrobials to combat this global issue and therapeutic agents with novel targets, rather than established synthesis pathways are desperately required (Cheng et al., 2016; Kohanski et al., 2010; Oldfield and Feng, 2014). Among all the targeting process, cell division is a fundamental and essential bacterial process multiply and colonize their niche: proteins underlying this could prove ideal candidates as novel antibiotic targets (Lock and Harry, 2008). Moreover, since the cell-division proteins conserved among many bacterial species are mostly absent from eukaryotic cells, the specific inhibitors will be safe without perturbing the human homologue (den Blaauwen et al., 2014; Sass and Brotz-Oesterhelt, 2013; Vollmer, 2008). In addition, the external location of cell division proteins and protein-protein interaction that is required for divisome assemble are more accessible for inhibitory compounds (Lock and Harry, 2008).

Novel agents that target bacterial cell division proteins has been accepted as promising strategies for antimicrobial attack, which have currently become a research hotspot (Sass and Brotz-Oesterhelt, 2013). For *Streptococcus*, which mainly consist of opportunistic pathogens that gain predominance in the host, targeting cell division is an effective and efficient strategy to prevent their proliferation and colonization, thus also raising interests of researchers (Sham et al., 2012).

Strategies concerning inhibition of cell division mainly target FtsZ, the central mediator of divisome. The first antibacterial agent that targets cell division was reported in 2008 by Haydon *et al.*, who identified PC190723 as an FtsZ assembly inhibitor against *Staphylococcus aureus* (Haydon et al., 2008). PC190723 has potent and selective bactericidal activity against drug resistant S. *aureus* with minimal inhibitory concentrations (MICs) in the range of 0.5 to 1.0 µg/ml (Haydon et al., 2008). However, PC190723 showed species-selectivity with no effect against *S. pneumoniae*. With the development of research on its structural, biochemical and biological properties, multiple antibacterial approaches that interfere with FtsZ assembly and disassembly dynamics, GTPase activity, degradation of FtsZ protein, and expression of *ftsZ,* have been validated as effective (Hurley et al., 2016). Natural products such as phenylacrylamide and 5-methylphenanthridium derivatives have also been shown to inhibit FtsZ activity in *Streptococcus pyogenes* (Liu et al., 2017). Other approaches focus on FtsZ regulators that modulate the stability of FtsZ. Acyldepsipeptides convert the ClpP peptidase to an uncontrolled protease that degrades FtsZ, resulting in cell division inhibition in *B. subtilis*, *S. aureus* and *S. pneumoniae* (Sass et al., 2011).

The contrast between the huge amount of FtsZ targeting strategies in model rods and *S. aureus* and the few in streptococcus may have resulted from the limited knowledge of the cell division biological process in *Streptococcus*. Nevertheless, the development of several strategies including fragment discovery approaches and phage display has been made to discover small-molecule inhibitors of protein–protein interactions in cell division. Moreover, the emergence of antisense nucleic acids and high-throughput screening, along with further research to uncover the mysteries of cell division will lead us towards the discovery of new approaches (Hurley et al., 2016; Zhang et al., 2018).

Future Perspectives

In this review we look into the progress made in cell division of *Streptococcus* over recent years, detailing major events that are similar to those observed in model rods. However, studies also highlighted the specificities and differences in streptococcal cell division compared to the model organisms. Although the precise mechanism of nucleoid occlusion and some functional proteins still remain unclear, the wealth of knowledge available has aided researchers towards the discovery of novel therapies targeting cell division. With the development of new available genetic and cytological tools, *Streptococcus* will rapidly become another model bacterial system for the study in cell division.

Acknowledgements

This work was supported by the National Natural Science Foundation of China (31200985, 81371135, and 31870065).

References

Adams, D.W., and Errington, J. (2009). Bacterial cell division: assembly, maintenance and disassembly of the Z ring. Nature reviews microbiology *7*, 642-653. https://dx.doi.org/10.1038/nrmicro2198.

Adams, D.W., Wu, L.J., and Errington, J. (2015). Nucleoid occlusion protein Noc recruits DNA to the bacterial cell membrane. The EMBO journal *34*, 491-501. https://dx.doi.org/10.15252/embj.201490177.

Arrigucci, R., and Pozzi, G. (2017). Identification of the chain-dispersing peptidoglycan hydrolase LytB of *Streptococcus gordonii*. PLoS ONE *12*, e0176117. https://dx.doi.org/10.1371/journal.pone.0176117.

Arumugam, S., Petrasek, Z., and Schwille, P. (2014). MinCDE exploits the dynamic nature of FtsZ filaments for its spatial regulation. Proceedings of the National Academy of Sciences of the United States of America *111*, E1192-1200. https://dx.doi.org/10.1073/pnas.1317764111.

Bai, X.H., Li, Q., Jiang, Y.L., Zhang, J.R., Chen, Y., and Zhou, C.Z. (2016). Structural Comparison and Simulation of Pneumococcal Peptidoglycan Hydrolase LytB. Methods in molecular biology (Clifton, NJ) *1440*, 271-283. https://dx.doi.org/10.1007/978-1-4939-3676-2_19.

Bartual, S.G., Straume, D., Stamsas, G.A., Munoz, I.G., Alfonso, C., Martinez-Ripoll, M., Havarstein, L.S., and Hermoso, J.A. (2014). Structural basis of PcsB-mediated cell separation in *Streptococcus pneumoniae*. Nature communications *5*, 3842. https://dx.doi.org/10.1038/ncomms4842.

Berg, K.H., Stamsas, G.A., Straume, D., and Havarstein, L.S. (2013). Effects of low PBP2b levels on cell morphology and peptidoglycan composition in *Streptococcus pneumoniae* R6. Journal of bacteriology *195*, 4342-4354. https://dx.doi.org/10.1128/jb.00184-13.

Bisicchia, P., Arumugam, S., Schwille, P., and Sherratt, D. (2013). MinC, MinD, and MinE drive counter-oscillation of early-cell-division proteins prior to *Escherichia coli* septum formation. mBio *4*, e00856-00813. https://dx.doi.org/10.1128/mBio.00856-13.

Boersma, M.J., Kuru, E., Rittichier, J.T., VanNieuwenhze, M.S., Brun, Y.V., and Winkler, M.E. (2015). Minimal Peptidoglycan (PG) Turnover in Wild-Type and PG Hydrolase and Cell Division Mutants of *Streptococcus pneumoniae* D39 Growing Planktonically and in Host-Relevant Biofilms. Journal of bacteriology *197*, 3472-3485. https://dx.doi.org/10.1128/jb.00541-15.

Bonny, M., Fischer-Friedrich, E., Loose, M., Schwille, P., and Kruse, K. (2013). Membrane binding of MinE allows for a comprehensive description of Min-protein pattern formation. PLoS computational biology *9*, e1003347. https://dx.doi.org/10.1371/journal.pcbi.1003347.

Bramkamp, M. (2015). Following the equator: division site selection in *Streptococcus pneumoniae*. Trends in microbiology *23*, 121-122. https://dx.doi.org/10.1016/j.tim.2015.02.001.

Bramkamp, M., Emmins, R., Weston, L., Donovan, C., Daniel, R.A., and Errington, J. (2008). A novel component of the division-site selection system of *Bacillus subtilis* and a new mode of action for the division inhibitor MinCD.

Molecular microbiology *70*, 1556-1569. https://dx.doi.org/10.1111/j.1365-2958.2008.06501.x.

Burnside, K., Lembo, A., Harrell, M.I., Gurney, M., Xue, L., BinhTran, N.T., Connelly, J.E., Jewell, K.A., Schmidt, B.Z., de los Reyes, M.*, et al.* (2011). Serine/threonine phosphatase Stp1 mediates post-transcriptional regulation of hemolysin, autolysis, and virulence of group B Streptococcus. The Journal of biological chemistry *286*, 44197-44210. https://dx.doi.org/10.1074/jbc.M111.313486.

Buske, P.J., and Levin, P.A. (2013). A flexible C-terminal linker is required for proper FtsZ assembly *in vitro* and cytokinetic ring formation *in vivo*. Molecular microbiology *89*, 249-263. https://dx.doi.org/10.1111/mmi.12272.

Cabre, E.J., Monterroso, B., Alfonso, C., Sanchez-Gorostiaga, A., Reija, B., Jimenez, M., Vicente, M., Zorrilla, S., and Rivas, G. (2015). The nucleoid occlusion SlmA protein accelerates the disassembly of the FtsZ protein polymers without affecting their GTPase activity. PloS ONE *10*, e0126434. https://dx.doi.org/10.1371/journal.pone.0126434.

Cheng, G., Dai, M., Ahmed, S., Hao, H., Wang, X., and Yuan, Z. (2016). Antimicrobial drugs in fighting against antimicrobial resistance. Frontiers in microbiology *7*, 470. https://dx.doi.org/10.3389/fmicb.2016.00470.

Claessen, D., Emmins, R., Hamoen, L.W., Daniel, R.A., Errington, J., and Edwards, D.H. (2008). Control of the cell elongation-division cycle by shuttling of PBP1 protein in *Bacillus subtilis*. Molecular microbiology *68*, 1029-1046. https://dx.doi.org/10.1111/j.1365-2958.2008.06210.x.

Cleverley, R.M., Barrett, J.R., Basle, A., Bui, N.K., Hewitt, L., Solovyova, A., Xu, Z.Q., Daniel, R.A., Dixon, N.E., Harry, E.J., *et al.* (2014). Structure and function of a spectrin-like regulator of bacterial cytokinesis. Nature communications *5*, 5421. https://dx.doi.org/10.1038/ncomms6421.

Cleverley, R.M., Rismondo, J., Lockhart-Cairns, M.P., Van Bentum, P.T., Egan, A.J., Vollmer, W., Halbedel, S., Baldock, C., Breukink, E., and Lewis, R.J. (2016). Subunit arrangement in GpsB, a regulator of cell wall biosynthesis. Microbial drug resistance (Larchmont, NY) *22*, 446-460. https://dx.doi.org/10.1089/mdr.2016.0050.

Duque, C., Stipp, R.N., Wang, B., Smith, D.J., Hofling, J.F., Kuramitsu, H.K., Duncan, M.J., and Mattos-Graner, R.O. (2011). Downregulation of GbpB, a component of the VicRK regulon, affects biofilm formation and cell surface characteristics of *Streptococcus mutans*. Infection and immunity *79*, 786-796. https://dx.doi.org/10.1128/IAI.00725-10.

de Boer, P., Crossley, R., and Rothfield, L. (1992). The essential bacterial cell-division protein FtsZ is a GTPase. Nature *359*, 254-256. https://dx.doi.org/10.1038/359254a0.

den Blaauwen, T., Andreu, J.M., and Monasterio, O. (2014). Bacterial cell division proteins as antibiotic targets. Bioorg Chem *55*, 27-38. https://dx.doi.org/10.1016/j.bioorg.2014.03.007.

Du, S., and Lutkenhaus, J. (2014). SlmA antagonism of FtsZ assembly employs a two-pronged mechanism like MinCD. PLoS genetics *10*, e1004460. https://dx.doi.org/10.1371/journal.pgen.1004460.

Du, S., and Lutkenhaus, J. (2017). Assembly and activation of the *Escherichia coli* divisome. Molecular microbiology *105*, 177-187. https://dx.doi.org/10.1111/mmi.13696.

Duman, R., Ishikawa, S., Celik, I., Strahl, H., Ogasawara, N., Troc, P., Lowe, J., and Hamoen, L.W. (2013). Structural and genetic analyses reveal the protein SepF as a new membrane anchor for the Z ring. Proceedings of the National Academy of Sciences of the United States of America *110*, E4601-E4610. https://dx.doi.org/10.1073/pnas.1313978110.

Dworkin, J. (2015). Ser/Thr phosphorylation as a regulatory mechanism in bacteria. Current opinion in microbiology *24*, 47-52. https://dx.doi.org/10.1016/j.mib.2015.01.005.

Egan, A.J., Cleverley, R.M., Peters, K., Lewis, R.J., and Vollmer, W. (2017). Regulation of bacterial cell wall growth. The FEBS journal *284*, 851-867. https://dx.doi.org/10.1111/febs.13959.

Egan, A.J., and Vollmer, W. (2013). The physiology of bacterial cell division. Annals of the New York Academy of Sciences *1277*, 8-28. https://dx.doi.org/10.1111/j.1749-6632.2012.06818.x.

Engholm, D.H., Kilian, M., Goodsell, D.S., Andersen, E.S., and Kjaergaard, R.S. (2017). A visual review of the human pathogen *Streptococcus pneumoniae*. FEMS microbiology reviews *41*, 854-879. https://dx.doi.org/10.1093/femsre/fux037.

Errington, J. (2015). Bacterial morphogenesis and the enigmatic MreB helix. Nature reviews microbiology *13*, 241-248. https://dx.doi.org/10.1038/nrmicro3398.

Eswaramoorthy, P., Erb, M.L., Gregory, J.A., Silverman, J., Pogliano, K., Pogliano, J., and Ramamurthi, K.S. (2011). Cellular architecture mediates DivIVA ultrastructure and regulates min activity in *Bacillus subtilis*. mBio 2, e00257-11. https://dx.doi.org/10.1128/mBio.00257-11.

Fadda, D., Santona, A., D'Ulisse, V., Ghelardini, P., Ennas, M.G., Whalen, M.B., and Massidda, O. (2007). *Streptococcus pneumoniae* DivIVA: localization and interactions in a MinCD-free context. Journal of bacteriology *189*, 1288-1298. https://dx.doi.org/10.1128/jb.01168-06.

Fenton, A.K., Manuse, S., Flores-Kim, J., Garcia, P.S., Mercy, C., Grangeasse, C., Bernhardt, T.G., and Rudner, D.Z. (2018). Phosphorylation-dependent activation of the cell wall synthase PBP2a in *Streptococcus pneumoniae* by MacP. Proceedings of the National Academy of Sciences of the United States of America *115*, 2812-2817. https://dx.doi.org/10.1073/pnas.1715218115.

Fenton, A.K., and Gerdes, K. (2013). Direct interaction of FtsZ and MreB is required for septum synthesis and cell division in *Escherichia coli*. The EMBO journal *32*, 1953-1965. https://dx.doi.org/10.1038/emboj.2013.129.

Fleurie, A., Lesterlin, C., Manuse, S., Zhao, C., Cluzel, C., Lavergne, J.P., Franz-Wachtel, M., Macek, B., Combet, C., Kuru, E., *et al.* (2014a). MapZ marks the division sites and positions FtsZ rings in *Streptococcus pneumoniae*. Nature *516*, 259-262. https://dx.doi.org/10.1038/nature13966.

Fleurie, A., Manuse, S., Zhao, C., Campo, N., Cluzel, C., Lavergne, J.P., Freton, C., Combet, C., Guiral, S., Soufi, B., *et al.* (2014b). Interplay of the serine/threonine-kinase StkP and the paralogs DivIVA and GpsB in pneumococcal cell elongation and division. PLoS genetics *10*, e1004275. https://dx.doi.org/10.1371/journal.pgen.1004275.

Garcia, P.S., Simorre, J.P., Brochier-Armanet, C., and Grangeasse, C. (2016). Cell division of *Streptococcus pneumoniae*: think positive! Current opinion in microbiology *34*, 18-23. https://dx.doi.org/10.1016/j.mib.2016.07.014.

Geissler, B., Elraheb, D., and Margolin, W. (2003). A gain-of-function mutation in *ftsA* bypasses the requirement for the essential cell division gene *zipA* in *Escherichia coli*. Proceedings of the National Academy of Sciences of the United States of America *100*, 4197-4202. https://dx.doi.org/10.1073/pnas.0635003100.

Gopalani, M., Dhiman, A., Rahi, A., Kandari, D., and Bhatnagar, R. (2016). Identification, functional characterization and regulon prediction of a novel two component system comprising BAS0540-BAS0541 of *Bacillus anthracis*. PLoS ONE *11*, e0158895. https://dx.doi.org/10.1371/journal.pone.0158895.

Grangeasse, C. (2016). Rewiring the pneumococcal cell cycle with serine/threonine- and tyrosine-kinases. Trends in microbiology *24*, 713-724. https://dx.doi.org/10.1016/j.tim.2016.04.004.

Gregory, J.A., Becker, E.C., and Pogliano, K. (2008). *Bacillus subtilis* MinC destabilizes FtsZ-rings at new cell poles and contributes to the timing of cell division. Genes & development *22*, 3475-3488. https://dx.doi.org/10.1101/gad.1732408.

Gupta, S., Banerjee, S.K., Chatterjee, A., Sharma, A.K., Kundu, M., and Basu, J.

(2015). Essential protein SepF of mycobacteria interacts with FtsZ and MurG to regulate cell growth and division. Microbiology (Reading, England) *161*, 1627-1638. https://dx.doi.org/10.1099/mic.0.000108.

Haeusser, D.P., and Margolin, W. (2016). Splitsville: structural and functional insights into the dynamic bacterial Z ring. Nature reviews microbiology *14*, 305-319. https://dx.doi.org/10.1038/nrmicro.2016.26.

Haeusser, D.P., Rowlett, V.W., and Margolin, W. (2015). A mutation in *Escherichia coli ftsZ* bypasses the requirement for the essential division gene *zipA* and confers resistance to FtsZ assembly inhibitors by stabilizing protofilament bundling. Molecular microbiology *97*, 988-1005. https://dx.doi.org/10.1111/mmi.13081.

Hale, C.A., and de Boer, P.A. (1997). Direct binding of FtsZ to ZipA, an essential component of the septal ring structure that mediates cell division in *E. coli*. Cell *88*, 175-185. https://dx.doi.org/10.1016/s0092-8674(00)81838-3.

Hamoen, L.W., Meile, J.C., de Jong, W., Noirot, P., and Errington, J. (2006). SepF, a novel FtsZ-interacting protein required for a late step in cell division. Molecular microbiology *59*, 989-999. https://dx.doi.org/10.1111/j.1365-2958.2005.04987.x.

Haydon, D.J., Stokes, N.R., Ure, R., Galbraith, G., Bennett, J.M., Brown, D.R.,

Baker, P.J., Barynin, V.V., Rice, D.W., Sedelnikova, S.E., *et al.* (2008). An inhibitor of FtsZ with potent and selective anti-staphylococcal activity. Science *321*, 1673-1675. https://dx.doi.org/10.1126/science.1159961.

Higgins, M.L., and Shockman, G.D. (1970). Model for cell wall growth of *Streptococcus faecalis*. Journal of bacteriology *101*, 643-648.

Higgins, M.L., and Shockman, G.D. (1976). Study of cycle of cell wall assembly in *Streptococcus faecalis* by three-dimensional reconstructions of thin sections of cells. Journal of bacteriology *127*, 1346-1358.

Holeckova, N., Doubravova, L., Massidda, O., Molle, V., Buriankova, K., Benada, O., Kofronova, O., Ulrych, A., and Branny, P. (2014). LocZ is a new cell division protein involved in proper septum placement in *Streptococcus pneumoniae*. mBio *6*, e01700-01714. https://dx.doi.org/10.1128/mBio.01700-14.

Hurley, K.A., Santos, T.M., Nepomuceno, G.M., Huynh, V., Shaw, J.T., and Weibel, D.B. (2016). Targeting the bacterial division protein FtsZ. Journal of medicinal chemistry *59*, 6975-6998. https://dx.doi.org/10.1021/acs.jmedchem.5b01098.

Jacq, M., Arthaud, C., Manuse, S., Mercy, C., Bellard, L., Peters, K., Gallet, B., Galindo, J., Doan, T., Vollmer, W., *et al.* (2018). The cell wall hydrolase Pmp23 is important for assembly and stability of the division ring in *Streptococcus pneumoniae*. Scientific reports *8*, 7591. https://dx.doi.org/10.1038/s41598-018-25882-y.

Jones, L.J., Carballido-Lopez, R., and Errington, J. (2001). Control of cell shape in bacteria: helical, actin-like filaments in *Bacillus subtilis*. Cell *104*, 913-922. https://dx.doi.org/10.1016/s0092-8674(01)00287-2.

Kiekebusch, D., Michie, K.A., Essen, L.O., Lowe, J., and Thanbichler, M. (2012). Localized dimerization and nucleoid binding drive gradient formation by the bacterial cell division inhibitor MipZ. Molecular cell *46*, 245-259. https://dx.doi.org/10.1016/j.molcel.2012.03.004.

Kocaoglu, O., Tsui, H.C., Winkler, M.E., and Carlson, E.E. (2015). Profiling of beta-lactam selectivity for penicillin-binding proteins in *Streptococcus pneumoniae* D39. Antimicrobial agents and chemotherapy *59*, 3548-3555. https://dx.doi.org/10.1128/aac.05142-14.

Kocaoglu, O., Calvo, R.A., Sham, L.T., Cozy, L.M., Lanning, B.R., Francis, S., Winkler, M.E., Kearns, D.B., and Carlson, E.E. (2012). Selective penicillin-binding protein imaging probes reveal substructure in bacterial cell division. ACS chemical biology *7*, 1746-1753. https://dx.doi.org/10.1021/cb300329r.

Kohanski, M.A., Dwyer, D.J., and Collins, J.J. (2010). How antibiotics kill bacteria:

from targets to networks. Nature reviews microbiology *8*, 423-435. https://dx.doi.org/10.1038/nrmicro2333.

Krupka, M., Cabre, E.J., Jimenez, M., Rivas, G., Rico, A.I., and Vicente, M. (2014). Role of the FtsA C terminus as a switch for polymerization and membrane association. mBio *5*, e02221. https://dx.doi.org/10.1128/mBio.02221-14.

Land, A.D., Luo, Q., and Levin, P.A. (2014). Functional domain analysis of the cell division inhibitor EzrA. PLoS ONE *9*, e102616. https://dx.doi.org/10.1371/journal.pone.0102616.

Land, A.D., Tsui, H.C., Kocaoglu, O., Vella, S.A., Shaw, S.L., Keen, S.K., Sham, L.T., Carlson, E.E., and Winkler, M.E. (2013). Requirement of essential Pbp2x and GpsB for septal ring closure in *Streptococcus pneumoniae* D39. Molecular microbiology *90*, 939-955. https://dx.doi.org/10.1111/mmi.12408.

Land, A.D., and Winkler, M.E. (2011). The requirement for pneumococcal MreC and MreD is relieved by inactivation of the gene encoding PBP1a. Journal of bacteriology *193*, 4166-4179. https://dx.doi.org/10.1128/jb.05245-11.

Lara, B., Rico, A.I., Petruzzelli, S., Santona, A., Dumas, J., Biton, J., Vicente, M., Mingorance, J., and Massidda, O. (2005). Cell division in cocci: localization and properties of the *Streptococcus pneumoniae* FtsA protein. Molecular microbiology *55*, 699-711. https://dx.doi.org/10.1111/j.1365-2958.2004.04432.x.

Le Gouellec, A., Roux, L., Fadda, D., Massidda, O., Vernet, T., and Zapun, A. (2008). Roles of pneumococcal DivIB in cell division. Journal of bacteriology *190*, 4501-4511. https://dx.doi.org/10.1128/jb.00376-08.

Lewis, R.J. (2017). The GpsB files: the truth is out there. Molecular microbiology *103*, 913-918. https://dx.doi.org/10.1111/mmi.13612.

Li, Y., Shao, S., Xu, X., Su, X., Sun, Y., and Wei, S. (2018). MapZ Forms a Stable Ring Structure That Acts As a Nanotrack for FtsZ Treadmilling in *Streptococcus mutans*. ACS nano. https://dx.doi.org/10.1021/acsnano.8b02469.

Liu, F., Venter, H., Bi, F., Semple, S.J., Liu, J., Jin, C., and Ma, S. (2017). Synthesis and antibacterial activity of 5-methylphenanthridium derivatives as FtsZ inhibitors. Bioorganic & medicinal chemistry letters *27*, 3399-3402. https://dx.doi.org/10.1016/j.bmcl.2017.06.005.

Lock, R.L., and Harry, E.J. (2008). Cell-division inhibitors: new insights for future antibiotics. Nature reviews drug Discovery *7*, 324-338. https://dx.doi.org/10.1038/nrd2510.

Loose, M., and Mitchison, T.J. (2014). The bacterial cell division proteins FtsA and FtsZ self-organize into dynamic cytoskeletal patterns. Nature Cell Biology *16*, 38-46. https://dx.doi.org/10.1038/ncb2885.

Lowe, J., and Amos, L.A. (1998). Crystal structure of the bacterial cell-division protein FtsZ. Nature *391*, 203-206. https://dx.doi.org/10.1038/34472.

Ma, X., Ehrhardt, D.W., and Margolin, W. (1996). Colocalization of cell division proteins FtsZ and FtsA to cytoskeletal structures in living *Escherichia coli* cells by using green fluorescent protein. Proceedings of the National Academy of Sciences of the United States of America *93*, 12998-13003. https://dx.doi.org/10.1073/pnas.93.23.12998.

Maggi, S., Massidda, O., Luzi, G., Fadda, D., Paolozzi, L., and Ghelardini, P. (2008). Division protein interaction web: identification of a phylogenetically conserved common interactome between *Streptococcus pneumoniae* and *Escherichia coli.* Microbiology 154, 3042-3052. https://dx.doi.org/10.1099/mic.0.2008/018697-0.

Manuse, S., Jean, N.L., Guinot, M., Lavergne, J.P., Laguri, C., Bougault, C.M., VanNieuwenhze, M.S., Grangeasse, C., and Simorre, J.P. (2016). Structure-function analysis of the extracellular domain of the pneumococcal cell division site positioning protein MapZ. Nature communications *7*, 12071. https://dx.doi.org/10.1038/ncomms12071.

Martin, J.E., Lisher, J.P., Winkler, M.E., and Giedroc, D.P. (2017). Perturbation of manganese metabolism disrupts cell division in *Streptococcus pneumoniae*.

Molecular microbiology *104*, 334-348. https://dx.doi.org/10.1111/mmi.13630.

Massidda, O., Novakova, L., and Vollmer, W. (2013). From models to pathogens: how much have we learned about *Streptococcus pneumoniae* cell division? Environmental microbiology *15*, 3133-3157. https://dx.doi.org/10.1111/1462-2920.12189.

Masson, S., Kern, T., Le Gouellec, A., Giustini, C., Simorre, J.P., Callow, P., Vernet, T., Gabel, F., and Zapun, A. (2009). Central domain of DivIB caps the C-terminal regions of the FtsL/DivIC coiled-coil rod. The Journal of biological chemistry 284, 27687-27700. https://dx.doi.org/10.1074/jbc.M109.019471.

Minnen, A., Attaiech, L., Thon, M., Gruber, S., and Veening, J.W. (2011). SMC is recruited to oriC by ParB and promotes chromosome segregation in *Streptococcus pneumoniae.* Molecular microbiology *81*, 676-688. https://dx.doi.org/10.1111/j.1365-2958.2011.07722.x.

Mitchell, T.J. (2003). The pathogenesis of streptococcal infections: from tooth decay to meningitis. Nature reviews microbiology *1*, 219-230. https://dx.doi.org/10.1038/nrmicro771.

Monahan, L.G., Liew, A.T., Bottomley, A.L., and Harry, E.J. (2014). Division site positioning in bacteria: one size does not fit all. Frontiers in microbiology *5*, 19. https://dx.doi.org/10.3389/fmicb.2014.00019.

Monds, R.D., Lee, T.K., Colavin, A., Ursell, T., Quan, S., Cooper, T.F., and Huang, K.C. (2014). Systematic perturbation of cytoskeletal function reveals a linear scaling relationship between cell geometry and fitness. Cell reports *9*, 1528-1537. https://dx.doi.org/10.1016/j.celrep.2014.10.040.

Morgenstein, R.M., Bratton, B.P., Nguyen, J.P., Ouzounov, N., Shaevitz, J.W., and Gitai, Z. (2015). RodZ links MreB to cell wall synthesis to mediate MreB rotation and robust morphogenesis. Proceedings of the National Academy of Sciences of the United States of America *112*, 12510-12515. https://dx.doi.org/10.1073/pnas.1509610112.

Morlot, C., Bayle, L., Jacq, M., Fleurie, A., Tourcier, G., Galisson, F., Vernet, T., Grangeasse, C., and Di Guilmi, A.M. (2013). Interaction of Penicillin-Binding Protein 2x and Ser/Thr protein kinase StkP, two key players in *Streptococcus pneumoniae* R6 morphogenesis. Molecular microbiology *90*, 88-102. https://dx.doi.org/10.1111/mmi.12348.

Morlot, C., Pernot, L., Le Gouellec, A., Di Guilmi, A.M., Vernet, T., Dideberg, O., and Dessen, A. (2005). Crystal structure of a peptidoglycan synthesis regulatory factor (PBP3) from *Streptococcus pneumoniae*. The Journal of biological chemistry *280*, 15984-15991. https://dx.doi.org/10.1074/jbc.M408446200.

Morlot, C., Noirclerc-Savoye, M., Zapun, A., Dideberg, O., and Vernet, T. (2004).

The D,D-carboxypeptidase PBP3 organizes the division process of *Streptococcus pneumoniae*. Molecular microbiology *51*, 1641-1648. https://dx.doi.org/10.1046/j.1365-2958.2003.03953.x

Morlot, C., Zapun, A., Dideberg, O., and Vernet, T. (2003). Growth and division of *Streptococcus pneumoniae*: localization of the high molecular weight penicillin-binding proteins during the cell cycle. Molecular microbiology *50*, 845-855. https://dx.doi.org/10.1046/j.1365-2958.2003.03767.x.

Mosyak, L., Zhang, Y., Glasfeld, E., Haney, S., Stahl, M., Seehra, J., and Somers, W.S. (2000). The bacterial cell-division protein ZipA and its interaction with an FtsZ fragment revealed by X-ray crystallography. The EMBO journal *19*, 3179-3191. https://dx.doi.org/10.1093/emboj/19.13.3179.

Mura, A., Fadda, D., Perez, A.J., Danforth, M.L., Musu, D., Rico, A.I., Krupka, M., Denapaite, D., Tsui, H.T., Winkler, M.E., *et al.* (2016). Roles of the essential protein FtsA in cell growth and division in *Streptococcus pneumoniae*. Journal of bacteriology. https://dx.doi.org/10.1128/jb.00608-16.

Ni, H., Fan, W., Li, C., Wu, Q., Hou, H., Hu, D., Zheng, F., Zhu, X., Wang, C., Cao, X., *et al.* (2018). *Streptococcus suis* DivIVA protein is a substrate of Ser/Thr kinase STK and involved in cell division regulation. Frontiers in cellular and infection microbiology *8*, 85. https://dx.doi.org/10.3389/fcimb.2018.00085.

Noirclerc-Savoye, M., Lantez, V., Signor, L., Philippe, J., Vernet, T., and Zapun, A. (2013). Reconstitution of membrane protein complexes involved in pneumococcal septal cell wall assembly. PLoS ONE *8*, e75522. https://dx.doi.org/10.1371/journal.pone.0075522.

Oldfield, E., and Feng, X. (2014). Resistance-resistant antibiotics. Trends in Pharmacological Sciences *35*, 664-674. https://dx.doi.org/10.1016/j.tips.2014.10.007.

Ortiz, C., Natale, P., Cueto, L., and Vicente, M. (2016). The keepers of the ring: regulators of FtsZ assembly. FEMS microbiology reviews *40*, 57-67. https://dx.doi.org/10.1093/femsre/fuv040.

Park, K.T., Villar, M.T., Artigues, A., and Lutkenhaus, J. (2017). MinE conformational dynamics regulate membrane binding, MinD interaction, and Min oscillation. Proceedings of the National Academy of Sciences of the United States of America *114*, 7497-7504. https://dx.doi.org/10.1073/pnas.1707385114.

Park, K.T., Wu, W., Battaile, K.P., Lovell, S., Holyoak, T., and Lutkenhaus, J. (2011). The Min oscillator uses MinD-dependent conformational changes in MinE to spatially regulate cytokinesis. Cell *146*, 396-407. https://dx.doi.org/10.1016/j.cell.2011.06.042.

Pazos, M., Natale, P., and Vicente, M. (2013). A specific role for the ZipA protein in cell division: stabilization of the FtsZ protein. The Journal of biological chemistry *288*, 3219-3226. https://dx.doi.org/10.1074/jbc.M112.434944.

Perez-Nunez, D., Briandet, R., David, B., Gautier, C., Renault, P., Hallet, B., Hols, P., Carballido-Lopez, R., and Guedon, E. (2011). A new morphogenesis pathway in bacteria: unbalanced activity of cell wall synthesis machineries leads to coccus-to-rod transition and filamentation in ovococci. Molecular microbiology *79*, 759-771. https://dx.doi.org/10.1111/j.1365-2958.2010.07483.x.

Peters, K., Schweizer, I., Beilharz, K., Stahlmann, C., Veening, J.W., Hakenbeck, R., and Denapaite, D. (2014). *Streptococcus pneumoniae* PBP2x mid-cell localization requires the C-terminal PASTA domains and is essential for cell shape maintenance. Molecular microbiology *92*, 733-755. https://dx.doi.org/10.1111/mmi.12588.

Philippe, J., Vernet, T., and Zapun, A. (2014). The elongation of ovococci. Microbial drug resistance (Larchmont, NY) *20*, 215-221. https://dx.doi.org/10.1089/mdr.2014.0032.

Pichoff, S., and Lutkenhaus, J. (2002). Unique and overlapping roles for ZipA and FtsA in septal ring assembly in *Escherichia coli*. The EMBO journal *21*, 685-693.

Pichoff, S., and Lutkenhaus, J. (2005). Tethering the Z ring to the membrane through a conserved membrane targeting sequence in FtsA. Molecular microbiology *55*, 1722-1734. https://dx.doi.org/10.1111/j.1365-2958.2005.04522.x.

Pinho, M.G., Kjos, M., and Veening, J.W. (2013). How to get (a)round: mechanisms controlling growth and division of coccoid bacteria. Nature reviews microbiology *11*, 601-614. https://dx.doi.org/10.1038/nrmicro3088.

Pompeo, F., Foulquier, E., Serrano, B., Grangeasse, C., and Galinier, A. (2015). Phosphorylation of the cell division protein GpsB regulates PrkC kinase activity through a negative feedback loop in *Bacillus subtilis*. Molecular microbiology *97*, 139-150. https://dx.doi.org/10.1111/mmi.13015.

Rismondo, J., Cleverley, R.M., Lane, H.V., Grosshennig, S., Steglich, A., Moller, L., Mannala, G.K., Hain, T., Lewis, R.J., and Halbedel, S. (2016). Structure of the bacterial cell division determinant GpsB and its interaction with penicillin-binding proteins. Molecular microbiology *99*, 978-998. https://dx.doi.org/10.1111/mmi.13279.

Rodrigues, C.D., and Harry, E.J. (2012). The Min system and nucleoid occlusion are not required for identifying the division site in *Bacillus subtilis* but ensure its efficient utilization. PLoS genetics *8*, e1002561. https://dx.doi.org/10.1371/journal.pgen.1002561.

Rowlett, V.W., and Margolin, W. (2015). The Min system and other nucleoid-independent regulators of Z ring positioning. Frontiers in microbiology *6*, 478. https://dx.doi.org/10.3389/fmicb.2015.00478.

Rued, B.E., Zheng, J.J., Mura, A., Tsui, H.T., Boersma, M.J., Mazny, J.L., Corona, F., Perez, A.J., Fadda, D., Doubravova, L., *et al.* (2017). Suppression and synthetic-lethal genetic relationships of DeltagpsB mutations indicate that GpsB mediates protein phosphorylation and penicillin-binding protein interactions in *Streptococcus pneumoniae* D39. Molecular microbiology *103*, 931-957. https://dx.doi.org/10.1111/mmi.13613.

Sass, P., and Brotz-Oesterhelt, H. (2013). Bacterial cell division as a target for new antibiotics. Current opinion in microbiology *16*, 522-530. https://dx.doi.org/10.1016/j.mib.2013.07.006.

Sass, P., Josten, M., Famulla, K., Schiffer, G., Sahl, H.G., Hamoen, L., and Brotz-Oesterhelt, H. (2011). Antibiotic acyldepsipeptides activate ClpP peptidase to degrade the cell division protein FtsZ. Proceedings of the National Academy of Sciences of the United States of America *108*, 17474-17479. https://dx.doi.org/10.1073/pnas.1110385108.

Sauvage, E., Kerff, F., Terrak, M., Ayala, J.A., and Charlier, P. (2008). The penicillin-binding proteins: structure and role in peptidoglycan biosynthesis.

FEMS microbiology reviews *32*, 234-258. https://dx.doi.org/10.1111/j.1574-6976.2008.00105.x.

Scheffers, D.J., de Wit, J.G., den Blaauwen, T., and Driessen, A.J. (2002). GTP hydrolysis of cell division protein FtsZ: evidence that the active site is formed by the association of monomers. Biochemistry *41*, 521-529. https://dx.doi.org/10.1021/bi011370i.

Schumacher, M.A., and Zeng, W. (2016). Structures of the nucleoid occlusion protein SlmA bound to DNA and the C-terminal domain of the cytoskeletal protein FtsZ. Proceedings of the National Academy of Sciences of the United States of America *113*, 4988-4993. https://dx.doi.org/10.1073/pnas.1602327113.

Shak, J.R., Vidal, J.E., and Klugman, K.P. (2013). Influence of bacterial interactions on pneumococcal colonization of the nasopharynx. Trends in microbiology *21*, 129-135. https://dx.doi.org/10.1016/j.tim.2012.11.005.

Sham, L.T., Barendt, S.M., Kopecky, K.E., and Winkler, M.E. (2011). Essential PcsB putative peptidoglycan hydrolase interacts with the essential FtsXSpn cell division protein in *Streptococcus pneumoniae* D39. Proceedings of the National Academy of Sciences of the United States of America *108*, E1061-1069. https://dx.doi.org/10.1073/pnas.1108323108.

Sham, L.T., Tsui, H.C., Land, A.D., Barendt, S.M., and Winkler, M.E. (2012).

Recent advances in pneumococcal peptidoglycan biosynthesis suggest new vaccine and antimicrobial targets. Current opinion in microbiology *15*, 194-203. https://dx.doi.org/10.1016/j.mib.2011.12.013.

Sham, L.T., Jensen, K.R., Bruce, K.E., and Winkler, M.E. (2013). Involvement of FtsE ATPase and FtsX extracellular loops 1 and 2 in FtsEX-PcsB complex function in cell division of *Streptococcus pneumoniae* D39. mBio 4. https://dx.doi.org/10.1128/mBio.00431-13.

Sharifzadeh, S., Boersma, M.J., Kocaoglu, O., Shokri, A., Brown, C.L., Shirley, J.D., Winkler, M.E., and Carlson, E.E. (2017). Novel Electrophilic Scaffold for Imaging of Essential Penicillin-Binding Proteins in *Streptococcus pneumoniae*. ACS chemical biology *12*, 2849-2857. https://dx.doi.org/10.1021/acschembio.7b00614.

Shiomi, D., and Margolin, W. (2007). Dimerization or oligomerization of the actin-like FtsA protein enhances the integrity of the cytokinetic Z ring. Molecular microbiology *66*, 1396-1415. https://dx.doi.org/10.1111/j.1365-2958.2007.05998.x.

Skoog, K., and Daley, D.O. (2012). The *Escherichia coli* cell division protein ZipA forms homodimers prior to association with FtsZ. Biochemistry *51*, 1407-1415. https://dx.doi.org/10.1021/bi2015647.

Stamsas, G.A., Straume, D., Ruud Winther, A., Kjos, M., Frantzen, C.A., and Havarstein, L.S. (2017). Identification of EloR (Spr1851) as a regulator of cell elongation in *Streptococcus pneumoniae*. Molecular microbiology *105*, 954-967. https://dx.doi.org/10.1111/mmi.13748.

Steele, V.R., Bottomley, A.L., Garcia-Lara, J., Kasturiarachchi, J., and Foster, S.J. (2011). Multiple essential roles for EzrA in cell division of *Staphylococcus aureus*. Molecular microbiology *80*, 542-555. https://dx.doi.org/10.1111/j.1365-2958.2011.07591.x.

Stock, A.M., Robinson, V.L., and Goudreau, P.N. (2000). Two-component signal transduction. Annual review of biochemistry *69*, 183-215. https://dx.doi.org/10.1146/annurev.biochem.69.1.183.

Straume, D., Stamsas, G.A., Berg, K.H., Salehian, Z., and Havarstein, L.S. (2017). Identification of pneumococcal proteins that are functionally linked to penicillin-binding protein 2b (PBP2b). Molecular microbiology *103*, 99-116. https://dx.doi.org/10.1111/mmi.13543.

Szwedziak, P., Wang, Q., Freund, S.M., and Lowe, J. (2012). FtsA forms actin-like protofilaments. The EMBO journal *31*, 2249-2260. https://dx.doi.org/10.1038/emboj.2012.76.

Thanassi, J.A., Hartman-Neumann, S.L., Dougherty, T.J., Dougherty, B.A., and

Pucci, M.J. (2002). Identification of 113 conserved essential genes using a high-throughput gene disruption system in *Streptococcus pneumoniae*. Nucleic acids research *30*, 3152-3162.

Thibessard, A., Fernandez, A., Gintz, B., Leblond-Bourget, N., and Decaris, B. (2002). Effects of rodA and pbp2b disruption on cell morphology and oxidative stress response of *Streptococcus thermophilus* CNRZ368. Journal of bacteriology *184*, 2821-2826. https://dx.doi.org/10.1128/jb.184.10.2821-2826.2002.

Tonthat, N.K., Milam, S.L., Chinnam, N., Whitfill, T., Margolin, W., and Schumacher, M.A. (2013). SlmA forms a higher-order structure on DNA that inhibits cytokinetic Z-ring formation over the nucleoid. Proceedings of the National Academy of Sciences of the United States of America *110*, 10586-10591. https://dx.doi.org/10.1073/pnas.1221036110.

Treuner-Lange, A., Aguiluz, K., van der Does, C., Gomez-Santos, N., Harms, A., Schumacher, D., Lenz, P., Hoppert, M., Kahnt, J., Munoz-Dorado, J., *et al.* (2013). PomZ, a ParA-like protein, regulates Z-ring formation and cell division in *Myxococcus xanthus*. Molecular microbiology *87*, 235-253. https://dx.doi.org/10.1111/mmi.12094.

Tsui, H.C., Zheng, J.J., Magallon, A.N., Ryan, J.D., Yunck, R., Rued, B.E.,

Bernhardt, T.G., and Winkler, M.E. (2016). Suppression of a deletion mutation in the gene encoding essential PBP2b reveals a new lytic transglycosylase involved in peripheral peptidoglycan synthesis in *Streptococcus pneumoniae* D39. Molecular microbiology *100*, 1039-1065. https://dx.doi.org/10.1111/mmi.13366.

Tsui, H.T., Boersma, M.J., Vella, S.A., Kocaoglu, O., Kuru, E., Peceny, J.K., Carlson, E.E., VanNieuwenhze, M.S., Brun, Y.V., Shaw, S.L., *et al.* (2014). Pbp2x localizes separately from Pbp2b and other peptidoglycan synthesis proteins during later stages of cell division of *Streptococcus pneumoniae* D39. Molecular microbiology *94*, 21-40. https://dx.doi.org/10.1111/mmi.12745.

Typas, A., Banzhaf, M., Gross, C.A., and Vollmer, W. (2011). From the regulation of peptidoglycan synthesis to bacterial growth and morphology. Nature reviews microbiology *10*, 123-136. https://dx.doi.org/10.1038/nrmicro2677.

van Baarle, S., and Bramkamp, M. (2010). The MinCDJ system in *Bacillus subtilis* prevents minicell formation by promoting divisome disassembly. PLoS one *5*, e9850. https://dx.doi.org/10.1371/journal.pone.0009850.

van Raaphorst, R., Kjos, M., and Veening, J.W. (2017). Chromosome segregation drives division site selection in *Streptococcus pneumoniae*. Proceedings of the National Academy of Sciences of the United States of

America *114*, E5959-e5968. https://dx.doi.org/10.1073/pnas.1620608114.

Vollmer, W. (2008). Targeting the bacterial Z-ring. Chemistry & biology *15*, 93-94. https://dx.doi.org/10.1016/j.chembiol.2008.02.001.

Wen, Z.T., Bitoun, J.P., and Liao, S. (2015). PBP1a-deficiency causes major defects in cell division, growth and biofilm formation by *Streptococcus mutans*. PLoS ONE *10*, e0124319. https://dx.doi.org/10.1371/journal.pone.0124319.

Wheeler, R., Mesnage, S., Boneca, I.G., Hobbs, J.K., and Foster, S.J. (2011). Super-resolution microscopy reveals cell wall dynamics and peptidoglycan architecture in ovococcal bacteria. Molecular microbiology *82*, 1096-1109. https://dx.doi.org/10.1111/j.1365-2958.2011.07871.x.

Willemse, J., Borst, J.W., de Waal, E., Bisseling, T., and van Wezel, G.P. (2011). Positive control of cell division: FtsZ is recruited by SsgB during sporulation of Streptomyces. Genes & development *25*, 89-99. https://dx.doi.org/10.1101/gad.600211.

Willis, L., and Huang, K.C. (2017). Sizing up the bacterial cell cycle. Nature reviews microbiology *15*, 606-620. https://dx.doi.org/10.1038/nrmicro.2017.79.

Wu, L.J., and Errington, J. (2011). Nucleoid occlusion and bacterial cell division. Nature reviews microbiology *10*, 8-12. https://dx.doi.org/10.1038/nrmicro2671.

Xu, P., Ge, X., Chen, L., Wang, X., Dou, Y., Xu, J.Z., Patel, J.R., Stone, V., Trinh,

M., Evans, K., *et al.* (2011). Genome-wide essential gene identification in *Streptococcus sanguinis*. Scientific reports *1*, 125. https://dx.doi.org/10.1038/srep00125.

Zapun, A., Vernet, T., and Pinho, M.G. (2008). The different shapes of cocci. FEMS microbiology reviews *32*, 345-360. https://dx.doi.org/10.1111/j.1574-6976.2007.00098.x.

Zhang, C., Sun, W., Tan, M., Dong, M., Liu, W., Gao, T., Li, L., Xu, Z., and Zhou, R. (2017). The Eukaryote-Like Serine/Threonine Kinase STK Regulates the Growth and Metabolism of Zoonotic *Streptococcus suis*. Frontiers in cellular and infection microbiology *7*, 66. https://dx.doi.org/10.3389/fcimb.2017.00066.

Zhang, Y., Ma, W., Zhu, Y., Shi, S., Li, Q., Mao, C., Zhao, D., Zhan, Y., Shi, J., Li, W., *et al.* (2018). Inhibiting Methicillin-Resistant *Staphylococcus aureus* by Tetrahedral DNA Nanostructure-Enabled Antisense Peptide Nucleic Acid Delivery. Nano Letter https://dx.doi.org/10.1021/acs.nanolett.8b02166.

Zhao, H., Patel, V., Helmann, J.D., and Dorr, T. (2017). Don't let sleeping dogmas lie: new views of peptidoglycan synthesis and its regulation. Molecular microbiology *106*, 847-860. https://dx.doi.org/10.1111/mmi.13853.

Zucchini, L., Mercy, C., Garcia, P.S., Cluzel, C., Gueguen-Chaignon, V., Galisson, F., Freton, C., Guiral, S., Brochier-Armanet, C., Gouet, P., *et al.* (2017). PASTA repeats of the protein kinase StkP interconnect cell constriction and separation of *Streptococcus pneumoniae*. Nature microbiology. https://dx.doi.org/10.1038/s41564-017-0069-3

https://doi.org/10.21775/9781912530229.08

Application of Omics and Bioinformatics Tools in *Streptococcus* Research

Min Liao[1,2#], Ting Tong[1,2#], Yawen Zong[1,2#], Xuedong Zhou[1], Lei Cheng[1,2], Ruijie Huang[1,3*], Biao Ren[1*], Gil Alterovitz[4]

1. State Key Laboratory of Oral Diseases and National Clinical Research Center for Oral Diseases, West China Hospital of Stomatology, Sichuan University
2. Department of Operative Dentistry and Endodontics, West China Hospital of Stomatology, Sichuan University
3. Department of Pediatric Dentistry, West China School of Stomatology, Sichuan University.
4. Harvard Medical School, Boston Children's Hospital, Boston, MA, USA

#: These authors contributed equally to this work.

*Corresponding authors

Ruijie Huang: hwangrj@163.com

Biao Ren: renbiao@scu.edu.cn

Abstract

Researchers used to focus on analyzing single gene or protein expression of the microbes. But recently, genome, transcriptome, proteome and metabolome have gained more and more attention. Based on technologies of omics, including genomics, transcriptomics and metabolomics, a large quantity of information about cells, microbes and human, such as the information about phylogeny, virulence, antibiotic resistance and other aspects, has been revealed. Genus *Streptococcus* is one of the most invasive groups of bacteria that cause both human and animal diseases, threatening public health. In this review, we summarize the application of omics to analyze this genus-*Streptococcus*.

Introduction

Over the past century, biotechnology has developed rapidly and provided possibilities to determine the function of micro composition of life and describe life phenomena in a micro or macro perspective. Omics, which is a multiple biotechnology based on the genomics, focuses on global and high-throughput analytical methods to investigate the mechanisms of life evolution and the role of genes (Yan et al., 2015). Saving genomics, omics includes proteomics, transcriptomics, and metabolomics. Meanwhile, a large quantity of information, such as the information about phylogeny, virulence, antibiotic resistance and other aspects of microbes, has been revealed about cells, microbes, and human in the past two decades. With abundant data and increasing accuracy, omics is applied in many life science aspects including microorganism. Genus *Streptococcus* is one of the most invasive groups of bacteria that causes both human and animal's disease, threatening human's health, especially *Streptococcus pneumoniae*, *Streptococcus pyogenes*, *Streptococcus agalactiae*, *Streptococcus mutans* and *Streptococcus suis*. Taking this genus as the research object, omics and bioinformatics tools provide a systematic path to cognize the law of biological activities of microorganisms.

In 1995, the first complete genomic sequence, as also the first completed bacterial genome sequence, was obtained from *Haemophilus influenzae* Rd (Fleischmann et al., 1995). Subsequently, researchers have completed sequencing of hundreds of microbes. With the improvement of sequencing technology and accuracy of computer model analysis, omics enables us to understand the mechanism of molecular function and interaction. As for *Streptococcus*, the known genome of Streptococcal strains varies in size from about 0.36 to 9.62 Mbp. The genus *Streptococcus* possesses relatively low GC content, which may attribute to mutation and various selections (Gao et al., 2014). Related to genomics, proteomics mainly focuses on the bacterial surface proteome, a major source for identifying vaccine targets and relating to serological assays. For example, Jimenez-Munguia et al. (2015) found that a kind of cell wall surface anchor family protein was significantly lower in pneumococcal pneumonia children compared to those in controls. However, Kreth et al. (2011) found that the balance of transcriptome expression of *Streptococcus* was regulated by special genes.

Overall, omics technology has been applied to many fields of microbe investigation, such as molecular function, identification and immune response to drugs. Simultaneously, the development of omics has advanced the development of researches on the origin of life, evolution, disease occurrence and drug industry. However, there are still plentiful challenges, such as the need of standard database and convenient analysis software, systematic research processes and macroscopic application. In this review, we summarize the systematic application of omics technology to analyze genus-*Streptococcus* (Figure 1).

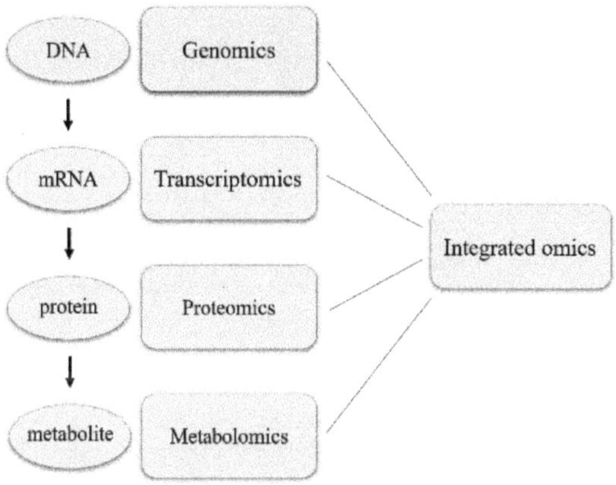

Figure 1. Schematic diagram of different omics tools targeting different molecular layers of *Streptococcus*.

Genomics

Genomics, a branch of genetics, is a popular subject based on whole gene sequencing (WGS) technology and bioinformatic tools. Since the first complete genome of free-living organism, *Haemophilus influenzae* Rd, was sequenced in 1995 (Fleischmann et al., 1995), DNA sequencing technology has experienced a rapid development over the next few decades. In 2005, a profound revolution had been brought about by a new technology in this filed, namely "next-generation sequencing (NGS)", also called high-throughput sequencing, which made genomic sequencing much more accessible with relatively lower expense and higher efficiency since then. Nowadays, many new companies and equipments in this field have sprung up, including Illumina, Oxford Nanopore, Pacific Biosciences, and etc. (Levy and Myers, 2016). They can not only provide genome organization of a selected subject, but also make a comprehensive analysis based on the sequencing data. Thanks to it, genomics can be widely implemented to explore the secret of different lives on earth, including microorganisms, human, plants, and etc.

Genome Characteristics of genus Streptococcus

With the development of sequencing, new whole genomes are sequenced completely and reported almost weekly with higher speed and lower cost. Today, the completed genomes of nearly 12,500 Streptococcal strains can be found at the National Center for Biotechnology Information (NCBI) website, with frequent data update. The genome of Streptococcal strains varies in size from about 0.36 to 9.62 Mbp. The genus *Streptococcus* possesses relatively low GC content, which may attribute to mutation and various selection (Gao et al., 2014). The pan-genome (all genes present in the genomes of the genus) size increases with the addition of newly sequenced strains, indicating that this genus has an open pan-genome. Conversely, the core-gene size (common genes shared by all Streptococcal strains) size tends to reduce (Gao et al., 2014; Liu et al., 2013b).

Comparative genomics

With an increasing number of sequenced Streptococcal whole genomes, comparative genomics is widely conducted to illustrate the functions of genes by comparing the genome of interested isolates with the reference genomes available in public database and using bioinformatic tools. Through comparing intraspecies variations or closely related species variations (Gao et al., 2014), it can provide us with comprehensive understanding of this genus and its connections between distinct phenotype and genetic characteristics, including the evolutionary relationship of *Streptococcus*, virulence determinants, as well as the genetic basis underlying adaptive changes to selective pressure and different hosts. Analysis of genome sequences reveal the dynamic nature of genetic material, and existence of multiple genetic events, such as genome decay, genome rearrangement and acquisition of new genes by horizontal gene transfer (HGT) (Fraser-Liggett, 2005; Lefébure and Stanhope, 2007; Muzzi and Donati, 2011), which have a profound influence on the evolutionary process and decide the phenotype to some extent.

Environmental adaptation

The environment that bacteria live in has a strong impact on the bacterial evolution over time, and genomics tools can help to investigate the genetic basis of their adaptation to habitat (Papadimitriou et al., 2014; Prajapati, 2013). A representative example of environmental adaptation is about the dairy starter strain and the nutritious dairy they live in. For instance, *Streptococcus thermophilus*, a common dairy starter with outstanding fermentation capacity, is widely used in dairy industry. The mechanism of fermentation property and food safety attracts significant attention and has been explored genetically. *S. thermophilus*, as a member of salivarius group, shows certain differences in genomic characteristics compared with other two genetically similar group members, *Streptococcus salivarius* and *Streptococcus vestibularis*. The *S. thermophilus* has a smaller genome size of about 1.7Mb and a relatively low level of nucleotide variability (Delorme et al., 2015). Notably, most Streptococcal virulence-associated genes are either presented as pseudogenes or missing in this species (Bolotin et al., 2004; Prajapati, 2013), suggesting that it is nonpathogenic. For instance, antibiotic resistance is an importance character of pathogenicity, but there is no antibiotic resistance genes detected in *S. thermophilus* genome. Similarly, genome decay and absence of classical Streptococcal virulent factors are detected in *Streptococcus infantarius* subsp. *infantarius* and the *Streptococcus macedonicus* isolated from fermented dairy (Jans et al., 2013; Papadimitriou et al., 2014). It seems that narrow and nutritious dairy habitat induces Streptococcal genome to experience a reductive evolution (Delorme et al., 2015; Papadimitriou et al., 2014). Meanwhile, *S. thermophilus* possesses some functional genes that seem to be beneficial to human when the bacteria adapt to milk. This species is reported to possess encoding genes for probiotic features, like adhesion, acid resistance, producing bacteriocin and utilizing lactose (Prajapati, 2013). Recently, an isolate that has a novel eps gene cluster can produce the highest level of exopolysaccharide (EPS) among this species and produce both capsular and ropy EPS, making the viscosity and texture of dairy better (Wu et al., 2014). Four separate clustered regularly interspaced short palindromic repeats (CRISPR)/Cas loci found in its genome enable it to resist unwanted bacteriophage infection. These researches support the use of *S. thermophilus* in manufacture of yogurt and cheese and probably wider use in biotechnological applications (Wu et al., 2014).

Conversely, the adaption to environment of pathogens always involves in complex regulation of virulence rather than merely attenuation of virulence. Lefebure et al. (2012) compared genomes of *S. pyogenes*, a common human pathogen, with several closely related species, *Streptococcus dysgalactiae*, *Streptococcus equi*, and *Streptococcus canis*. It turned out that the *S. pyogenes* gained 113 genes during evolution and about 14 genes were related to virulence, including *sal* antibiotic locus, *speB* gene, *rgg* gene, and etc. The *rgg* can regulate the expression of *speB* and interact with it, and this interaction may be beneficial to *S. pyogenes* colonization in human pharynx without dispersing to cause invasive diseases (Dmitriev et al., 2006). Both the acquisition of virulence genes and the establishment of regulation network contribute to the bacterial adaption to human host (Lefebure et al., 2012).

Virulence factors

Many species of genus *Streptococcus* are critical pathogens in both human and animal diseases, and their virulence plays an important role in infection. Through comparing the genome of interested strains with the annotated sequences in public database by using bioinformatics tools, a number of virulence genes may be detected in the interested strains. The identified conserved Streptococcal virulence determinates are mainly involved in adhesion, invasion, spreading, and other related regulatory genes, which can be inferred in virulence-related public database, such as Virulence Factors of the Pathogenic Bacteria Database (Gao et al., 2014; Olson et al., 2013). Some other genetic elements derived from HGT also belong to Streptococcal virulence. For instance, 12 characterized prophages in *S. suis* probably contribute to the pathogenicity of *S. suis* for carrying putative virulence-related genes, like toxin–antitoxin system, Clp protease and DNA methyltransferase (Tang et al., 2013). A virulent *Streptococcus zooepidemicus* strain, Sz35246, is reported to acquire four pathogenicity islands (PAIs) through HGT, including three toxin-antitoxin (TA) systems PAIs and a restriction modification system (RM system) PAI (Ma et al., 2013).

A recent research compared genomes of 3615 M1 Group A Streptococcus (GAS) strains isolated within about the last 100 years all around the world, revealing that the contemporary epidemic strain experienced a series genetic evolutionary events to form a high virulent strain. The precursor cell just contained one phage carrying an extracellular DNase gene, and then it acquired a phage encoding SpeA1. After that, it experienced a single-nucleotide change to form SpeA2 variant. Next, the species obtained a 36-kb region encoding secreted toxins NAD+-glycohydrolase and streptolysin O through HGT, shaping the high-virulent strain, MGAS5005-like M1 strains that spread worldwide recently (Nasser, 2014). However, Fiebig et al. (2015) found that the hypervirulence of *S. pyogenes* M1 is not restricted to the MGAS5005-like genotype. They compared a MGAS5005-like strain, 5884, with AP1. Although the AP1 is not MGAS5005-like strain, it can cause systemic infection and a quick death in infectious mouse (Oehmcke et al., 2009; Oehmcke et al., 2013). Compared to the 5448 strain, the higher virulent strain AP1 had its CRISPR-Cas system destroyed by the phage insertion, which attenuated its defense capacity against intrusive genetic element and caused an enrichment of phages. Besides, the AP1 genome has an additional protein H encoding gene and a mutation of *rofA* gene, which may influence its regulatory networks and virulence, but whether these characteristics contribute to its virulence deserves further researches (Fiebig et al., 2015). All these results suggest that the virulence determinants are dynamics and influenced extensively by the gene gain, gene loss and genetic recombination events.

Genomics analysis can be used to compare genome sequence focusing on different virulence determinants and may explain the genetic basis of distinct virulence and pathogenesis of different strains to some extent. However, this does not work every time. Sometimes, different clinical outcomes cannot be explained by the bacterial genetic differences associated with virulence. For instance, no clear difference of gene content was detected from The Streptococcus Anginosus Group (SAG) strains isolated from both respiratory and invasive infections (Olson et al., 2013). Likewise, Argimon et al. (2014) compared the genome of *S. mutans* from severe early childhood caries (S-ECC) and caries-free (CF) child using a new genome-scale in silico subtractive hybridization. Although they quickly identified the genetic loci unique to strains associated with S-ECC, they failed to find any specific genes related to caries, suggesting that there is no virulence determinants difference of *S. mutans* from S-ECC or SF children responsible for the distinct clinical outcomes. Moreover, Hossain et al. (2015) compared genomes of virulent and avirulent *Streptococcus uberis* strains and found that the potential virulence genes are detected in both strains and highly variable with high level of SNPs, indicating that the presence of virulence genes merely is hardly to explain the virulence capacity and infection status. There is a more complex dynamic that determines the infection status and clinical outcomes, and this is more than one simple genetic component. More comprehensive analysis of multiple factors need to be carried out, including genetic transcription, protein translation, environmental effect, interaction of various microorganisms, and etc.

Antibiotic resistance

With the increasing use of antibiotics, many strains that are susceptible to antibiotics primitively become resistant and widespread, making it difficult to control of the epidemics of infection. A bacterial resistance surveillance in USA shows that the *S. pneumoniae* isolated from patient infection site shows increased rates of insusceptibility to several antibiotics, including amoxicillin/clavulanate, erythromycin, and levofloxacin (Pfaller et al., 2012). Analysis of bacterial genome sequences enables us to identify the putative antibiotic resistance genes and may provide guidance for clinical treatment. Several antibiotic resistance gene databases have been established, including ARGO, MvirDB, ARDB, Resfinder, CARD and ARG-ANNOT, of which the last three are relatively detailed and comprehensive. The powerful and versatile ARG-ANNOT, released in 2013, contains more than 1600 antibiotic resistance genes and is widely used in bacterial drug resistance researches (Gupta et al., 2014). Metcalf et al. (2017) used short-read whole genome sequencing combined with ARG-ANNOT and ResFinder databases to detect multiple resistance determinants in invasive group B Streptococci (iGBS), and this was highly accurate, economical and practical for large-scale analysis compared to phenotypic testing.

Bacteria can obtain resistance genes mainly through mutations and HGT from other microbes over the course of evolution, especially under the selection pressure of antibiotics (Blair et al., 2015; Huddleston, 2014). The main types of HGT include conjugation, natural transformation and transduction (Huddleston, 2014). *S. pneumoniae* is a common pathogen that shows high rate of recombination. The Molecular Epidemiology Network clone 1 (PMEN1), a multi-drug resistant *S. pneumoniae* lineage, is reported to possess multiple genetic elements related to antibiotic resistance, including integrative and conjugative element (ICE), plasmids and other element with resistance genes (Croucher et al., 2009). Croucher et al. (2011) analyzed 240 sequenced strains of this lineage from all over the world and found distinguished different ways of resistance genes acquisition. The fluoroquinolone and rifampicin resistance mainly resulted from gene mutation, while resistance to non-β-lactam derived from recombination. Three different genetic elements were integrated into the Tn916 transposon, including Tn917 transposon (containing an *ermB* gene), mega element (containing a *mef/mel* gene) and omega element (containing an *ermB* gene and an aminoglycoside phosphotransferase). Those elements encode products involved in methylation of target ribosomal RNA, drug efflux pump system, and enabling the PMEN1 insensitive to macrolide. These elements are acquired by the strains across the established phylogeny more than once, and the macrolide-sensitive strains are replaced by the insusceptible strains, suggesting a genetic plasticity of this lineage during adaption to the selection pressure of clinical intervention.

Phylogeny

Phylogenetic analysis uses substantial genomic data and various analysis software to build a phylogenetic tree, describing the phylogenetic relationships of target species visually. In general, phylogenetic tree is built based on certain genetic loci, such as housekeeping genes or some genes of interest (Olson et al., 2013; Teng et al., 2014) as well as multiple genetic loci or even the whole genome sequences. The tree built on single gene, such as virulence genes, and this can throw light on the evolutionary history of the specific factor at species or genus level (Argimon et al., 2013; Gao et al., 2014). Glucosyltransferases (Gtfs) is an important enzyme in *Streptococcus* to catalyze the synthesis of glucans. The GtfB catalysate, a type of water-insoluble glucans (WIG), is beneficial for the formation of dental plaque and the adhesion of bacteria. Thus, it influences the development of dental caries and periodontitis (Mattos-Graner et al., 2004; Munro et al., 1991). Researchers conducted phylogenetic analysis among the *gtfs* genes of strains isolated from oral cavity of human, chimpanzees and macaque monkeys. They found that the gene duplication of *gtfB* and *gtfC* and the divergency of primates took place almost at the same period. Besides, the divergency of viviparous occurred before the expression of *S. mutans gtfB* and *gtfC*, which regulates the WIG synthesis. This phenomenon indicates that the g*tfs* diversification precedes human, and it is unrelated to the changes in human diet (Argimon et al., 2013).

16S rRNA is a popular marker used in phylogenetic analysis, and the majority of Streptococcal species are clustered into six major clades based on the 16S rRNA gene phylogenetic tree accordingly. Those species are classified into six "species group", named: "Pyogenic", "Mitis", "Anginosus", "Bovis", "Mutans" and "Salivarius" (Gao et al., 2014; Richards et al., 2014; Stackebrandt et al., 2002). Yet, there are limitations in this analysis and classification. For instance, sometimes different species may share identical 16S rRNA sequence, like the *Streptococcus mitis* and *Streptococcus oralis* (Suzuki et al., 2005). This identical 16S rRNA sequence makes it difficult to distinguish different species, and besides, the exist of horizontal transfer of segments of the 16S rRNA genes among species of the anginosus group makes the result inaccurate (Schouls et al., 2003). Hence, an established database based on larger volume of molecular data is needed. Phylogenomic analysis, a preferable method, is widely adopted by scientists. Phylogenomic analysis is based on multiple locus and possesses advantages of discriminate different species over other single locus analysis. Olson et al. (2013) compared the phylogenetic trees based on single locus 16S rRNA, housekeeping genes (*rpoB*, *cpn60*) and in-house core-SNP pipeline, respectively. Although these trees have similar overall topology, the multilocus approaches acquired longer branch lengths and a greater degree of statistical certainty; conversely, the 16S rRNA possesses a lower discriminatory power. Moreover, the multilocus analysis is particularly suitable for defining location of some ambiguous species. For instance, *Streptococcus sinensis* turned out to be equally close to both the anginosus and mitis groups in the phylogenetic tree built on 16S rRNA gene (Teng et al., 2014). The accurate location of *S. sinensis* on the phylogenetic trees remains ambiguous merely according to the single locus analysis. Teng et al. (2014) used phylogenomic analysis based on concatenated genes and whole genomes, and at the same time, they also used matrix-assisted laser desorption ionization-time of flight mass spectrometry to analyze the phylogenetic clade of *S. sinensis*, *Streptococcus oligofermentans* and *Streptococcus cristatus*. By these approaches, they found those three species formed a distinct phylogenetic clade, thus they proposed a new group of the three species, named "sinensis group". These results indicate that the multilocus phylogenomic analysis plays an important role in the taxonomy and identification of *Streptococcus,* and the multilocus approach possesses distinct advantages over the single genetic locus approach.

The phylogenomic tools and the genomic data have also been extensively used in the genetic and evolutionary relationship investigation at both strain and species levels. Watanabe et al. (2013) used BEAST with a post-probabilistic approach to construct a phylogenetic tree based on whole genomes of *Streptococcus dysgalactiae* subsp. *equisimilis* (SDSE) strains and estimated the ages of the subclade, and the result revealed that the most recent common ancestor of the selected SDSE strains appeared 446 years ago. Meanwhile, the results of analysis based on a core set of genes after omitting the phage regions exhibit a remarkable consistency. A recently published global phylogenomic analysis of core genome identified distinct lineages among nonencapsulated *S. pneumoniae* (non-Ec-*Sp*) and encapsulated *S. pneumoniae* (Ec-*Sp*) strains. Most non-Ec-*Sp* isolates clustered to form a separate (deep-branching classic) lineage and few non-Ec-*Sp* strains sporadically joined in the lineage of Ec-*Sp* on the phylogenetic tree. Most isolates from this separate non-Ec-*Sp* lineage showed highly insensitive to antibiotics and possessed more mobile elements that might contribute to the increased ability of adherence, revealing that non-Ec-*Sp* isolates react differently under the selection pressure of clinical intervention, such as the use of vaccines and antimicrobial (Hilty et al., 2014).

Genome editing

CRISPR/Cas system was first characterized over a decade ago as an adaptive immune system in genomes of various bacteria and archaea to eliminate extrinsic plasmid or phage (Barrangou and Horvath, 2017). The host is able to integrate the invading genes on its CRISPR loci as “spacer”, although the mechanism of incorporation remains unclear (Marraffini, 2016). Afterwards, this sequence is transcribed to RNA, namely small-interfering CRISPR RNA (crRNA). The crRNA hybridizes with the transactivating CRISPR RNA (tracrRNA) to form a guide RNA (gRNA), guiding the Cas nuclease to specifically cleave invading cognate sequences with the help of protospacer-adjacent motif (PAM) located downstream of the target sequence (Makarova et al., 2015; Steinert et al., 2015; Xu et al., 2015). To date, it has developed into a versatile and powerful tool widely applied to manipulate almost any genomes in biology, including yeast, human cells, plant, mice, drosophila, zebrafish, and etc. (DiCarlo et al., 2013; Gratz et al., 2013; Hwang et al., 2013; Li et al., 2013b; Muller et al., 2016; Sander and Joung, 2014). Realization of gene editing including specific gene knock-out as well as integration of desired genes requires introducing or expressing the Cas9 and engineered gRNA in the target cells or organisms. (Sander and Joung, 2014). The most frequently studied and used type of this system is the *S. pyogenes* CRISPR-Cas9 (SpCas9). It can generate sequence-specific double stand DNA break (DSB) at multiple sites more precisely, compared to other editing tools such as zinc-finger or TALE nucleases (Kleinstiver et al., 2015; Marraffini, 2016). CRISPR/Cas9 systems from different bacteria possess different Cas proteins of distinct sizes and various PAM requirements (Steinert et al., 2015). The other widely used Cas9 system is from *S. thermophilus*, which recognizes longer PAM sequences and shows relatively lower off-target mutagenesis compared to the SpCas9 system (Muller et al., 2016).

Although the CRISPR/Cas system has been widely and successfully used, the system still has drawbacks that need to be optimized, particularly, the high frequency off-target effect that may lead to DNA cleavage at wrong sites. A recently published research used T7 endonuclease I (T7EI) assay to unveil multiple off-target sites of the *VEGFA*, *EMX1* genes and its high frequency at genomic level in human U2OS.EGFP cells. The off-target cleavage can lead to various mutations, including insertions, deletion and point mutations, and the rate of mutations is reported to reach up to 125% (Cradick et al., 2013; Fu et al., 2013). Therefore, several approaches of reducing the off-target effect have been designed, including using cas9 nickase (Shen et al., 2014), truncating the sgRNA (Fu et al., 2014), constructing fusions of catalytically inactive Cas9 and FokI nuclease (fCas9) (Wishart, 2016), and etc. However, these methods can only solve part of the problem. Recently, researches constructed a SpCas9-HF1 variant, of which four Spcas9 residues (N497A, R661A, Q695A, and Q926A) were substituted by alanine. The SpCas9-HF1 was able to significantly minimized the genome-wide off-target effects to undetectable levels with the on-target activities remaining high compared to wide type, and this might provide a general approach to optimize other CRISPR-RNA-guide nucleases (Kleinstiver et al., 2016).

Metagenomics

Oral microenvironment is a pool of more than 700 species of microbes. It not only affects the oral health and diseases, but also affects systematic diseases, such as infective endocarditis, bacterial pneumonia, and etc. (Brown, 2007; Paik et al., 2005; Xu et al., 2007; Xu and Gunsolley, 2014). To gain insight in the microbial related diseases requires comprehensive study of entirety of microbes in a niche. Metagenomics has emerged in recent years and has been applied in analyzing the total genomes in a selected microenvironment (Figure 2). It is a preferred method of investigating the microbial community and its diversity, without isolating and cultivating the enormous organisms (Martínez-Porchas and Vargas-Albores, 2017). Belda-Ferre et al. (2012) analyzed the oral metagenome under different oral health conditions by 454 pyrosequencing. Interestingly, they found that *Streptococcus sanguinis* and *Aggregatibacter* were abundant in the caries-free individuals, while a plenty of *Streptococcus gordonii* and *Leptotrichia buccalis* were detected in the caries affected individuals. Besides, *S. mutans* was not prominent in the carious cavities, although it was thought to be the principle pathogenic factor of caries. Instead, there was an intricate bacterial community, indicating that the caries is a polymicrobial disease, and the *S. mutans* may trigger the carious and cooperate with other bacteria to promote the development of caries.

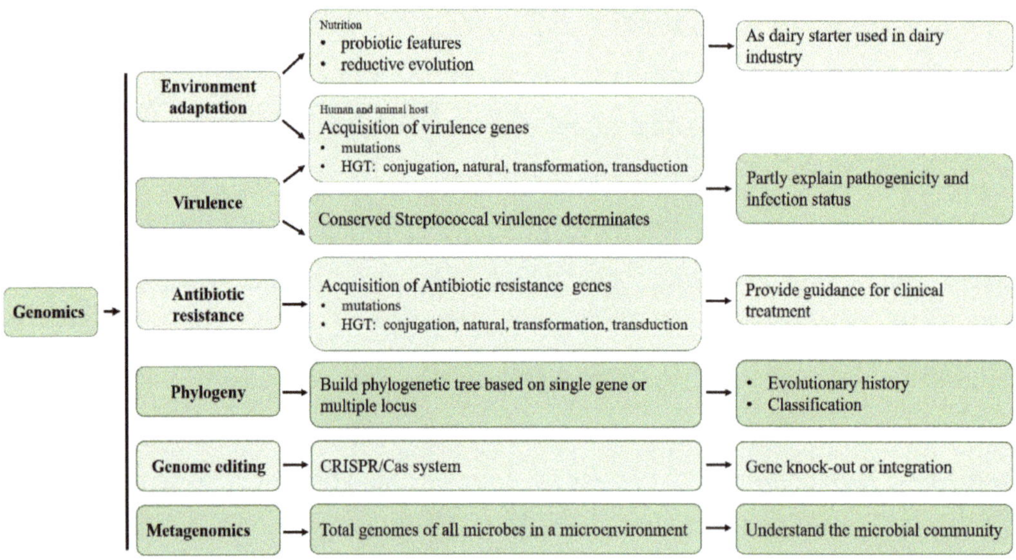

Figure 2. Schematic view of application of genomics in *Streptococcus*. By comparing with the genome of genetically similar group members in database, *S. thermophilus* shows a lack of most Streptococcal virulence-associated genes. Besides, it possesses encoding genes for probiotic features, enabling it to serve as diary starter. Genomics can also unveil the conserved virulence

genes and the putative virulence-associated genes acquired over the course of evolution, which may be attribute to the environment adaptation and partly explain the pathogenicity and infection status of host. Some genes acquired is related to antibiotic resistance, which might provide guidance for clinical treatment. Through phylogenetic analysis, a phylogenetic tree based on single gene or multiple genes can be made to illustrate the evolutionary history of the *Streptococcus* and contribute to the classification of some species. Moreover, the CRISPR/Cas system of *Streptococcus* is a powerful genome editing tool to realize gene knock-out or integration. As for a niche, the metagenomics can be used to analyze the total genomes of all microorganisms in the microenvironment and gain insight in the microbial community and its diversity.

Transcriptomics

Transcriptomics is a discipline that analyzes the transcription of genes in cells and the laws of transcription regulation on the whole. Briefly, transcriptomics studies the gene expression at the RNA level. Transcriptomics plays an important role in bridging genomics and proteomics. The transcriptome is defined as the sum total of all mRNA molecules of an organism (McGettigan, 2013). Since a human EST (expression sequence tag) database that retained 609 cDNA clones was firstly reported by a group from National Institutes of Health (NIH) in 1991 (Adams et al., 1991), transcriptomics, which has become an important method in the field of cell phenotype and function research, enables us to unveil the gene expression under a certain condition and provides us information about the strength of the gene expression relative to a reference (Afzal et al., 2015a).

For many years, researchers thought that the structure of bacterial transcriptome was less complicated compared to the structure of a eukaryote (Sharma and Vogel, 2014; Stazic and Voss, 2016). However, with the development of sequencing technologies, the interests in bacterial transcriptomics has increased recently due to the discovery of dense transcriptions on the bacterial chromosome, abundant small regulatory RNAs (sRNAs) and antisense transcripts (Sorek and Cossart, 2010; Thomason and Storz, 2010). Based on the next-generation sequencing technologies, the process of bacterial transcriptomics includes total RNA extraction, target RNA enrichment, library preparation and sequencing, reads mapping, analysis, and etc. It is acknowledged that microarray, containing abundant probes that can perform multiple genetic tests at the same time, is a reliable, rapid and comparatively economical method in bacterial transcriptomics (Afzal et al., 2015a; Conway and Schoolnik, 2003). However, compared with microarrays, the advantages of high-throughput RNA sequencing (RNA-seq) includes single base pair resolution, low background signal, a large dynamic range of expression levels, higher levels of reproducibility, smaller sample requirements for starting RNA and no limitation in detecting. Those advantages make it a powerful approach for assaying bacterial transcriptomes (McClure et al., 2013; Wang et al., 2009). Although it is still a developing technology, transcriptomics provides a novel approach of studying bacterial diversification at the genetic level.

Advances in Streptococcus transcriptomics

Virulence and Pathogenicity

In transcriptomic studies, a large number of genes have been shown to be involved in the expression of *Streptococcus* virulence and pathogenicity. For instance, in the absence of the two-component control of virulence regulator-sensor (CovRS) of Group A Streptococcus, an increased expression of genes associated with antiphagocytosis, toxin gene abrogation and destruction of some gene products was detected, unlike previous studies. Moreover, a genome-scale metabolic model was established and 16 nonredundant metabolic gene modules were found to be essential for pathogen growth (Yun-Juan et al., 2015). Meanwhile, Kreth et al. (2011) found that the global regulator CodY could control the regulation behavior of CovRS according to nutritional status, which indicated that there was a counteractive balance of transcriptome expression. Kajfasz et al. (2011) studied the mechanism of how the ClpXP proteolytic complex controlled the key virulence properties of *S. mutans,* and they demonstrated that genes encoding enzymes were required for the biosynthesis of intracellular polysaccharides and malolactic fermentation was upregulated. In addition, environment plays a key role in pathogenicity of *Streptococcus* as well. Pettigrew et al. (2014) pointed out that genes, associated with carbohydrate metabolism and production of bacteriocins, were upregulated. Besides, genes with colonization were downregulated. Those gene expression changes resulted in higher virulence when *S. pneumoniae* exposed to influenza A virus (IAV) infection. In the entire pathogenic process, Virtaneva et al. (2005) reported that there were some temporal changes in the Group A Streptococcus transcriptome. In particular, prophage virulence gene expression and prophage induction occurred predominantly when high pathogen cell densities and acute inflammation presented. As for the interaction with the host microenvironment, Geng et al. (2012) found that genes related to overall pathogenic capacity of *Streptococcus parasanguinis* expressed differently to enable bacteria to avoid host immune clearance, to colonize in host tissues and to survive within oral biofilms.

The main content of *Streptococcus* transcriptomics is centered on its virulence and pathogenicity, which are comprehensively affected by physiological characteristics such as competence and physiological development. Genes associated with those characteristics may influence the final characterization of bacteria, and usually, virulence and pathogenicity are mostly affected. The advances of these researches are as follows.

Competence

The competence is important for *Streptococcus* to survive and behave in the society. Referring to the regulatory links in the competence of *S. mutans*, Khan et al. (2016) found that the XIP-encoding gene *comS* determined the entire late response to competence-stimulating peptides (CSP) and the response regulator ComE determined the immediate transcriptional response to CSP. They also compared their data with published transcriptome data, permitting the identification of all of the operons in each regulon and identified a core set of 27 pan streptococcal competence genes within the SigX regulon. However, combined with the bacterial flow-cytometric sorting of cells, Lemme et al. (2011) concluded that the activation of bacteriocin-related genes was the main approach of cells' response to CSP, and then, some of these cells would be separated into two subpopulations of which only one would become competent.

Physiological development and biofilms

The different expression of transcriptome of *Streptococcus* in its different growth phases might be associated with its host infection capacity. Geng et al. (2012) indicated that genomic islands (GIs) and various open reading frames outside the GIs expressed differently in early exponential and early stationary growth phases, contributing to its pathogenic capacity. As for biofilm cells, transcriptome of *Streptococcus* affects its phenotype and response to microenvironment. Klein et al. (2010) reported that the expression of genes involved in sugar metabolism was mostly significantly affected in the *S. mutans* biofilm formation. And different expression of other related genes, such as genes associated with two-component systems, fermentation/glycolysis and iron transport, may lead to its pathogenicity. Liu et al. (2013a) created a hyperosmotic condition for *S. mutans* as an environmental stress and found that biofilm was dispersed due to *gtfB* and *comC* genes down-regulation and genes involved in carbohydrate metabolism up-regulation. In addition, the upregulation of some specific genes related to other environmental stress may suggest that there was a potential cross-talk between these stimuli. Wen et al. (2011) analyzed the growth of LuxS-deficient *S. mutans* in biofilm and found that some genes, such as those related to sugar-specific enzymes II of the phosphotransferase (PTS) system and the ATP operon, were upregulated and the ability of confronting environmental stress was compromised. These results may reveal the relationship between the select gene, LuxS, and biofilm formation and virulence of *S. mutans*. Polymicrobial biofilm is a more reality biofilm mode compared to monomicrobial biofilm in *Streptococcus* research. Transcriptomics plays an important role in interspecies interactions for their physiology and virulence. Sztajer et al. (2011) co-cultivated the *S. mutans* and *Candida albicans* and found the communication between the two pathogens. They used gas chromatography–mass spectrometry and transcriptome analysis and revealed a relatively lower level of EPS expression of *S. mutans* in dual-species biofilm. Activation of *sigX*, the alternative sigma factor of *S. mutans* induced by quorum sensing signals, was detected only when cultured in mixed biofilms or treated with conditioned media from co-cultivated-biofilms, not in the single-species biofilms or with culture supernatants from single-species biofilms.

These results demonstrate that *C. albicans* is able to stimulate the quorum sensing system of *S. mutans*, leading to a change of pathogenicity of *S. mutans*.

Novel gene function of *Streptococcus*

Transcriptomics makes it possible to discover new gene functions. For example, Yuzenkova et al. (2014) suggested that gene expression in *S. pneumoniae* may regulate transcription elongation. Afzal et al. (2015b) found that the transcriptional regulator NanR acted as a transcriptional activator of nan operon I and the *nanA* gene regulated the presence of sialic acid in *S. pneumoniae*.

Identification

In addition, transcriptomics, as an important part of bacterial omics, can also help to identify the genetic distance and reveal the extent of genetic relatedness between closely related strains. Park et al. (2012) proved this point of view by comparing transcriptomes of *Streptococcus pseudopneumoniae* with *S. pneumoniae*, *S. mitis* and *S. oralis*. With genes up- or down-regulation, this novel species was demonstrated to be closest related to *S. pneumoniae*.

Other aspects

Transcriptomics is also used in exploring the effect of some specific factors on *Streptococcus* transcriptome, such as metal ions (Manzoor et al., 2015), biologically active substances (Feng et al., 2013; Haase et al., 2015; Zeng and Burne, 2015), drug ingredients (Feng et al., 2011; Ferrándiz and de, 2014), and special environmental conditions (Kalpana et al., 2016; Richards et al., 2013; Sitkiewicz et al., 2009). In addition, transcriptomics was used in studying transcriptome changes of host immune organs after the infection of *Streptococcus* in animal models (Gaur et al., 2014; Sitkiewicz et al., 2009; Wu et al., 2010).

Macro transcriptomics of Streptococcus

In a microenvironment, transcriptomics of microbiome becomes a novel research direction. Taking oral microenvironment as an example, Peterson et al. (2014) used RNA-seq to analyze dental plaque microbiota and they found that 45%

transcription was derived from *Streptococcus*. And Benitez-Paez et al. (2014) found that expression of genes, which were associated with bacterial adhesion to tooth surface, changed significantly at the the early biofilm stage compared to the mature biofilm stage.

Transcriptomics enables us to understand bacterial physiological activities, competence, interaction with the host and other characteristics of *Streptococcus* at gene level. And it also provides novel clues of the relationship of bacterial genus and the microenvironment that bacteria live. RNA-seq becomes the most frequently used method in transcriptomics, although there are still problems in analyzing excessive data of bacterial genes. The future directions of new sequencing technologies are on the premise of ensuring safety to determine the transcriptome of a certain bacteria quickly and simply.

Proteomics

With the development of omics, proteomics technology is widely used and it plays an important role in microbiology research. Based on proteome, proteomics is applied in the studies of the composition and protein expressions of cells, tissues and organisms. Compared to genomics, proteomics provides more information about proteins, such as protein localization, interactions among proteins and the occurrence of post-translational modifications. Genes and transcripts are both important and useful in characterizing the activity of microorganisms, because they control the protein expression. However, protein studies provide more direct information of the functional activity of the cell, organism or community (Franzosa et al., 2015). Microbes are special organisms that they live in human, animal and the environment around us. Global analysis of microbial proteome has been widely used in investigating the pathogenesis of important human bacterial pathogens (Bittaye and Cash, 2015). Analyzing the proteome expressed by microbes is essential to demonstrate the relationship of microbes and human diseases, and the pathogenic mechanism of microbial virulence. Here, the proteomics researches of *Streptococcus* are summarized as below.

Streptococcus genus is one of the most invasive bacterial groups and has been identified as sources of invasive infections in human. *S. pneumoniae*, *S. pyogenes*, *S. agalactiae*, *S. mutans* and *S. suis* are considered to cause infections. Therefore, revealing the virulence of *Streptococcus* genus and the mutation of *Streptococcus* is a worldwide object for human's health. This requires more information about the pathogen, especially the proteome data (Bittaye and Cash, 2015; Olaya-Abril et al., 2014). As we know, penicillin and other antibiotics are widely used in treating infections of *Streptococcus*, meanwhile, the antibiotic resistance emerges worldwide and becomes a serious problem that threatens public health. The efficiency of vaccines and antibiotics are compromised because of the super virulent strains and multidrug resistant strains, such as *S. pneumoniae* mutant. Protein vaccines have many advantages compared to the current polysaccharide-based vaccines, which are serotype-dependent and subjected to losing efficacy because of serotype replacement and high manufacturing complexity (Olaya-Abril et al., 2013). Proteomics plays a vital role in the researches about microbes, such as *Streptococcus.*

Virulence of Streptococcus

Proteins are closely related to the virulence of microbes. Through analyzing different expression of proteins especially through proteomics, researchers could provide more information to treat or avoid the infection of microbes. For example, protein vaccines are higher immunogenicity and more cost-effective than polysaccharide-based vaccines (Donnarumma et al., 2016). Therefore, current researches focus on developing vaccine candidates of several surface-associated proteins that elicit protection against more *Streptococcus* strains, especially *Pneumoniae.* Searching for potential microbial therapeutic targets and providing microbial protein-makers are also the objectives. Through proteomics researches, such as functional and comparative proteomics, more and more proteins are revealed (Bogaert et al., 2004).

Functional proteomics related to *Streptococcus*

Streptococcus pneumoniae

As one of the most common kinds of pneumonia, bacterial pneumonia usually happens in children and the elderly, and it is responsible for high morbidity and mortality rates in both developed and developing countries. *S. pneumoniae* is a leading pathogen that causes invasive diseases such as pneumonia, even sepsis, meningitis, and etc. Current controlling strategies include the use of antibiotics for treating ongoing infections together with pneumococcal vaccines for the prevention of further pneumococcal infection (Bittaye and Cash, 2015; Bogaert et al., 2004; Olaya-Abril et al., 2014).

Two-dimensional gel electrophoresis (2DGE) has been frequently used in many researches to investigate the proteome of *S. pneumoniae*. Recent development in global proteome analysis by label-free liquid chromatography-tandem mass spectrometry (LC-MS/MS) -based technologies also helps to provide more information about many microbial proteomes. Studies on the global proteome of *S. pneumoniae* were very limited before, but with the effort of numerous scientists, extensive information about the bacterial surface proteome has been collected, providing a major source for identifying vaccine targets and relating to serological assays. On the bases of proteomic experimental identification followed by LC-MS/MS analysis, Jimenez-Munguia et al. (2015) selected 64 recombined pneumococcal proteins and finally found that acute serum IgG levels against RrgB, a kind of cell wall surface anchor family protein, were significantly lower in less than 4 years old children with pneumococcal pneumonia than those in controls. This result helps the scientists to design a multiplex bead-based platform to assess natural IgG antibodies against pneumococcal protein antigens in children. In addition, Olaya-Abril et al. (2013) identified potential new protein vaccines through pan-surfomic analysis of pneumococcal clinical isolates from adults. In this study, four novel surface proteins (Spr0012, Spr0328, Spr0561 and SP670_2141) are selected for vaccine test, because of their capacity to raise immune response in infected patients and induce an IgM response. IgM should be a priori and good indicator of antigen exposure.

Recently, extracellular vesicles have gained more attention. They are produced by pathogens with various microbial virulence factors. Therefore, synthesizing and releasing of extracellular vesicles derived from the outer surface are in favor of bacterial pathogens to combat the defense responses of invaded hosts. Based on proteomics and fatty acid analysis of *S. pneumoniae*, it has been demonstrated that these vesicles and the plasma membrane are similar in essential aspects, but there are some differences. The vesicles are more enriched in lipoproteins and short-chain fatty acids, and they act as vectors of surface proteins and virulence factors (Olaya-Abril et al., 2014). Studies on the composition of extracellular vesicles are essential to reveal the virulence of microbes and to provide information to explore vaccine candidates.

In order to cause infection, pathogens such as *S. pneumoniae* need to capture iron from the host environment. In Streptococcal species, there are three known iron-uptake systems responsible for iron acquisition from the host, the ABC transporters PiaABC, PiuABC and PitABC. Because of further analysis by the integrative transcriptomics and proteomics, rich information and insightful clues have been provided for the investigations of iron-transporting mechanism in bacteria, the interplay between Streptococcal iron availability and the biological metabolic pathways (Yang et al., 2016a). By using immuno-proteomic, N-acetylglucosamine-6-phosphate deacetylase (NagA) was identified in a protein mixture secreted by *S. pneumoniae* and its strong immunogenicity was confirmed. On account of high immunogenicity and immunospecificity, NagA may be a novel diagnostic marker for *S. pneumoniae* (Choi et al., 2013).

Streptococcus suis

S. suis is an emerging zoonotic pathogen that causes severe human infections, such as meningitis, septicemia, endocarditis and pneumonia. Among these strains, *S. suis* serotype 2 (SS2) infection is a leading cause of septicemia and meningitis in pigs and humans. And it is often accompanied by bacteremia, the infection of which has become increasingly potent, especially in the southeast Asian countries like Thailand, Vietnam and China. Fibrinogen-binding surface proteins of *S. suis* (SsFBPs) can bind to human fibrinogen (hFg) to help the microbe to survive in the host, to evade host immunity and to survive and disseminate in blood. By combining the results of proteomic and immune blot, novel FBPs of SS2 were identified, especially, two SsFBPs (MRP and Enolase) were identified as important FBPs of SS2 (Pian et al., 2015). Haas and Grenier (Haas and Grenier, 2015) applied analyzed proteomics of membrane vesicles (MVs), which range in diameter from 13 to 130 nm and appear to be coated by capsular material. They revealed that in the MVs there were 46 proteins, 9 of which are considered as proven or suspected virulence factors. The results implied that *S. suis* MVs may play a role as a virulence factor in the pathogenesis of *S. suis* infections. Zhou et al. (2015) demonstrated that substrate binding protein (Sbp) might play an important role in the pathogenesis of SS2. Identified by immunoproteomic method, four SS2 membrane associated proteins, including L-lactate dehydrogenase (Ldh), Dihydrolipoamide dehydrogenase (Dldh), Pyruvate dehydrogenase E1 component (Pec) and amino acid ABC Sbp, were cloned and expressed as recombinant proteins with Histag. These findings may provide foundations for future researches of SS2 pathogenic mechanisms and targets, provide foundations for excellent candidates for vaccine development, and are important for novel therapies for targeting SS2-induced sepsis.

Streptococcus pyogenes

S. pyogenes, a kind of human pathogens, can cause various infectious diseases, ranging from mild pharyngitis and impetigo to severe necrotizing fasciitis and autoimmune sequelae. Acidic stimulus and growth phase influence the secretion of *S. pyogenes*-associated virulence factors, which are ultimately associated with *S. pyogenes* invasive capability. Proteomics, electron microscopy and Fluorescence Activated Cell Sorting (FACS) analysis demonstrate that *S. pyogenes* can release lipoproteins (Lpps) in lipoprotein-rich membrane vesicles (LMVs), a vesicle-like structure, and reveal that Lpps are the major components of the LMVs. LMVs represent as an excellent vector for antigen delivery via their potential intrinsic adjuvanticity (Biagini et al., 2015). The "surface interactome" of *S. pyogenes* was characterized by protein array technology (Galeotti et al., 2012). The proteins were proved to get involved in protein folding and transportation.

The proteins are OppA, DppA, PrsA and TlpA, all of which are localized at the septum. Besides, scientists have detected interactions among the metal transporters AdcA, Lmb and SpeI, a major player in *S. pyogenes* pathogenesis. Those findings might indicate the way SpeI obtains the metal from the environment. Through applying selected reaction monitoring mass spectrometry (SRM-MS), a method to identify and quantify proteins with high sensitivity, specificity and high reproducibility, Karlsson et al. (2012) provided a proteome-wide selected reaction monitoring (SRM) assay repository resource for *S. pyogenes* to facilitate SRM-MS analysis for this bacterium.

Other *Streptococcus*

S. mutans is one of the major pathogens that causes dental caries. *S. mutans* and sucrose are responsible for the development of virulent-cariogenic biofilms. Klein et al. (2012) used high-throughput quantitative proteomics to examine relevant proteins that were produced by *S. mutans* to facilitate its growth and optimal survival during mixed-species biofilm development induced by sucrose. They found that the protein synthesis and gene expression were usually augmented when *S. mutans* grew in mixed-species biofilms compared to in single-species biofilms. The specific genes, which were upregulated, were associated with glucan synthesis and remodeling (gtfBC, dexA) and glucan binding (gbpB). This result indicated that the presence of other organisms, which were fundamentally different in the matrix assembly, survival and biofilm maintenance, increased the virulence of *S. mutans*.

S. thermophiles, as a starter for dairy productions, is widely used in food industry. In order to contribute to future investigations on protein complex assembly and composition or search for bacterial strains with specific biotechnological applications, Salzano et al. (2013) applied proteomic procedures to characterize the heteromultimeric and homomeric protein complexes from the membrane fraction of *S. thermophilus*. Moreover, researchers analyzed proteome of extracted cytoplasmic proteins from *S. sanguinis* and revealed O-Glycosylation of the N-terminal region of the serine-rich adhesin Srr1 of *S. agalactiae* by Mass Spectrometry (Chaze et al., 2014; El-Rami et al., 2017).

Comparative proteomics

Isobaric tagging for relative and absolute quantification (iTRAQ), a high-throughput gel-free proteomic approach, allows high resolution quantitative comparisons of protein profiles between multiple phenotypes. By iTRAQ, Allan et al. (2014) studied the growth and metabolic proteins of *S. pneumoniae*, a clinical serotype 14 strain, in planktonic and biofilm status to represent target candidates for future vaccine development. They found several ABC transporter system proteins were significantly up-regulated in pneumococcal biofilms. In one study, iTRAQ-based quantitative proteomics was also applied to analyze the protein alterations in *S. pneumoniae* in response to Sodium new houttuyfonate (SNH) treatment. It was proved that SNH killed *S. pneumoniae* in a dose-dependent manner accompanied by the increase of H_2O_2 (Yang et al., 2016b). Yang et al. (2015) performed 2-DGE-based proteomic analysis to characterize the protein alterations in *S. pneumoniae* after being treated with polypyridylruthenium (II) (Ru (II)) complex X-03, a metal compound. They found that it could obstruct bacterial fatty acid synthesis and oxidation-reduction process in order to suppress the growth of *S. pneumoniae* and interfere iron-acquisition pathway in the bacterium. Vancomycin is one of the most important drugs currently used in the treatment of gram-positive bacterial infections. Yang et al. (2017) applied 2-DGE and LM-MS/MS to analyze differentially expressed proteins (DEPs) of whole cellular proteins extracted from *S. pneumoniae* D39 with or without vancomycin treatment. They found that 27 proteins were upregulated and four proteins were downregulated in vancomycin-treated *S. pneumoniae*. These findings provide important information for future investigations of vancomycin tolerance mechanisms of *S. pneumoniae*. Through the iTRAQ strategy, Wang et al. (2016a) compared the protein expression profiles of *S. suis* growing in sub-MIC tylosin versus no tylosin, and finally 1501 proteins were identified. These proteins were associated with *S. suis* metabolic regulation and virulence, and appear to be important in biofilm growth at sub-MIC tylosin level. Through analyzing acid-influenced secretome of *S. pyogenes* in acidic and neutral conditions, it was demonstrated that acidic stimuli and growth-phase cues were crucial for classical protein secretion in *S. pyogenes*. Histidine triad protein A (HtpA), as a novel acid-induced virulence factor, was proved to improve the anti-phagocytosis ability for causing necrotizing fasciitis (Wen et al., 2014). By using immunoproteomics to analyze and compare the transcriptomes and proteomes of *S. pyogenes* biofilms, Freiberg et al. (2016) found that the *S. pyogenes in vivo* expressed proteome matches the proteome *in vitro* biofilm better, in compared with *in vitro* planktonic proteome.

With the help of different proteomics analysis, it was demonstrated that virulence factors of *S. mutant* expressed differently. By a 2DGE-based proteome reference map of the cytoplasmic and extracellular proteins from *S. mutans*, Li et al. (2013a) identified 239 protein spots that represented 192 different cytoplasmic proteins, which were regulated by the carolacton isolated from the myxobacterium *Sorangium cellulosum*. Finally, carolacton was found to inhibit *S. mutans* by disturbing the peptidoglycan biosynthesis and degradation, thus causing damages to the integrity of the cell envelope and disturbing *S. mutans* biofilm viability. Thereby this function makes carolacton a potential anti-biofilm drug.

Streptococcus coaggregates with *Porphyromonas gingivalis*, a major periodontal pathogen, and assists it with the establishment of subgingival biofilm. *S. gordonii*, *S. oralis*, *S. mutans* and *S. sanguinis* have been proved to interact with *P. gingivalis*. By using shotgun proteomics, Maeda et al. (2015) examined the molecular basis of mixed-culture biofilm formation by *P. gingivalis* with *S. oralis*. Compared to the profile in the respective monoculture biofilm, 31 proteins of *P. gingivalis* and two proteins of *S. oralis* were down-regulated in the mixed-culture biofilm. The Glyceraldehyde-3-phosphate dehydrogenase (GAPDH) of *S. oralis* (GapC) as well as the FimA of *P. gingivalis* were up-regulated in mixed-culture, which indicated that the GapC might interacted with FimA to promote the formation of mixed-culture biofilm. The housekeeping proteins GyrB and RpoD increased in the mixed-culture biofilm and reflected the activity or viability of bacterial cells.

Interaction between Streptococcus and other microbes/human

A recent study based on LC/Q-TOF mass spectrometry proteomics analyzed the changing protein profiles of *Helicobacter pylori* and *S. mitis* in mixed-culture biofilm, and the results indicated that the interaction between *H. pylori* and *S. mitis* may enhance bacterial survival rate and reduce carcinogenesis. The *S. mitis* in mixed-culture biofilm has a high-level expression of posphoglycerate kinase (PGK), a major enzyme involved in glycolysis, which is beneficial to the survival of the bacterium. Meanwhile, some oxidative stress related proteins of *H. pylori* decreased in the multi-species setting, including thioredoxin (TrxA) and other redox-regulating enzymes, and this was thought to be protective and less pathogenic (Khosravi et al., 2016). *S. gordonii* is a pioneer colonizer of the tooth surface. By 2-DE, Yoshida et al. (2015) analyzed the protein expression profiles of *S. mutans* in the biofilm, with or without the presence of *S. gordonii*. It was suggested that Dps-like peroxide resistant protein (Dpr) might be one of the essential proteins for *S. mutans* survival on teeth in the presence of early colonizing oral *streptococcus*.

Sjöholm et al. (2014) used a combination of MS based techniques to comprehensively quantify the components of the *S. pyogenes*-plasma protein interaction network. They subdivided the interacting protein list by using a SRM depending on the level of enrichment and protein concentration on the bacterial surface. And their results provided important information about preventing or treating the infection of *Streptococcus*.

Metabolomics

Since the metabolite is the ultimate output of the biological hierarchy and is more stable than mRNA or proteins (Patel and Ahmed, 2015), the process of metabolism and the metabolite is always directly related to the bacterial growth, pathogenicity, evading host immune system, interaction with other organisms, research of antibiotics, vaccine development, and etc., therefore, metabolite seems to be important for disease prevention and control (Fei et al., 2016; Schulz and Hammerschmidt, 2013; Takahashi et al., 2012; Wang et al., 2016b). For example, the acid produced by *S. mutans* and other bacteria in oral cavity plays a vital role in the development of caries (Takahashi and Nyvad, 2011). Metabolomics, serving as a complement of other omics, provide us with an integral view of the metabolism of the living cells and can be used to explore some key products and potential mechanisms by comparing metabolic differences under different circumstance (Schulz and Hammerschmidt, 2013; Tang, 2011).

Streptococcus iniae is an important zoonotic pathogen that causes infections in various fish including tilapia (Lau et al., 2006). Recently, a series of gas chromatography-mass spectrometer (GC/MS) based metabolomic researches revealed a regulatory mechanism of tilapia against *S. iniae* infection. The L-leucine in the survival fish infected by *S. iniae* was relatively higher than that in the dead fish, indicating the potential modulator role of L-leucine in the *S. iniae* infection. Furthermore, the addition of exogenous L-leucine can raise the survival capacity of infected tilapia, and this validated the result of metabolomic analysis (Ma et al., 2015). The GC/MS-based metabolomic analysis revealed the mechanism of how L-leucine works. Elevated level of L-serine triggers further up-regulation of the *IL-1β* and *IL-8* mRNA expression (Du et al., 2017). The metabolic profile of bacteria can also reflect the bacterial adaption changes by comparing the final outputs. *S. agalactiae* is a common pathogen that causes systemic infections in human and fish, and it is serum-resistant and capable to escape from the host immune system. By using GC/MS-based metabolomics analysis, a recent research illuminated a metabolic trick of *S. agalactiae* and explained the key pathways in response to plasma of yellow grouper, including an increase of adenosine and a decrease of malic acid that decreased the sensitivity to plasma (Wang et al., 2016b).

Drug discovery

Genomics, proteomics and metabolomics are the three fields conventionally and effectively applied in drug development and clinical application, especially when confronted with the increased antibiotic resistance. Genomics has been applied in multiple areas, including identification of biomarker (osteoarthritis), expression profiling of tissue, toxicogenomics, and etc. (Meulenbelt et al., 2011; Russell et al., 2013). The proteomics owns its unique advantages over genomics, because the proteomics reflected the cellular events after gene transcriptions. The employment of it can be found nearly throughout the pipeline of drug development, from target identification to clinical trial of drugs (Burbaum and Tobal, 2002; Russell et al., 2013). Metabolomics make it clear about the metabolic pathways and all of the products of both pathogens and the host, revealing how the pathogenic factors act and host defense responses, helping us to understand the mechanisms of diseases at the molecular level and enabling us to fight with them. There are multiple approaches implemented in the development of antibiotics against Streptococcal infections. The comparative genomics can be used to identify the essential genes for S. pneumoniae growth as potential target (Song et al., 2005). A more valid approach is the computational subtractive genomics analysis. The complete proteome of S. pneumonia is downscaling gradually to a small number, which are related to pathogenicity and involved in pathogen metabolic pathways that are essential for the survival of the bacteria. These selected proteins may be potential drug targets. Simultaneously, this approach also encompasses the use of BLAST tool to recognize the non-homologous against human genomes. Eventually, two candidates can guide new drug discovery and reduce the side effect on human (Wadood et al., 2018). A recent published research used an integrated proteomic and metabolomic analysis has evaluated the mechanism of how rhodomyrtone antagonizes S. pneumoniae. A significant down-regulate of two enzymes (glycosyltransferase and UTP-glucose-1-phosphate Uridylyltransferase) and three metabolites (uridine 5′-diphosphoglucuronic acid (UDP-glucuronic acid), UDP-glucose, and UDP-N-acetyl-D-Galactosamine) related to the pneumococcal capsule synthesis was observed in the S. pneumoniae stimulated by rhodomyrtone. This result, confirmed the antimicrobial activity and potential use of rhodomyrtone as antibiotic, thus, providing an alternative antibiotic for treatment against S. pneumoniae (Mitsuwan et al., 2017).

Integrated omics

Omics is a powerful tool used in covering genomics, transcriptomics, proteomics and metabolomics to gain insight of the biological features of *Streptococcus* as well as other creatures. Each of these omics tools can provide a relatively comprehensive understanding of the given object, but sometimes it is not enough to explain more complex phenomena based on a single method. For example, Argimon et al. (2014) used genomics tools and failed to find critical virulence genes of *S. mutans* to explain caries status. But a combination of multiple omics can provide ample evidence of genes, mRNA, proteins and other metabolites at the molecular level. For example, a transcriptomic analysis revealed an up regulation of salvage and de novo nucleotide synthesis pathway of *Streptococcus intermedius* in the aerobic culture compared to anaerobic environment, which was validated by the metabolomic results (Fei et al., 2016). The integration of different omics can imply the “missing link” between gene transcription, translation of protein, the following modification and other process, revealing the complicated regulatory mechanisms and networks behind certain phenotype. For example, the matrix metalloproteinases 2 (MMP2) mRNA expression of oral epithelia stimulated by *C. albicans* is up-regulated, but the MMP2 protein expression is down-regulated, which is remained confused (Claveau et al., 2004). An integrated omics may help to solve these problems.

A recent published research used an integrated proteomic and metabolomic analysis has revealed the mechanism of rhodomyrtone antibacterial activity against *S. pneumoniae.* A significant decrease of two enzymes (glycosyltransferase and UTP-glucose-1-phosphate Uridylyltransferase) and three metabolites (uridine 5′-diphosphoglucuronic acid (UDP-glucuronic acid), UDP-glucose, and UDP-N-acetyl-D-Galactosamine) related to the pneumococcal capsule synthesis was observed in the *S. pneumoniae* stimulated by rhodomyrtone. This result confirmed the antimicrobial activity and potential use of rhodomyrtone as an antibiotic at both transcription and protein levels (Mitsuwan et al., 2017).

The putative genes that regulate iron-uptake of *S. pneumoniae* are ATP-binding cassette (ABC) transporters, PiaABC, PiuABC, and PitABC. Those genes are critical to the growth and pathogenicity of the bacteria. Recently, SPD_0090 and SPD_1609 were identified as potential novel iron-bunding proteins by using translatomics and proteomics, and they were upregulated as compensation in the known ABC transporters mutant *S. pneumoniae* (Yang et al., 2016a). The integration of different omics can provide more detailed and comprehensive evidence to throw light on the potential mechanisms and pathways of certain phenotype. The progress in bioinformation and sequencing technology makes it more practical and economic to carry out integrated omics.

References

Adams, M.D., Kelley, J.M., Gocayne, J.D., Dubnick, M., Polymeropoulos, M.H., Xiao, H., Merril, C.R., Wu, A., Olde, B., Moreno, R.F., *et al.* (1991). Complementary DNA sequencing: expressed sequence tags and human genome project. *Science 252*, 1651-1656.

Afzal, M., Manzoor, I., and Kuipers, O.P. (2015a). A Fast and Reliable Pipeline for Bacterial Transcriptome Analysis Case study: Serine-dependent Gene Regulation in Streptococcus pneumoniae. *J Vis Exp*. doi:10.3791/52649

Afzal, M., Shafeeq, S., Ahmed, H., and Kuipers, O.P. (2015b). Sialic acid-mediated gene expression in Streptococcus pneumoniae and role of NanR as a transcriptional activator of the nan gene cluster. *Applied and environmental microbiology 81*, 3121-3131. doi:10.1128/AEM.00499-15

Allan, R.N., Skipp, P., Jefferies, J., Clarke, S.C., Faust, S.N., Hall-Stoodley, L., and Webb, J. (2014). Pronounced metabolic changes in adaptation to biofilm growth by Streptococcus pneumoniae. *PLoS ONE 9*, e107015. doi:10.1371/journal.pone.0107015

Argimon, S., Alekseyenko, A.V., DeSalle, R., and Caufield, P.W. (2013). Phylogenetic analysis of glucosyltransferases and implications for the coevolution of mutans streptococci with their mammalian hosts. *PLoS One 8*, e56305. doi:10.1371/journal.pone.0056305

Argimon, S., Konganti, K., Chen, H., Alekseyenko, A.V., Brown, S., and Caufield, P.W. (2014). Comparative genomics of oral isolates of Streptococcus mutans by in silico genome subtraction does not reveal accessory DNA associated with severe early childhood caries. *Infect Genet Evol 21*, 269-278. doi:10.1016/j.meegid.2013.11.003

Barrangou, R., and Horvath, P. (2017). A decade of discovery: CRISPR functions and applications. *Nat Microbiol* 2, 17092. doi:10.1038/nmicrobiol.2017.92

Belda-Ferre, P., Alcaraz, L.D., Cabrera-Rubio, R., Romero, H., Simon-Soro, A., Pignatelli, M., and Mira, A. (2012). The oral metagenome in health and disease. *ISME J 6*, 46-56. doi:10.1038/ismej.2011.85

Benitez-Paez, A., Belda-Ferre, P., Simon-Soro, A., and Mira, A. (2014). Microbiota diversity and gene expression dynamics in human oral biofilms. *BMC Genomics*

15, 311. doi:10.1186/1471-2164-15-311

Biagini, M., Garibaldi, M., Aprea, S., Pezzicoli, A., Doro, F., Becherelli, M., Taddei, A.R., Tani, C., Tavarini, S., Mora, M., *et al.* (2015). The Human Pathogen Streptococcus pyogenes Releases Lipoproteins as Lipoprotein-rich Membrane Vesicles. *Molecular & Cellular Proteomics : MCP 14*, 2138-2149. doi:10.1074/mcp.M114.045880

Bittaye, M., and Cash, P. (2015). Streptococcus pneumoniae proteomics: determinants of pathogenesis and vaccine development. *Expert review of proteomics 12*, 607-621. doi:10.1586/14789450.2015.1108844

Blair, J.M., Webber, M.A., Baylay, A.J., Ogbolu, D.O., and Piddock, L.J. (2015). Molecular mechanisms of antibiotic resistance. *Nat Rev Microbiol 13*, 42-51. doi:10.1038/nrmicro3380

Bogaert, D., De Groot, R., and Hermans, P.W. (2004). Streptococcus pneumoniae colonisation: the key to pneumococcal disease. *The Lancet Infectious diseases 4*, 144-154. doi:10.1016/s1473-3099(04)00938-7

Bolotin, A., Quinquis, B., Renault, P., Sorokin, A., Ehrlich, S.D., Kulakauskas, S., Lapidus, A., Goltsman, E., Mazur, M., Pusch, G.D., *et al.* (2004). Complete sequence and comparative genome analysis of the dairy bacterium Streptococcus thermophilus. *Nat Biotechnol 22*, 1554-1558. doi:10.1038/nbt1034

Brown, J.S. (2007). Oral biofilms, periodontitis and pulmonary infections. *Oral Diseases 13*, 513-514. doi:10.1111/j.1601-0825.2007.01410b.x

Burbaum, J., and Tobal, G.M. (2002). Proteomics in drug discovery. *Current Opinion in Chemical Biology 6*, 427-433. doi:https://doi.org/10.1016/S1367-5931(02)00337-X

Chaze, T., Guillot, A., Valot, B., Langella, O., Chamot-Rooke, J., Di Guilmi, A.M., Trieu-Cuot, P., Dramsi, S., and Mistou, M.Y. (2014). O-Glycosylation of the N-terminal region of the serine-rich adhesin Srr1 of Streptococcus agalactiae explored by mass spectrometry. *Molecular & Cellular Proteomics : MCP 13*, 2168-2182. doi:10.1074/mcp.M114.038075

Choi, C.W., An, H.Y., Lee, Y.J., Lee, Y.G., Yun, S.H., Park, E.C., Hong, Y., Kim, G.H., Park, J.E., Baek, S.J., *et al.* (2013). Characterization of Streptococcus pneumoniae N-acetylglucosamine-6-phosphate deacetylase as a novel diagnostic marker. *Journal of microbiology (Seoul, Korea) 51*, 659-664. doi:10.1007/s12275-013-3451-8

Claveau, I., Mostefaoui, Y., and Rouabhia, M. (2004). Basement membrane protein and matrix metalloproteinase deregulation in engineered human oral mucosa following infection with Candida albicans. *Matrix Biol 23*, 477-486. doi:10.1016/j.matbio.2004.08.006

Conway, T., and Schoolnik, G.K. (2003). Microarray expression profiling: capturing a genome-wide portrait of the transcriptome. *Mol Microbiol 47*, 879-889.

Cradick, T.J., Fine, E.J., Antico, C.J., and Bao, G. (2013). CRISPR/Cas9 systems targeting beta-globin and CCR5 genes have substantial off-target activity. *Nucleic Acids Res 41*, 9584-9592. doi:10.1093/nar/gkt714

Croucher, N.J., Harris, S.R., Fraser, C., Quail, M.A., Burton, J., van der Linden, M., McGee, L., von Gottberg, A., Song, J.H., Ko, K.S., *et al.* (2011). Rapid Pneumococcal Evolution in Response to Clinical Interventions. *Science 331*, 430.

Croucher, N.J., Walker, D., Romero, P., Lennard, N., Paterson, G.K., Bason, N.C., Mitchell, A.M., Quail, M.A., Andrew, P.W., Parkhill, J., *et al.* (2009). Role of Conjugative Elements in the Evolution of the Multidrug-Resistant Pandemic Clone Streptococcus pneumoniae(Spain23F) ST81. *Journal of Bacteriology 191*, 1480-1489. doi:10.1128/JB.01343-08

Delorme, C., Abraham, A.L., Renault, P., and Guedon, E. (2015). Genomics of Streptococcus salivarius, a major human commensal. *Infect Genet Evol 33*, 381-392. doi:10.1016/j.meegid.2014.10.001

DiCarlo, J.E., Norville, J.E., Mali, P., Rios, X., Aach, J., and Church, G.M. (2013). Genome engineering in Saccharomyces cerevisiae using CRISPR-Cas systems. *Nucleic Acids Res 41*, 4336-4343. doi:10.1093/nar/gkt135

Dmitriev, A.V., McDowell, E.J., Kappeler, K.V., Chaussee, M.A., Rieck, L.D., and Chaussee, M.S. (2006). The Rgg regulator of Streptococcus pyogenes influences utilization of nonglucose carbohydrates, prophage induction, and expression of the NAD-glycohydrolase virulence operon. *Journal of bacteriology 188*, 7230-7241. doi:10.1128/jb.00877-06

Donnarumma, D., Faleri, A., Costantino, P., Rappuoli, R., and Norais, N. (2016). The role of structural proteomics in vaccine development: recent advances and future prospects. *Expert review of proteomics 13*, 55-68. doi:10.1586/14789450.2016.1121113

Du, C.C., Yang, M.J., Li, M.Y., Yang, J., Peng, B., Li, H., and Peng, X.X. (2017). Metabolic Mechanism for l-Leucine-Induced Metabolome To Eliminate Streptococcus iniae. *J Proteome Res 16*, 1880-1889. doi:10.1021/acs.jproteome.6b00944

El-Rami, F., Nelson, K., and Xu, P. (2017). Proteomic Approach for Extracting Cytoplasmic Proteins from Streptococcus sanguinis using Mass Spectrometry. *Journal of molecular biology research 7*, 50-57. doi:10.5539/jmbr.v7n1p50

Fei, F., Mendonca, M.L., McCarry, B.E., Bowdish, D.M.E., and Surette, M.G. (2016). Metabolic and transcriptomic profiling of Streptococcus intermedius during aerobic and anaerobic growth. *Metabolomics 12*. doi:10.1007/s11306-016-0966-0

Feng, G., Klein, M., Gregoire, S., Singh, A., Vorsa, N., and Koo, H. (2013). The specific degree-of-polymerization of A-type proanthocyanidin oligomers impacts Streptococcus mutans glucan-mediated adhesion and transcriptome responses within biofilms. *Biofouling 29*, 629-640.

Feng, J., Billal, D.S., Lupien, A., Racine, G., Winstall, E., Legare, D., Leprohon, P., and Ouellette, M. (2011). Proteomic and transcriptomic analysis of linezolid resistance in Streptococcus pneumoniae. *J Proteome Res 10*, 4439-4452. doi:10.1021/pr200221s

Ferrándiz, M., and de, l., Campa, AG. (2014). The fluoroquinolone levofloxacin triggers the transcriptional activation of iron transport genes that contribute to cell death in Streptococcus pneumoniae. *Antimicrob Agents Chemother 58*, 247-257.

Fiebig, A., Loof, T.G., Babbar, A., Itzek, A., Koehorst, J.J., Schaap, P.J., and Nitsche-Schmitz, D.P. (2015). Comparative genomics of Streptococcus pyogenes M1 isolates differing in virulence and propensity to cause systemic infection in mice. *Int J Med Microbiol 305*, 532-543. doi:10.1016/j.ijmm.2015.06.002

Fleischmann, R., Adams, M., White, O., Clayton, R., Kirkness, E., Kerlavage, A., Bult, C., Tomb, J., Dougherty, B., Merrick, J., *et al.* (1995). Whole-genome random sequencing and assembly of Haemophilus influenzae Rd. *Science 269*, 496-512.

Franzosa, E.A., Hsu, T., Sirota-Madi, A., Shafquat, A., Abu-Ali, G., Morgan, X.C., and Huttenhower, C. (2015). Sequencing and beyond: integrating molecular 'omics' for microbial community profiling. *Nature reviews Microbiology 13*, 360-372. doi:10.1038/nrmicro3451

Fraser-Liggett, C.M. (2005). Insights on biology and evolution from microbial genome sequencing. *Genome research 15*, 1603-1610. doi:10.1101/gr.3724205

Freiberg, J.A., Le Breton, Y., Tran, B.Q., Scott, A.J., Harro, J.M., Ernst, R.K., Goo, Y.A., Mongodin, E.F., Goodlett, D.R., McIver, K.S., *et al.* (2016). Global Analysis and Comparison of the Transcriptomes and Proteomes of Group A Streptococcus Biofilms. *mSystems 1*. doi:10.1128/mSystems.00149-16

Fu, Y., Foden, J.A., Khayter, C., Maeder, M.L., Reyon, D., Joung, J.K., and Sander, J.D. (2013). High-frequency off-target mutagenesis induced by CRISPR-Cas nucleases in human cells. *Nat Biotechnol 31*, 822-826. doi:10.1038/nbt.2623

Fu, Y., Sander, J.D., Reyon, D., Cascio, V.M., and Joung, J.K. (2014). Improving CRISPR-Cas nuclease specificity using truncated guide RNAs. *Nature Biotechnology 32*, 279. doi:10.1038/nbt.2808
https://www.nature.com/articles/nbt.2808#supplementary-information

Galeotti, C.L., Bove, E., Pezzicoli, A., Nogarotto, R., Norais, N., Pileri, S., Lelli, B., Falugi, F., Balloni, S., Tedde, V., *et al.* (2012). Surface Interactome in Streptococcus pyogenes*. *Molecular & Cellular Proteomics : MCP 11*. doi:10.1074/mcp.M111.015206

Gao, X.Y., Zhi, X.Y., Li, H.W., Klenk, H.P., and Li, W.J. (2014). Comparative genomics of the bacterial genus Streptococcus illuminates evolutionary implications of species groups. *PLoS One 9*, e101229. doi:10.1371/journal.pone.0101229

Gaur, U., Xiong, Y.Y., Luo, Q.P., Yuan, F.Y., Wu, H.Y., Qiao, M., Wimmers, K., Li, K., Mei, S.Q., and Liu, G.S. (2014). Breed-specific transcriptome response of spleen from six to eight week old piglet after infection with Streptococcus suis type 2. *Mol Biol Rep 41*, 7865-7873. doi:10.1007/s11033-014-3680-x

Geng, J., Chiu, C.H., Tang, P., Chen, Y., Shieh, H.R., Hu, S., and Chen, Y.Y. (2012). Complete genome and transcriptomes of Streptococcus parasanguinis FW213: phylogenic relations and potential virulence mechanisms. *PLoS One 7*, e34769. doi:10.1371/journal.pone.0034769

Gratz, S.J., Cummings, A.M., Nguyen, J.N., Hamm, D.C., Donohue, L.K., Harrison, M.M., Wildonger, J., and O'Connor-Giles, K.M. (2013). Genome engineering of Drosophila with the CRISPR RNA-guided Cas9 nuclease. *Genetics 194*, 1029-1035. doi:10.1534/genetics.113.152710

Gupta, S.K., Padmanabhan, B.R., Diene, S.M., Lopez-Rojas, R., Kempf, M., Landraud, L., and Rolain, J.-M. (2014). ARG-ANNOT, a new bioinformatic tool to discover antibiotic resistance genes in bacterial genomes. *Antimicrobial agents and chemotherapy 58*, 212-220. doi:10.1128/aac.01310-13

Haas, B., and Grenier, D. (2015). Isolation, Characterization and Biological Properties of Membrane Vesicles Produced by the Swine Pathogen Streptococcus suis. *PLoS ONE 10*. doi:10.1371/journal.pone.0130528

Haase, E., Feng, X., Pan, J., Miecznikowski, J., and Scannapieco, F. (2015). Dynamics of the Streptococcus gordonii Transcriptome in Response to Medium, Salivary α-Amylase, and Starch. *Appl Environ Microbiol 81*, 5363-5374.

Hilty, M., Wuthrich, D., Salter, S.J., Engel, H., Campbell, S., Sa-Leao, R., de Lencastre, H., Hermans, P., Sadowy, E., Turner, P., *et al.* (2014). Global phylogenomic analysis of nonencapsulated Streptococcus pneumoniae reveals a deep-branching classic lineage that is distinct from multiple sporadic lineages. *Genome Biol Evol 6*, 3281-3294. doi:10.1093/gbe/evu263

Hossain, M., Egan, S.A., Coffey, T., Ward, P.N., Wilson, R., Leigh, J.A., and Emes, R.D. (2015). Virulence related sequences; insights provided by comparative genomics of Streptococcus uberis of differing virulence. *BMC Genomics 16*, 334. doi:10.1186/s12864-015-1512-6

Huddleston, J.R. (2014). Horizontal gene transfer in the human gastrointestinal tract: potential spread of antibiotic resistance genes. *Infect Drug Resist 7*, 167-176. doi:10.2147/IDR.S48820

Hwang, W.Y., Fu, Y., Reyon, D., Maeder, M.L., Tsai, S.Q., Sander, J.D., Peterson, R.T., Yeh, J.R.J., and Joung, J.K. (2013). Efficient genome editing in zebrafish using a CRISPR-Cas system. *Nature Biotechnology 31*, 227. doi:10.1038/nbt.2501
https://www.nature.com/articles/nbt.2501#supplementary-information

Jans, C., Follador, R., Hochstrasser, M., Lacroix, C., Meile, L., and Stevens, M.J.A.

(2013). Comparative genome analysis of Streptococcus infantarius subsp. infantariusCJ18, an African fermented camel milk isolate with adaptations to dairy environment. *BMC Genomics 14*, 200. doi:10.1186/1471-2164-14-200

Jimenez-Munguia, I., van Wamel, W.J., Olaya-Abril, A., Garcia-Cabrera, E., Rodriguez-Ortega, M.J., and Obando, I. (2015). Proteomics-driven design of a multiplex bead-based platform to assess natural IgG antibodies to pneumococcal protein antigens in children. *Journal of proteomics 126*, 228-233. doi:10.1016/j.jprot.2015.06.011

Kajfasz, J., Abranches, J., and Lemos, J. (2011). Transcriptome analysis reveals that ClpXP proteolysis controls key virulence properties of Streptococcus mutans. *Microbiology 157*, 2800-2890.

Kalpana, D., Im, C., and Lee, Y.S. (2016). Comparative growth, cross stress resistance, transcriptomics of Streptococcus pyogenes cultured under low shear modeled microgravity and normal gravity. *Saudi J Biol Sci 23*, 24-33. doi:10.1016/j.sjbs.2015.02.004

Karlsson, C., Malmström, L., Aebersold, R., and Malmström, J. (2012). Proteome-wide selected reaction monitoring assays for the human pathogen Streptococcus pyogenes. *Nature Communications 3*, 1301-. doi:10.1038/ncomms2297

Khan, R., Rukke, H.V., Hovik, H., Amdal, H.A., Chen, T., Morrison, D.A., and Petersen, F.C. (2016). Comprehensive Transcriptome Profiles of Streptococcus mutans UA159 Map Core Streptococcal Competence Genes. *mSystems 1*. doi:10.1128/mSystems.00038-15

Khosravi, Y., Loke, M.F., Goh, K.L., and Vadivelu, J. (2016). Proteomics Analysis Revealed that Crosstalk between Helicobacter pylori and Streptococcus mitis May Enhance Bacterial Survival and Reduces Carcinogenesis. *Frontiers in microbiology 7*, 1462. doi:10.3389/fmicb.2016.01462

Klein, M.I., DeBaz, L., Agidi, S., Lee, H., Xie, G., Lin, A.H., Hamaker, B.R., Lemos, J.A., and Koo, H. (2010). Dynamics of Streptococcus mutans transcriptome in response to starch and sucrose during biofilm development. *PLoS One 5*, e13478. doi:10.1371/journal.pone.0013478

Klein, M.I., Xiao, J., Lu, B., Delahunty, C.M., Yates, J.R., 3rd, and Koo, H. (2012). Streptococcus mutans protein synthesis during mixed-species biofilm development by high-throughput quantitative proteomics. *PLoS ONE 7*, e45795. doi:10.1371/journal.pone.0045795

Kleinstiver, B.P., Pattanayak, V., Prew, M.S., Tsai, S.Q., Nguyen, N.T., Zheng, Z., and Joung, J.K. (2016). High-fidelity CRISPR-Cas9 nucleases with no detectable genome-wide off-target effects. *Nature 529*, 490-495. doi:10.1038/nature16526

Kleinstiver, B.P., Prew, M.S., Topkar, V.V., Tsai, S.Q., and Joung, J.K. (2015). 58. Engineered Cas9 Variants with Novel PAM Specificities Expand the Targeting Range of CRISPR/Cas Nucleases. *Molecular Therapy 23*, S26. doi:10.1016/S1525-0016(16)33663-2

Kreth, J., Chen, Z., Ferretti, J., and Malke, H. (2011). Counteractive balancing of transcriptome expression involving CodY and CovRS in Streptococcus pyogenes. *J Bacteriol 193*, 4153-4165. doi:10.1128/JB.00061-11

Lau, S.K.P., Woo, P.C.Y., Luk, W.-k., Fung, A.M.Y., Hui, W.-t., Fong, A.H.C., Chow, C.-w., Wong, S.S.Y., and Yuen, K.-y. (2006). Clinical isolates of Streptococcus iniae from Asia are more mucoid and β-hemolytic than those from North America. *Diagnostic Microbiology and Infectious Disease 54*, 177-181. doi:https://doi.org/10.1016/j.diagmicrobio.2005.09.012

Lefebure, T., Richards, V.P., Lang, P., Pavinski-Bitar, P., and Stanhope, M.J. (2012). Gene repertoire evolution of Streptococcus pyogenes inferred from phylogenomic analysis with Streptococcus canis and Streptococcus dysgalactiae. *PLoS One 7*, e37607. doi:10.1371/journal.pone.0037607

Lefébure, T., and Stanhope, M.J. (2007). Evolution of the core and pan-genome of Streptococcus: positive selection, recombination, and genome composition. *Genome Biology 8*, R71. doi:10.1186/gb-2007-8-5-r71

Lemme, A., Grobe, L., Reck, M., Tomasch, J., and Wagner-Dobler, I. (2011). Subpopulation-specific transcriptome analysis of competence-stimulating-peptide-induced Streptococcus mutans. *J Bacteriol 193*, 1863-1877. doi:10.1128/JB.01363-10

Levy, S.E., and Myers, R.M. (2016). Advancements in Next-Generation Sequencing. *Annu Rev Genomics Hum Genet 17*, 95-115. doi:10.1146/annurev-genom-083115-022413

Li, J., Wang, W., Wang, Y., and Zeng, A.P. (2013a). Two-dimensional gel-based proteomic of the caries causative bacterium Streptococcus mutans UA159 and insight into the inhibitory effect of carolacton. *Proteomics 13*, 3470-3477. doi:10.1002/pmic.201300077

Li, W., Teng, F., Li, T., and Zhou, Q. (2013b). Simultaneous generation and germline transmission of multiple gene mutations in rat using CRISPR-Cas systems. *Nat Biotechnol 31*, 684-686. doi:10.1038/nbt.2652

Liu, C., Niu, Y., Zhou, X., Zhang, K., Cheng, L., Li, M., Li, Y., Wang, R., Yang, Y., and Xu, X. (2013a). Hyperosmotic response of streptococcus mutans: from microscopic physiology to transcriptomic profile. *BMC Microbiol 13*, 275. doi:10.1186/1471-2180-13-275

Liu, G., Zhang, W., and Lu, C. (2013b). Comparative genomics analysis of Streptococcus agalactiae reveals that isolates from cultured tilapia in China are closely related to the human strain A909. *BMC Genomics 14*, 775. doi:10.1186/1471-2164-14-775

Ma, Y.M., Yang, M.J., Wang, S., Li, H., and Peng, X.X. (2015). Liver functional metabolomics discloses an action of L-leucine against Streptococcus iniae infection in tilapias. *Fish Shellfish Immunol 45*, 414-421. doi:10.1016/j.fsi.2015.04.037

Ma, Z., Geng, J., Yi, L., Xu, B., Jia, R., Li, Y., Meng, Q., Fan, H., and Hu, S. (2013). Insight into the specific virulence related genes and toxin-antitoxin virulent pathogenicity islands in swine streptococcosis pathogen Streptococcus equi ssp. zooepidemicus strain ATCC35246. *BMC Genomics 14*, 377. doi:10.1186/1471-2164-14-377

Maeda, K., Nagata, H., Ojima, M., and Amano, A. (2015). Proteomic and transcriptional analysis of interaction between oral microbiota Porphyromonas gingivalis and Streptococcus oralis. *Journal of proteome research 14*, 82-94. doi:10.1021/pr500848e

Makarova, K.S., Wolf, Y.I., Alkhnbashi, O.S., Costa, F., Shah, S.A., Saunders, S.J., Barrangou, R., Brouns, S.J., Charpentier, E., Haft, D.H., *et al.* (2015). An updated evolutionary classification of CRISPR-Cas systems. *Nat Rev Microbiol 13*, 722-736. doi:10.1038/nrmicro3569

Manzoor, I., Shafeeq, S., and Kuipers, O.P. (2015). Transcriptome analysis of

Streptococcus pneumoniae D39 in the presence of cobalt. *Genom Data 6*, 151-153. doi:10.1016/j.gdata.2015.08.033

Marraffini, L.A. (2016). The CRISPR-Cas system of Streptococcus pyogenes: function and applications (University of Oklahoma Health Sciences Center, Oklahoma City (OK)).

Martínez-Porchas, M., and Vargas-Albores, F. (2017). Microbial metagenomics in aquaculture: a potential tool for a deeper insight into the activity. *Reviews in Aquaculture 9*, 42-56. doi:10.1111/raq.12102

Mattos-Graner, R.O., Napimoga, M.H., Fukushima, K., Duncan, M.J., and Smith, D.J. (2004). Comparative Analysis of Gtf Isozyme Production and Diversity in Isolates of Streptococcus mutans with Different Biofilm Growth Phenotypes. *Journal of Clinical Microbiology 42*, 4586-4592. doi:10.1128/JCM.42.10.4586-4592.2004

McClure, R., Balasubramanian, D., Sun, Y., Bobrovskyy, M., Sumby, P., Genco, C.A., Vanderpool, C.K., and Tjaden, B. (2013). Computational analysis of bacterial RNA-Seq data. *Nucleic Acids Res 41*, e140. doi:10.1093/nar/gkt444

McGettigan, P.A. (2013). Transcriptomics in the RNA-seq era. *Curr Opin Chem Biol 17*, 4-11. doi:10.1016/j.cbpa.2012.12.008

Metcalf, B.J., Chochua, S., Gertz, R.E., Jr., Hawkins, P.A., Ricaldi, J., Li, Z., Walker, H., Tran, T., Rivers, J., Mathis, S., *et al.* (2017). Short-read whole genome sequencing for determination of antimicrobial resistance mechanisms and capsular serotypes of current invasive Streptococcus agalactiae recovered in the USA. *Clin Microbiol Infect 23*, 574 e577-574 e514. doi:10.1016/j.cmi.2017.02.021

Meulenbelt, I., Kraus, V.B., Sandell, L.J., and Loughlin, J. (2011). Summary of the OA biomarkers workshop 2010 – genetics and genomics: new targets in OA. *Osteoarthritis and Cartilage 19*, 1091-1094. doi:https://doi.org/10.1016/j.joca.2011.06.002

Mitsuwan, W., Olaya-Abril, A., Calderon-Santiago, M., Jimenez-Munguia, I., Gonzalez-Reyes, J.A., Priego-Capote, F., Voravuthikunchai, S.P., and Rodriguez-Ortega, M.J. (2017). Integrated proteomic and metabolomic analysis reveals that rhodomyrtone reduces the capsule in Streptococcus pneumoniae. *Sci Rep 7*, 2715. doi:10.1038/s41598-017-02996-3

Muller, M., Lee, C.M., Gasiunas, G., Davis, T.H., Cradick, T.J., Siksnys, V., Bao, G., Cathomen, T., and Mussolino, C. (2016). Streptococcus thermophilus CRISPR-Cas9 Systems Enable Specific Editing of the Human Genome. *Mol Ther 24*, 636-644. doi:10.1038/mt.2015.218

Munro, C., Michalek, S.M., and Macrina, F.L. (1991). Cariogenicity of Streptococcus mutans V403 glucosyltransferase and fructosyltransferase mutants constructed by allelic exchange. *Infection and immunity 59*, 2316-2323.

Muzzi, A., and Donati, C. (2011). Population genetics and evolution of the pan-genome of Streptococcus pneumoniae. *Int J Med Microbiol 301*, 619-622. doi:10.1016/j.ijmm.2011.09.008

Nasser, W. (2014). Evolutionary pathway to increased virulence and epidemic group A Streptococcus disease derived from 3,615 genome sequences. *Proceedings of the National Academy of Sciences of the United States of America v. 111*, pp. E1768-E1776-2014 v.1111 no.1717. doi:10.1073/pnas.1403138111

Oehmcke, S., Shannon, O., von Köckritz-Blickwede, M., Mörgelin, M., Linder, A., Olin, A.I., Björck, L., and Herwald, H. (2009). Treatment of invasive streptococcal infection with a peptide derived from human high-molecular weight kininogen. *Blood 114*, 444-451. doi:10.1182/blood-2008-10-182527

Oehmcke, S., Westman, J., Malmström, J., Mörgelin, M., Olin, A.I., Kreikemeyer, B., and Herwald, H. (2013). A Novel Role for Pro-Coagulant Microvesicles in the Early Host Defense against Streptococcus pyogenes. *PLOS Pathogens 9*, e1003529. doi:10.1371/journal.ppat.1003529

Olaya-Abril, A., Jimenez-Munguia, I., Gomez-Gascon, L., Obando, I., and Rodriguez-Ortega, M.J. (2013). Identification of potential new protein vaccine candidates through pan-surfomic analysis of pneumococcal clinical isolates from adults. *PLoS ONE 8*, e70365. doi:10.1371/journal.pone.0070365

Olaya-Abril, A., Prados-Rosales, R., McConnell, M.J., Martin-Pena, R., Gonzalez-Reyes, J.A., Jimenez-Munguia, I., Gomez-Gascon, L., Fernandez, J., Luque-Garcia, J.L., Garcia-Lidon, C., *et al.* (2014). Characterization of protective extracellular membrane-derived vesicles produced by Streptococcus pneumoniae. *Journal of proteomics 106*, 46-60. doi:10.1016/j.jprot.2014.04.023

Olson, A.B., Kent, H., Sibley, C.D., Grinwis, M.E., Mabon, P., Ouellette, C., Tyson,

S., Graham, M., Tyler, S.D., Van Domselaar, G., *et al.* (2013). Phylogenetic relationship and virulence inference of Streptococcus Anginosus Group: curated annotation and whole-genome comparative analysis support distinct species designation. *BMC Genomics 14*, 895. doi:10.1186/1471-2164-14-895

Paik, S., Senty, L., Das, S., Noe, J.C., Munro, C.L., and Kitten, T. (2005). Identification of Virulence Determinants for Endocarditis in Streptococcus sanguinis by Signature-Tagged Mutagenesis. *Infection and Immunity 73*, 6064-6074. doi:10.1128/IAI.73.9.6064-6074.2005

Papadimitriou, K., Anastasiou, R., Mavrogonatou, E., Blom, J., Papandreou, N.C., Hamodrakas, S.J., Ferreira, S., Renault, P., Supply, P., Pot, B., *et al.* (2014). Comparative genomics of the dairy isolate Streptococcus macedonicus ACA-DC 198 against related members of the Streptococcus bovis/Streptococcus equinus complex. *BMC Genomics 15*, 272. doi:10.1186/1471-2164-15-272

Park, H.K., Myung, S.C., and Kim, W. (2012). Comparative transcriptomic analysis of streptococcus pseudopneumoniae with viridans group streptococci. *BMC Microbiol 12*, 77. doi:10.1186/1471-2180-12-77

Patel, S., and Ahmed, S. (2015). Emerging field of metabolomics: Big promise for cancer biomarker identification and drug discovery. *Journal of Pharmaceutical and Biomedical Analysis 107*, 63-74. doi:https://doi.org/10.1016/j.jpba.2014.12.020

Peterson, S.N., Meissner, T., Su, A.I., Snesrud, E., Ong, A.C., Schork, N.J., and Bretz, W.A. (2014). Functional expression of dental plaque microbiota. *Front Cell Infect Microbiol 4*, 108. doi:10.3389/fcimb.2014.00108

Pettigrew, M., Marks, L., Kong, Y., Gent, J., Roche-Hakansson, H., and Hakansson, A. (2014). Dynamic changes in the Streptococcus pneumoniae transcriptome during transition from biofilm formation to invasive disease upon influenza A virus infection. *Infect Immun 82*, 4607-4619.

Pfaller, M.A., Farrell, D.J., Sader, H.S., and Jones, R.N. (2012). AWARE Ceftaroline Surveillance Program (2008-2010): trends in resistance patterns among Streptococcus pneumoniae, Haemophilus influenzae, and Moraxella catarrhalis in the United States. *Clin Infect Dis 55 Suppl 3*, S187-193. doi:10.1093/cid/cis561

Pian, Y., Wang, P., Liu, P., Zheng, Y., Zhu, L., Wang, H., Xu, B., Yuan, Y., and Jiang, Y. (2015). Proteomics identification of novel fibrinogen-binding proteins of Streptococcus suis contributing to antiphagocytosis. *Frontiers in cellular and infection microbiology 5*, 19. doi:10.3389/fcimb.2015.00019

Prajapati, J.B. (2013). Genomic Analysis of Dairy Starter Culture Streptococcus thermophilus MTCC 5461. *Journal of Microbiology and Biotechnology 23*, 459-466. doi:10.4014/jmb.1210.10030

Richards, V.P., Choi, S.C., Pavinski Bitar, P.D., Gurjar, A.A., and Stanhope, M.J. (2013). Transcriptomic and genomic evidence for Streptococcus agalactiae adaptation to the bovine environment. *BMC Genomics 14*, 920. doi:10.1186/1471-2164-14-920

Richards, V.P., Palmer, S.R., Pavinski Bitar, P.D., Qin, X., Weinstock, G.M., Highlander, S.K., Town, C.D., Burne, R.A., and Stanhope, M.J. (2014). Phylogenomics and the dynamic genome evolution of the genus Streptococcus. *Genome Biol Evol 6*, 741-753. doi:10.1093/gbe/evu048

Russell, C., Rahman, A., and Mohammed, A.R. (2013). Application of genomics, proteomics and metabolomics in drug discovery, development and clinic. *Therapeutic Delivery 4*, 395-413. doi:10.4155/tde.13.4

Salzano, A.M., Novi, G., Arioli, S., Corona, S., Mora, D., and Scaloni, A. (2013). Mono-dimensional blue native-PAGE and bi-dimensional blue native/urea-PAGE or/SDS-PAGE combined with nLC-ESI-LIT-MS/MS unveil membrane protein heteromeric and homomeric complexes in Streptococcus thermophilus. *Journal of proteomics 94*, 240-261. doi:10.1016/j.jprot.2013.09.007

Sander, J.D., and Joung, J.K. (2014). CRISPR-Cas systems for editing, regulating and targeting genomes. *Nat Biotechnol 32*, 347-355. doi:10.1038/nbt.2842

Schouls, L.M., Schot, C.S., and Jacobs, J.A. (2003). Horizontal Transfer of Segments of the 16S rRNA Genes between Species of the Streptococcus anginosus Group. *Journal of Bacteriology 185*, 7241-7246. doi:10.1128/JB.185.24.7241-7246.2003

Schulz, C., and Hammerschmidt, S. (2013). Exploitation of physiology and metabolomics to identify pneumococcal vaccine candidates. *Expert Rev Vaccines 12*, 1061-1075. doi:10.1586/14760584.2013.824708

Sharma, C.M., and Vogel, J. (2014). Differential RNA-seq: the approach behind and the biological insight gained. *Curr Opin Microbiol 19*, 97-105. doi:10.1016/j.mib.2014.06.010

Shen, B., Zhang, W., Zhang, J., Zhou, J., Wang, J., Chen, L., Wang, L., Hodgkins, A., Iyer, V., Huang, X., *et al.* (2014). Efficient genome modification by CRISPR-Cas9 nickase with minimal off-target effects. *Nat Methods 11*, 399-402. doi:10.1038/nmeth.2857

Sitkiewicz, I., Green, N.M., Guo, N., Bongiovanni, A.M., Witkin, S.S., and Musser, J.M. (2009). Transcriptome adaptation of group B Streptococcus to growth in human amniotic fluid. *PLoS One 4*, e6114. doi:10.1371/journal.pone.0006114

Sjöholm, K., Karlsson, C., Linder, A., and Malmström, J. (2014). A comprehensive analysis of the Streptococcus pyogenes and human plasma protein interaction network. *Molecular bioSystems 10*, 1698-1708. doi:10.1039/c3mb70555b

Song, J.-H., Ko, K.S., Lee, J.-Y., Yang Baek, J., Sup Oh, W., Sik Yoon, H., Jeong, J.-Y., and Chun, J. (2005). Identification of essential genes in Streptococcus pneumoniae by allelic replacement mutagenesis, Vol 19.

Sorek, R., and Cossart, P. (2010). Prokaryotic transcriptomics: a new view on regulation, physiology and pathogenicity. *Nat Rev Genet 11*, 9-16. doi:10.1038/nrg2695

Stackebrandt, E., Frederiksen, W., Garrity, G.M., Grimont, P.A.D., Kämpfer, P., Maiden, M.C.J., Nesme, X., Rosselló-Mora, R., Swings, J., Trüper, H.G., *et al.* (2002). Report of the ad hoc committee for the re-evaluation of the species definition in bacteriology. *International journal of systematic and evolutionary microbiology 52*, 1043-1047. doi:10.1099/ijs.0.02360-0

Stazic, D., and Voss, B. (2016). The complexity of bacterial transcriptomes. *J Biotechnol 232*, 69-78. doi:10.1016/j.jbiotec.2015.09.041

Steinert, J., Schiml, S., Fauser, F., and Puchta, H. (2015). Highly efficient heritable plant genome engineering using Cas9 orthologues from Streptococcus thermophilus and Staphylococcus aureus. *Plant J 84*, 1295-1305. doi:10.1111/tpj.13078

Suzuki, N., Seki, M., Nakano, Y., Kiyoura, Y., Maeno, M., and Yamashita, Y. (2005). Discrimination of Streptococcus pneumoniae from Viridans Group Streptococci by Genomic Subtractive Hybridization. *Journal of Clinical Microbiology 43*, 4528-4534. doi:10.1128/JCM.43.9.4528-4534.2005

Sztajer, H., Szafranski, S.P., Tomasch, J., Reck, M., Nimtz, M., Rohde, M., and Wagner-Dobler, I. (2014). Cross-feeding and interkingdom communication in dual-species biofilms of Streptococcus mutans and Candida albicans. *ISME J 8*, 2256-2271. doi:10.1038/ismej.2014.73

Takahashi, N., and Nyvad, B. (2011). The role of bacteria in the caries process: ecological perspectives. *J Dent Res 90*, 294-303. doi:10.1177/0022034510379602

Takahashi, N., Washio, J., and Mayanagi, G. (2012). Metabolomic approach to oral biofilm characterization—A future direction of biofilm research. *Journal of Oral Biosciences 54*, 138-143. doi:10.1016/j.job.2012.02.005

Tang, F., Bossers, A., Harders, F., Lu, C., and Smith, H. (2013). Comparative genomic analysis of twelve Streptococcus suis (pro)phages. *Genomics 101*, 336-344. doi:10.1016/j.ygeno.2013.04.005

Tang, J. (2011). Microbial Metabolomics. *Current Genomics 12*, 391-403. doi:10.2174/138920211797248619

Teng, J.L., Huang, Y., Tse, H., Chen, J.H., Tang, Y., Lau, S.K., and Woo, P.C. (2014). Phylogenomic and MALDI-TOF MS analysis of Streptococcus sinensis HKU4T reveals a distinct phylogenetic clade in the genus Streptococcus. *Genome Biol Evol 6*, 2930-2943. doi:10.1093/gbe/evu232

Thomason, M., and Storz, G. (2010). Bacterial antisense RNAs: how many are there, and what are they doing? *Annul Rev Genet 44*, 167-188.

Virtaneva, K., Porcella, S.F., Graham, M.R., Ireland, R.M., Johnson, C.A., Ricklefs, S.M., Babar, I., Parkins, L.D., Romero, R.A., Corn, G.J., *et al.* (2005). Longitudinal analysis of the group A Streptococcus transcriptome in experimental pharyngitis in cynomolgus macaques. *Proc Natl Acad Sci U S A 102*, 9014-9019. doi:10.1073/pnas.0503671102

Wadood, A., Jamal, A., Riaz, M., Khan, A., Uddin, R., Jelani, M., and Azam, S.S. (2018). Subtractive genome analysis for in silico identification and characterization of novel drug targets in Streptococcus pneumonia strain JJA. *Microbial Pathogenesis 115*, 194-198.

doi:https://doi.org/10.1016/j.micpath.2017.12.063

Wang, S., Yang, Y., Zhao, Y., Zhao, H., Bai, J., Chen, J., Zhou, Y., Wang, C., and Li, Y. (2016a). Sub-MIC Tylosin Inhibits Streptococcus suis Biofilm Formation and Results in Differential Protein Expression. *Frontiers in microbiology 7*, 384. doi:10.3389/fmicb.2016.00384

Wang, Z., Gerstein, M., and Snyder, M. (2009). RNA-Seq: a revolutionary tool for transcriptomics. *Nat Rev Genet 10*, 57-63. doi:10.1038/nrg2484

Wang, Z., Li, M.Y., Peng, B., Cheng, Z.X., Li, H., and Peng, X.X. (2016b). GC-MS-Based Metabolome and Metabolite Regulation in Serum-Resistant Streptococcus agalactiae. *J Proteome Res 15*, 2246-2253. doi:10.1021/acs.jproteome.6b00215

Watanabe, S., Kirikae, T., and Miyoshi-Akiyama, T. (2013). Complete genome sequence of Streptococcus dysgalactiae subsp. equisimilis 167 carrying Lancefield group C antigen and comparative genomics of S. dysgalactiae subsp. equisimilis strains. *Genome Biol Evol 5*, 1644-1651. doi:10.1093/gbe/evt117

Wen, Y.T., Wang, J.S., Tsai, S.H., Chuan, C.N., Wu, J.J., and Liao, P.C. (2014). Label-free proteomic analysis of environmental acidification-influenced Streptococcus pyogenes secretome reveals a novel acid-induced protein histidine triad protein A (HtpA) involved in necrotizing fasciitis. *Journal of proteomics 109*, 90-103. doi:10.1016/j.jprot.2014.06.026

Wen, Z.T., Nguyen, A.H., Bitoun, J.P., Abranches, J., Baker, H.V., and Burne, R.A. (2011). Transcriptome analysis of LuxS-deficient Streptococcus mutans grown in biofilms. *Mol Oral Microbiol 26*, 2-18. doi:10.1111/j.2041-1014.2010.00581.x

Wishart, D.S. (2016). Emerging applications of metabolomics in drug discovery and precision medicine. *Nat Rev Drug Discov 15*, 473-484. doi:10.1038/nrd.2016.32

Wu, Q., Tun, H.M., Leung, F.C., and Shah, N.P. (2014). Genomic insights into high exopolysaccharide-producing dairy starter bacterium Streptococcus thermophilus ASCC 1275. *Sci Rep 4*, 4974. doi:10.1038/srep04974

Wu, Z., Zhang, W., Lu, Y., and Lu, C. (2010). Transcriptome profiling of zebrafish infected with Streptococcus suis. *Microb Pathog 48*, 178-187. doi:10.1016/j.micpath.2010.02.007

Xu, K., Ren, C., Liu, Z., Zhang, T., Zhang, T., Li, D., Wang, L., Yan, Q., Guo, L., Shen, J., *et al.* (2015). Efficient genome engineering in eukaryotes using Cas9 from Streptococcus thermophilus. *Cell Mol Life Sci 72*, 383-399. doi:10.1007/s00018-014-1679-z

Xu, P., Alves, J.M., Kitten, T., Brown, A., Chen, Z., Ozaki, L.S., Manque, P., Ge, X., Serrano, M.G., Puiu, D., *et al.* (2007). Genome of the opportunistic pathogen Streptococcus sanguinis. *Journal of bacteriology 189*, 3166-3175. doi:10.1128/jb.01808-06

Xu, P., and Gunsolley, J. (2014). Application of metagenomics in understanding oral health and disease. *Virulence 5*, 424-432. doi:10.4161/viru.28532

Yan, S.K., Liu, R.H., Jin, H.Z., Liu, X.R., Ye, J., Shan, L., and Zhang, W.D. (2015). "Omics" in pharmaceutical research: overview, applications, challenges, and future perspectives. *Chin J Nat Med 13*, 3-21. doi:10.1016/S1875-5364(15)60002-4

Yang, X.Y., He, K., Du, G., Wu, X., Yu, G., Pan, Y., Zhang, G., Sun, X., and He, Q.Y. (2016a). Integrated Translatomics with Proteomics to Identify Novel Iron-Transporting Proteins in Streptococcus pneumoniae. *Front Microbiol 7*, 78. doi:10.3389/fmicb.2016.00078

Yang, X.Y., Shi, T., Du, G., Liu, W., Yin, X.F., Sun, X., Pan, Y., and He, Q.Y. (2016b). iTRAQ-Based Proteomics Revealed the Bactericidal Mechanism of Sodium New Houttuyfonate against Streptococcus pneumoniae. *Journal of agricultural and food chemistry 64*, 6375-6382. doi:10.1021/acs.jafc.6b02147

Yang, X.Y., Xu, J.Y., Wei, Q.X., Sun, X., and He, Q.Y. (2017). Comparative Proteomics of Streptococcus pneumoniae Response to Vancomycin Treatment. *Omics : a journal of integrative biology 21*, 531-539.

Yang, X.Y., Zhang, L., Liu, J., Li, N., Yu, G., Cao, K., Han, J., Zeng, G., Pan, Y., Sun, X., *et al.* (2015). Proteomic analysis on the antibacterial activity of a Ru(II) complex against Streptococcus pneumoniae. *Journal of proteomics 115*, 107-116. doi:10.1016/j.jprot.2014.11.018

Yoshida, A., Niki, M., Yamamoto, Y., Yasunaga, A., and Ansai, T. (2015). Proteome analysis identifies the Dpr protein of Streptococcus mutans as an important factor in the presence of early streptococcal colonizers of tooth surfaces. *PLoS ONE 10*, e0121176. doi:10.1371/journal.pone.0121176

Yun-Juan, B., Zhong, L., Jeffrey, A.M., Shaun, W.L., Victoria, A.P., and Francis, J.C. (2015). CovRS-Regulated Transcriptome Analysis of a Hypervirulent M23 Strain of Group A Streptococcus pyogenes Provides New Insights into Virulence Determinants. *J Bacteriol 197*, 3191-3205.

Yuzenkova, Y., Gamba, P., Herber, M., Attaiech, L., Shafeeq, S., Kuipers, O.P., Klumpp, S., Zenkin, N., and Veening, J.W. (2014). Control of transcription elongation by GreA determines rate of gene expression in Streptococcus pneumoniae. *Nucleic Acids Res 42*, 10987-10999. doi:10.1093/nar/gku790

Zeng, L., and Burne, R. (2015). Sucrose- and Fructose-Specific Effects on the Transcriptome of Streptococcus mutans, as Determined by RNA Sequencing. *Appl Environ Microbiol 82*, 146-156.

Zhou, Y., Wang, Y., Deng, L., Zheng, C., Yuan, F., Chen, H., Bei, W., and Li, J. (2015). Evaluation of the protective efficacy of four novel identified membrane associated proteins of Streptococcus suis serotype 2. *Vaccine 33*, 2254-2260. doi:10.1016/j.vaccine.2015.03.038

Chapter 9

https://doi.org/10.21775/9781912530229.09

Intrageneric and Intergeneric Interactions Developed by Oral Streptococci: Pivotal Role in the Pathogenesis of Oral Diseases

Yandi Chen, Yuqing Li*, Jing Zou*

State Key Laboratory of Oral Diseases, National Clinical Research Center for Oral Diseases, West China Hospital of Stomatology, Sichuan University, Chengdu, PR China.

*Corresponding authors: liyuqing@scu.edu.cn, zoujing1970@126.com

Abstract

Oral streptococci are among the most abundant genera present in the oral cavity. They are usually the first colonizers of oral surfaces and they develop extensive microbial interactions, playing a fundamental role in the pathogenesis of oral diseases such as dental caries and periodontitis. In addition to physical adherence, streptococcal cells also exchange messages with cells from another *Streptococcus spp.* and other microorganisms in the form of metabolites and signaling molecules. In this review, we focused on these intrageneric and intergeneric interactions, and their association with oral diseases.

Introduction

Dental biofilms, microbial interactions and polymicrobial diseases

As we have known so far, human oral cavity contains diverse range of microorganisms that exist as biofilms on dental and mucosal surfaces (Marsh, 2016). Dental plaque is one of the most widely studied biofilms. It is a complex system, not merely a simple sum of its bacterial components, but a sophisticated microbial community which may result in novel functions other than the ones of planktonic cells. These novel functions are essential for the biofilm architecture and microbial physiology (Kolenbrander et al., 2002; Marsh, 2005). Notably, dental biofilms are not formed by random simultaneous colonization. The process of colonization is selective, reproducible and sequential (Diaz et al., 2006). Along with *Actinomyces spp.* and a few other genera, oral streptococci are present in high numbers and serve as pioneer colonizers in dental biofilm development by binding to complementary salivary receptors in the acquired pellicle coating the tooth surface (Jakubovics et al., 2014; Parashar et al., 2015; Ichinosawa et al., 2017). Later inhabitants, such as *Porphyromonas gingivalis* and other pathogenic species, are capable of binding to the antecedent organisms. All of these microorganisms are embedded in a self-produced matrix of hydrated extracellular polymeric substances (EPS) that form their immediate environment, where they are protected against environmental threats such as antibiotics, and develop an advantage over planktonic bacteria (Flemming and Wingender, 2010; Marsh, 2016).

Throughout the process of biofilm formation, adherent bacteria can sense their neighbors and give appropriate responses (Parashar et al., 2015). It is generally accepted that synergistic, mutualistic, and antagonistic interactions occur between microorganisms and contribute to the development of polymicrobial biofilm communities (Kuramitsu et al., 2007). Traits such as high species diversity, physical contact between adjacent cells and large metabolic activity all make the oral microbial community a typical model for studying microbial interactions (Guo et al., 2014).

Oral microbial communities are in a dynamic equilibrium with their environment, offering certain benefits to the host by protecting the epithelial cells from damage and enhancing the digestion of some substrates (Vasudevan, 2017). However, under some circumstances, microbial communities may have negative effects on the host. The etiology of oral diseases such as dental caries, went through a paradigm shift from Koch´s postulates focusing on pathogen isolation, to the concept of a polymicrobial disease (Simónsoro and Mira, 2015; Marsh, 2016). Dysbiosis of a mixed-species community is prone to break down the synergistic relationship between host and microbiota.

The overall incidence of caries and periodontal diseases in the population has not significantly decreased in multiple decades. They continue to be among the most common human diseases worldwide (Kassebaum et al., 2015). Interactions among different oral microorganisms may influence the balance between health and dysbiosis (Thompson et al., 2016). Thus, understanding these interactions is of great importance to evaluate the cause of these diseases.

In the present review, we aim to provide an overview of the intrageneric, and intergeneric interactions developed by oral streptococci, and the consequent influence on caries and periodontitis pathogenesis.

Interactions in the microbial community

Coaggregation and coadhesion

Coaggregation is a specific cell-to-cell recognition and binding reaction that occurs between genetically distinct bacterial cells. It is one of the most critical mechanisms for the temporary retention and eventual colonization of bacteria on dental surfaces, and contributes to dental biofilm formation (Hojo et al., 2009). Coaggregated partners are both suspended in the planktonic phase, whereas coadhesion usually refers to one microorganism immobilized on a surface while the other is suspended (Ruhl et al., 2014). Coaggregation and coadhesion among oral microorganisms play an important role in the development of biofilms and formation of microbial communities. Either of the processes can be synergistic, since one microorganism may generate a niche and facilitate the retention of the other, which may be a pathogenic microorganism. Notably, some "bridging" species represented by *Fusobacterium nucleatum* can coaggregate with both early colonizers and late colonizers, serving as a coordinator that allows connections of species that are not coaggregation partners (Kolenbrander et al., 2002; Denes and Barraud, 2016). Extensive coaggregation and coadhesion allow a close cell-to-cell contact, facilitating metabolic communication, signal transmission and reception, and genetic exchange (Hojo et al., 2009).

Coaggregation and coadhesion between distinct microorganisms involve the specific recognition of macromolecules on the surface of one cell by macromolecular surface components of a partner organism (Jakubovics et al., 2014). Bacterial surfaces are rich in adhesins, some of which are proteins while others are lectins (Marsh, 2016). Antigen (Ag) I/II polypeptides are the best characterized adhesins expressed by most indigenous streptococci and are involved in binding a wide range of host receptors, mediating biofilm formation and coaggregating with oral microbial partners (Back et al., 2015; Ito et al., 2017). AgI/II proteins of oral streptococci have a molecular mass of approximately 160–180 kDa and have a conserved linear structure consisting of several clearly defined domains (Jakubovics et al., 2014). Notably, distinct domains mediate adherence to various species, making the recognition of interacting partner species-specific. Representative AgI/II proteins such as *S. gordonii* SspA and SspB are important in microbial interactions. These proteins differ in binding specificities and can coaggregate with distinct groups of actinomyces (Egland et al., 2001). Not only AgI/II, some streptococci also harbor other adhesins such as CshA and CshB which were discovered in *S. gordonii DL1* (Mcnab et al., 1999). The CshA polypeptide is a structural and functional component of fibrils and CshA-like fibrillar protein also occurs on the surfaces of other members of the oral *Streptococcus mitis* group (Elliott et al., 2003). Interestingly, many AgI/II proteins require accessory adhesins to enable coaggregation. For instance, the AgI/II family adhesion SpaP participates in binding to certain actinomyces when expressed in *S. gordonii*, but this interaction also requires the presence of CshA (Guo et al., 2016).

Streptococcus spp. effectively utilizes dietary sucrose to synthesize extracellular polysaccharides as a scaffold for its biofilm (Lei et al., 2015). These extracellular polysaccharides can mediate coaggregation interactions. The receptor polysaccharide (RPS) recognizes lectin-like adhesins found on actinomyces, veillonellae and other streptococci (Hsu et al., 1994; Cisar et al., 1997). For instance, *S. mutans* is capable of effectively converting dietary sucrose into acids and producing extracellular glucans using exoenzymes termed glucosyltransferases (Gtfs) (Bowen and Koo, 2011). Specifically, insoluble glucans produced by glucosyltransferase B (GtfB) provide bacterial binding sites and form the core of the extracellular matrix of cariogenic dental plaque biofilm *in vivo*. Besides, the enzyme glyceraldehyde-3-phosphate dehydrogenase (GAPDH) expressed on the surface of streptococcal cells is also responsible for a number of interactions (Jakubovics et al., 2014), predominantly participating in the interactions between streptococci and *P. gingivalis* (Enersen et al., 2013; Wright et al., 2014).

Metabolic communications

In addition to the physical attachment mentioned above, microbial metabolites play important roles in the establishment of stable oral biofilm communities. Metabolic communications may occur where the excretion of a metabolite by one organism is used as a nutrient by other organisms, or breakdown of a substrate by enzymatic activity of one organism creates available substrates for different organisms (Hojo et al., 2009). A good example is represented by oral streptococci producing short-chain fatty acids, which are an essential carbon source for certain oral bacteria. Furthermore, the production of acids can lower the pH in dental biofilm and inhibit the growth of less aciduric microorganisms. Therefore, short-chain fatty acids have a role in competitive and/or mutualistic interactions and bacterial communication. They could even take a part in quorum sensing (Yuce et al., 2017).

To compete for restricted nutrient supply and limited foothold with other bacteria, oral streptococci produce a plethora of general and specific antimicrobial agents (James, 2014). Bacteriocins are defined as proteinaceous bactericidal substances and their production is controlled by many genetic and environmental factors such as cell density, pH, nutrient source and oxygen (Hojo et al., 2009; Merritt and Qi, 2012). Unlike traditional antibiotics, bacteriocins have a narrow killing spectrum. Competition through bacteriocins commonly occurs and these activities enable bacteria to select their neighbors, promote the establishment of a community with specific bacterial species, and influence the ecological balance of the oral ecosystem. In addition, bacteriocins exert autolytic cell destruction and release of DNA to transform or stabilize the biofilm matrix (Jakubovics et al., 2014). Among all oral bacteria, streptococci possess the greatest bacteriocin production. *S. mutans* produces several bacteriocins named mutacins, including lantibiotics and nonlantibiotics, which inhibit the growth of other bacteria in close proximity such as *S. sanguinis* (Merritt and Qi, 2012). Lantibiotics contain either a lanthionine or methyllanthionine ring structure as well as dehydrated amino acids, while nonlantibiotics are unmodified peptides (Mohammad Shahnoor and Indranil, 2011). *S. salivarius* also has strains that are able to release lantibiotics in large amounts to eliminate harmful bacteria (Burton et al., 2013).

Another thing to note is that oral streptococci are able to produce growth-inhibiting amounts of hydrogen peroxide (H_2O_2) as byproduct of aerobic metabolism (Zhu and Kreth, 2012). Actually, it is more than a simple byproduct and functions in several aspects of oral bacterial biofilm ecology. Many species such as *S. sanguinis* and *S. gordonii* produce H_2O_2 as a weapon to compete over other species by limiting their carbon source availability and causing oxidative stress . The activities thereby help these species in structuring the initial biofilm on tooth surfaces. Furthermore, H_2O_2 can serve as a signaling molecule to regulate gene expression, as shown by *Aggregatibacter actinomycetemcomitans* (Stacy et al., 2014). However, H_2O_2 production depends on the presence of oxygen. When these microorganisms are grown in an anaerobic environment, the lack of oxygen will lead to diminished H_2O_2 production (Kreth et al., 2008). As a result of H_2O_2 action, extracellular DNA released from cells is crucial in the biofilm development and stabilization. It also serves as the source for horizontal gene transfer between oral streptococci (Zhu and Kreth, 2012).

A selected group of oral bacteria produce ammonia (NH_3) via the arginine deiminase system (ADS) to increase intracellular pH and the pH of oral biofilms (Nascimento and Burne, 2014; Huang et al., 2015). Typically, the expression of ADS genes is inducible by arginine and is sensitive to carbohydrate catabolite repression. The gene expression can be highly variable within and between species due to both constitutional and environmental differences (Huang et al., 2015; Huang et al., 2017). It can be enhanced by conditions such as low pH and anaerobic environments. Other than inhibiting tooth demineralization by neutralizing glycolytic acids and by suppressing the emergence of a cariogenic pathogens (Nascimento and Burne, 2014), more recently arginine has been shown to influence the architecture and physicochemical properties of the biofilm matrix (He et al., 2016).

Quorum sensing

Quorum-sensing (QS) is the ability to detect and respond to population density by regulating gene expression through a sophisticated intercellular chemical signaling pathway (Ng and Bassler, 2009; Basavaraju et al., 2016). As bacteria grow, they produce and release a series of molecules called autoinducers (AI) into the external environment at a low basal level (Mashima and Nakazawa, 2015). While the bacterial cells assess cell density until the population reaches a certain scale, accumulated molecules reach a certain threshold level. Subsequently, different sets of target genes are activated to allow bacteria adaptation to environmental changes. By means of QS, pathogens can regulate many aspects, including their virulence factor production, virulence-related behaviors, biofilm formation and motility (Vendeville et al., 2005; Irie and Parsek, 2008; Plančak et al., 2015). QS can be briefly described as a bacterial synchronizing behavior on a community-wide scale through communication.

QS systems in bacteria are generally divided into at least three classes (Sudheer et al., 2015): (1) LuxI/LuxR-type QS in Gram-negative bacteria, using acyl-homoserine lactones as signal molecules; (2) oligopeptide-two-component-type QS in Gram-positive bacteria, using small peptides as signal molecules; and (3) LuxS-encoded AI-2 QS in both Gram-negative and Gram-positive bacteria. There are two types of QS systems in Streptococci. One is ComCDE system that generates competence stimulating peptide (CSP), which is synthesized in the cell and released into the extracellular medium, enabling intraspecies cell-to-cell communication and modulating the expression of many genes in the same bacterial species (Li et al., 2001; Parashar et al., 2015).The other is LuxS/AI-2 system, which acts as a universal signaling molecule that can mediate interspecies interactions in the multispecies plaque community (Xiao et al., 2017).

Genetic exchanges

Gene exchanges certainly occur between oral streptococci and other microorganisms due to the easy contact among neighboring cells. Extracellular DNA can be released from cells as a result of lysis or active secretion, and it may contribute to the structural integrity of the biofilm (Jakubovics et al., 2013; Okshevsky and Meyer, 2015). Conjugation, transduction and transformation are the 3 basic ways for possible exchange of DNA among bacterial cells. This section will be amply discussed in another chapter.

Interactions between oral streptococci and caries-associated microorganisms.

In the human oral cavity, certain streptococci such as *S. mutans* are associated with the onset and progression of caries. *S. mutans* shows multiple virulences such as the capacity to produce large quantities of organic acids (acidogenicity), the ability to tolerate low pH in the environment (aciduricity), and the ability to synthesize multiple secreted proteins and water-insoluble glucan exopolysaccharides (Lemos et al., 2013; Huang et al., 2017). However, despite its important role in caries pathogenesis, *S. mutans* is not an etiologic factor in caries (Giacaman et al., 2015). To understand the dysbiosis of the biofilm, not only should we focus on selected caries-associated species, but modulation of the biofilm as a whole, including species in the core community and species that can balance the acid production after normal dietary intakes of carbohydrates (Tanner et al., 2018). Thus, the microbial interactions involving caries pathogenesis is of great significance.

Intrageneric interactions among oral streptococci

In terms of intrageneric interactions among oral streptococci, the relationship between *S. sanguinis* and *S. mutans* associated to their coexistence and competition is considered as a pertinent model. Both species share the same ecological niche and are similar in metabolic requirements, thus one may inhibit the other and compete for tooth colonization. The cycle of early colonization by *S. sanguinis* likely occurs in every human after tooth eruption or extensive cleaning (Kreth et al., 2016). It is worth noting that high levels of *S. sanguinis* in the mouth is associated to a delayed colonization by *S. mutans* and similarly, *S. mutans* teeth colonization is associated to low levels of *S. sanguinis* (Caufield et al., 2000). A recent study using next-generation sequencing approach also confirmed the opposite relationship between these two species (Richards et al., 2017). Clinically, the predominance of *S. sanguinis* over *S. mutans* in the dental biofilm may be associated with lower caries prevalence in both children and adults (Ge et al., 2008; Giacaman et al., 2015), and *S. sanguinis* seems to be among the species that are more prevalent in subjects with periodontal health (Mason et al., 2015).

As previously reported, oxygen availability is a crucial factor in the intrageneric competition. *S. sanguinis* can outcompete *S. mutans* via production of H_2O_2, since *S. mutans* does not produce significant amounts of H_2O_2 itself and is highly susceptible to the H_2O_2 antimicrobial activity. Valdebenito et al. speculated that the competitive advantage of *S. sanguinis* over *S. mutans* is mainly attributed to its glutathione peroxidase and capacity to undergo gluconeogenesis, as demonstrated by in silico analysis (Valdebenito et al., 2017). Glutathione peroxidase endows *S. sanguinis* with the capacity to resist its own H_2O_2 production, while gluconeogenesis occurs under low-nutrient conditions to allow continuous production of *S. mutans*-toxic peroxide. Interestingly, the inhibitory effect of *S. sanguinis* over *S. mutans* is accompanied by an inhibition of several *S. mutans* genes related with virulence. Therefore, H_2O_2 may be implicated not only in direct killing, but also in modulating the expression of *S. mutans* virulence genes (Wen et al., 2010). Importantly, ammonia production via arginine metabolism is another strategy used by many oral streptococci including *S. sanguinis* to compete with *S. mutans* and survive the acidification of oral biofilms (Huang et al., 2017). However, in retaliation, the acidogenicity of *S. mutans* can help lower the environmental pH to the point of inhibition of the growth of *S. sanguinis*. Under conditions where bacterial cells have enough energy to compete but not enough food for optimal growth, *S. mutans* produces mutacins I and IV that inhibit *S. sanguinis* (Kreth et al., 2005).

S. gordonii is also an ADS-positive bacterium, producing alkali in the form of ammonia that neutralize glycolytic acids and create an environment which is compatible with dental health (Nascimento and Burne, 2014). Besides, *S. gordonii* produces H_2O_2 that effectively affects *S. mutans* survival (Kreth et al., 2008). Unlike H_2O_2 production by *S. sanguinis*, *S. gordonii* producing H_2O_2 seems to be inversely correlated with carbohydrate availability (Zhu and Kreth, 2012). By contrast, the production of challisin is a more recognized feature of *S. gordonii*. Some strains of *S. gordonii* utilize this protease to reduce mutacin production and biofilm colonization of *S. mutans* by reducing the levels of the stimulating factor CSP (Kuramitsu et al., 2007; Wang et al., 2011). However, *S. gordonii* is sensitive to the mutacins produced by *S. mutans* , especially mutacin IV which is specifically active against members of the mitis group of oral streptococci (Kuramitsu et al., 2007).

Another species that needs attention is *Streptococcus oligofermentans*. Discovered in caries-free subjects, *S. oligofermentans* show relatively specific inhibitory effect on *S. mutans* without causing collateral damage to other streptococci due to its H_2O_2 production from lactic acid produced by *S. mutans* through lactate oxidase (Tong et al., 2007). The inhibition on *S. mutans* is available in biofilms at both neutral pH and cariogenic conditions (Bao et al., 2015). *S. oligofermentans* is thereby considered as a prominent probiotic candidate to compete against *S. mutans* at sites prone to caries.

Oral streptococci interactions with Actinomyces spp.

Several *Actinomyces spp.* have the cariogenic ability of acid production and acid tolerance (Tanner et al., 2018), and are implicated as pathogens of root surface caries (Dame-Teixeira et al., 2016). Nevertheless, actinomyces such as *Actinomyces naeslundii* metabolize carbohydrates into relatively weak acids (such as acetate) under aerobic conditions and also degrades lactate produced by other cohabitants into weak acids, thereby neutralizing dental biofilm pH (Oliveira et al., 2015).

Like oral streptococci, actinomyces are early colonizers of the tooth surface. Either physical or metabolic interactions between the two genera have profound influence on early plaque development and subsequent bacterial adhesion. Xiao et al. discovered that, the presence of *A. naeslundii* can enhance the expression of *gtfB/gtfC* genes in *S. mutans*, mediating the establishment of an EPS-rich matrix and forming more biomass (Jin Xiao, 2012). Thus, there is a potential interaction between *S. mutans* and *A. naelsundii*, which affects biofilms' glucose metabolism.

Lectin-like interactions between actinomyces and streptococci occur in both directions. *Actinomyces oris* provides both adhesins and receptors for coaggregation with oral streptococci. The type 2 fimbriae expressed by actinomyces target on streptococci cell wall phosphopolysaccharides containing the linkages GalNAcb1-3Gal or Galb1-3GalNAc (the principal mechanism of these intergeneric bindings), while the polysaccharide receptor on *A. oris* is recognized by *S. sanguinis* (Yang et al., 2014; Back et al., 2015)

Mutual assistance has been demonstrated between these early colonizers. Sialidase activity in *A. naeslundii*, along with the glycolytic and proteolytic activities in *S. gordonii*, provide nutrients supply for each other (Bradshaw et al., 1994). H_2O_2 secreted by *S. gordonii* probably serve as a key factor influencing interaction dynamics between these two species. As has been previously noted, *S. gordonii* can produce H_2O_2 at concentrations sufficient to kill many oral bacteria, but it cannot produce catalase to tolerate H_2O_2. *A. naeslundii* may help remove H_2O_2 from coaggregate cultures, protecting *S. gordonii* from oxidative damage (Jakubovics et al., 2008). Another problem is that H_2O_2 production is likely to deplete the intracellular arginine pool in *S. gordonii* and increase its requirement for arginine. Coaggregation with *A.naeslundii* can stabilize arginine biosynthesis in *S. gordonii*, overcoming the requirement and enabling the growth of *S. gordonii* in the absence of arginine (Jakubovics et al., 2008). In exchange, the ability of *A. naeslundii* to bind to *S. gordonii* contributes to its retention in biofilms under flowing saliva (Jr et al., 2001).

Oral streptococci interactions with Veillonella spp.

Oral *Veillonella spp.*, especially *V. parvula*, is associated with caries and intraradicular infections (Mashima and Nakazawa, 2015). Veillonellae are among the most predominant species in the oral cavity and coaggregate with many initial, early, middle and late colonizers. They are considered as bridging species similar to oral fusobacteria (Zhou et al., 2015), and usually coexist with their streptococcal coaggregation partners in specific parts of the mouth. Their coaggregation and the subsequent metabolic cooperation are of major importance in biofilm formation, and are also key elements in facilitating the succession of species in developing dental plaque (Periasamy and Kolenbrander, 2010).

Veillonellae were unable to establish monoinfections. Mcbride et al discovered that when rats were monoinfected by *S. mutans* firstly and subsequently infected with veillonellae, the number of veillonellae in coinfected animals' teeth increased significantly (Mcbride and Van der Hoeven, 1981). This experiment demonstrated that veillonellae and oral streptococci are metabolically linked. Although veillonellae are unable to ferment sugars, streptococcal fermentation of sugars to lactic acid can serve as a favored carbon source for them. A food chain can thus be developed between these bacteria with the end-product of one organism serving as the energy source for the other (Egland et al., 2004; Chalmers et al., 2008).

The synergistic relationship between veillonellae and oral streptococci was demonstrated early in the 1970s, consequently resulting in reduced caries activity and enamel demineralization (Van der Hoeven et al., 1978). But an opposite conclusion was made in later studies. A convincing reason could be that, although veillonellae form propionic and acetic acids (weaker than lactic acid) from metabolism of lactic acid, and propionic and acetic acids less likely dissolve the enamel of the teeth, veillonellae have been detected in high proportions in progressing incipient lesions (Milnes and Bowden, 1985). Consistent results acquired by molecular identification methods, demonstrated that significantly more veillonellae in dentinal lesions were detected than at any other site (Becker et al., 2002). More recently, significantly higher levels of veillonellae were detected in subjects with caries than those free of caries (Xu et al., 2014a; Thuy et al., 2015). Numerous findings indicated a strong association between *Veillonella spp.* and dental caries, thus the *Veillonella spp.* level in an individual is considered as a sensitive biologic indicator and early warning sign of dental caries. Gross et al. stated that, among children without history of caries, the presence of veillonellae helped to foresee a possible development of caries (Gross et al., 2012). Clinical data also demonstrated high *Veillonella spp.* levels are in association with more caries (Lima et al., 2011; Tanner et al., 2011).

The combination of veillonellae and *S.mutans* leads to more acid production and greater demineralization than the production and demineralization by *S. mutans* alone, as shown in an *in vitro* study (Noorda et al., 1988). The elevated acid production is probably facilitated by the lactate removal from the environment by veillonellae, creating a higher pH microenvironment (Marsh, 1994). An *in vivo* study showed that *Veillonella spp.* can mitigate the inhibitory effects of *S. gordonii* on *S. mutans* sugar metabolism, suggesting a specific interaction between *S.mutans* and *Veillonella spp.* that may be more complex than pH (Kreth et al., 2009; Liu et al., 2011). Promotion of acid production ensures better nutrition supply for veillonellae. In return to the utilization of lactic acid, *Veillonella spp.* also provides protection to *S. mutans*. Luppens et al. reported that *S. mutans* grow in a dual-species biofilm together with *V. parvula* is subjected to a physiological change, and acquires an advantage in its ability to survive under antimicrobial treatment (Luppens et al., 2008).

Various surface molecules and the streptococcal transcription factor catabolite control protein A are required for the interspecies interaction between *S. gordonii* and *Veillonella atypical* (Johnson et al., 2009). The adhesins and/or transcription factors may also involve in the interaction and biofilm formation of other veilloella-streptococcal pairs, as suggested by Mashima et al. (Mashima and Nakazawa, 2014). They also stated that, signaling molecules between bacterial cells should be considered as an important way of communication, in that *Veillonella tobetsuensis* promoted *S. gordonii* biofilm formation to the greatest extent without intergeneric coaggregation (Mashima and Nakazawa, 2014). They hypothesized that a small molecule such as AI produced by *V. tobetsuensis* may stimulate *S. gordonii* biofilm formation. In a following study, AI-1 and AI-2 were detected in the culture supernatants of *V. tobetsuensis* and the researchers concluded that these molecules (mainly AI-2) may play key roles in facilitating biofilm formation of *S. gordonii* (Mashima and Nakazawa, 2015). Besides, Zhou et al discovered that Hag 1, a multivalent hemagglutinin in *V.atypica*, is involved in its adherence to oral streptococci, *P. gingivalis* and human oral buccal cells (Zhou et al., 2015).

Oral streptococci interactions with Candida Albicans

Candida Albicans is able to survive as a commensal in several anatomically distinct sites. However, under certain circumstances, *C. albicans* can cause infections that range from superficial infections of the skin to life-threatening systemic infections (Mayer et al., 2013). *C. albicans* can exhibit two forms under different environmental conditions. The yeast form colonizes predominantly surfaces, whereas the hyphal form confers invasiveness to *C. albicans*, which can thus cause serious damages to human tissues (Gow et al., 2012; Ashkanane et al., 2017). Moreover, the hyphal form can provide structural integrity to biofilms (Banerjee et al., 2013). Oral candidiasis usually includes pseudomembraneous stomatitis, erythematous stomatitis and hyperplastic lesions.

C. albicans can coaggregate with a variety of oral commensal bacteria, especially in the range of oral streptococci. Oral streptococci were once believed to protect humans against oral candidiasis (Liljemark and Gibbons, 1973). However, with the development of research, the streptococcal species considered as avirulent can show its pathogenicity when coaggregated with *C. albicans*. Within the biofilms, fungal and bacterial cells use metabolites or cell contact-mediated signals to communicate with each other, further influencing gene expression, host responses and progression of the disease (Xu et al., 2017).

Interactions between *C. albicans* and commensal oral streptococci are bidirectional and generally considered as mutualistic beneficial. The synergistic relationship has been demonstrated in several aspects. In addition to providing adhesion sites, streptococci can also excrete lactate as a carbon source for yeast growth. In turn, yeast cells can reduce the oxygen tension to more preferable levels for streptococci, providing growth stimulatory factors for them (Diaz et al., 2012; Metwalli et al., 2013). In practical terms, *C. albicans* promotes the ability of streptococci to form biofilms in oral environment, and oral streptococci in turn enhances the growth of *C. albicans* biofilm and oral mucosa invasion (manifested by enhanced hyphal production and increased biomass) (Bamford et al., 2009).

S. gordonii cell wall-associated polypeptides SspA, SspB, CshA, and EPS-mediated interactions play important roles in binding to *C. albicans* (Xu et al., 2014b). Along with the secretion and/or modulation of QS molecules, these mechanisms together lead to synergism for their survival as mixed species biofilms (Diaz et al., 2012). *S. gordonii* provides an adherent surface to *C. albicans* thus facilitating its colonization on oral tissues. Moreover, *S. gordonii* can prevent *C. albicans* from detecting farnesol, a quorum sensor produced by the fungus that functions as a self-restriction signal (Bamford et al., 2009). The inhibition of farnesol detection leads to production of more robust biofilms, increasing pathogenicity and higher levels of antimicrobial resistance (Daniel et al., 2016).This aspect is of clinical significance particularly for immunocompromised individuals, because this synergism can subsequently develop into a fungal infection.

In another study focused on *S. oralis*, the introduction of *C. albicans* enhances mucosal biofilm formation by *S. oralis* (which lacks the ability to form robust mucosal biofilms), and the co-infection significantly increases the frequency and size of the oral thrush lesions in mice (Xu et al., 2014c). Xu et al. demonstrated that this synergism can activate host enzymes that cleave epithelial junction proteins, increasing fungal invasion (Xu et al., 2016).

The correlation between *C. albicans* and dental caries has also been highlighted. Indeed, high candidal presence in dental plaque and saliva are correlated with caries experience and severity (De-La-Torre et al., 2016; Fragkou et al., 2016; Moraga et al., 2016). With regard to caries etiology, in addition to the ability of *C. albicans* to produce and tolerate acids, much of the attention has focused on the relationship between *S. mutans* and *C. albicans* (Falsetta et al., 2014; Ellepola et al., 2017; Pereira et al., 2017a). A symbiotic relationship between these two species has been demonstrated to enhance the virulence of cospecies plaque biofilms, ultimately amplifying the severity of caries (Falsetta et al., 2014). As have been shown in an animal model, coinfection of rats with *S. mutans* and *C. albicans* enhances the colonization and carriage of both organisms *in vivo* and dramatically amplifies the virulence of plaque biofilms formed on rodent dentition, leading to the development of rampant carious lesions (Falsetta et al., 2014).

Unlike the binding mechanism between *S. gordonii* and *C. albicans*, which is sucrose-independent, *S. mutans* poorly binds to the fungal surface without sucrose. Gtf-derived EPS is a key mediator of cospecies biofilm development (Hwang et al., 2015; Ellepola et al., 2017), and it binds to the mannan layer of *C. albicans* independent of hyphae or other known major cell surface adhesins (Hwang et al., 2017). Moreover, Kim et al. discovered that bacterial-fungal derived metabolites increases the growth of *S. mutans* and its Gtf activity, furtherly altering the biofilm architecture into enlarged and densely packed bacterial cell-clusters (Kim et al., 2017). Co-cultivation with *C. albicans* also influences carbohydrate utilization by *S. mutans*, as was revealed by He et al. via

RNA sequencing demonstrating that the majority of up-regulated genes are related to carbohydrate transport and metabolic/catabolic processes (He et al., 2017).

Some streptococci exert an inhibition on fungal growth. *S. sanguinis* has an antagonistic effect on *C. albicans* growth due to its production of *S. sanguinis* bacteriocin. The bacteriocin can change the cell shape of *C. albicans*, increase the fungal cell membrane permeability, and reduce its adhesion ability (Ma et al., 2014; Ma et al., 2015; Ma et al., 2017) .

Besides, Ishijima et al. found that *S. salivarius* is effectively working against *C. albicans* growth and exerts a protective effect against candidiasis, as shown in a candidiasis model (Ishijima et al., 2012). *S. salivarius* is able to directly bond candida cells, thus inhibiting the adhesion of *C. albicans* to a plastic Petri dish and reducing *C. albicans*' ability to maintain its blastospore shape. MacDonald also showed that two strains of *S. salivarius* do not reduce yeast growth but inhibit its hyphae formation and adhesion to surfaces (Macdonald, 2015). The inhibition effect is likely to be attributed to protein secretion of *S. salivarius* (Fairiska et al., 2017).

Interactions between oral streptococci and periodontal pathogens

Periodontitis is a common infection worldwide and affects a large population (Kassebaum et al., 2014). This disease not only compromises the integrity of the tooth supporting tissues (gingiva, periodontal ligament and alveolar bone), but it is also associated with severe systemic conditions such as coronary artery disease, rheumatoid arthritis, and diabetes (Hajishengallis, 2015). Oral streptococci can develop extensive communication with periodontal pathogens. *S. gordonii* is metabolically compatible with definite periodontal pathogens such as *P. gingivalis*, *F. nucleatum* and *A. actinomycetemcomitans*. It has been proposed as one of the "Helper Pathogens Facilitating the Formation of Periodontal Disease" (Whitmore and Lamont, 2011). Unlike *S. gordonii*, some *Streptococcus spp.* such as *S. sanguinis* is often found in subgingival biofilm. It is correlated with a delay in colonization by periodontal pathogens, and it antagonizes a variety of periodontal pathogens (Lee, 2015; Herrero et al., 2016). In the following part several representative periodontal pathogens were selected and their relationships with oral streptococci were discussed.

Oral streptococci interactions with Porphyromonas gingivalis

P. gingivalis is a major pathogenic species involved in gingivitis and periodontitis. It has been designated as a "keystone pathogen" which induces dysbiosis through the manipulation of the host innate immune response, leading to uncontrolled inflammation and tissue damage (Hajishengallis and Lamont, 2014; Kalia et al., 2017). A high level of *P. gingivalis* in the subgingival oral biofilm is attributed to its extensive interactions with other gram-negative obligate and facultative anaerobes, such as *F. nucleatum*, *Treponema denticola* and *Tannerella forsythus* (Daep et al., 2008). However, its initial colonization in the oral cavity is in the supragingival biofilm (Wright et al., 2013). The fimbriae of *P. gingivalis* can attach to streptococci and both major and minor fimbriae are involved. The major fimbria (FimA) of *P. gingivalis* is linked to GAPDH located on the surface of streptococci, and the minor fimbria (Mfa) engages streptococcal cell surface protein SspA/B (Enersen et al., 2013; Wright et al., 2013; Wright et al., 2014).

Adherence of *P. gingivalis* to *Streptococcus spp.* such as *S. gordonii* is considered as an initial event that facilitates *P. gingivalis* colonization in the oral cavity (Wright et al., 2013; Kalia et al., 2017). Interbacterial coaggregation or coadhesion between *P. gingivalis* and *S. gordonii* enhances the colonization of both species in a biofilm model, where *S. gordonii* outcompetes *P. gingivalis* for attachment sites in the salivary pellicle, and substantial mixed bacterial biofilms develop on saliva-coated glass slides only when *S. gordonii* cells are plated to provide an attachment substrate for *P. gingivalis* (Cook et al., 1998). The synergistic interaction between *S. gordonii* and *P. gingivalis* is multifaceted, and metabolic interchanges between the two species have also been characterized. *S. gordonii* may deplete oxidants to allow the survival of *P. gingivalis*, and *P. gingivalis* secretes several proteases that may breakdown peptides for *S. gordonii* metabolism (Jenkinson, 2011). The proteolytic activity of *P. gingivalis* is increased when *S. gordonii* is present, and a community composed of *P. gingivalis* and *S. gordonii* is more pathogenic in animal models of periodontal diseases compared to each species alone (Daep et al., 2011; Whitmore and Lamont, 2011). Therefore, this synergy may play an important role in the development of bacterial populations associated with the onset and progression of severe periodontal disease forms (Forsgren et al., 2010).

In addition to an increased periodontal pathogenicity of the biofilm, the coadhesion of *P. gingivalis* with streptococci such as *S. gordonii* and *S. mutans* is important in the invasion of dentinal tubules, inducing infections of the root canal system (etiology of pulpal and periapical diseases) (Love et al., 2000). The recognition of collagen type I present within the tubules by streptococcal antigen

I/II polypeptides is essential for the bacterial invasion and for the intratubular growth (Love et al., 1997). Although *P. gingivalis* can also bind to collagen type I deposited onto hydroxylapatite surfaces (Naito et al., 2010), its ability to bind type I collagen within dentinal tubules alone is not sufficient to promote the invasion of tubules. The binding of *P. gingivalis* to intratubular collagen just probably help the invasion process (Love et al., 2000). Due to its noninvasive nature, *P. gingivalis* is able to penetrate tubules only after coadhering to the invasive partner streptococci.

A synergy in biofilm formation between *P. gingivalis* and *S. oralis*, another species of the mitis group, has also been uncovered by Maeda et al. via proteomic and transcriptional analysis. They observed an overexpression of *P. gingivalis* FimA and *S. oralis* GAPDH in mixed-biofilm and concluded that *S. oralis* regulates the transcriptional activity of *P. gingivalis luxS* (Maeda et al., 2015).

Not all *Streptococcus spp.* seem to offer help when confronted with this pathogen. Known that *S. sanguinis* have measures to limit the overgrowth of *P. gingivalis* via its aforementioned H_2O_2 production ability, the study conducted by Ma et al. also showed significant inhibitory effects on *P. intermedia* exerted by intracellular proteins extracted from *S. sanguinis* (Ma et al., 2014). Besides, Lee demonstrated that *S. sanguinis* peptidoglycan inhibited the cytokine expression triggered by lipopolysaccharide in *P. gingivalis* and may alleviate inflammatory responses (Lee, 2015).

Another species which may hamper the pathogenesis of *P. gingivalis* is

Streptococcus cristatus. It has been reviewed that there is an opposite relationship between the number of *S. cristatus* and *P. gingivalis* in the dental plaque isolated from periodontitis subjects, suggesting that *S. cristatus* may be beneficial to the host by antagonizing the colonization and accumulation of *P. gingivalis* (Wang et al., 2009). The arginine deiminase of *S. cristatus* represses the expression of FimA in *P. gingivalis* (Christopher et al., 2010), and also represses the expression of several well-known virulence genes involved in the production of gingipains (Ho et al., 2017a, b).

Oral streptococci interactions with Fusobacteria nucleatum

Fusobacteria are the most frequent Gram-negative bacteria in dental plaque, binding to a diverse array of microbial species. Fusobacteria are present in both supragingival and subgingival plaque, and are predominantly involved in both caries and periodontal diseases (Haffajee et al., 2008; Lima et al., 2011; Aruni et al., 2015). Most worthy of mention is *F. nucleatum*, which is described as a “bridge” connecting early and late colonizers, making great contributions to biofilm formation and architecture (Denes and Barraud, 2016; Lima et al., 2017). Regarded as a periodontal pathogenic bacteria, *F. nucleatum* is consistently associated with, and increased in number in periodontitis sites (Signat et al., 2011).

Within dental plaque, *F. nucleatum* is often found with streptococci in “corncob” formations (Lancy et al., 1983). The outer membrane protein RadD and CmpA, two arginine-inhibitable adhesins, are major fusobacterial adhesins allowing the physical attachment to *S. sanguinis and S.gordonii* (Kaplan et al., 2009; Lima et al., 2017). By adhering to *S. sanguinis*, *F. nucleatum* is able to mask the surface components and evade detection by antagonistic oral bacteria, thereby overcoming the colonization resistance (He et al., 2012). The adherence also triggers a specific cellular response by *F. nucleatum* and results in increased resistance to environmental stress (He et al., 2012). By this way, *F. nucleatum* integration into Gram-positive bacteria dominating supragingival microbial community is greatly facilitated.

In addition to physical contact mediated by the RadD and CmpA adhesins, in a more recent study, Sakanaka et al. discovered that *S. gordonii* arginine-ornithine antiporter-mediated ornithine efflux is indispensable for a successful colonization by *F. nucleatum*, where *F. nucleatum* utilizes the ornithine released by *S. gordonii* antiporter as a substrate of ornithine decarboxylase (Sakanaka et al., 2015). This leads to *S. gordonii-F. nucleatum* community development and finally result in the formation of a middle-stage periodontopathic biofilm.

Unlike the adhesion to *S. sanguinis* or *S. gordonii,* the specific interaction between *F. nucleatum* and *S. mutans* is mediated by the RadD-SpaP adhesin pair (Guo et al., 2016). The interaction can broaden the possibility of integration into oral streptococci biofilm, and is potentially beneficial, since *F. nucleatum* was found to have acid-neutralizing abilities (Takahashi et al., 1997), which may help decrease the risk of dental caries.

The binding of another early colonizer, *S.cristatus*, to *F. nucleatum*, which is mediated by the *S.cristatus* fibrillar tufts, arouse researchers' concern. Since *Fusobacteria spp.* can penetrate the epithelium (during periodontitis) while *S.cristatus* cannot, the adhesion to invasive *F. nucleatum* helps noninvasive *S.*

cristatus to enter epithelial cells (Edwards et al., 2006). As a consequence of that, *S. cristatus* attenuates the expression of a number of inflammatory cytokines induced by *F. nucleatum*, and upregulates several anti-inflammatory mediators, reducing the proinflammatory effect of *F. nucleatum* (Zhang and Rudney, 2011). This is of clinical significance because the mucosal hyporesponsiveness to invasive bacteria may also provide a possibility for these pathogens to later colonize the gingival crevice and remote locations (Zhang and Rudney, 2011), which is related to periodontal pathogenesis.

However, not all oral streptococci develop a mutualistic beneficial relationship with *F. nucleatum.* Jang et al. discovered that *F. nucleatum* AI-2 stimulated *S. gordonii* biofilm growth and aggregation of *F. nucleatum* with *S. gordonii*, while inhibited *S. oralis* biofilm growth and aggregation of *F. nucleatum* with *S. oralis*, indicating that initial colonizing streptococci affects sequential colonization of *F. nucleatum* in dental biofilms and enrichment of *S. oralis* in initial biofilm may reduce *F. nucleatum* attachment (Jang et al., 2013).

Oral streptococci interactions with Prevotella intermedia

Prevotella intermedia is strongly associated with the acute necrotizing ulcerative gingivitis and pregnancy-induced gingivitis (Chung et al., 1983). It also plays an important role in producing volatile sulfur compounds (Tanaka et al., 2004), causing halitosis.

S. salivarius is considered as a probiotic bacteria due to its ability to compete with the colonization of bacteria that increase volatile sulfur compounds (Burton et al., 2006a; Hyink et al., 2007; Burton et al., 2010). Therefore, its influence on *P.*

intermedia is of great clinical significance. The coaggregation between *S. salivarius* and *P. intermedia* is special. When tested by a fimbriae-negative mutant of *S. salivarius*, coaggregation is observed with *F. nucleatum* and with *P. gingivalis*, but not with *P. intermedia*. Only fimbriaed *S. salivarius* cells coaggregated with *P. intermedia*, suggesting that *S. salivarius-P. intermedia* coaggregation is mediated by *S. salivarius* fimbriae (Lévesque et al., 2003). *S. salivarius K12* showed an antibacterial effect against *P. intermedia* when exceeds a certain concentration (70%), leading to significant decrease of volatile sulfur compounds both *in vitro* and in human mouth (Burton et al., 2006b; Moon et al., 2016).

Oral streptococci interactions with Aggregatibacter actinomycetemcomitans

A. actinomycetemcomitans can be found in healthy individuals as part of the normal flora, and it is the predominant pathogen associated with localized aggressive periodontitis (Gholizadeh et al., 2017; Vaniabella et al., 2017). The most prominent pathogenic traits of *A. actinomycetemcomitans* should be its role in the decline of host immune response and degradation of gingival epithelial attachment on periodontal tissues (Vaniabella et al., 2017). Aggregatibacteria and streptococci are found in plaque samples as dense aggregates at the tips of abundant hedgehog structures (Mark Welch et al., 2016). Similar to veillonellae, *A. actinomycetemcomitans* entirely depends on other oral microorganisms to grow in saliva (Kolenbrander, 2011).

S. gordonii excretes lactate acid as a carbon source for *A. actinomycetemcomitans* and in exchange, the latter detoxifies H_2O_2 that *S. gordonii* produces. Their mutual benefit is also shown in a murine abscess model, where both species grow more virulent when they are together than when they are alone (Stacy et al., 2014; Stacy et al., 2016). Moreover, when cocultured with *S. gordonii* and senses the streptococcal metabolite H_2O_2, *A. actinomycetemcomitans* displays an enhanced resistance of being killed by the host innate immunity (Ramsey and Whiteley, 2009).

A. actinomycetemcomitans also interacts with *S. mutans*. In dual-species biofilms grown on artificial saliva, *A.actinomycetemcomitans* triggers the expression of the QS regulon of *S. mutans*, resulting in an up-regulation of the transformasome and mutacin related genes, and down-regulation of oxidative stress related genes (Szafrański et al., 2017). It can be concluded that *A. actinomycetemcomitans* protects *S. mutans* from oxidative stress possibly by the aerobic respiration. Simultaneously, *A. actinomycetemcomitans* grows in a highly virulent form but down-regulates genes important for its escape from the host immune response.

Not all *Streptococcus sp.* develop a synergistic relationship with *A. actinomycetemcomitans*. Species such as *S. sanguinis*, *S. mitis*, and *S. salivarius* show prominent inhibitory effects on *A. actinomycetemcomitans* recovery and colonization but without a bactericidal activity (Teughels et al., 2007), and the growth of tested streptococcal strains are also affected by *A. actinomycetemcomitans*. In a recent study, the inhibitory effect of *S. salivarius* on the growth of *A. actinomycetemcomitans* was further verified and the researchers proposed that the production of lactic acid and lantibiotics by *S. salivarius* should be the major cause (Vaniabella et al., 2017): The lactic acid produced by *S. salivarius* can increase the permeability of the outer membrane of *A. actinomycetemcomitans* and thus increase its sensitization of lantibiotics.

Conclusion and strategies for prevention and therapy regarding microbial interactions

Oral streptococci play a pivotal role in dental biofilms formation and interact with multiple microorganisms. These microbial interactions, either cooperative or competitive, may promote the establishment of a pathogenic community, causing problems such as dental caries and periodontal diseases. Ultimately, the goal of investigating these microbial interactions is to open the way for controlling oral diseases. Taking dental caries as an example, the removal of dental plaque and the application of fluoride and antimicrobial agents have always been the mainstay through the years. However, the fast accumulation of dental plaque, continual debate on the safety of fluoride use and increasingly severe antimicrobial resistance are problems we must face at this stage. There is an urgent need to develop new preventive and therapeutic methods against oral diseases. Since many oral diseases originate from the dysbiosis of dental biofilms, oral care strategies should place emphasis on maintaining the composition and activity of these biofilms at levels compatible with oral health rather than trying to eliminate them (Marsh, 2016). Microbial interactions within dental biofilms, as have been discussed in the present review, can help assembly and convert distinct bacterial cells to a pathogenic biofilm. Either utilizing or fighting against these interactions should be considered as feasible strategies to cure oral diseases.

One of the strategies being closely studied is the adoption of probiotics. Probiotics are defined as live microorganisms that, when administered in adequate amounts, confer a health benefit on the host (Gruner et al., 2016). Probiotic bacteria can bind and compete with already coaggregated pathogenic bacteria for nutritive sources or produce chemical substances that inhibit the development of pathogenic bacteria, in order to promote oral health without negatively impacting the normal oral macrobiotic of the host (Burton et al., 2013; Zambori et al., 2016; Pereira et al., 2017b). A range of bacteria (most of them being acidogenic such as lactobacilli, streptococci or bifidobacteria) exert such effects (Gruner et al., 2016). Taking probiotic lactobacilli as an example, they coaggregate with *S. mutans* and other caries-associated strains, inhibiting the growth of these microorganisms (Lin et al., 2018). By comparing the efficacy of milk supplemented with *Lactobacillus rhamnosus* with standard milk in preschool children, a recent clinical trial demonstrated a significantly lower increment of caries in the study group than the control group after 10 months of intervention (Rodríguez et al., 2016). In addition, a probiotic blend containing *S. salivarius*, *L. reuteri* and *L. paracasei* is a useful adjunct to scaling and root planing in chronic periodontitis patients (Mani et al., 2017).

Another approach, which is the interference with the cell-cell communication system targeting the QS signaling pathways, has also been considered as a possible solution to oral diseases, especially in this time of increasing antibiotic resistance and treatment failures. Because of the significant role of signaling molecules in coordinating gene expression and promoting biofilm formation, there is an impetus to investigate the potential of inhibitory analogues to disrupt these networks, thereby providing mechanisms to control or influence the development of dental plaque (Parashar et al., 2015; Pérez et al., 2018).To date, QS inhibitors and quorum quenching enzymes have been investigated for their

QS interfering capabilities (Fong et al., 2018). furthermore, addition of exogenous CSP or QS-modifying compounds showed effects on maintaining a healthy microbial ecology in dental plaque (Philip et al., 2018).

The specifically targeted antimicrobial peptide (STAMP), a synthetic fusion peptide, is also an ecological approach. It is based on the addition of a targeting peptide to an existing broad spectrum antimicrobial peptide (AMP), making it selective for a particular bacterial species or strain (He et al., 2010). It specifically targets pathogens such as *S. mutans* from multispecies biofilms and show membrane-disrupting activities (Philip et al., 2018). As a result, not only the pathogens are eliminated, but a more benign oral microbial community is established (Guo et al., 2015). A clinical study demonstrated the efficacy of a mouth rinse containing a STAMP (C16G2) resulting in a significantly lower levels of *S. mutans* in the plaque and saliva samples after rinsing (Sullivan et al., 2011). Furthermore, this rinse is effective in preventing *S. mutans* regrowth in spite of frequent exposure to sugar.

As mentioned above, new strategies are emerging and changing the way of preventing oral diseases from widespread killing to mediation of microbial communications. We can say that we are embracing a new era with more effective disease control and better oral health.

Acknowledgements

This work was supported by the National Natural Science Foundation of China (31200985, 81470035, and 31870065).

References

Aruni, A.W., Dou, Y., Mishra, A., and Fletcher, H.M. (2015). The Biofilm Community-Rebels with a Cause. Curr Oral Health Rep, 2, 48-56. DOI: 10.1007/s40496-014-0044-5.

Ashkanane, A., Gomez, G.F., Levon, J., Windsor, L.J., Eckert, G.J., and Gregory, R.L. (2017). Nicotine Upregulates Coaggregation of *Candida albicans* and *Streptococcus mutans*. J Prosthodont. DOI:10.1111/jopr.12643.

Back, C.R., Douglas, S.K., Emerson, J.E., Nobbs, A.H., and Jenkinson, H.F. (2015). *Streptococcus gordonii* DL1 adhesin SspB V-region mediates coaggregation via receptor polysaccharide of *Actinomyces oris* T14V. Mol Oral Microbiol, 30, 411-424. DOI: 10.1111/omi.12106.

Bamford, C.V., D'Mello, A., Nobbs, A.H., Dutton, L.C., Vickerman, M.M., and Jenkinson, H.F. (2009). *Streptococcus gordonii* Modulates *Candida albicans* Biofilm Formation through Intergeneric Communication. Infect Immun, 77, 3696-3704. DOI: 10.1128/IAI.00438-09.

Banerjee, M., Uppuluri, P., Zhao, X.R., Carlisle, P.L., Vipulanandan, G., Villar, C.C., Lópezribot, J.L., and Kadosh, D. (2013). Expression of *UME6*, a key regulator of *Candida albicans* hyphal development, enhances biofilm formation via Hgc1- and Sun41-dependent mechanisms. Eukaryot Cell, 12, 224-232. DOI: 10.1128/EC.00163-12.

Bao, X., de Soet, J.J., Tong, H., Gao, X., He, L., Van, L.C., and Deng, D.M. (2015). *Streptococcus oligofermentans* Inhibits *Streptococcus mutans* in Biofilms at Both Neutral pH and Cariogenic Conditions. PLoS One, 10, e0130962. DOI: 10.1371/journal.pone.0130962.

Basavaraju, M., Sisnity, V.S., Palaparthy, R., and Addanki, P.K. (2016). Quorum quenching: Signal jamming in dental plaque biofilms. J Dent Sci, 11, 349-352.

DOI: 10.1016/j.jds.2016.02.002.

Becker, M.R., Paster, B.J., Leys, E.J., Moeschberger, M.L., Kenyon, S.G., Galvin, J.L., Boches, S.K., Dewhirst, F.E., and Griffen, A.L. (2002). Molecular analysis of bacterial species associated with childhood caries. J Clin Microbiol, 40, 1001-1009. DOI: 10.1128/JCM.40.3.1001-1009.2002.

Bowen, W.H., and Koo, H. (2011). Biology of *Streptococcus mutans*-derived glucosyltransferases: role in extracellular matrix formation of cariogenic biofilms. Caries Res, 45, 69-86. DOI: 10.1159/000324598.

Bradshaw, D.J., Homer, K.A., Marsh, P.D., and Beighton, D. (1994). Metabolic cooperation in oral microbial communities during growth on mucin. Microbiology, 140, 3407-3412. DOI: 10.1099/13500872-140-12-3407.

Burton, J.P., Wescombe, P.A., Moore, C.J., Chilcott, C.N., and Tagg, J.R. (2006a). Safety assessment of the oral cavity probiotic *Streptococcus salivarius* K12. Appl Environ Microbiol, 72, 3050-3053. DOI: 10.1128/AEM.72.4.3050-3053.2006.

Burton, J.P., Chilcott, C.N., Moore, C.J., Speiser, G., and Tagg, J.R. (2006b). A preliminary study of the effect of probiotic *Streptococcus salivarius* K12 on oral malodour parameters. J Appl Microbiol, 100, 754–764. DOI: 10.1111/j.1365-2672.2006.02837.x.

Burton, J.P., Chilcott, C.N., and Tagg, J.R. (2010). The rationale and potential for the reduction of oral malodour using *Streptococcus salivarius* probiotics. Oral Dis, 11, 29-31. DOI: 10.1111/j.1601-0825.2005.01084.x.

Burton, J.P., Wescombe, P.A., Macklaim, J.M., Chai, M.H.C., Kyle, M.D., Hale, J.D.F., John, T., Gregor, R., Gloor, G.B., and Cadieux, P.A. (2013). Persistence of the Oral Probiotic *Streptococcus salivarius* M18 Is Dose Dependent and Megaplasmid Transfer Can Augment Their Bacteriocin Production and Adhesion Characteristics. PLoS One, 8, e65991. DOI: 10.1371/journal.pone.0065991.

Caufield, P.W., Dasanayake, A.P., Li, Y., Pan, Y., Hsu, J., and Hardin, J.M. (2000). Natural history of *Streptococcus sanguinis* in the oral cavity of infants: evidence for a discrete window of infectivity. Infect Immun, 68, 4018-4023.

Chalmers, N.I., Jr, R.J.P., Cisar, J.O., and Kolenbrander, P.E. (2008). Characterization of a *Streptococcus sp.-Veillonella sp.* Community Micromanipulated from Dental Plaque. J Bacteriol, 190, 8145-8154. DOI: 10.1128/JB.00983-08.

Christopher, A.B., Arndt, A., Cugini, C., and Davey, M.E. (2010). A streptococcal effector protein that inhibits *Porphyromonas gingivalis* biofilm development. Microbiology, 156, 3469-3477. DOI: 10.1099/mic.0.042671-0.

Chung, C.P., Nisengard, R.J., Slots, J., and Genco, R.J. (1983). Bacterial IgG and IgM antibody titers in acute necrotizing ulcerative gingivitis. J Periodontol, 54, 557-562. DOI: 10.1902/jop.1983.54.9.557.

Cisar, J.O., Sandberg, A.L., Reddy, G.P., Abeygunawardana, C., and Bush, C.A. (1997). Structural and antigenic types of cell wall polysaccharides from viridans group streptococci with receptors for oral actinomyces and streptococcal lectins. Infect Immun, 65, 5035-5041.

Cook, G.S., Costerton, J.W., and Lamont, R.J. (1998). Biofilm formation by *Porphyromonas gingivalis* and *Streptococcus gordonii*. J Periodontal Res, 33, 323-327.

Daep, C.A., Lamont, R.J., and Demuth, D.R. (2008). Interaction of *Porphyromonas gingivalis* with oral streptococci requires a motif that resembles the eukaryotic nuclear receptor box protein-protein interaction domain. Infect Immun, 6, 3273-3280. DOI: 10.1128/IAI.00366-08.

Daep, C.A., Novak, E.A., Lamont, R.J., and Demuth, D.R. (2011). Structural dissection and in vivo effectiveness of a peptide inhibitor of *Porphyromonas gingivalis* adherence to *Streptococcus gordonii.* Infect Immun, 79, 67-74. doi: 10.1128/IAI.00361-10.

Dame-Teixeira, N., Parolo, C.C.F., Maltz, M., Tugnait, A., Devine, D., and Do, T. (2016). *Actinomyces spp.* gene expression in root caries lesions. J Oral Microbiol, 8:1, 32383, DOI: 10.3402/jom.v8.32383.

Daniel, M.J., Anand, S., Ramasubramanian, A.K., and Lopez-Ribot, J.L. (2016). An In Vitro Model for Oral Mixed Biofilms of *Candida albicans* and *Streptococcus gordonii* in Synthetic Saliva. Front Microbiol, 7, 686. DOI: 10.3389/fmicb.2016.00686.

De-La-Torre, J., Marichalar-Mendia, X., Varona-Barquin, A., Marcos-Arias, C., Eraso, E., Aguirre-Urizar, J.M., and Quindós, G. (2016). Caries and Candida colonisation in adult patients in Basque Country (Spain). Mycoses, 59, 234-240. DOI: 10.1111/myc.12453.

Denes, E., and Barraud, O. (2016). *Fusobacterium nucleatum* infections: clinical spectrum and bacteriological features of 78 cases. Infection, 44, 475-481. DOI: 10.1007/s15010-015-0871-x.

Diaz, P.I., Chalmers, N.I., Rickard, A.H., Kong, C., Milburn, C.L., Jr, P.R., and Kolenbrander, P.E. (2006). Molecular characterization of subject-specific oral microflora during initial colonization of enamel. Appl Environ Microbiol, 72, 2837-2848. DOI: 10.1128/AEM.72.4.2837-2848.2006.

Diaz, P.I., Xie, Z., Sobue, T., Thompson, A., Biyikoglu, B., Ricker, A., Ikonomou, L., and Dongaribagtzoglou, A. (2012). Synergistic interaction between *Candida albicans* and commensal oral streptococci in a novel in vitro mucosal model. Infect Immun, 80, 620-632. DOI: 10.1128/IAI.05896-11.

Edwards, A.M., Grossman, T.J., and Rudney, J.D. (2006). *Fusobacterium nucleatum* transports noninvasive *Streptococcus cristatus* into human epithelial cells. Infect Immun, 74, 654-662. DOI: 10.1128/IAI.01559-06.

Egland, P.G., Dû, L.D., and Kolenbrander, P.E. (2001). Identification of Independent *Streptococcus gordonii* SspA and SspB Functions in Coaggregation with *Actinomyces naeslundii.* Infect Immun, 69, 7512-7516. DOI: 10.1128/IAI.69.12.7512-7516.2001.

Egland, P.G., Palmer, R.J., and Kolenbrander, P.E. (2004). Interspecies communication in *Streptococcus gordonii-Veillonella atypica* biofilms: signaling in flow conditions requires juxtaposition. Proc Natl Acad Sci U S A, 101, 16917-16922. DOI: 10.1073/pnas.0407457101.

Ellepola, K., Liu, Y., Cao, T., Koo, H., and Seneviratne, C.J. (2017). Bacterial GtfB Augments *Candida albicans* Accumulation in Cross-Kingdom Biofilms. J Dent Res, 96, 1129-1135. DOI: 10.1177/0022034517714414.

Elliott, D., Harrison, E., Handley, P.S., Ford, S.K., Jaffray, E., Mordan, N., and Mcnab, R. (2003). Prevalence of Csh-like fibrillar surface proteins among mitis group oral streptococci. Oral Microbiol Immunol, 18, 114-120.

Enersen, M., Nakano, K., and Amano, A. (2013). *Porphyromonas gingivalis* fimbriae. J Oral Microbiol, 5. DOI: 10.3402/jom.v5i0.20265.

Fairiska, N., Bachtiar, B.M., and Gultom, F.P. (2017). Analysis of the Inhibitory Potential of *Streptococcus Salivarius* Isolated from Adult Saliva and the Tongue Dorsum for the Growth of *Candida Albicans*. Int J Appl Pharm, 9, 3-7. DOI: http://dx.doi.org/10.22159/ijap.2017.v9s1.05.

Falsetta, M.L., Klein, M.I., Colonne, P.M., Scott-Anne, K., Gregoire, S., Pai, C.H., Gonzalez-Begne, M., Watson, G., Krysan, D.J., and Bowen, W.H. (2014). Symbiotic relationship between *Streptococcus mutans* and *Candida albicans* synergizes virulence of plaque biofilms in vivo. Infect Immun, 82, 1968-1981. DOI: 10.1128/IAI.00087-14.

Flemming, H.C., and Wingender, J. (2010). The biofilm matrix. Nat Rev Microbiol, 8, 623-633. DOI: 10.1038/nrmicro2415.

Fong, J., Zhang, C., Yang, R., Boo, Z.Z., Tan, S.K., Nielsen, T.E., Givskov, M., Liu, X.W., Bin, W., and Su, H. (2018). Combination Therapy Strategy of Quorum Quenching Enzyme and Quorum Sensing Inhibitor in Suppressing Multiple Quorum Sensing Pathways of P. aeruginosa. Sci Rep, 8, 1155. DOI: 10.1038/s41598-018-19504-w.

Forsgren, N., Lamont, R.J., and Persson, K. (2010). Two intramolecular isopeptide bonds are identified in the crystal structure of the *Streptococcus gordonii* SspB C-terminal domain. J Mol Biol, 397, 740-751. DOI: 10.1016/j.jmb.2010.01.065.

Fragkou, S., Balasouli, C., Tsuzukibashi, O., Argyropoulou, A., Menexes, G., Kotsanos, N., and Kalfas, S. (2016). *Streptococcus mutans*, *Streptococcus sobrinus* and *Candida albicans* in oral samples from caries-free and caries-active children. Eur Arch Paediatr Dent, 17, 1-9. DOI: 10.1007/s40368-016-0239-7.

Ge, Y., Caufield, P.W., Fisch, G.S., and Li, Y. (2008). *Streptococcus mutans* and *Streptococcus sanguinis* Colonization Correlated with Caries Experience in Children. Caries Res, 42, 444-448. DOI: 10.1159/000159608.

Gholizadeh, P., Pormohammad, A., Eslami, H., Shokouhi, B., Fakhrzadeh, V., and Kafil, H.S. (2017). Oral pathogenesis of *Aggregatibacter actinomycetemcomitans*. Microb Pathog 113, 303-311. DOI: 10.1016/j.micpath.2017.11.001.

Giacaman, R.A., Torres, S., Gómez, Y., Muñoz-Sandoval, C., and Kreth, J. (2015). Correlation of *Streptococcus mutans* and *Streptococcus sanguinis* colonization and ex vivo hydrogen peroxide production in carious lesion-free and high caries adults. Arch Oral Biol, 60, 154-159. DOI:

10.1016/j.archoralbio.2014.09.007.

Gow, N.A., Fl, V.D.V., Brown, A.J., and Netea, M.G. (2012). *Candida albicans* morphogenesis and host defence: discriminating invasion from colonization. Nat Rev Microbiol, 10, 112-122. DOI: 10.1038/nrmicro2711.

Gross, E.L., Beall, C.J., Kutsch, S.R., Firestone, N.D., Leys, E.J., and Griffen, A.L. (2012). Beyond *Streptococcus mutans*: Dental Caries Onset Linked to Multiple Species by 16S rRNA Community Analysis. PLoS One, 7, e47722. DOI: 10.1371/journal.pone.0047722.

Gruner, D., Paris, S., and Schwendicke, F. (2016). Probiotics for managing caries and periodontitis: Systematic review and meta-analysis. J Dent 48, 16-25. DOI: 10.1016/j.jdent.2016.03.002.

Guo, L., He, X., and Shi, W. (2014). Intercellular communications in multispecies oral microbial communities. Front Microbiol, 5, 328. DOI: 10.3389/fmicb.2014.00328.

Guo, L., Mclean, J.S., Yang, Y., Eckert, R., Kaplan, C.W., Kyme, P., Sheikh, O., Varnum, B., Lux, R., and Shi, W. (2015). Precision-guided antimicrobial peptide as a targeted modulator of human microbial ecology. Proc Natl Acad Sci U S A, 112, 7569-7574. DOI: 10.1073/pnas.1506207112.

Guo, L., Shokeen, B., He, X., Shi, W., and Lux, R. (2016). *Streptococcus mutans* SpaP binds to RadD of *Fusobacterium nucleatum* ssp. polymorphum. Mol Oral Microbiol, 32, 355-364. DOI: 10.1111/omi.12177.

Haffajee, A.D., Socransky, S.S., Patel, M.R., and Song, X. (2008). Microbial complexes in supragingival plaque. Oral Microbiol Immunol, 23, 196-205. DOI: 10.1111/j.1399-302X.2007.00411.x.

Hajishengallis, G. (2015). Periodontitis: from microbial immune subversion to systemic inflammation. Nat Rev Immunol., 15, 30-44. DOI: 10.1038/nri3785.

Hajishengallis, G., and Lamont, R.J. (2014). Breaking bad: Manipulation of the host response by *Porphyromonas gingivalis*. Eur J Immunol, 44, 328-338. DOI: 10.1002/eji.201344202.

He, J., Hwang, G., Liu, Y., Gao, L., Kilpatrickliverman, L.T., Santarpia, P., Zhou, X., and Koo, H. (2016). l-Arginine Modifies the Exopolysaccharide Matrix and Thwarts *Streptococcus mutans* Outgrowth within Mixed-Species Oral Biofilms. J Bacteriol., 198, 2651-2661. DOI: 10.1128/JB.00021-16.

He, J., Kim, D., Zhou, X., Ahn, S.J., Burne, R., Richards, V., and Koo, H. (2017). RNA-Seq Reveals Enhanced Sugar Metabolism in *Streptococcus mutans* Co-cultured with *Candida albicans* within Mixed-Species Biofilms. Front Microbiol, 8, 1036. DOI: 10.3389/fmicb.2017.01036.

He, J., Yarbrough, D.K., Kreth, J., Anderson, M.H., Shi, W.Y., and Eckert, R. (2010). Systematic approach to optimizing specifically targeted antimicrobial peptides against *Streptococcus mutans*. Antimicrob Agents Chemother, 54, 2143-2151. DOI: 10.1128/AAC.01391-09.

He, X., Hu, W., Kaplan, C.W., Guo, L., Shi, W., and Lux, R. (2012). Adherence to Streptococci Facilitates *Fusobacterium nucleatum* Integration into an Oral Microbial Community. Microb Ecol, 63, 532-542. DOI: 10.1007/s00248-011-9989-2.

Herrero, E.R., Slomka, V., Bernaerts, K., Boon, N., Hernandez-Sanabria, E., Passoni, B.B., Quirynen, M., and Teughels, W. (2016). Antimicrobial effects of commensal oral species are regulated by environmental factors. J Dent, 47, 23-33. DOI: 10.1016/j.jdent.2016.02.007.

Ho, M.H., Lamont, R.J., and Xie, H. (2017a). Identification of *Streptococcus cristatus* peptides that repress expression of virulence genes in *Porphyromonas gingivalis*. Sci Rep, 7, 1413. DOI: 10.1038/s41598-017-01551-4.

Ho, M.H., Lamont, R.J., and Xie, H. (2017b). A novel peptidic inhibitor derived from *Streptococcus cristatus* ArcA attenuates virulence potential of *Porphyromonas gingivalis*. Sci Rep, 7, 16217. DOI: 10.1038/s41598-017-16522-y

Hojo, K., Nagaoka, S., Ohshima, T., and Maeda, N. (2009). Bacterial interactions in dental biofilm development. J Dent Res, 88, 982-990. DOI: 10.1177/0022034509346811.

Hsu, S.D., Cisar, J.O., Sandberg, A.L., and Kilian, M. (1994). Adhesive Properties of Viridans Streptoccocal Species. Microb Ecol Health Dis, 7, 125-137. DOI：10.3109/08910609409141342.

Huang, X., Browngardt, C.M., Jiang, M., Ahn, S.J., Burne, R.A., and Nascimento, M.M. (2017). Diversity in Antagonistic Interactions between Commensal Oral Streptococci and Streptococcus mutans. Caries Res, 52, 88-101. DOI: 10.1159/000479091.

Huang, X., Schulte, R.M., Burne, R.A., and Nascimento, M.M. (2015). Characterization of the Arginolytic Microflora Provides Insights into pH Homeostasis in Human Oral Biofilms. Caries Res, 49, 165-176. DOI: 10.1159/000365296.

Hwang, G., Liu, Y., Kim, D., Li, Y., Krysan, D.J., and Koo, H. (2017). *Candida albicans* mannans mediate *Streptococcus mutans* exoenzyme GtfB binding to modulate cross-kingdom biofilm development in vivo. PLoS Pathog, 13, e1006407. DOI: 10.1371/journal.ppat.1006407.

Hwang, G., Marsh, G., Gao, L., Waugh, R., and Koo, H. (2015). Binding Force Dynamics of *Streptococcus mutans*–glucosyltransferase B to *Candida albicans*. J Dent Res, 94, 1310-1317. DOI: 10.1177/0022034515592859.

Hyink, O., Wescombe, P.A., Upton, M., Ragland, N., Burton, J.P., and Tagg, J.R.

(2007). Salivaricin A2 and the novel lantibiotic salivaricin B are encoded at adjacent loci on a 190-kilobase transmissible megaplasmid in the oral probiotic strain Streptococcus salivarius K12. Appl Environ Microbiol, 73, 1107-1113. DOI: 10.1128/AEM.02265-06.

Ichinosawa, T., Ito, T., Yonezawa, H., Senpuku, H., and Shimizu, T. (2017). Molecular Interaction of the Analogous Peptide SspB (390-T400K-402) derived from *Streptococcus gordonii* Surface Protein Peptide with Periodontal Bacteria. Int J Oral-Med Sci, 15, 160-167.

Irie, Y., and Parsek, M.R. (2008). Quorum Sensing and Microbial Biofilms. Curr Top Microbiol Immunol, 322, 67-84.

Ishijima, S.A., Hayama, K., Burton, J.P., Reid, G., Okada, M., Matsushita, Y., and Abe, S. (2012). Effect of *Streptococcus salivarius* K12 on the in vitro growth of *Candida albicans* and its protective effect in an oral candidiasis model. Appl Environ Microbiol, 78, 2190-2199. DOI: 10.1128/AEM.07055-11.

Ito, T., Yoshida, Y., Shiota, Y., Ito, Y., Yamamoto, T., and Takashiba, S. (2017). Effects of Lectins on initial attachment of cariogenic *Streptococcus mutans*. Glycoconj J, 35, 41-51. DOI: 10.1007/s10719-017-9795-2.

Jakubovics, N.S., Gill, S.R., Vickerman, M.M., and Kolenbrander, P.E. (2008). Role of hydrogen peroxide in competition and cooperation between *Streptococcus gordonii* and *Actinomyces naeslundii*. FEMS Microbiol Ecol, 66, 637–644. DOI: 10.1111/j.1574-6941.2008.00585.x.

Jakubovics, N.S., Shields, R.C., Rajarajan, N., and Burgess, J.G. (2013). Life after death: the critical role of extracellular DNA in microbial biofilms. Lett Appl Microbiol, 57, 467-475. DOI: 10.1111/lam.12134.

Jakubovics, N.S., Yassin, S.A., and Rickard, A.H. (2014). Community interactions of oral streptococci. Adv Appl Microbiol, 87, 43-110. DOI:

10.1016/B978-0-12-800261-2.00002-5.

James, C.E. (2014). Recent Advances in Studies of Polymicrobial Interactions in Oral Biofilms. Curr Oral Health Rep, 1, 59-69.

Jang, Y.J., Sim, J., Jun, H.K., and Choi, B.K. (2013). Differential effect of autoinducer 2 of Fusobacterium nucleatum on oral streptococci. Arch Oral Biol, 58, 1594-1602. DOI: 10.1016/j.archoralbio.2013.08.006.

Jenkinson, H.F. (2011). Beyond the oral microbiome. Environ Microbiol, 13, 3077-3087.

Jin Xiao, M.I.K., Megan L. Falsetta, Bingwen Lu, Claire M. Delahunty, John R. Yates, III, Arne Heydorn, Hyun Koo (2012). The Exopolysaccharide Matrix Modulates the Interaction between 3D Architecture and Virulence of a Mixed-Species Oral Biofilm. PLoS Pathog, 8, e1002623. DOI: 10.1371/journal.ppat.1002623.

Johnson, B.P., Jensen, B.J., Ransom, E.M., Heinemann, K.A., Vannatta, K.M., Egland, K.A., and Egland, P.G. (2009). Interspecies signaling between *Veillonella atypica* and *Streptococcus gordonii* requires the transcription factor CcpA. J Bacteriol, 191, 5563-5565. DOI: 10.1128/JB.01226-08.

Jr, P.R., Kazmerzak, K., Hansen, M.C., and Kolenbrander, P.E. (2001). Mutualism versus independence: strategies of mixed-species oral biofilms in vitro using saliva as the sole nutrient source. Infect Immun, 69, 5794-5804. DOI: 10.1128/IAI.69.9.5794-5804.2001.

Kalia, P., Jain, A., Krishnan, R.R., Demuth, D.R., and Steinbachrankins, J.M. (2017). Peptide-modified nanoparticles inhibit formation of *Porphyromonas gingivalis* biofilms with *Streptococcus gordonii*. Int J Nanomedicine, 12, 4553-4562. DOI: 10.2147/IJN.S139178.

Kaplan, C.W., Lux, R., Haake, S.K., and Shi, W. (2009). The *Fusobacterium nucleatum* outer membrane protein RadD is an arginine - inhibitable adhesin required for inter - species adherence and the structured architecture of multispecies biofilm. Mol Microbiol, 71, 35-47. DOI: 10.1111/j.1365-2958.2008.06503.x.

Kassebaum, N.J., Bernabé, E., Dahiya, M., Bhandari, B., Murray, C.J., and Marcenes, W. (2015). Global burden of untreated caries: a systematic review and metaregression. J Dent Res, 94, 650-658. DOI: 10.1177/0022034515573272.

Kassebaum, N.J., Bernabé, E., Dahiya, M., Bhandari, B., Murray, C.J.L., and Marcenes, W. (2014). Global Burden of Severe Periodontitis in 1990-2010. J Dent Res, 93, 1045-1053.

Kim, D., Sengupta, A., Niepa, T.H., Lee, B.H., Weljie, A., Freitas-Blanco, V.S., Murata, R.M., Stebe, K.J., Lee, D., and Koo, H. (2017). Candida albicans stimulates Streptococcus mutans microcolony development via cross-kingdom biofilm-derived metabolites. Sci Rep, 7, 41332. DOI: 10.1038/srep41332.

Kolenbrander, P.E. (2011). Multispecies communities:interspecies interactions influence growth on saliva as sole nutritional source. Int J Oral Sci, 3, 49-54. DOI: 10.4248/IJOS11025.

Kolenbrander, P.E., Andersen, R.N., Blehert, D.S., Egland, P.G., Foster, J.S., and Jr, R.J.P. (2002). Communication among Oral Bacteria. Microbiol Mol Biol Rev, 66, 486-505.

Kreth, J., Giacaman, R.A., Raghavan, R., and Merritt, J. (2016). The road less traveled – defining molecular commensalism with *Streptococcus sanguinis*. Mol Oral Microbiol, 32, 181-196. DOI: 10.1111/omi.12170.

Kreth, J., Merritt, J., Shi, W., and Qi, F. (2005). Competition and Coexistence between *Streptococcus mutans* and *Streptococcus sanguinis* in the Dental

Biofilm. J Bacteriol, 187, 7193-7203. DOI: 10.1128/JB.187.21.7193-7203.2005.

Kreth, J., Vu, H., Zhang, Y., and Herzberg, M.C. (2009). Characterization of hydrogen peroxide-induced DNA release by *Streptococcus sanguinis* and *Streptococcus gordonii*. J Bacteriol, 191, 6281-6291. DOI: 10.1128/JB.00906-09.

Kreth, J., Zhang, Y., and Herzberg, M.C. (2008). Streptococcal Antagonism in Oral Biofilms: *Streptococcus sanguinis* and *Streptococcus gordonii* Interference with Streptococcus mutans. J Bacteriol, 190, 4632-4640. DOI: 10.1128/JB.00276-08.

Kuramitsu, H.K., He, X., Lux, R., Anderson, M.H., Shi, and Wenyuan (2007). Interspecies interactions within oral microbial communities. Microbiol Mol Biol Rev, 71, 653-670. DOI: 10.1128/MMBR.00024-07.

Lévesque, C., Lamothe, J., and Frenette, M. (2003). Coaggregation of *Streptococcus salivarius* with periodontopathogens: evidence for involvement of fimbriae in the interaction with Prevotella intermedia. Oral Microbiol Immunol, 18, 333–337.

Lancy, P., ., Dirienzo, J.M., Appelbaum, B., ., Rosan, B., ., and Holt, S.C. (1983). Corncob formation between *Fusobacterium nucleatum* and *Streptococcus sanguis*. Infect Immun, 40, 303-309.

Lee, S.H. (2015). Antagonistic effect of peptidoglycan of *Streptococcus sanguinis* on lipopolysaccharide of major periodontal pathogens. J Microbiol, 53, 553-560. DOI: 10.1007/s12275-015-5319-6.

Lei, L., Yang, Y., Mao, M., Hong, L., Meng, L., Yan, Y., Yin, J., and Tao, H. (2015). Modulation of Biofilm Exopolysaccharides by the *Streptococcus mutans vicX* Gene. Front Microbiol, 6, 1432. doi: 10.3389/fmicb.2015.01432.

Lemos, J.A., Jr, Q.R., Koo, H., and Abranches, J. (2013). *Streptococcus mutans*: a new Gram-positive paradigm? Microbiology 159, 436-445.

DOI: 10.1099/mic.0.066134-0.

Li, Y.H., Hanna, M.N., Svensäter, G., Ellen, R.P., and Cvitkovitch, D.G. (2001). Cell density modulates acid adaptation in *Streptococcus mutans*: implications for survival in biofilms. J Bacteriol, 183, 6875-6884. DOI: 10.1128/JB.183.23.6875-6884.2001.

Liljemark, W.F., and Gibbons, R.J. (1973). Suppression of *Candida albicans* by human oral streptococci in gnotobiotic mice. Infect Immun, 8, 846-849.

Lima, B.P., Shi, W., and Lux, R. (2017). Identification and characterization of a novel *Fusobacterium nucleatum* adhesin involved in physical interaction and biofilm formation with Streptococcus gordonii. Microbiologyopen, 6, e00444. DOI: 10.1002/mbo3.444.

Lima, K.C., Coelho, L.T., Pinheiro, I.V.A., Rôças, I.N., and Jr, J.F.S. (2011). Microbiota of Dentinal Caries as Assessed by Reverse-Capture Checkerboard Analysis. Caries Res, 45, 21-30. DOI: 10.1159/000322299.

Lin, T.H., Lin, C.H., and Pan, T.M. (2018). The implication of probiotics in the prevention of dental caries. Appl Microbiol Biotechnol, 102, 577-586. DOI: 10.1007/s00253-017-8664-z.

Liu, J., Wu, C., Huang, I.H., Merritt, J., and Qi, F. (2011). Differential response of *Streptococcus mutans* towards friend and foe in mixed-species cultures. Microbiology 157, 2433-2444. DOI: 10.1099/mic.0.048314-0.

Love, R.M., Mcmillan, M.D., and Jenkinson, H.F. (1997). Invasion of dentinal tubules by oral streptococci is associated with collagen recognition mediated by the antigen I/II family of polypeptides. Infect Immun, 65, 5157-5164.

Love, R.M., Mcmillan, M.D., Park, Y., and Jenkinson, H.F. (2000). Coinvasion of dentinal tubules by *Porphyromonas gingivalis* and *Streptococcus gordonii* depends upon binding specificity of streptococcal antigen I/II adhesin. Infect

Immun, 68, 1359-1365.

Luppens, S.B.I., Kara, D., Bandounas, L., Jonker, M.J., Wittink, F.R.A., Bruning, O., Breit, T.M., Cate, J.M.T., and Crielaard, W. (2008). Effect of *Veillonella parvula* on the antimicrobial resistance and gene expression of *Streptococcus mutans* grown in a dual-species biofilm. Oral Microbiol Immunol, 23, 183-189. DOI: 10.1111/j.1399-302X.2007.00409.x.

Ma, S., Ge, W., Yan, Y., Huang, X., Ma, L., Li, C., Yu, S., and Chen, C. (2017). Effects of *Streptococcus sanguinis* Bacteriocin on Deformation, Adhesion Ability, and Young's Modulus of *Candida albicans*. Biomed Res Int, 2017, 5291486. DOI: 10.1155/2017/5291486.

Ma, S., Li, H., Yan, C., Wang, D., Li, H., Xia, X., Dong, X., Zhao, Y., Sun, T., and Hu, P. (2014). Antagonistic effect of protein extracts from *Streptococcus sanguinis* on pathogenic bacteria and fungi of the oral cavity. Exp Ther Med, 7, 1486-1494. DOI: 10.3892/etm.2014.1618.

Ma, S., Zhao, Y., Xia, X., Dong, X., Ge, W., and Li, H. (2015). Effects of *Streptococcus sanguinis* Bacteriocin on Cell Surface Hydrophobicity, Membrane Permeability, and Ultrastructure of *Candida* Thallus. Biomed Res Int, 2015, 514152. DOI: 10.1155/2015/514152.

Macdonald, K.W. (2015). The Role of *Streptococcus salivarius* as a Modulator of Homeostasis in the Oral Cavity. Electr Thes Diss Rep. 2816. https://ir.lib.uwo.ca/etd/2816.

Maeda, K., Nagata, H., Ojima, M., and Amano, A. (2015). Proteomic and transcriptional analysis of interaction between oral microbiota *Porphyromonas gingivalis* and *Streptococcus oralis*. J Proteome Res, 14, 82-94. DOI: 10.1021/pr500848e.

Mani, A., Saini, R., and Saini, S.R. (2017). Efficacy of Oral Probiotics as an

Adjunct to Scaling and Root Planing in Nonsurgical Treatment Outcome of Generalized Chronic Periodontitis Patients: A Clinico-Microbiological Study. Int J Exp Dent Sci, 6, 6-13.

Mark Welch, J.L., Rossetti, B.J., Rieken, C.W., Dewhirst, F.E., and Borisy, G.G. (2016). Biogeography of a human oral microbiome at the micron scale. Proc Nat Acad Sci USA, 113, 201522149. DOI: 10.1073/pnas.1522149113.

Marsh, P.D. (1994). Microbial ecology of dental plaque and its significance in health and disease. Adv Dent Res, 8, 263-271. DOI: 10.1177/08959374940080022001.

Marsh, P.D. (2005). Dental plaque: biological significance of a biofilm and community life-style. J Clin Periodontol 32 Suppl 6, 7-15. DOI: 10.1111/j.1600-051X.2005.00790.x.

Marsh, P.D. (2016). Dental Biofilms in Health and Disease (Springer International Publishing), pp.41-52. DOI: https://doi.org/10.1007/978-3-319-30552-3_5.

Mashima, I., and Nakazawa, F. (2014). The influence of oral Veillonella species on biofilms formed by Streptococcus species. Anaerobe, 28, 54-61. DOI: 10.1016/j.anaerobe.2014.05.003.

Mashima, I., and Nakazawa, F. (2015). The interaction between *Streptococcus spp.* and *Veillonella tobetsuensis* in the early stages of oral biofilm formation. J Bacteriol, 197, 13-54. DOI: 10.1128/JB.02512-14.

Mason, M.R., Preshaw, P.M., Nagaraja, H.N., Dabdoub, S.M., Rahman, A., and Kumar, P.S. (2015). The subgingival microbiome of clinically healthy current and never smokers. ISME J, 9, 268-272. DOI: 10.1038/ismej.2014.114.

Mayer, F.L., Wilson, D., and Hube, B. (2013). Candida albicans pathogenicity mechanisms. Virulence, 4, 119-128. DOI: 10.4161/viru.22913.

Mcbride, B.C., and Js, V.D.H. (1981). Role of interbacterial adherence in colonization of the oral cavities of gnotobiotic rats infected with *Streptococcus mutans* and *Veillonella alcalescens*. Infect Immun, 33, 467-472.

Mcnab, R., Forbes, H., Handley, P.S., Loach, D.M., Tannock, G.W., and Jenkinson, H.F. (1999). Cell wall-anchored CshA polypeptide (259 kilodaltons) in *Streptococcus gordonii* forms surface fibrils that confer hydrophobic and adhesive properties. J Bacteriol, 181, 3087-3095.

Merritt, J., and Qi, F. (2012). The Mutacins of *Streptococcus mutans*: Regulation and Ecology. Mol Oral Microbiol, 27, 57-69. DOI: 10.1111/j.2041-1014.2011.00634.x.

Metwalli, K.H., Khan, S.A., Krom, B.P., and Jabrarizk, M.A. (2013). Streptococcus mutans, Candida albicans, and the human mouth: a sticky situation. PLoS Pathog, 9, e1003616. DOI: 10.1371/journal.ppat.1003616.

Milnes, A.R., and Bowden, G.H. (1985). The microflora associated with developing lesions of nursing caries. Caries Res, 19, 289-297. DOI: 10.1159/000260858.

Mohammad Shahnoor, H., and Indranil, B. (2011). Mutacins from *Streptococcus mutans* UA159 are active against multiple streptococcal species. Appl Environ Microbiol, 77, 2428-2834. DOI: 10.1128/AEM.02320-10.

Moon, J.E., Moon, Y.M., and Cho, J.W. (2016). The Effect of *Streptococcus salivarius* K12 against Prevotella intermedia on the Reduction of Oral Malodor. Int J Clin Prev Dent, 12, 153-161. DOI: 10.15236/ijcpd.2016.12.3.153.

Moraga, C.P.L., Puente, C.A.L., Bozo, I.C.M., and Orellana, B.R.U.a. (2016). Prevalence of *Candida albicans* and carriage of *Candida* non-*albicans* in the saliva of preschool children, according to their caries status. Acta Odontol Scand, 75, 30-35. DOI: 10.1080/00016357.2016.1244560.

Naito, Y., Tohda, H., Okuda, K., and Takazoe, I. (2010). Adherence and hydrophobicity of invasive and noninvasive strains of *Porphyromonas gingivalis*.

Oral Microbiol Immunol., 8, 195-202.

Nascimento, M.M., and Burne, R.A. (2014). Caries Prevention by Arginine Metabolism in Oral Biofilms: Translating Science into Clinical Success. Curr Oral Health Rep, 1, 79-85. DOI: 10.1007/s40496-013-0007-2.

Ng, W., and Bassler, B.L. (2009). Bacterial Quorum-Sensing Network Architectures. Annu Rev Genet, 43, 197-222. DOI: 10.1146/annurev-genet-102108-134304.

Noorda, W.D., Purdell-Lewis, D.J., van Montfort, A.M., and Weerkamp, A.H. (1988). Monobacterial and mixed bacterial plaques of *Streptococcus mutans* and *Veillonella alcalescens* in an artificial mouth: development, metabolism, and effect on human dental enamel. Caries Res, 22, 342-347. DOI: 10.1159/000261134.

Okshevsky, M., and Meyer, R.L. (2015). The role of extracellular DNA in the establishment, maintenance and perpetuation of bacterial biofilms. Crit Rev Microbiol, 41, 341-352. DOI: 10.3109/1040841X.2013.841639.

Oliveira, R.V.D.D., Albuquerque, Y.E., Koga-Ito, C.Y., and Brighenti, F.L. (2015). Influence of different fermentable carbohydrates on dual-species biofilms of *S. mutans* and *A. naeslundii*: a pilot study. Braz Dent Sci, 18, 82-88. doi: 10.14295/bds.2015.v18i2.1100.

Pérez, T.M., Ratia, K., Wang, D.S., Gogos, A., Driver, T.G., and Federle, M.J. (2018). A novel chemical inducer of *Streptococcus* quorum sensing acts by inhibiting the pheromone-degrading endopeptidase PepO. J Biol Chem, 293, 931-940. DOI: 10.1074/jbc.M117.810994.

Parashar, A., Parashar, S., Zingade, A., Gupta, S., and Sanikop, S. (2015). Interspecies communication in oral biofilm: An ocean of information. Oral Sci Int, 12, 37-42. DOI: 10.1016/S1348-8643(15)00016-6.

Pereira, D., Seneviratne, C.J., Koga-Ito, C.Y., and Samaranayake, L.P. (2017a). Is the oral fungal pathogen *Candida albicans* a cariogen? Oral Dis, 24, 518-526. DOI: 10.1111/odi.12691.

Carneiro Pereira, A.L., de Lima, L.G., da Silva Maia, S.É., Melo de Matos, J.M., Melo de Matos, J.M., and Luiza de Aguiar Rocha Martins, A. (2017b). Co-aggregation of Probiotics in the Dental Biofilm and Inhibition of Bacterial Growth in Caries Prevention. Int J Oral Health Med Res, 4, 4-7.

Periasamy, S., and Kolenbrander, P.E. (2010). Central role of the early colonizer *Veillonella sp.* in establishing multispecies biofilm communities with initial, middle, and late colonizers of enamel. J Bacteriol, 192, 2965-2972. DOI: 10.1128/JB.01631-09.

Philip, N., Suneja, B., and Walsh, L.J. (2018). Ecological Approaches to Dental Caries Prevention: Paradigm Shift or Shibboleth? Caries Res, 52, 153-165. DOI: 10.1159/000484985.

Plančak, D., Musić, L., and Puhar, I. (2015). Quorum Sensing of Periodontal Pathogens. Acta Stomatol Croat, 49, 234-241. DOI: 10.15644/asc49/3/6.

Ramsey, M.M., and Whiteley, M. (2009). Polymicrobial interactions stimulate resistance to host innate immunity through metabolite perception. Proc Natl Acad Sci U S A, 106, 1578-1583. DOI: 10.1073/pnas.0809533106.

Richards, V.P., Alvarez, A.J., Luce, A.R., Bedenbaugh, M., Mitchell, M.L., Burne, R.A., and Nascimento, M.M. (2017). Microbiomes of Site-Specific Dental Plaques from Children with Different Caries Status. Infect Immun, 85, e00106-17. DOI: 10.1128/IAI.00106-17.

Rodríguez, G., Ruiz, B., Faleiros, S., Vistoso, A., Marró, M.L., Sánchez, J., Urzúa, I., and Cabello, R. (2016). Probiotic compared with standard milk for high-caries children: a cluster randomized trial. J Dent Res, 95, 402-407. DOI:

10.1177/0022034515623935.

Ruhl, S., Eidt, A., Melzl, H., Reischl, U., and Cisar, J.O. (2014). Probing of microbial biofilm communities for coadhesion partners. Appl Environ Microbiol, 80, 6583-6590. DOI: 10.1128/AEM.01826-14.

Sakanaka, A., Kuboniwa, M., Takeuchi, H., Hashino, E., and Amano, A. (2015). Arginine-ornithine Antiporter ArcD Controls Arginine Metabolism and Interspecies Biofilm Development of *Streptococcus gordonii*. J Biol Chem., 290, 21185-21198. DOI: 10.1074/jbc.M115.644401.

Signat, B., Roques, C., Poulet, P., and Duffaut, D. (2011). *Fusobacterium nucleatum* in periodontal health and disease. Curr Issues Mol Biol , 13, 25-36.

Simónsoro, A., and Mira, A. (2015). Solving the etiology of dental caries. Trends Microbiol, 23, 76-82. DOI: 10.1016/j.tim.2014.10.010.

Stacy, A., Everett, J., Jorth, P., Trivedi, U., Rumbaugh, K.P., and Whiteley, M. (2014). Bacterial fight-and-flight responses enhance virulence in a polymicrobial infection. Proc Natl Acad Sci U S A., 111, 7819-7824. DOI: 10.1073/pnas.1400586111.

Stacy, A., Mcnally, L., Darch, S.E., Brown, S.P., and Whiteley, M. (2016). The biogeography of polymicrobial infection. Nat Rev Microbiol, 14, 93-105. DOI: 10.1038/nrmicro.2015.8.

Sudheer, Y., Kamalesh, B., Siddharth, S., Indra, G., and Swetha, R.K. (2015). Quorum Sensing Inhibition, Relevance to Periodontics. J Int Oral Health, 7, 67-69.

Sullivan, R., Santarpia, P., Lavender, S., Gittins, E., Liu, Z., Anderson, M.H., He, J., Shi, W., and Eckert, R. (2011). Clinical efficacy of a specifically targeted antimicrobial peptide mouth rinse: targeted elimination of Streptococcus mutans and prevention of demineralization. Caries Res, 45, 415-428. DOI:

10.1159/000330510.

Szafrański, S.P., Deng, Z.L., Tomasch, J., Jarek, M., Bhuju, S., Rohde, M., Sztajer, H., and Wagnerdöbler, I. (2017). Quorum sensing of *Streptococcus mutans* is activated by *Aggregatibacter actinomycetemcomitans* and by the periodontal microbiome. BMC Genomics, 18, 238. DOI: 10.1186/s12864-017-3618-5.

Takahashi, N., Saito, K., Schachtele, C.F., and Yamada, T. (1997). Acid tolerance and acid-neutralizing activity of *Porphyromonas gingivalis*, *Prevotella intermedia* and *Fusobacterium nucleatum*. Oral Microbiol Immunol, 12, 323-328.

Tanaka, M., Yamamoto, Y., Kuboniwa, M., Nonaka, A., Nishida, N., Maeda, K., Kataoka, K., Nagata, H., and Shizukuishi, S. (2004). Contribution of periodontal pathogens on tongue dorsa analyzed with real-time PCR to oral malodor. Microbes Infect, 6, 1078-1083. DOI: 10.1016/j.micinf.2004.05.021.

Tanner, A.C.R., Kressirer, C.A., Rothmiller, S., Johansson, I., and Chalmers, N.I. (2018). The Caries Microbiome: Implications for Reversing Dysbiosis. Adv Dent Res, 29, 78-85. DOI: 10.1177/0022034517736496.

Tanner, A.C.R., Mathney, J.M.J., Kent, R.L., Chalmers, N.I., Hughes, C.V., Loo, C.Y., Pradhan, N., Kanasi, E., Hwang, J., and Dahlan, M.A. (2011). Cultivable Anaerobic Microbiota of Severe Early Childhood Caries. J Clin Microbiol, 49, 1464-1474. DOI: 10.1128/JCM.02427-10.

Teughels, W., Kinder, H.S., Sliepen, I., Pauwels, M., Van, E.J., Cassiman, J.J., and Quirynen, M. (2007). Bacteria interfere with *A. actinomycetemcomitans* colonization. J Dent Res, 86, 611-617. DOI: 10.1177/154405910708600706.

Thompson, J.A., Oliveira, R.A., and Xavier, K.B. (2016). Chemical conversations in the gut microbiota. Gut Microbes, 7, 163-170. DOI: 10.1080/19490976.2016.1145374.

Thuy, D., Evelyn, S., Tonnie, K.M., Francis, J.H., and David, B. (2015). Transcriptomic analysis of three *Veillonella spp.* present in carious dentine and in the saliva of caries-free individuals. Front Cell Infect Microbiol, 5, 25. DOI: 10.3389/fcimb.2015.00025.

Tong, H., Chen, W., Merritt, J., Qi, F., Shi, W., and Dong, X. (2007). *Streptococcus oligofermentans* inhibits *Streptococcus mutans* through conversion of lactic acid into inhibitory H_2O_2: a possible counteroffensive strategy for interspecies competition. Mol Microbiol, 63, 872-880. DOI: 10.1111/j.1365-2958.2006.05546.x.

Valdebenito, B., Tullume-Vergara, P.O., González, W., Kreth, J., and Giacaman, R.A. (2017). In silico analysis of the competition between *Streptococcus sanguinis* and *Streptococcus mutans* in the dental biofilm. Mol Oral Microbiol, 33, 168-180. DOI: 10.1111/omi.12209.

Van der Hoeven, J.S., Toorop, A.I., and Mikx, F.H.M. (1978). Symbiotic Relationship of *Veillonella alcalescens* and *Streptococcus mutans* in Dental Plaque in Gnotobiotic Rats. Caries Res, 12, 142-147. DOI: 10.1159/000260324.

Vaniabella, R., Joenoes, H., and Bachtiar, B.M. (2017). Analysis on the inhibition of *Aggregatibacter Actinomycetemcomitans* by *Streptococcus Salivarius* isolated from saliva and tongue dorsum of adults. Asian J Pharm Clin Res, 10, 6-10. DOI: 10.22159/ajpcr.2017.v10s5.23081.

Vasudevan, R. (2017). Dental Plaques: Microbial Community of the Oral Cavity. J Microbiol Exp, 4, 00100.

Vendeville, A., Winzer, K., Heurlier, K., Tang, C.M., and Hardie, K.R. (2005). Making 'sense' of metabolism: autoinducer-2, LUXS and pathogenic bacteria. Nat Rev Microbiol, 3, 383-396. DOI: 10.1038/nrmicro1146.

Wang, B.Y., Deutch, A., ., Hong, J., ., and Kuramitsu, H.K. (2011). Proteases of an early colonizer can hinder *Streptococcus mutans* colonization in vitro. J Dent Res, 90, 501-505. DOI: 10.1177/0022034510388808.

Wang, B.Y., Wu, J., Lamont, R.J., Lin, X., and Xie, H. (2009). Negative correlation of distributions of Streptococcus cristatus and Porphyromonas gingivalis in subgingival plaque. J Clin Microbiol, 47, 3902-3906. DOI: 10.1128/JCM.00072-09.

Wen, Z.T., Yates, D., Ahn, S.J., and Burne, R.A. (2010). Biofilm formation and virulence expression by *Streptococcus mutans* are altered when grown in dual-species model. BMC Microbiol, 10, 111. DOI: 10.1186/1471-2180-10-111.

Whitmore, S.E., and Lamont, R.J. (2011). The pathogenic persona of community associated oral streptococci. Mol Microbiol, 81, 305-314. DOI: 10.1111/j.1365-2958.2011.07707.x.

Wright, C.J., Burns, L.H., Jack, A.A., Back, C.R., Dutton, L.C., Nobbs, A.H., Lamont, R.J., and Jenkinson, H.F. (2013). Microbial interactions in building of communities. Mol Oral Microbiol, 28, 83-101. DOI: 10.1111/omi.12012.

Wright, C.J., Xue, P., Hirano, T., Liu, C., Whitmore, S.E., Hackett, M., and Lamont, R.J. (2014). Characterization of a bacterial tyrosine kinase in *Porphyromonas gingivalis* involved in polymicrobial synergy. Microbiologyopen 3, 383-394. DOI: 10.1002/mbo3.177.

Xiao, W., Li, X., and Ling, J. (2017). Streptococcus gordonii LuxS/autoinducer - 2 quorum - sensing system modulates the dual - species biofilm formation with Streptococcus mutans. J Basic Microbiol, 57, 605-616. DOI: 10.1002/jobm.201700010.

Xu, H., Hao, W., Zhou, Q., Wang, W., Xia, Z., Liu, C., Chen, X., Qin, M., and Chen, F. (2014a). Plaque bacterial microbiome diversity in children younger than 30 months with or without caries prior to eruption of second primary molars.

PLoS One, 9, e89269. DOI: 10.1371/journal.pone.0089269.

Xu, H., Jenkinson, H.F., and Dongaribagtzoglou, A. (2014b). Innocent until proven guilty: mechanisms and roles of *Streptococcus–Candida* interactions in oral health and disease. Mol Oral Microbiol, 29, 99-116.

Xu, H., Sobue, T., Bertolini, M., Thompson, A., and Dongaribagtzoglou, A. (2016). *Streptococcus oralis* and *Candida albicans* Synergistically Activate μ-Calpain to Degrade E-cadherin From Oral Epithelial Junctions. J Infect Dis, 214, 925-934. DOI: 10.1093/infdis/jiw201.

Xu, H., Sobue, T., Bertolini, M., Thompson, A., Vickerman, M., Nobile, C.J., and Dongari-Bagtzoglou, A. (2017). *S. oralis* activates the Efg1 filamentation pathway in *C. albicans* to promote cross-kingdom interactions and mucosal biofilms. Virulence, 8, 1602-1617. DOI: 10.1080/21505594.2017.1326438.

Xu, H., Sobue, T., Thompson, A., Xie, Z., Poon, K., Ricker, A., Cervantes, J., Diaz, P.I., and DongariBagtzoglou, A. (2014c). Streptococcal co-infection augments *Candida* pathogenicity by amplifying the mucosal inflammatory response. Cell Microbiol, 16, 214–231. DOI: 10.1111/cmi.12216.

Yang, J., Yoshida, Y., and Cisar, J.O. (2014). Genetic basis of coaggregation receptor polysaccharide biosynthesis in *Streptococcus sanguinis* and related species. Mol Oral Microbiol, 29, 24-31. DOI: 10.1111/omi.12042.

Yuce, H., Tulu, F., Inis, S., Karaman, I., Yuce, H., Tulu, F., Inis, S., and Karaman, I. (2017). Growth behavior of *Eikenella corrodens* and *Streptococcus gordonii* in response to a short chain fatty acid metabolite-acetic acid. J Turgut Ozal Med Cent, 24, 396-400. DOI: 10.5455/jtomc.2017.04.057.

Zambori, C., Morvay, A.A., Sala, C., Licker, M., Gurban, C., Tanasie, G., and Tirziu, E. (2016). Antimicrobial effect of probiotics on bacterial species from dental plaque. J Infect Dev Ctries, 10, 214-221. DOI: 10.3855/jidc.6800.

Zhang, G., and Rudney, J.D. (2011). *Streptococcus cristatus* attenuates *Fusobacterium nucleatum*-induced cytokine expression by influencing pathways converging on nuclear factor-κB. Mol Oral Microbiol, 26, 150-163. DOI: 10.1111/j.2041-1014.2010.00600.x.

Zhou, P., Liu, J., Merritt, J., and Qi, F. (2015). A YadA - like autotransporter, Hag1 in *Veillonella atypica* is a multivalent hemagglutinin involved in adherence to oral streptococci, *Porphyromonas gingivalis*, and human oral buccal cells. Mol Oral Microbiol, 30, 269-279. DOI: 10.1111/omi.12091.

Zhu, L., and Kreth, J. (2012). The role of hydrogen peroxide in environmental adaptation of oral microbial communities. Oxid Med Cell Longev, 2012, 2012, 717843. DOI: 10.1155/2012/717843.

Chapter 10 https://doi.org/10.21775/9781912530229.10

The Emergence of Hypervirulent M1T1 Clone of Group A *Streptococcus* via Genetic Recombination and Host Selection

Guanghui Liu[1*]

[1]Division of Medical Protein Chemistry, Lund University, 21428 Malmö, Sweden
*Corresponding email: ghliu83@gmail.com

Abstract

Streptococcus pyogenes (Group A Streptococcus, GAS) is a strictly human bacterial pathogen. Since the mid-1980s, GAS M1T1 clone has been the most prevalent and globally disseminated serotype and is the culprit causing invasive and severe streptococcal infections, urging a better understanding of the emergence of hypervirulent M1T1 clone from an evolutionary perspective. This review highlights the molecular and evolutionary events leading to pandemic M1T1 strains, and discusses the pressure driving the genetic acquisition of novel virulence genes and the selection of hypervirulent isolates in host. By understanding the evolutionary selection and pressures that select and shape the pandemic M1T1 clone, we could potentially develop new therapeutic strategies to tackle challenges when dealing with the globally disseminated M1T1 GAS clone.

Introduction

Streptococcus pyogenes is a Gram-positive human-restricted pathogen causing β-hemolysis of red blood cells and also known as Group A *Streptococcus* (GAS) due to the Lancefield group A antigen expressed on its surface (Lancefield, 1933). GAS primarily colonizes the nasopharynx and the skin of healthy people including children, without causing symptoms of diseases (Kline and Runge, 1994; Shaikh et al., 2010). However, when epithelial barrier damage occurs or our immunity becomes weakened, GAS is able to pass through the nasopharynx or skin barrier, and causes a wide variety of infections in human (Carapetis et al., 2005; Ralph and Carapetis, 2013). These infections result in mild diseases such as pharyngitis (strep throat) and impetigo (skin infection), and severe and life-threatening diseases such as necrotizing fasciitis (the flesh-eating disease) and streptococcal toxic shock syndrome (Carapetis et al., 2005; Ralph and Carapetis, 2013). Moreover, people infected by GAS have a risk to develop autoimmune complications, *e.g.*, rheumatic fever and acute postinfectious glomerulonephritis (Carapetis et al., 2005; Ralph and Carapetis, 2013). It is estimated that 700 million GAS infections occur worldwide each year, of which 2.9% of cases are severe and invasive, leading to more than 500,000 deaths per year (Carapetis et al., 2005).

In concert with its strikingly capacity to cause diseases, GAS has been well equipped to colonize various hostile niches and to evade host protective immunity (Cole et al., 2011; Lappin and Ferguson, 2009; Olsen and Musser, 2010). In addition to the nasopharynx and the skin, GAS has been identified as an infectious agent in muscle, kidney, lung, heart, and brain (Cole et al., 2011; Lappin and Ferguson, 2009; Olsen and Musser, 2010). To colonize and invade these tissues and organs, GAS needs to overcome tissue barriers and the immune defense elicited in various niches. To cause necrotizing fasciitis, GAS exhibits sophisticated virulence mechanisms that initiate host cell adhesion, soft tissue entry, immune evasion and tissue destruction (Olsen and Musser, 2010). In patients with streptococcal toxic shock syndrome, GAS releases superantigens to bypass normal antigen processing, causing a cytokine storm that induces an acute multi-organ failure (Lappin and Ferguson, 2009).

The most known virulence factor of GAS is the M protein, which interacts with its host ligands, *e.g.*, complement, immunoglobulins and fibrinogen, playing relevant immunomodulatory functions (Smeesters et al., 2010). As a highly variable molecule, M protein together with surface T antigen are employed to classify GAS serotypes (Lancefield, 1962). Among the >200 GAS known serotypes, M1 is the most frequently identified serotype associated with severe and invasive infections (Steer et al., 2009; Sumby et al., 2005b). For example, M1 isolates share 18.3% of all GAS isolates worldwide, and particularly dominate in high-income countries, Asia, and Latin America (Steer et al., 2009). However, since the mid-1980s, the subclone of M1 GAS, M1T1, has been rising to be the most prevalent and globally disseminated serotype and deserved greater attention (Aziz and Kotb, 2008; Maamary et al., 2012; Nasser et al., 2014; Sumby et al., 2005b).

In recent years, omics studies have provided more evolutionary and molecular based evidence for the success of M1T1 clone. This clone evolves through a series of genetic recombination events that acquire genetic regions encoding significant streptococcal virulence factors (Aziz and Kotb, 2008; Nasser et al., 2014; Steer et al., 2009; Sumby et al., 2005b). Furthermore, the M1T1 clone frequently underdoes spontaneous mutations in its major regulatory genes, leading to the emergence of hypervirulent isolates with distinctive switching of phenotypes (Aziz et al., 2004b; Ikebe et al., 2010). This review aims to outline the molecular and evolutionary events leading to pandemic M1T1 strains, and to pinpoint the mechanisms governing the phenotypic shift of M1T1 GAS clone, which underlie the selection of hypervirulent isolates in the host.

I. Genetic acquisition of novel virulence genes determines the success of M1T1 GAS clone

The past 30 years have witnessed the emergence of the highly successful M1T1 GAS clone. Comparing to other non-M1 and noninvasive M1 types of GAS, M1T1 clone shows more advantages in host adaptation, survival, persistence and dissemination (Maamary et al., 2012; Nasser et al., 2014). These M1T1 clones including widely studied MGAS2221 and strain 5448, are more virulent in mouse and primate model of pharyngitis and invasive infections (Aziz et al., 2005; Sumby et al., 2005b). How M1T1 clone developed such advantages remained a mystery until high-throughput sequencing technologies were adopted to comparatively analyze genetic compositions of pandemic M1T1 clonal isolates. The comparative genetic analysis reveals that the emergence of M1T1 clone is attributed to two genetic recombination events that acquire regions encoding significant streptococcal virulence factors (Aziz and Kotb, 2008; Maamary et al., 2012; Nasser et al., 2014; Sumby et al., 2005b). The acquisition of these genes in a stepwise fashion confer the versatile capacity of M1T1 GAS in host adaptation and invasiveness through synergistic effects.

1. The first event: acquisition of bacteriophage-encoded DNase Sda1 and superantigen SpeA2

Genomic analysis demonstrates the presence of novel bacteriophages or prophages encoding the extracellular deoxyribonuclease (Sda1, also called SdaD2) and the streptococcal pyrogenic exotoxin A (SpeA2) in the genome of M1T1 clones (Aziz et al., 2005; Banks et al., 2004). Although GAS harbors remarkable bacteriophages, which account for more than 10% of the total genome in certain M serotypes (Beres et al., 2002; Ferretti et al., 2001), the novel bacteriophages encoding DNase Sda1 and superantigen SpeA2 are exclusively found in the invasive strains, particularly in the epidemic M1T1 clone. Early at 1996, Cleary et al. observed that the pandemic M1T1 subclone differs from its closely related M1 subtypes by nearly 70 kb of two prophage DNA (Cleary et al., 1998). Through analysis of a M1T1 genome, Aziz et al. reported that these unique prophages encoding Sda1 and SpeA2 were introduced horizontally into the M1T1 global strain, and proposed that such a genetic recombination could contribute to the emergence and diversification of the M1T1 subclone (Aziz et al., 2005; Aziz and Kotb, 2008). The precise genetic recombination was later elucidated by Maamary and colleagues (Maamary et al., 2012). They showed that the acquisition of Sda1 occurred relatively early in the molecular evolutionary history of M1T1 strains, followed by acquisition of the superantigen SpeA2 (Maamary et al., 2012; Nasser et al., 2014). Notably, the acquisition of Sda1 and SpeA2 does not necessarily improve the ability of M1 GAS to survive the killing of human neutrophils *in vitro* (Maamary et al., 2012). Instead, Sda1 and SpeA2 increase GAS virulence in the subcutaneous infection model of humanized plasminogen transgenic *AlbPLG1* mice (Maamary et al., 2012). However, M1 strain SF370, which contains four different bacteriophages encoding several superantigen-like proteins but not Sda1 and SpeA2, is less virulent and noninvasive, and genetically viewed as a pre-epidemic M1 serotype (Ferretti et al., 2001). To some extent, the acquisition of Sda1 and SpeA2 can be defined as evolutionary markers that distinguish epidemic M1T1 from non-invasive M1T1 and M1 serotypes.

1.1 DNase Sda1 and its role in GAS virulence

M1T1 GAS clone secrets several distinct DNases. Of these DNases, Sda1, was originally discovered by Aziz et al. in 2004 in culture supernatant of a M1T1 clinical isolate. Sda1 was determined to be bacteriophage-encoded (Aziz et al., 2004a) and found to be highly similar in amino-acid composition to streptodornase D from M49 GAS. Recombinant Sda1 can degrade both streptococcal and mammalian DNA, and its nuclease activity depends on the availability of divalent metal ions such as Ca^{++} and Mg^{++} (Aziz et al., 2004a). The crystal structure of Sda1 has been recently solved, revealing surface residues of Sda1 such as Arg12 and Asn211 involved in DNA binding and catalysis (Moon et al., 2016). It has been shown that Sda1 expression is regulated by the peroxide responsive regulator PerR under oxidative stress (Wang et al., 2013). However, the cysteine proteinase SpeB secreted by GAS seems to play a key role in controlling Sda1 expression by degrading Sda1 (Nelson et al., 2011; Walker et al., 2007).

Sda1 is a virulence factor of GAS with multi-faceted functions. Sumby et al. demonstrated that Sda1 alone is sufficient to induce wild-type levels of skin lesions in mouse model of skin infection, indicating that Sda1 but not other DNases (namely Spd and Spd3) is the major DNase that contributes to virulence of MGAS5005 in mouse models (Sumby et al., 2005a). Furthermore, the virulence of Sda1 was extensively characterized by Buchanan et al. using M1T1 strain 5448

(Buchanan et al., 2006). They found that Sda1 promotes GAS neutrophil resistance and virulence through degrading neutrophil extracellular traps (NETs), which is a significant innate immune structure composed of chromatin and granule proteins (Brinkmann et al., 2004; Buchanan et al., 2006). A recent study further suggested that Sda1 could also cleave bacterial DNA to prevent TLR9-mediated recognition of GAS by host innate immune cells, which damped the secretion of inflammatory cytokines, resulting in GAS survival in a murine model of necrotizing fasciitis (Uchiyama et al., 2012).

DNase as a bacterial virulence factor is not limited to GAS alone. Sda1 homologues can be found in various virulent pathogens, such as *Plasmodium falciparum*, *Staphylococcus aureus* and *Streptococcus pneumoniae* (Aziz et al., 2004a; Beiter et al., 2006; Thammavongsa et al., 2013). These DNases share the similar function in degrading NETs, suggesting their potential as a therapeutic target. Inhibition of DNase activity or development of therapeutic antibodies neutralizing the activity of these enzymes may facilitate the resolution of pathogenic infections. As an example, mice immunized with recombinant *Plasmodium* DNase exhibit increased immunity against lethal challenge of *P. falciparum* (Chang et al., 2016).

1.2 The superantigen SpeA2 – a potent scarlet fever toxin

GAS produces numerous pyrogenic toxins called superantigens. Unlike the normal antigens that are presented within the major histocompatibility complex class II (MHC II) of antigen-presenting cells and the T cell receptor (TCR), superantigens are capable to crosslink MHC II and TCR outside the peptide-binding groove, causing non-specific activation of T cells and an immense cytokine release (Fraser and Proft, 2008; Li et al., 1999). Therefore, superantigens have been implicated as the major culprit of acute infections such as sepsis, necrotizing fasciitis, and toxic shock syndrome (Norrby-Teglund et al., 2001; Proft et al., 2003; Salgado-Pabon et al., 2013; Sriskandan et al., 1996). As a representative of superantigens, SpeA has been known as a scarlet fever toxin, preferentially stimulating human T cells bearing specific TCR β-chain variable domains (Abe et al., 1991; Braun et al., 1993; Leonard et al., 1991). Accordingly, SpeA has been shown enhancing acute nasopharyngeal infection by GAS in mice expressing human MHC II, while vaccination with inactive SpeA or SpeA devoid of MHC II binding protected mice from GAS infection (Kasper et al., 2014; Zeppa et al., 2017).

GAS encodes four SpeA variants, namely SpeA1 through SpeA4. SpeA2 differs from SpeA1 by a single amino acid substitution at position 80 from glycine to serine, which is close to the zinc-binding site in SpeA for recognition of the MHC II molecules (Earhart et al., 2000; Kline and Collins, 1996, 1997; Nelson et al., 1991; Papageorgiou et al., 1999). Importantly, all contemporary M1 isolates exclusively harbor the SpeA2 allele (Musser et al., 1991; Musser et al., 1993). In fact, a more recent research performed by Nasser et al., after analysis of genome sequences from 3,615 M1 strains, indicate that SpeA2 variant subsequently evolved from the old SpeA1 variant via a single-nucleotide change leading to glycine-to-serine substitution (Nasser et al., 2014). Although only one single amino acid difference from SpeA1, recombinant SpeA2 has a higher binding affinity for the MHC II molecules (Abe et al., 1991; Kline and Collins, 1996, 1997; Papageorgiou et al., 1999). This increased binding affinity may lead to more severe effects in the host, and could explain why SpeA2 has been detected in a majority of M1T1 GAS strains isolated from patients with streptococcal toxic shock syndrome (Musser et al., 1991; Musser et al., 1993).

SpeA2 expression is controlled by multiple factors, either from GAS or from host. Kazmi et al. initially failed to induce SpeA expression in invasive M1T1 isolates at varying culture conditions, whereas detected SpeA2 production from mouse-passaged isolates (Kazmi et al., 2001). The expression of SpeA2 was shown to correlate reciprocally and temporally to the downregulation of SpeB, and it was regulated negatively by the global regulator, the control of virulence two-component system CovRS (also known as CsrRS) in M1T1 (Kazmi et al., 2001; Sumby et al., 2006). Interestingly, SpeA2 expression can be induced in cocultures with human peripheral blood mononuclear cells *ex vivo* (Kansal et al., 2005). The host signals, such as transferrin and lactoferrin, play a role in regulation of SpeA2 expression, and may be implicated in GAS invasive infections (Kansal et al., 2005).

2. The second event: replacement through recombination of streptolysin O and NAD⁺-glycohydrolase

The second genetic recombination event occuring is that M1T1 clone acquires a 36-kb chromosomal region, which encodes streptolysin O (SLO) and NAD⁺-glycohydrolase (NADase, also called *Streptococcus pyogenes* NAD-glycohydrolase, SPN). SLO is an oxygen-labile streptococcal hemolytic exotoxin that belongs to a large family of cholesterol-dependent cytolysins (Tweten, 2005), while NADase is an enzyme that cleaves β-NAD⁺ to produce ADP-ribose and nicotinamide (Ghosh et al., 2010; Grushoff et al., 1975). SLO and NADase, known as well-characterized secreted toxins of GAS, have been widely found in isolates from patients with toxic shock syndrome and other severe infections. For instance, Muller-Alouf et al. reported all 212 isolates from patients with severe invasive GAS infections secreted SLO (Muller-Alouf et al., 1997). Furthermore, a strong association has been proposed between isolates causing severe invasive GAS infections and the production of NADase, and all M1 strains isolated after 1988 were positive for NADase (Stevens et al., 2000; Tatsuno et al., 2007). Compared to bacteriophage-encoded Sda1 and SpeA2, which are exclusively found in the epidemic M1T1 clones, SLO and NADase are chromosome-encoded and present in many invasive M types of GAS such as M1, M3 and M12 (Ikebe et al., 2010; Muller-Alouf et al., 1997). However, by analyzing single-nucleotide polymorphism of contemporary GAS isolates, Sumby et al. identified a considerable difference between genomes of pre-epidemic M1 strain SF370 and M1T1 isolate MGAS5005 (Sumby et al., 2005b). The latter contains a ~36 kb chromosomal region encoding SLO and NADase, which was suggested to be acquired from a M12 GAS strain by horizontal gene transfer and recombinational replacement (Sumby et al., 2005b). It is believed that acquisition of this 36-kb region together with the integration of bacteriophage-encoded Sda1 and SpeA2, cooperatively leads to the emergence and rapid global spread of the contemporary epidemic M1T1 clone (Nasser et al., 2014; Wilkening and Federle, 2017; Zhu et al., 2015).

2.1 SLO – the potent pore-forming toxin

Nearly all GAS clinical isolates possess a highly conserved *slo* gene encoding SLO, which is detectable during exponential and early stationary growth phases of GAS isolates in the culture medium (Alouf, 1980). Moreover, SLO expression is highly upregulated in epidemic M1T1 GAS strains (Sumby et al., 2006; Zhu et al., 2015). It is also suggested that the expression of *slo* gene is upregulated by cathelicidin antimicrobial peptide LL-37, preventing GAS to be killed by human cells (Love et al., 2012).

As a potent pore-forming toxin, SLO has been shown to lyze numerous host cells through destructing cytoplasmic membrane integrity. SLO causes lysis in eukaryotic cells by binding to membrane cholesterol in a monomeric form, followed by oligomerization and pore formation, a mechanism similarly to that mediated by staphylococcal α-toxin and lytic membrane attack complex of complement (Bhakdi et al., 1985; Duncan and Schlegel, 1975; Hugo et al., 1986; Niedermeyer, 1985; Otto, 2014; Tegla et al., 2011). Cells, such as erythrocytes, neutrophils, macrophages, platelets, and keratinocytes, are the reported targets of SLO (Duncan and Schlegel, 1975; Launay and Alouf, 1979; Sierig et al., 2003). Similar with streptolysin S, SLO contributes to β-hemolytic activity of GAS on the blood agar (Fontaine et al., 2003; Sierig et al., 2003). However, its hemolytic activity is inactivated by oxygen, while streptolysin S is oxygen-stable.

SLO contributes to GAS virulence with multiple functions. It has been demonstrated that *slo*-deficient GAS mutants decreased their virulence in various murine models of invasive diseases (Ato et al., 2008; Fontaine et al., 2003; Limbago et al., 2000; Timmer et al., 2009). More in details, studies suggested that SLO contributes to GAS virulence in the early stage of infections when GAS encounters epithelial barrier and phagocytes including neutrophils and macrophages (Ato et al., 2008; Fontaine et al., 2003; Limbago et al., 2000;

Timmer et al., 2009). In term of neutrophils, sublytic SLO sufficiently suppresses neutrophil oxidative burst, which could be reversed by free cholesterol and anti-SLO antibodies (Uchiyama et al., 2015). Furthermore, SLO impairs neutrophil degranulation, interleukin-8 secretion and antibacterial activity (Uchiyama et al., 2015). Interestingly, SLO at sublytic concentrations also inhibits neutrophil chemotaxis, possibly due to that SLO induces a rapid and massive shedding of L-selectin, the cell adhesion molecule guiding leukocytes to sites of inflammation (Walcheck et al., 1996; Walev et al., 2000; Van Epps and Andersen, 1974). However, in epidemic M1T1 clones, SLO functionally induces neutrophil lysis and cell death (Timmer et al., 2009). It could be explained by the observation that epidemic M1T1 clones typically produce a high level of SLO (Sumby et al., 2006). During neutrophil lysis, SLO apparently activates neutrophils through a p38 MAPK signaling pathway, resulting in the release of heparin-binding protein, LL-37, defensins, and elastase (Andersen and Duncan, 1980; Nilsson et al., 2006). Moreover, SLO induces the coaggregation of neutrophils and platelets in rat model, which is largely mediated by platelet P-selectin and may contributes to the vascular dysfunction and ischemic destruction observed in patients with toxic shock syndrome (Bryant et al., 2005).

Macrophages, as similar with neutrophils, response to the stimulation of recombinant SLO through activation of the NLRP3 inflammasome, leading to IL-1β secretion and pyroptosis (Keyel et al., 2013; Mitsui et al., 2002). Notably, GAS also requires SLO to activate intracellular NLRP3 inflammasome in macrophages in a caspase-1 dependent manner, independently of TLR signaling that is critical for macrophages exposed to certain bacterial pore-forming toxins (Harder et al., 2009; Mariathasan et al., 2006). This finding was constitutively mimicked by Timmer et al. who showed an accelerated, caspase-dependent SLO-induced macrophage apoptosis contributed to M1T1 GAS immune evasion and virulence (Timmer et al., 2009). Interestingly, not only caspase-1, caspases-3 and -7 are also involved in SLO-induced macrophage apoptosis, indicating that SLO rather activates multiple inflammatory programmed cell death pathways (Goldmann et al., 2009; Timmer et al., 2009). In addition to the lytic activity and modulation of inflammation, SLO prevents internalization of GAS into lysosomes (Hakansson et al., 2005; Logsdon et al., 2011). SLO-deficient GAS was internalized directly via a clathrin-dependent pathway and rapidly transported into lysosomes, and killed through the lysosomal degradation (Hakansson et al., 2005; Logsdon et al., 2011). In contrast, SLO prevented lysosomal colocalization with intracellular GAS and acidification of GAS-containing vacuoles, and conferred GAS resist internalization and intracellular killing by pharyngeal epithelial cells (Hakansson et al., 2005; Logsdon et al., 2011). Collectively, these suggest a lysis-independent function of SLO, which is being a translocator delivering GAS toxins, in particular NADase into cell cytoplasm (Magassa et al., 2010).

2.2 NADase – an intracellular enzyme and toxin

GAS NADase or SPN is a secreted enzyme that catalyzes the hydrolysis of the nicotinamide-ribose bond of NAD^+ to yield nicotinamide and ADP-ribose (Ghosh et al., 2010; Grushoff et al., 1975). Because cyclic ADP-ribose is a potent second messenger for intracellular calcium homeostasis (Guse, 2004; Lee, 2001), it has been suggested that NADase could be translocated into the cytosol of host cells to deplete cellular NAD pools and disrupting cellular metabolism, leading to host cell injury and death (Bricker et al., 2005; Bricker et al., 2002; Madden et al., 2001; Michos et al., 2006).

NADase is directly toxic for a wide variety of host cells, and toxic to *E. coli* and yeast when they were employed for expression of recombinant GAS NADase (Ghosh et al., 2010; Tatsuno et al., 2007). However, NADase has no toxic effects on GAS, suggesting an alternative factor from GAS that essentially regulates the toxicity of NADase within GAS. This factor, designated as immunity factor for SPN (IFS) by Meehl et al., is an endogenous inhibitor of NADase activity (Kimoto et al., 2006; Meehl et al., 2005). IFS is co-expressed as a small protein of 161 amino acids with NADase and forms a stable complex with the later at a 1:1 molar ratio in the cytoplasm of GAS. Binding of IFS to NADase in a form of complex blocks its access to the substrate β-NAD^+, therefore competitively inhibiting activities of NADase and protecting GAS from exhaustion of NAD^+ (Kimoto et al., 2006; Meehl et al., 2005; Smith et al., 2011). Furthermore, recombinant NADase can be normally produced in *E. coli* when co-expressed with the inhibitor IFS, and dissociated from IFS by treatment with sodium thiocyanate (Kimoto et al., 2006; Meehl et al., 2005). However, to be active and toxic to host cells, NADase need to be released from the stable complex with IFS during secretion by GAS, for which the mechanism remains to be elucidated.

NADase acts specifically as an intracellular toxin to the host. The cytotoxic effects of NADase on host requires SLO that delivers NADase to the cytoplasm of human cells, making it difficult to distinguish toxic effects of NADase from those of SLO (Karasawa et al., 1995; Magassa et al., 2010; Sharma et al., 2016). Alternatively, Sharma et al. fused active NADase to anthrax toxin, which mediates delivery of NADase into human oropharyngeal keratinocytes, independently of SLO (Sharma et al., 2016). This active NADase, but not the inactive form of NADase, induced cell death of keratinocytes. Moreover, NADase/SLO-deficient GAS had a defective intracellular survival in human keratinocytes, however, this defective survival could be recused by delivery of anthrax toxin-fused NADase into keratinocytes (Sharma et al., 2016). These indicate that NADase acts specifically as an intracellular toxin, mediating cytotoxicity and promoting intracellular survival of GAS in host cells (Bricker et al., 2002; Sharma et al., 2016). Furthermore, the toxic effects of NADase in the host are also demonstrated *in vivo* using mouse models of invasive infections, which showed a significantly attenuation of virulence in GAS lacking NADase activity (Bricker et al., 2005; Zhu et al., 2017). NADase also has an immune regulatory function through inhibiting inflammasome-dependent IL-1β release from infected macrophages, independent on SLO-mediated translocation of NADase into the host cell cytosol (Hancz et al., 2017).

2.3 SLO and NADase – One plus one equals more than two

As two potent toxins of GAS, SLO and NADase play relevant roles in GAS pathogenesis. More importantly, both toxins act synergistically and closely associate with each other. For example, the genes encoding SLO and NADase are co-transcribed, while two proteins can be co-purified from the culture supernatants of GAS (Shany et al., 1973). SLO mediates the translocation of NADase into human epithelial cells, while NADase binds to SLO and enhances its activity (Michos et al., 2006; Velarde et al., 2017). Interaction between SLO and NADase seems to be a win-win event that stabilizes both toxins and increases their activities (Michos et al., 2006; Velarde et al., 2017). Furthetmore, GAS with both SLO and NADase induces an augmented cell injury of human keratinocytes (Michos et al., 2006; Velarde et al., 2017). Interestingly, SLO stimulates xenophagy in pharyngeal keratinocytes. However, SLO and NADase coordinately prevent maturation of GAS-containing autophagosomes, thereby prolonging GAS intracellular survival (O'Seaghdha and Wessels, 2013). Similarly, both toxins also mediate GAS intracellular survival and cytotoxicity for macrophages. They do not inhibit maturation of GAS-containing lysophagosomes, but rather prevent phagolysosome acidification in macrophages that may serve as reservoirs or vehicles to enable persistent GAS infection in the host (Bastiat-Sempe et al., 2014).

The synergistical effects of SLO and NADase have been elucidated not only in host cells *ex vivo* but also in various mouse models of invasive GAS infections. Bricker et al. showed that mutants deficient in SLO and NADase activity were significantly attenuated for virulence compared with the wild-type parent M3 GAS strain in mouse models of invasive soft-tissue infection and septicemia (Bricker et al., 2005). In an epidemic M1T1 GAS background, SLO and NADase had equally capacities to cause necrotizing myositis (Zhu et al., 2017). Moreover, mutants lacking SLO and NADase are impaired to resist killing by human neutrophils, and are significantly attenuated in the bacteremia and soft tissue infection models (Zhu et al., 2017). Therefore, recombinational replacement of SLO and NADase-encoded genetic region from M12 GAS contributes significantly to the emergence of contemporary epidemic M1T1 isolates.

II. Selection of hypervirulent M1T1 GAS clone in the host

Phenotypic switching refers to a switch between two phenotypic states with different gene expression patterns, which leads to variation in morphology, physiology, and pathogenicity of bacterial pathogens. For example, pathogen *S. aureus* switches from normal colony phenotype to a slow-growing, small colony variant phenotype that facilitates bacterial persistent in host cells and recurrent infections (Proctor et al., 2006). GAS, particularly the invasive M1T1 and M3 types, also undergoes a switch from a SpeB-positive and SpeA negative ($SpeB^+/SpeA^-$) phenotype to a $SpeB^-/SpeA^+$ phenotype when they are recovered from infected human or passaged in mouse, which is termed as *in vivo* selection (Ikebe et al., 2010; Kazmi et al., 2001; Walker et al., 2007). The *in vivo* selection, characterized as the loss of SpeB activity, has been associated with the mutations in global regulatory proteins of GAS (Carroll and Musser, 2011). GAS possesses various global regulators including two-component regulatory systems and stand-alone regulators (Kreikemeyer et al., 2003). Of these global regulators, CovRS has been well characterized (Graham et al., 2002; Sumby et al., 2006). CovRS represents a sophisticated regulatory network negatively controlling expression of multiple virulence factors including SpeB (Graham et al., 2002). However, CovRS is highly prone to spontaneous mutation in the host, which results in downregulation of SpeB production but enhancement of virulence, leading to the emergence of variants with hypervirulent/invasive capacities (Graham et al., 2002; Ikebe et al., 2010; Trevino et al., 2009).

1. CovRS – the control of GAS virulence

CovRS consists of a membrane-bound histidine kinase CovS and a cognate response regulator CovR. As a typical sensor histidine kinase, CovS contains a surface-exposed extracellular domain, two membrane-spanning domains, a cytosolic HAMP domain (present in Histidine kinases, Adenyl cyclases, Methyl-accepting proteins and Phosphatases), and a histidine kinase-like ATPase (HATPase) domain (Tran-Winkler et al., 2011). The extracellular domain acts to sense various environmental stimuli or stresses, *e.g.*, Mg^{2+} concentration, external pH and temperature, iron starvation and LL-37 (Dalton and Scott, 2004; Froehlich et al., 2009; Gryllos et al., 2007; Gryllos et al., 2003; Gryllos et al., 2008). It is presumed that the environmental signals could influence the histidine autophosphorylation of HATPase domain in CovS and subsequent phosphotransfer to the aspartate residue of CovR, which determines CovR phosphorylation status (Dalton and Scott, 2004; Tran-Winkler et al., 2011). CovR acts primarily as a repressor of gene expression whose phosphorylation enhances its binding to regulated promoters (Dalton and Scott, 2004; Federle and Scott, 2002; Gusa et al., 2006; Gusa and Scott, 2005; Horstmann et al., 2015; Horstmann et al., 2014; Tran-Winkler et al., 2011), mediating the cellular responses through altering distinct transcriptional profiling of GAS (Churchward, 2007; Dalton et al., 2006; Graham et al., 2002; Gryllos et al., 2007; Gryllos et al., 2003; Gusa et al., 2006; Horstmann et al., 2015; Horstmann et al., 2014; Tran-Winkler et al., 2011; Trevino et al., 2009).

In response to specific environmental stimuli, CovRS mediates distinct signal transduction and strikingly different transcriptomes. Human cationic LL-37, being a well-characterized stimulus of GAS CovRS, exhibits a broad spectrum of antimicrobial activity as well as immunomodulatory activity (Durr et al., 2006). M1T1 GAS clone is more resistant to the antimicrobial activity of human LL-37 through several mechanisms, for example, sequestering and neutralizing its activity by M1 protein, and inactivating LL-37 by the streptococcal inhibitor of complement and secreted streptococcal proteins (Durr et al., 2006; Frick et al., 2003; Johansson et al., 2008; LaRock et al., 2015; Nyberg et al., 2004). Notably, LL-37 binds directly to the extracellular domain of CovS, which dephosphorylate CovR and subsequently release the CovR-mediated repression of expression of multiple virulence factors including Sda1, SpeA2, SLO, NADase and the hyaluronic acid capsule (Gryllos et al., 2008; Love et al., 2012; Tran-Winkler et al., 2011; Velarde et al., 2014). The LL-37 signaling through CovRS leads to a marked increase of GAS virulence by promoting GAS resistance to the killing by human oropharyngeal keratinocytes, neutrophils, and macrophages (Gryllos et al., 2008; Love et al., 2012; Tran-Winkler et al., 2011). Interestingly, LL-37 signaling-upregulated GAS virulence can be adversely damped by elevated extracellular Mg^{2+}, although it may not be physiologically relevant (Gryllos et al., 2007; Gryllos et al., 2003; Gryllos et al., 2008; Tran-Winkler et al., 2011). The opposite effects on expression of CovRS-regulated genes through LL-37 and Mg^{2+} signaling, as an example, suggest the complexity of CovRS regulatory system, especially when invasive GAS survives in host.

2. The spontaneous mutation of CovRS in the host

In concert with CovRS-controlled enhanced GAS virulence elicited by host LL-37, CovRS undergoes spontaneous mutations more frequently in patients with severe GAS infections, leading to the emergence of hypervirulent/invasive variants. This naturally occurring CovRS mutation has been reported in various M types of GAS including not only M1 but also M3, M12, M28, M49 and M89 GAS clinical isolates (Ikebe et al., 2010; Uchiyama et al., 2012). However, CovRS mutations occur with a higher frequency in the invasive M1T1 isolates, which contributes to the phenotype of $SpeB^-/SpeA^+$ (Aziz et al., 2004b; Engleberg et al., 2001; Ikebe et al., 2010; Kansal et al., 2010). The mutations in most cases, either in CovR or CovS, in the form of nucleotide deletions, point mutations, or insertions, eventually cause dysfunction of CovRS, leading to enhanced virulence in mouse models of invasive disease (Ikebe et al., 2010; Li et al., 2014; Mayfield et al., 2014). CovRS mutation variants, for example, either generated in the laboratory or recovered after passage *in vivo*, produce elevated levels of CovRS-targeted virulent factors, especially that are involved in GAS dissemination into deep tissue and organs in the cases of necrotizing fasciitis and toxic shock syndrome (Engleberg et al., 2001; Graham et al., 2002; Horstmann et al., 2011; Kansal et al., 2010; Mayfield et al., 2014; Miyoshi-Akiyama et al., 2006; Sumby et al., 2006; Tatsuno et al., 2013; Trevino et al., 2009).

It is worth noting that CovRS mutations not only mediate the upregulation of virulence, but also modulate the cellular responses of GAS, *e.g.*, nutrient metabolism, protein biosynthesis, and environmental adaptation. More fundamentally, CovRS mutation changes the transcriptome profile of M1T1 GAS clone from a phenotype favoring epithelial colonization to the one that promotes immune evasion and dissemination of GAS (Graham et al., 2002; Hollands et al., 2010; Sumby et al., 2006). It suggests, potentially, that CovRS mutations may shape GAS virulence in combination with other transcriptional regulators that regulate carbon source utilization and nitrogen metabolism during GAS infection (Shelburne et al., 2010). Furthermore, CovRS mutations facilitate virulence through dismissing the production of SpeB, which proteolytically degrades multiple virulence factors (Aziz et al., 2004b; Walker et al., 2007).

3. The selection pressure for CovRS mutation variants

What pressure or factors in the host drive the occurrence of mutations in CovRS has not been well understood. However, several studies have provided clues that help us understand the *in vivo* selection of CovRS mutations. Two types of selection pressure have been suggested to govern the mutation of CovRS in the host, one from host, another from the bacteria itself.

The selective force from the host is navigated to the innate immunity where neutrophils play a dominant role (Li et al., 2014). Neutrophils are the most abundant innate immune cells, killing invading pathogens through a comprehensive mechanism consisting of phagocytosis, respiratory burst and antimicrobial molecules-mediated killing, and NETs formation (Dohrmann et al., 2016). Compared to other bacteria, GAS is more successful to counteract the antimicrobial activities of neutrophils, which could be attributed to the observation that neutrophils provide selection pressure for the phenotypic switch of M1T1 GAS via the mutation of CovRS in the mouse infection models (Kobayashi et al., 2003; Li et al., 2014). Accordingly, such a selection was inhibited in mice whose neutrophils were depleted by specific monoclonal antibodies or when neutrophil migration was impaired during infection due to the deficiency in chemokine receptor 2 (Li et al., 2014). The recovered CovRS mutation variants from mice showed a hypervirulent phenotype as similarly as found in the clinical isolates from patients with toxic shock syndrome and other severe infections (Ikebe et al., 2010; Li et al., 2014).

GAS virulence factors also act as selective pressures that determine the CovRS mutation in the host. At least two virulence factors, M1 protein and hyaluronic acid capsule, have been proposed being involved (Cole et al., 2010). M protein, as the major virulence factor of GAS, mediates GAS colonization of host epithelium, promotes GAS resistance to phagocytosis, inactivates LL-37 activity, as well as facilitates bacterial survival in neutrophils and NETs (Carlsson et al., 2003; Horstmann et al., 1988; Horstmann et al., 1992; LaRock et al., 2015; Lauth et al., 2009; Oehmcke et al., 2010; Staali et al., 2006). Similarly, hyaluronic acid capsule increases GAS resistance to phagocytosis and survival within NETs (Cole et al., 2010; Dale et al., 1996; Wessels et al., 1991). Both M1 protein and hyaluronic acid capsule share a common capacity that is to resist neutrophil-mediated killing (Buchanan et al., 2006; Lauth et al., 2009; Staali et al., 2003; Wessels et al., 1991). Consequently, GAS deficient in M1 protein or hyaluronic acid capsule failed to undergo phenotypic switching and was rapidly cleared by the host immune responses (Cole et al., 2010; Liu et al., 2015). Therefore, the occurrence of CovRS mutations in the host is rather a strategy for GAS to counteract host immunity, particularly the antimicrobial activities of neutrophils.

Sda1 has been also regarded as a selective force for the emergence of hypervirulent isolates in M1T1 GAS 5448 (Walker et al., 2007). Similarly as M1 protein and hyaluronic acid capsule, Sda1 confers GAS virulence through degrading NETs and increasing GAS survival in the host (Brinkmann et al., 2004; Buchanan et al., 2006; Uchiyama et al., 2012). However, horizontal transfer of the Sda1-encoding bacteriophage into the genome of M1 SF370 didn't confer GAS switch to hypervirulent *covRS* mutants in the mouse model as the phage donor strain 5448 did (Venturini et al., 2013). Moreover, we found that M1 protein, instead of Sda1, contributed to the CovRS mutation of two typical M1T1 GAS isolates in the host (Liu et al., 2015).

In a word, the selection of CovRS mutations in the host, theoretically, fits the paradigm of "Stress-Response" (Figure 1). To survive in host detrimental circumstance ("Stress"), GAS must adapt itself to evade immune attack ("Response"). It seems that GAS virulence factors including M1 protein and the capsule but not Sda1, provide primary and synergistical protection when the

M1T1 clone colonizes and invades host niches, where it survives and has opportunities to adjust the transcriptional profile through mutating in the global regulatory proteins (Cole et al., 2010; Feng et al., 2016; Liu et al., 2015; Venturini et al., 2013). Furthermore, when the detrimental circumstance or the selection pressure do not exist, GAS tends to attenuate virulence at the expense of downregulation of M protein (Zhou et al., 2013).

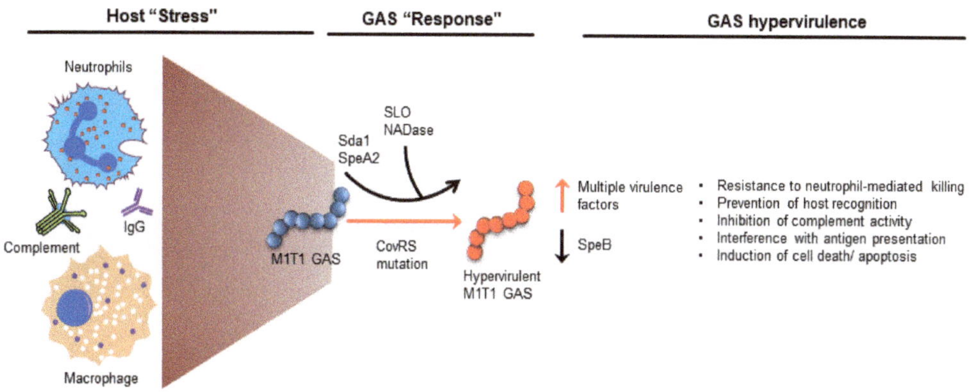

Figure 1. Schematic paradigm of "Stress-Response" driving the emergence of hypervirulent M1T1 GAS clone. Host innate immunity consisting of immune cells, complement and antibodies provides a "Stress" environment for GAS. To survive such a detrimental circumstance, GAS must adapt itself in response to immune stress. By acquiring genetic regions encoding Sda1, SpeA2, SLO and NADase and mutating its global regulatory proteins such as CovRS, GAS changes the transcriptional profile that promotes its hypervirulence, leading to the successful emergence of pandemic M1T1 clone.

Concluding Remarks

The emergence of highly successful and globally disseminated M1T1 GAS clone has been identified as a threat to human health. This review highlights the importance of genetic recombination events, through which M1T1 GAS obtains virulence factors including Sda1, SpeA2, SLO and NADase. Further, this review highlights the significance of selection pressures from either neutrophils or bacteria, that modulate the upregulation of GAS virulence via the mutation of CovRS. Finally, genetic recombination alongside with the selective force cooperatively drive the emergence of successful pandemic M1T1 isolates. By understanding the evolutionary pressures that select and shape the hypervirulent M1T1 clone, we could potentially develop new therapeutic strategies to tackle challenges when dealing with the globally disseminated M1T1 GAS clone.

Acknowledgment

The author would like to thank Ms. Kinga Balogh Sivars (AstraZeneca AB R&D, Sweden) for expert proofreading.

Disclosures

The author is currently employed by AstraZeneca AB R&D, Sweden. AstraZeneca has no financial conflict of interest.

References

Abe, J., Forrester, J., Nakahara, T., Lafferty, J.A., Kotzin, B.L., and Leung, D.Y. (1991). Selective stimulation of human T cells with streptococcal erythrogenic toxins A and B. J Immunol *146*, 3747-3750.

Alouf, J.E. (1980). Streptococcal toxins (streptolysin O, streptolysin S, erythrogenic toxin). Pharmacol Ther *11*, 661-717.

Andersen, B.R., and Duncan, J.L. (1980). Activation of human neutrophil metabolism by streptolysin O. J Infect Dis *141*, 680-685.

Ato, M., Ikebe, T., Kawabata, H., Takemori, T., and Watanabe, H. (2008). Incompetence of neutrophils to invasive group A *streptococcus* is attributed to induction of plural virulence factors by dysfunction of a regulator. PLoS One *3*, e3455.

Aziz, R.K., Edwards, R.A., Taylor, W.W., Low, D.E., McGeer, A., and Kotb, M. (2005). Mosaic prophages with horizontally acquired genes account for the emergence and diversification of the globally disseminated M1T1 clone of *Streptococcus pyogenes*. J Bacteriol *187*, 3311-3318.

Aziz, R.K., Ismail, S.A., Park, H.W., and Kotb, M. (2004a). Post-proteomic identification of a novel phage-encoded streptodornase, Sda1, in invasive M1T1 *Streptococcus pyogenes*. Mol Microbiol *54*, 184-197.

Aziz, R.K., and Kotb, M. (2008). Rise and persistence of global M1T1 clone of *Streptococcus pyogenes*. Emerg Infect Dis *14*, 1511-1517.

Aziz, R.K., Pabst, M.J., Jeng, A., Kansal, R., Low, D.E., Nizet, V., and Kotb, M. (2004b). Invasive M1T1 group A *Streptococcus* undergoes a phase-shift *in vivo* to prevent proteolytic degradation of multiple virulence factors by SpeB. Mol Microbiol *51*, 123-134.

Banks, D.J., Porcella, S.F., Barbian, K.D., Beres, S.B., Philips, L.E., Voyich, J.M., DeLeo, F.R., Martin, J.M., Somerville, G.A., and Musser, J.M. (2004). Progress toward characterization of the group A *Streptococcus* metagenome: complete genome sequence of a macrolide-resistant serotype M6 strain. J Infect Dis *190*, 727-738.

Bastiat-Sempe, B., Love, J.F., Lomayesva, N., and Wessels, M.R. (2014). Streptolysin O and NAD-glycohydrolase prevent phagolysosome acidification and promote group A *Streptococcus* survival in macrophages. MBio *5*, e01690-01614.

Beiter, K., Wartha, F., Albiger, B., Normark, S., Zychlinsky, A., and Henriques-Normark, B. (2006). An endonuclease allows *Streptococcus pneumoniae* to escape from neutrophil extracellular traps. Curr Biol *16*, 401-407.

Beres, S.B., Sylva, G.L., Barbian, K.D., Lei, B., Hoff, J.S., Mammarella, N.D., Liu, M.Y., Smoot, J.C., Porcella, S.F., Parkins, L.D., *et al.* (2002). Genome sequence of a serotype M3 strain of group A *Streptococcus*: phage-encoded toxins, the high-virulence phenotype, and clone emergence. Proc Natl Acad Sci U S A *99*, 10078-10083.

Bhakdi, S., Tranum-Jensen, J., and Sziegoleit, A. (1985). Mechanism of membrane damage by streptolysin-O. Infect Immun *47*, 52-60.

Braun, M.A., Gerlach, D., Hartwig, U.F., Ozegowski, J.H., Romagne, F., Carrel, S., Kohler, W., and Fleischer, B. (1993). Stimulation of human T cells by streptococcal "superantigen" erythrogenic toxins (scarlet fever toxins). J Immunol *150*, 2457-2466.

Bricker, A.L., Carey, V.J., and Wessels, M.R. (2005). Role of NADase in virulence in experimental invasive group A streptococcal infection. Infect Immun *73*, 6562-6566.

Bricker, A.L., Cywes, C., Ashbaugh, C.D., and Wessels, M.R. (2002). NAD+-glycohydrolase acts as an intracellular toxin to enhance the extracellular survival of group A streptococci. Mol Microbiol *44*, 257-269.

Brinkmann, V., Reichard, U., Goosmann, C., Fauler, B., Uhlemann, Y., Weiss, D.S., Weinrauch, Y., and Zychlinsky, A. (2004). Neutrophil extracellular traps kill bacteria. Science *303*, 1532-1535.

Bryant, A.E., Bayer, C.R., Chen, R.Y., Guth, P.H., Wallace, R.J., and Stevens, D.L. (2005). Vascular dysfunction and ischemic destruction of tissue in *Streptococcus pyogenes* infection: the role of streptolysin O-induced platelet/neutrophil complexes. J Infect Dis *192*, 1014-1022.

Buchanan, J.T., Simpson, A.J., Aziz, R.K., Liu, G.Y., Kristian, S.A., Kotb, M., Feramisco, J., and Nizet, V. (2006). DNase expression allows the pathogen group A *Streptococcus* to escape killing in neutrophil extracellular traps. Curr Biol *16*, 396-400.

Carapetis, J.R., Steer, A.C., Mulholland, E.K., and Weber, M. (2005). The global burden of group A streptococcal diseases. Lancet Infect Dis *5*, 685-694.

Carlsson, F., Berggard, K., Stalhammar-Carlemalm, M., and Lindahl, G. (2003). Evasion of phagocytosis through cooperation between two ligand-binding regions in *Streptococcus pyogenes* M protein. J Exp Med *198*, 1057-1068.

Carroll, R.K., and Musser, J.M. (2011). From transcription to activation: how group A streptococcus, the flesh-eating pathogen, regulates SpeB cysteine protease production. Mol Microbiol *81*, 588-601.

Chang, Z., Jiang, N., Zhang, Y., Lu, H., Yin, J., Wahlgren, M., Cheng, X., Cao, Y., and Chen, Q. (2016). The TatD-like DNase of Plasmodium is a virulence factor and a potential malaria vaccine candidate. Nat Commun *7*, 11537.

Churchward, G. (2007). The two faces of Janus: virulence gene regulation by CovR/S in group A streptococci. Mol Microbiol *64*, 34-41.

Cleary, P.P., LaPenta, D., Vessela, R., Lam, H., and Cue, D. (1998). A globally disseminated M1 subclone of group A streptococci differs from other subclones by 70 kilobases of prophage DNA and capacity for high-frequency intracellular invasion. Infect Immun *66*, 5592-5597.

Cole, J.N., Barnett, T.C., Nizet, V., and Walker, M.J. (2011). Molecular insight into invasive group A streptococcal disease. Nat Rev Microbiol *9*, 724-736.

Cole, J.N., Pence, M.A., von Kockritz-Blickwede, M., Hollands, A., Gallo, R.L., Walker, M.J., and Nizet, V. (2010). M protein and hyaluronic acid capsule are essential for *in vivo* selection of covRS mutations characteristic of invasive serotype M1T1 group A *Streptococcus*. MBio *1* (4):e00191-10.

Dale, J.B., Washburn, R.G., Marques, M.B., and Wessels, M.R. (1996). Hyaluronate capsule and surface M protein in resistance to opsonization of group A streptococci. Infect Immun *64*, 1495-1501.

Dalton, T.L., Collins, J.T., Barnett, T.C., and Scott, J.R. (2006). RscA, a member of the MDR1 family of transporters, is repressed by CovR and required for growth of *Streptococcus pyogenes* under heat stress. J Bacteriol *188*, 77-85.

Dalton, T.L., and Scott, J.R. (2004). CovS inactivates CovR and is required for growth under conditions of general stress in *Streptococcus pyogenes*. J Bacteriol *186*, 3928-3937.

Dohrmann, S., Cole, J.N., and Nizet, V. (2016). Conquering Neutrophils. PLoS Pathog *12*, e1005682.

Duncan, J.L., and Schlegel, R. (1975). Effect of streptolysin O on erythrocyte membranes, liposomes, and lipid dispersions. A protein-cholesterol interaction. J Cell Biol *67*, 160-174.

Durr, U.H., Sudheendra, U.S., and Ramamoorthy, A. (2006). LL-37, the only human member of the cathelicidin family of antimicrobial peptides. Biochim Biophys Acta *1758*, 1408-1425.

Earhart, C.A., Vath, G.M., Roggiani, M., Schlievert, P.M., and Ohlendorf, D.H. (2000). Structure of streptococcal pyrogenic exotoxin A reveals a novel metal cluster. Protein Sci *9*, 1847-1851.

Engleberg, N.C., Heath, A., Miller, A., Rivera, C., and DiRita, V.J. (2001). Spontaneous mutations in the CsrRS two-component regulatory system of *Streptococcus pyogenes* result in enhanced virulence in a murine model of skin and soft tissue infection. J Infect Dis *183*, 1043-1054.

Federle, M.J., and Scott, J.R. (2002). Identification of binding sites for the group A streptococcal global regulator CovR. Mol Microbiol *43*, 1161-1172.

Feng, W., Liu, M., Chen, D.G., Yiu, R., Fang, F.C., and Lei, B. (2016). Contemporary Pharyngeal and Invasive emm1 and Invasive emm12 Group A *Streptococcus* Isolates Exhibit Similar In Vivo Selection for CovRS Mutants in Mice. PLoS One *11*, e0162742.

Ferretti, J.J., McShan, W.M., Ajdic, D., Savic, D.J., Savic, G., Lyon, K., Primeaux, C., Sezate, S., Suvorov, A.N., Kenton, S., *et al.* (2001). Complete genome sequence of an M1 strain of *Streptococcus pyogenes*. Proc Natl Acad Sci U S A *98*, 4658-4663.

Fontaine, M.C., Lee, J.J., and Kehoe, M.A. (2003). Combined contributions of streptolysin O and streptolysin S to virulence of serotype M5 *Streptococcus pyogenes* strain Manfredo. Infect Immun *71*, 3857-3865.

Fraser, J.D., and Proft, T. (2008). The bacterial superantigen and superantigen-like proteins. Immunol Rev *225*, 226-243.

Frick, I.M., Akesson, P., Rasmussen, M., Schmidtchen, A., and Bjorck, L. (2003). SIC, a secreted protein of *Streptococcus pyogenes* that inactivates antibacterial peptides. J Biol Chem *278*, 16561-16566.

Froehlich, B.J., Bates, C., and Scott, J.R. (2009). *Streptococcus pyogenes* CovRS mediates growth in iron starvation and in the presence of the human cationic antimicrobial peptide LL-37. J Bacteriol *191*, 673-677.

Ghosh, J., Anderson, P.J., Chandrasekaran, S., and Caparon, M.G. (2010). Characterization of *Streptococcus pyogenes* beta-NAD+ glycohydrolase: re-evaluation of enzymatic properties associated with pathogenesis. J Biol Chem *285*, 5683-5694.

Goldmann, O., Sastalla, I., Wos-Oxley, M., Rohde, M., and Medina, E. (2009). *Streptococcus pyogenes* induces oncosis in macrophages through the activation of an inflammatory programmed cell death pathway. Cell Microbiol *11*, 138-155.

Graham, M.R., Smoot, L.M., Migliaccio, C.A., Virtaneva, K., Sturdevant, D.E., Porcella, S.F., Federle, M.J., Adams, G.J., Scott, J.R., and Musser, J.M. (2002). Virulence control in group A *Streptococcus* by a two-component gene regulatory system: global expression profiling and in vivo infection modeling. Proc Natl Acad Sci U S A *99*, 13855-13860.

Grushoff, P.S., Shany, S., and Bernheimer, A.W. (1975). Purification and properties of streptococcal nicotinamide adenine dinucleotide glycohydrolase. J Bacteriol *122*, 599-605.

Gryllos, I., Grifantini, R., Colaprico, A., Jiang, S., Deforce, E., Hakansson, A., Telford, J.L., Grandi, G., and Wessels, M.R. (2007). Mg(2+) signalling defines the group A streptococcal CsrRS (CovRS) regulon. Mol Microbiol *65*, 671-683.

Gryllos, I., Levin, J.C., and Wessels, M.R. (2003). The CsrR/CsrS two-component system of group A *Streptococcus* responds to environmental Mg^{2+}. Proc Natl Acad Sci U S A *100*, 4227-4232.

Gryllos, I., Tran-Winkler, H.J., Cheng, M.F., Chung, H., Bolcome, R., 3rd, Lu, W., Lehrer, R.I., and Wessels, M.R. (2008). Induction of group A *Streptococcus* virulence by a human antimicrobial peptide. Proc Natl Acad Sci U S A *105*, 16755-16760.

Gusa, A.A., Gao, J., Stringer, V., Churchward, G., and Scott, J.R. (2006). Phosphorylation of the group A Streptococcal CovR response regulator causes dimerization and promoter-specific recruitment by RNA polymerase. J Bacteriol *188*, 4620-4626.

Gusa, A.A., and Scott, J.R. (2005). The CovR response regulator of group A streptococcus (GAS) acts directly to repress its own promoter. Mol Microbiol *56*, 1195-1207.

Guse, A.H. (2004). Regulation of calcium signaling by the second messenger cyclic adenosine diphosphoribose (cADPR). Curr Mol Med *4*, 239-248.

Hakansson, A., Bentley, C.C., Shakhnovic, E.A., and Wessels, M.R. (2005). Cytolysin-dependent evasion of lysosomal killing. Proc Natl Acad Sci U S A *102*, 5192-5197.

Hancz, D., Westerlund, E., Bastiat-Sempe, B., Sharma, O., Valfridsson, C., Meyer, L., Love, J.F., O'Seaghdha, M., Wessels, M.R., and Persson, J.J. (2017). Inhibition of Inflammasome-Dependent Interleukin 1beta Production by Streptococcal NAD(+)-Glycohydrolase: Evidence for Extracellular Activity. MBio *8* (4): e00756-17.

Harder, J., Franchi, L., Munoz-Planillo, R., Park, J.H., Reimer, T., and Nunez, G. (2009). Activation of the Nlrp3 inflammasome by *Streptococcus pyogenes* requires streptolysin O and NF-kappa B activation but proceeds independently of TLR signaling and P2X7 receptor. J Immunol *183*, 5823-5829.

Hollands, A., Pence, M.A., Timmer, A.M., Osvath, S.R., Turnbull, L., Whitchurch, C.B., Walker, M.J., and Nizet, V. (2010). Genetic switch to hypervirulence reduces colonization phenotypes of the globally disseminated group A streptococcus M1T1 clone. J Infect Dis *202*, 11-19.

Horstmann, N., Sahasrabhojane, P., Saldana, M., Ajami, N.J., Flores, A.R., Sumby, P., Liu, C.G., Yao, H., Su, X., Thompson, E., *et al.* (2015). Characterization of the effect of the histidine kinase CovS on response regulator phosphorylation in group A Streptococcus. Infect Immun *83*, 1068-1077.

Horstmann, N., Sahasrabhojane, P., Suber, B., Kumaraswami, M., Olsen, R.J., Flores, A., Musser, J.M., Brennan, R.G., and Shelburne, S.A., 3rd (2011). Distinct single amino acid replacements in the control of virulence regulator protein differentially impact streptococcal pathogenesis. PLoS Pathog *7*, e1002311.

Horstmann, N., Saldana, M., Sahasrabhojane, P., Yao, H., Su, X., Thompson, E., Koller, A., and Shelburne, S.A., 3rd (2014). Dual-site phosphorylation of the control of virulence regulator impacts group a streptococcal global gene expression and pathogenesis. PLoS Pathog *10*, e1004088.

Horstmann, R.D., Sievertsen, H.J., Knobloch, J., and Fischetti, V.A. (1988). Antiphagocytic activity of streptococcal M protein: selective binding of complement control protein factor H. Proc Natl Acad Sci U S A *85*, 1657-1661.

Horstmann, R.D., Sievertsen, H.J., Leippe, M., and Fischetti, V.A. (1992). Role of fibrinogen in complement inhibition by streptococcal M protein. Infect Immun *60*, 5036-5041.

Hugo, F., Reichwein, J., Arvand, M., Kramer, S., and Bhakdi, S. (1986). Use of a monoclonal antibody to determine the mode of transmembrane pore formation by streptolysin O. Infect Immun *54*, 641-645.

Ikebe, T., Ato, M., Matsumura, T., Hasegawa, H., Sata, T., Kobayashi, K., and Watanabe, H. (2010). Highly frequent mutations in negative regulators of multiple virulence genes in group A streptococcal toxic shock syndrome isolates. PLoS Pathog *6*, e1000832.

Johansson, L., Thulin, P., Sendi, P., Hertzen, E., Linder, A., Akesson, P., Low, D.E., Agerberth, B., and Norrby-Teglund, A. (2008). Cathelicidin LL-37 in severe *Streptococcus pyogenes* soft tissue infections in humans. Infect Immun *76*, 3399-3404.

Kansal, R.G., Aziz, R.K., and Kotb, M. (2005). Modulation of expression of superantigens by human transferrin and lactoferrin: a novel mechanism in host-Streptococcus interactions. J Infect Dis *191*, 2121-2129.

Kansal, R.G., Datta, V., Aziz, R.K., Abdeltawab, N.F., Rowe, S., and Kotb, M. (2010). Dissection of the molecular basis for hypervirulence of an in vivo-selected phenotype of the widely disseminated M1T1 strain of group A *Streptococcus* bacteria. J Infect Dis *201*, 855-865.

Karasawa, T., Takasawa, S., Yamakawa, K., Yonekura, H., Okamoto, H., and Nakamura, S. (1995). NAD(+)-glycohydrolase from *Streptococcus pyogenes* shows cyclic ADP-ribose forming activity. FEMS Microbiol Lett *130*, 201-204.

Kasper, K.J., Zeppa, J.J., Wakabayashi, A.T., Xu, S.X., Mazzuca, D.M., Welch, I., Baroja, M.L., Kotb, M., Cairns, E., Cleary, P.P., *et al.* (2014). Bacterial superantigens promote acute nasopharyngeal infection by *Streptococcus pyogenes* in a human MHC Class II-dependent manner. PLoS Pathog *10*, e1004155.

Kazmi, S.U., Kansal, R., Aziz, R.K., Hooshdaran, M., Norrby-Teglund, A., Low, D.E., Halim, A.B., and Kotb, M. (2001). Reciprocal, temporal expression of SpeA and SpeB by invasive M1T1 group a streptococcal isolates in vivo. Infect Immun *69*, 4988-4995.

Keyel, P.A., Roth, R., Yokoyama, W.M., Heuser, J.E., and Salter, R.D. (2013). Reduction of streptolysin O (SLO) pore-forming activity enhances inflammasome activation. Toxins (Basel) *5*, 1105-1118.

Kimoto, H., Fujii, Y., Hirano, S., Yokota, Y., and Taketo, A. (2006). Genetic and biochemical properties of streptococcal NAD-glycohydrolase inhibitor. J Biol Chem *281*, 9181-9189.

Kline, J.A., and Runge, J.W. (1994). Streptococcal pharyngitis: a review of pathophysiology, diagnosis, and management. J Emerg Med *12*, 665-680.

Kline, J.B., and Collins, C.M. (1996). Analysis of the superantigenic activity of mutant and allelic forms of streptococcal pyrogenic exotoxin A. Infect Immun *64*, 861-869.

Kline, J.B., and Collins, C.M. (1997). Analysis of the interaction between the bacterial superantigen streptococcal pyrogenic exotoxin A (SpeA) and the human T-cell receptor. Mol Microbiol *24*, 191-202.

Kobayashi, S.D., Braughton, K.R., Whitney, A.R., Voyich, J.M., Schwan, T.G., Musser, J.M., and DeLeo, F.R. (2003). Bacterial pathogens modulate an apoptosis differentiation program in human neutrophils. Proc Natl Acad Sci U S A *100*, 10948-10953.

Kreikemeyer, B., McIver, K.S., and Podbielski, A. (2003). Virulence factor regulation and regulatory networks in *Streptococcus pyogenes* and their impact on pathogen-host interactions. Trends Microbiol *11*, 224-232.

Lancefield, R.C. (1933). A Serological Differentiation of Human and Other Groups of Hemolytic Streptococci. J Exp Med *57*, 571-595.

Lancefield, R.C. (1962). Current knowledge of type-specific M antigens of group A streptococci. J Immunol *89*, 307-313.

Lappin, E., and Ferguson, A.J. (2009). Gram-positive toxic shock syndromes. Lancet Infect Dis *9*, 281-290.

LaRock, C.N., Dohrmann, S., Todd, J., Corriden, R., Olson, J., Johannssen, T., Lepenies, B., Gallo, R.L., Ghosh, P., and Nizet, V. (2015). Group A Streptococcal M1 Protein Sequesters Cathelicidin to Evade Innate Immune Killing. Cell Host Microbe *18*, 471-477.

Launay, J.M., and Alouf, J.E. (1979). Biochemical and ultrastructural study of the disruption of blood platelets by streptolysin O. Biochim Biophys Acta *556*, 278-291.

Lauth, X., von Kockritz-Blickwede, M., McNamara, C.W., Myskowski, S., Zinkernagel, A.S., Beall, B., Ghosh, P., Gallo, R.L., and Nizet, V. (2009). M1 protein allows Group A streptococcal survival in phagocyte extracellular traps through cathelicidin inhibition. J Innate Immun *1*, 202-214.

Lee, H.C. (2001). Physiological functions of cyclic ADP-ribose and NAADP as calcium messengers. Annu Rev Pharmacol Toxicol *41*, 317-345.

Leonard, B.A., Lee, P.K., Jenkins, M.K., and Schlievert, P.M. (1991). Cell and receptor requirements for streptococcal pyrogenic exotoxin T-cell mitogenicity. Infect Immun *59*, 1210-1214.

Li, H., Llera, A., Malchiodi, E.L., and Mariuzza, R.A. (1999). The structural basis of T cell activation by superantigens. Annu Rev Immunol *17*, 435-466.

Li, J., Liu, G., Feng, W., Zhou, Y., Liu, M., Wiley, J.A., and Lei, B. (2014). Neutrophils select hypervirulent CovRS mutants of M1T1 group A *Streptococcus* during subcutaneous infection of mice. Infect Immun *82*, 1579-1590.

Limbago, B., Penumalli, V., Weinrick, B., and Scott, J.R. (2000). Role of streptolysin O in a mouse model of invasive group A streptococcal disease. Infect Immun *68*, 6384-6390.

Liu, G., Feng, W., Li, D., Liu, M., Nelson, D.C., and Lei, B. (2015). The Mga Regulon but Not Deoxyribonuclease Sda1 of Invasive M1T1 Group A *Streptococcus* Contributes to In Vivo Selection of CovRS Mutations and Resistance to Innate Immune Killing Mechanisms. Infect Immun *83*, 4293-4303.

Logsdon, L.K., Hakansson, A.P., Cortes, G., and Wessels, M.R. (2011). Streptolysin O inhibits clathrin-dependent internalization of group A *Streptococcus*. MBio *2*, e00332-00310.

Love, J.F., Tran-Winkler, H.J., and Wessels, M.R. (2012). Vitamin D and the human antimicrobial peptide LL-37 enhance group a streptococcus resistance to killing by human cells. MBio *3*(5): e00394-12.

Maamary, P.G., Ben Zakour, N.L., Cole, J.N., Hollands, A., Aziz, R.K., Barnett, T.C., Cork, A.J., Henningham, A., Sanderson-Smith, M., McArthur, J.D., *et al.* (2012). Tracing the evolutionary history of the pandemic group A streptococcal M1T1 clone. FASEB J *26*, 4675-4684.

Madden, J.C., Ruiz, N., and Caparon, M. (2001). Cytolysin-mediated translocation (CMT): a functional equivalent of type III secretion in gram-positive bacteria. Cell *104*, 143-152.

Magassa, N., Chandrasekaran, S., and Caparon, M.G. (2010). *Streptococcus pyogenes* cytolysin-mediated translocation does not require pore formation by streptolysin O. EMBO Rep *11*, 400-405.

Mariathasan, S., Weiss, D.S., Newton, K., McBride, J., O'Rourke, K., Roose-Girma, M., Lee, W.P., Weinrauch, Y., Monack, D.M., and Dixit, V.M. (2006). Cryopyrin activates the inflammasome in response to toxins and ATP. Nature *440*, 228-232.

Mayfield, J.A., Liang, Z., Agrahari, G., Lee, S.W., Donahue, D.L., Ploplis, V.A., and Castellino, F.J. (2014). Mutations in the control of virulence sensor gene from *Streptococcus pyogenes* after infection in mice lead to clonal bacterial variants with altered gene regulatory activity and virulence. PLoS One *9*, e100698.

Meehl, M.A., Pinkner, J.S., Anderson, P.J., Hultgren, S.J., and Caparon, M.G. (2005). A novel endogenous inhibitor of the secreted streptococcal NAD-glycohydrolase. PLoS Pathog *1*, e35.

Michos, A., Gryllos, I., Hakansson, A., Srivastava, A., Kokkotou, E., and Wessels, M.R. (2006). Enhancement of streptolysin O activity and intrinsic cytotoxic effects of the group A streptococcal toxin, NAD-glycohydrolase. J Biol Chem *281*, 8216-8223.

Mitsui, K., Takano, K., Nakatani, S., Nambu, H., Shibata, F., and Nakagawa, H. (2002). Chemokine production by rat macrophages stimulated with streptolysin O from *Streptococcus pyogenes*. Microbiol Immunol *46*, 37-45.

Miyoshi-Akiyama, T., Ikebe, T., Watanabe, H., Uchiyama, T., Kirikae, T., and Kawamura, Y. (2006). Use of DNA arrays to identify a mutation in the negative regulator, csrR, responsible for the high virulence of a naturally occurring type M3 group A streptococcus clinical isolate. J Infect Dis *193*, 1677-1684.

Moon, A.F., Krahn, J.M., Lu, X., Cuneo, M.J., and Pedersen, L.C. (2016). Structural characterization of the virulence factor Sda1 nuclease from *Streptococcus pyogenes*. Nucleic Acids Res *44*, 3946-3957.

Muller-Alouf, H., Geoffroy, C., Geslin, P., Bouvet, A., Felten, A., Gunther, E., Ozegowski, J.H., and Alouf, J.E. (1997). Streptococcal pyrogenic exotoxin A, streptolysin O, exoenzymes, serotype and biotype profiles of *Streptococcus pyogenes* isolates from patients with toxic shock syndrome and other severe infections. Zentralbl Bakteriol *286*, 421-433.

Musser, J.M., Hauser, A.R., Kim, M.H., Schlievert, P.M., Nelson, K., and Selander, R.K. (1991). *Streptococcus pyogenes* causing toxic-shock-like syndrome and other invasive diseases: clonal diversity and pyrogenic exotoxin expression. Proc Natl Acad Sci U S A *88*, 2668-2672.

Musser, J.M., Kapur, V., Kanjilal, S., Shah, U., Musher, D.M., Barg, N.L., Johnston, K.H., Schlievert, P.M., Henrichsen, J., Gerlach, D., *et al.* (1993). Geographic and temporal distribution and molecular characterization of two highly pathogenic clones of *Streptococcus pyogenes* expressing allelic variants of pyrogenic exotoxin A (Scarlet fever toxin). J Infect Dis *167*, 337-346.

Nasser, W., Beres, S.B., Olsen, R.J., Dean, M.A., Rice, K.A., Long, S.W., Kristinsson, K.G., Gottfredsson, M., Vuopio, J., Raisanen, K., *et al.* (2014). Evolutionary pathway to increased virulence and epidemic group A *Streptococcus* disease derived from 3,615 genome sequences. Proc Natl Acad Sci U S A *111*, E1768-1776.

Nelson, D.C., Garbe, J., and Collin, M. (2011). Cysteine proteinase SpeB from *Streptococcus pyogenes* - a potent modifier of immunologically important host and bacterial proteins. Biol Chem *392*, 1077-1088.

Nelson, K., Schlievert, P.M., Selander, R.K., and Musser, J.M. (1991). Characterization and clonal distribution of four alleles of the speA gene encoding pyrogenic exotoxin A (scarlet fever toxin) in *Streptococcus pyogenes*. J Exp Med *174*, 1271-1274.

Niedermeyer, W. (1985). Interaction of streptolysin-O with biomembranes: kinetic and morphological studies on erythrocyte membranes. Toxicon *23*, 425-439.

Nilsson, M., Sorensen, O.E., Morgelin, M., Weineisen, M., Sjobring, U., and Herwald, H. (2006). Activation of human polymorphonuclear neutrophils by streptolysin O from *Streptococcus pyogenes* leads to the release of proinflammatory mediators. Thromb Haemost *95*, 982-990.

Norrby-Teglund, A., Thulin, P., Gan, B.S., Kotb, M., McGeer, A., Andersson, J., and Low, D.E. (2001). Evidence for superantigen involvement in severe group a streptococcal tissue infections. J Infect Dis *184*, 853-860.

Nyberg, P., Rasmussen, M., and Bjorck, L. (2004). alpha2-Macroglobulin-proteinase complexes protect *Streptococcus pyogenes* from killing by the antimicrobial peptide LL-37. J Biol Chem *279*, 52820-52823.

O'Seaghdha, M., and Wessels, M.R. (2013). Streptolysin O and its co-toxin NAD-glycohydrolase protect group A *Streptococcus* from Xenophagic killing. PLoS Pathog *9*, e1003394.

Oehmcke, S., Shannon, O., Morgelin, M., and Herwald, H. (2010). Streptococcal M proteins and their role as virulence determinants. Clin Chim Acta *411*, 1172-1180.

Olsen, R.J., and Musser, J.M. (2010). Molecular pathogenesis of necrotizing fasciitis. Annu Rev Pathol *5*, 1-31.

Otto, M. (2014). Staphylococcus aureus toxins. Curr Opin Microbiol *17*, 32-37.

Papageorgiou, A.C., Collins, C.M., Gutman, D.M., Kline, J.B., O'Brien, S.M., Tranter, H.S., and Acharya, K.R. (1999). Structural basis for the recognition of superantigen streptococcal pyrogenic exotoxin A (SpeA1) by MHC class II molecules and T-cell receptors. EMBO J *18*, 9-21.

Proctor, R.A., von Eiff, C., Kahl, B.C., Becker, K., McNamara, P., Herrmann, M., and Peters, G. (2006). Small colony variants: a pathogenic form of bacteria that facilitates persistent and recurrent infections. Nat Rev Microbiol *4*, 295-305.

Proft, T., Sriskandan, S., Yang, L., and Fraser, J.D. (2003). Superantigens and streptococcal toxic shock syndrome. Emerg Infect Dis *9*, 1211-1218.

Ralph, A.P., and Carapetis, J.R. (2013). Group a streptococcal diseases and their global burden. Curr Top Microbiol Immunol *368*, 1-27.

Salgado-Pabon, W., Breshears, L., Spaulding, A.R., Merriman, J.A., Stach, C.S., Horswill, A.R., Peterson, M.L., and Schlievert, P.M. (2013). Superantigens are critical for *Staphylococcus aureus* Infective endocarditis, sepsis, and acute kidney injury. MBio *4*.

Shaikh, N., Leonard, E., and Martin, J.M. (2010). Prevalence of streptococcal pharyngitis and streptococcal carriage in children: a meta-analysis. Pediatrics *126*, e557-564.

Shany, S., Grushoff, P.S., and Bernheimer, A.W. (1973). Physical separation of streptococcal nicotinamide adenine dinucleotide glycohydrolase from streptolysin O. Infect Immun *7*, 731-734.

Sharma, O., O'Seaghdha, M., Velarde, J.J., and Wessels, M.R. (2016). NAD+-Glycohydrolase Promotes Intracellular Survival of Group A *Streptococcus*. PLoS Pathog *12*, e1005468.

Shelburne, S.A., Olsen, R.J., Suber, B., Sahasrabhojane, P., Sumby, P., Brennan, R.G., and Musser, J.M. (2010). A combination of independent transcriptional regulators shapes bacterial virulence gene expression during infection. PLoS Pathog *6*, e1000817.

Sierig, G., Cywes, C., Wessels, M.R., and Ashbaugh, C.D. (2003). Cytotoxic effects of streptolysin o and streptolysin s enhance the virulence of poorly encapsulated group a streptococci. Infect Immun *71*, 446-455.

Smeesters, P.R., McMillan, D.J., and Sriprakash, K.S. (2010). The streptococcal M protein: a highly versatile molecule. Trends Microbiol *18*, 275-282.

Smith, C.L., Ghosh, J., Elam, J.S., Pinkner, J.S., Hultgren, S.J., Caparon, M.G., and Ellenberger, T. (2011). Structural basis of *Streptococcus pyogenes* immunity to its NAD+ glycohydrolase toxin. Structure *19*, 192-202.

Sriskandan, S., Moyes, D., and Cohen, J. (1996). Detection of circulating bacterial superantigen and lymphotoxin-alpha in patients with streptococcal toxic-shock syndrome. Lancet *348*, 1315-1316.

Staali, L., Bauer, S., Morgelin, M., Bjorck, L., and Tapper, H. (2006). *Streptococcus pyogenes* bacteria modulate membrane traffic in human neutrophils and selectively inhibit azurophilic granule fusion with phagosomes. Cell Microbiol *8*, 690-703.

Staali, L., Morgelin, M., Bjorck, L., and Tapper, H. (2003). *Streptococcus pyogenes* expressing M and M-like surface proteins are phagocytosed but survive inside human neutrophils. Cell Microbiol *5*, 253-265.

Steer, A.C., Law, I., Matatolu, L., Beall, B.W., and Carapetis, J.R. (2009). Global emm type distribution of group A streptococci: systematic review and implications for vaccine development. Lancet Infect Dis *9*, 611-616.

Stevens, D.L., Salmi, D.B., McIndoo, E.R., and Bryant, A.E. (2000). Molecular epidemiology of nga and NAD glycohydrolase/ADP-ribosyltransferase activity among *Streptococcus pyogenes* causing streptococcal toxic shock syndrome. J Infect Dis *182*, 1117-1128.

Sumby, P., Barbian, K.D., Gardner, D.J., Whitney, A.R., Welty, D.M., Long, R.D., Bailey, J.R., Parnell, M.J., Hoe, N.P., Adams, G.G., *et al.* (2005a). Extracellular deoxyribonuclease made by group A *Streptococcus* assists pathogenesis by enhancing evasion of the innate immune response. Proc Natl Acad Sci U S A *102*, 1679-1684.

Sumby, P., Porcella, S.F., Madrigal, A.G., Barbian, K.D., Virtaneva, K., Ricklefs, S.M., Sturdevant, D.E., Graham, M.R., Vuopio-Varkila, J., Hoe, N.P., *et al.* (2005b). Evolutionary origin and emergence of a highly successful clone of serotype M1 group a *Streptococcus* involved multiple horizontal gene transfer events. J Infect Dis *192*, 771-782.

Sumby, P., Whitney, A.R., Graviss, E.A., DeLeo, F.R., and Musser, J.M. (2006). Genome-wide analysis of group a streptococci reveals a mutation that modulates global phenotype and disease specificity. PLoS Pathog *2*, e5.

Tatsuno, I., Okada, R., Zhang, Y., Isaka, M., and Hasegawa, T. (2013). Partial loss of CovS function in *Streptococcus pyogenes* causes severe invasive disease. BMC Res Notes *6*, 126.

Tatsuno, I., Sawai, J., Okamoto, A., Matsumoto, M., Minami, M., Isaka, M., Ohta, M., and Hasegawa, T. (2007). Characterization of the NAD-glycohydrolase in streptococcal strains. Microbiology *153*, 4253-4260.

Tegla, C.A., Cudrici, C., Patel, S., Trippe, R., 3rd, Rus, V., Niculescu, F., and Rus, H. (2011). Membrane attack by complement: the assembly and biology of terminal complement complexes. Immunol Res *51*, 45-60.

Thammavongsa, V., Missiakas, D.M., and Schneewind, O. (2013). *Staphylococcus aureus* degrades neutrophil extracellular traps to promote immune cell death. Science *342*, 863-866.

Timmer, A.M., Timmer, J.C., Pence, M.A., Hsu, L.C., Ghochani, M., Frey, T.G., Karin, M., Salvesen, G.S., and Nizet, V. (2009). Streptolysin O promotes group A

Streptococcus immune evasion by accelerated macrophage apoptosis. J Biol Chem *284*, 862-871.

Tran-Winkler, H.J., Love, J.F., Gryllos, I., and Wessels, M.R. (2011). Signal transduction through CsrRS confers an invasive phenotype in group A *Streptococcus*. PLoS Pathog *7*, e1002361.

Trevino, J., Perez, N., Ramirez-Pena, E., Liu, Z., Shelburne, S.A., 3rd, Musser, J.M., and Sumby, P. (2009). CovS simultaneously activates and inhibits the CovR-mediated repression of distinct subsets of group A *Streptococcus* virulence factor-encoding genes. Infect Immun *77*, 3141-3149.

Tweten, R.K. (2005). Cholesterol-dependent cytolysins, a family of versatile pore-forming toxins. Infect Immun *73*, 6199-6209.

Uchiyama, S., Andreoni, F., Schuepbach, R.A., Nizet, V., and Zinkernagel, A.S. (2012). DNase Sda1 allows invasive M1T1 Group A *Streptococcus* to prevent TLR9-dependent recognition. PLoS Pathog *8*, e1002736.

Uchiyama, S., Dohrmann, S., Timmer, A.M., Dixit, N., Ghochani, M., Bhandari, T., Timmer, J.C., Sprague, K., Bubeck-Wardenburg, J., Simon, S.I., *et al.* (2015). Streptolysin O Rapidly Impairs Neutrophil Oxidative Burst and Antibacterial Responses to Group A *Streptococcus*. Front Immunol *6*, 581.

Van Epps, D.E., and Andersen, B.R. (1974). Streptolysin O inhibition of neutrophil chemotaxis and mobility: nonimmune phenomenon with species specificity. Infect Immun *9*, 27-33.

Velarde, J.J., Ashbaugh, M., and Wessels, M.R. (2014). The human antimicrobial peptide LL-37 binds directly to CsrS, a sensor histidine kinase of group A *Streptococcus*, to activate expression of virulence factors. J Biol Chem *289*, 36315-36324.

Velarde, J.J., O'Seaghdha, M., Baddal, B., Bastiat-Sempe, B., and Wessels, M.R. (2017). Binding of NAD(+)-Glycohydrolase to Streptolysin O Stabilizes Both Toxins and Promotes Virulence of Group A *Streptococcus*. MBio *8* (5): e01382-17.

Venturini, C., Ong, C.L., Gillen, C.M., Ben-Zakour, N.L., Maamary, P.G., Nizet, V., Beatson, S.A., and Walker, M.J. (2013). Acquisition of the Sda1-encoding bacteriophage does not enhance virulence of the serotype M1 *Streptococcus pyogenes* strain SF370. Infect Immun *81*, 2062-2069.

Walcheck, B., Kahn, J., Fisher, J.M., Wang, B.B., Fisk, R.S., Payan, D.G., Feehan, C., Betageri, R., Darlak, K., Spatola, A.F., *et al.* (1996). Neutrophil rolling altered by inhibition of L-selectin shedding in vitro. Nature *380*, 720-723.

Walev, I., Tappe, D., Gulbins, E., and Bhakdi, S. (2000). Streptolysin O-permeabilized granulocytes shed L-selectin concomitantly with ceramide generation via neutral sphingomyelinase. J Leukoc Biol *68*, 865-872.

Walker, M.J., Hollands, A., Sanderson-Smith, M.L., Cole, J.N., Kirk, J.K., Henningham, A., McArthur, J.D., Dinkla, K., Aziz, R.K., Kansal, R.G., *et al.* (2007). DNase Sda1 provides selection pressure for a switch to invasive group A streptococcal infection. Nat Med *13*, 981-985.

Wang, C.H., Chiang-Ni, C., Kuo, H.T., Zheng, P.X., Tsou, C.C., Wang, S., Tsai, P.J., Chuang, W.J., Lin, Y.S., Liu, C.C., *et al.* (2013). Peroxide responsive regulator PerR of group A *Streptococcus* is required for the expression of phage-associated DNase Sda1 under oxidative stress. PLoS One *8*, e81882.

Wessels, M.R., Moses, A.E., Goldberg, J.B., and DiCesare, T.J. (1991). Hyaluronic acid capsule is a virulence factor for mucoid group A streptococci. Proc Natl Acad Sci U S A *88*, 8317-8321.

Wilkening, R.V., and Federle, M.J. (2017). Evolutionary Constraints Shaping *Streptococcus pyogenes*-Host Interactions. Trends Microbiol *25*, 562-572.

Zeppa, J.J., Kasper, K.J., Mohorovic, I., Mazzuca, D.M., Haeryfar, S.M.M., and McCormick, J.K. (2017). Nasopharyngeal infection by *Streptococcus pyogenes* requires superantigen-responsive Vbeta-specific T cells. Proc Natl Acad Sci U S A *114* (38) 10226-10231.

Zhou, Y., Hanks, T.S., Feng, W., Li, J., Liu, G., Liu, M., and Lei, B. (2013). The sagA/pel locus does not regulate the expression of the M protein of the M1T1 lineage of group A *Streptococcus*. Virulence *4*, 698-706.

Zhu, L., Olsen, R.J., Lee, J.D., Porter, A.R., DeLeo, F.R., and Musser, J.M. (2017). Contribution of Secreted NADase and Streptolysin O to the Pathogenesis of Epidemic Serotype M1 *Streptococcus pyogenes* Infections. Am J Pathol *187*, 605-613.

Zhu, L., Olsen, R.J., Nasser, W., Beres, S.B., Vuopio, J., Kristinsson, K.G., Gottfredsson, M., Porter, A.R., DeLeo, F.R., and Musser, J.M. (2015). A molecular trigger for intercontinental epidemics of group A *Streptococcus*. J Clin Invest *125*, 3545-3559.

Chapter 11

https://doi.org/10.21775/9781912530229.11

Recent Proceedings on Prevalence and Pathogenesis of *Streptococcus suis*

Chen Tan[1,2,3,4,5], Anding Zhang[1,2,3,4,5], Huanchun Chen[1,2,3,4,5] and Rui Zhou[1,2,3,4,5,*]

[1] State Key Laboratory of Agricultural Microbiology, College of Veterinary Medicine, Huazhong Agricultural University, Wuhan 430070, China.

[2] Key Laboratory of Preventive Veterinary Medicine in Hubei Province, Wuhan, 430070, China.

[3] Cooperative Innovation Center for Sustainable Pig Production, Wuhan, 430070, China.

[4] Key Laboratory of Development of Veterinary Diagnostic Products(MOA), Wuhan, 430070, China.

[5] International Research Center for Animal Disease (MOST), Wuhan, 430070, China.

[*]Correspondence:rzhou@mail.hzau.edu.cn

Abstract

Streptococcus suis (S. suis) is an important zoonotic pathogen that causes huge economic losses in the pig industry, as well as severe illness and even death in humans. The outbreak of human infection of *S. suis* in China in 2005 led to significant human morbidity and death, prompting an increase in global studies of *S. suis*. In recent years, important advances have been made regarding the etiology, genomics, excavation of virulence genes, and vaccine research in *S. suis*. A number of countries and regions have identified their predominantly serotypes. The development of genome sequencing technology has laid an important foundation for the study of pathogenic mechanisms. For example, 89K PAI was found in representative virulence strains in China, and several studies have been carried out to confirm multiple genes which carries are closely related to virulence. Also, the functions of some regulatory genes represented by the two-component signal transduction system have been analyzed. The development of inactivated vaccines, natural avirulent vaccines, gene-deletion attenuated vaccines, subunit vaccines, and glycoconjugate vaccines have greatly contributed to the prevention and control of the disease in the future. This article aims to summarize the research progress to provide directions for future research and the prevention of *S. suis*.

Introduction

Streptococcus suis is an important zoonotic pathogen that causes enormous economic losses in the global swine industry. It is also considered as an emerging zoonotic pathogen with the potential to cause a wide variety of diseases including septicemia, meningitis, pneumonia, endocarditis, and arthritis in both pigs and humans (Segura *et al.*, 2014). A total of 29 serotypes of *S. suis* have been identified on the basis of various capsular antigens (Nomoto *et al.*, 2015; Okura *et al.*, 2016; Oh *et al.*, 2017; Tohya *et al.*, 2017), and some serotypes such as serotype 1, 2, 5, 9, 14, 16, 28, and 31 can infect humans (Lun *et al.*, 2007; Gottschalk *et al.*, 2010; Nutravong *et al.*, 2014; Taniyama *et al.*, 2016). Since the first human case of *S. suis* was reported in Denmark in 1968, over 1,000 cases have been reported in over 30 countries with intensive pig production. Most cases were European, Asian, and North American countries and regions. From the 29 serotypes of *S. suis*, serotype 2 is the most dominant serotype infecting humans, causing serious illness and streptococcal toxic shock syndrome (STSS), which was the main cause of death in the 2005 outbreak in Sichuan province of China (Feng *et al.*, 2010).

Epidemiology

Numerous *S. suis* serotypes are pathogenic, but there are differences in the major serotypes prevalent in different countries (Oh *et al.*, 2017). For example, in China, serotype 2 was the prevailing serotype, followed by serotype 1 and serotype 3 (Wei *et al.*, 2009; Li *et al.*, 2012). In Canada, serotype 3 was the most common serotype until 2009, but serotype 2 was the most common in 2011 (Gottschalk *et al.*, 2013). In South Korea, serotypes 7 and 21 were most commonly isolated from pigs during the 2010-2013 period (Gurung *et al.*, 2015). In contrast, in Thailand, serotype 23 was the most common, followed by serotypes 9, 7, and 2 (Thongkamkoon *et al.*, 2017).

In the last 20 years, *S. suis* has caused huge losses to the swine industry in many Asian countries, and is the most frequently isolated pathogenic bacterial pathogen in most pig farms in China. Two large outbreaks of *S. suis* in China have been identified since the 1990s. The first outbreak occurred in 1998 in the Jiangsu Province, in which 25 humans were infected, 14 died, and approximately 80,000 pigs were infected (Tang *et al.*, 2006). Sick pigs showed a high fever, shortness of breath, head, neck, and abdominal bleeding. All 25 patients were adult males who had contact with dead or sick pigs, 16 of whom presented with STSS and nine presented with meningitis (Zhu *et al.*, 2001). The second outbreak took place in 2005 in Sichuan Province, resulting in 215 human infections and 39 deaths (Yu *et al.*, 2006). All infected people were mainly adult male farmers with recent exposure to carcasses of pigs which died of unknown causes or sick pigs. Three distinct clinical symptom settings have been observed, as 61 had STSS and the others were sepsis, meningitis, or both (Yu *et al.*, 2006). Both outbreaks were caused by the sequence type 7 (ST7), an emerging highly virulent *S. suis* clone (Ye *et al.*, 2008).

In the northern and northeastern regions of Thailand, *S. suis* is a major pathogen causing a public health concern (Kerdsin *et al.*, 2011). The transmission of this pathogen in humans in these areas is due to specific ethnic practices, such as the consumption of raw pork and meat products (Fongcom *et al.*, 2001). There was a serious *S. suis* outbreak in northern Thailand in 2010, involving 171 human cases (Thongkamkoon *et al.*, 2017).

In Vietnam, *S. suis* type 2 is the most commonly detected organism causing acute bacterial meningitis in adults. Mai et al. studied 450 patients with suspected bacterial meningitis and showed *S. suis* was the etiologic pathogen in 151 of the patients (Mai *et al.*, 2008). In 2007, Wertheim et al. identified *S. suis* from 43 meningitis patients from a national hospital in Hanoi (Wertheim *et al.*, 2009). The risk factors of *S. suis* infection in humans were investigated by a different research group. Nghia et al. conducted a case–control study and determined that eating "high-risk" dishes popular in parts of Asia, occupational exposure to pigs and pig products, and preparation of pork in the presence of skin lesions are the most important risk factors associated with *S. suis* bacterial meningitis (Nghia *et al.*, 2011).

Healthy pigs also carry *S. suis* on their tonsils which significantly impacts swine productivity, animal welfare, and human health. In collaboration with the University of Cambridge, we have investigated the prevalence and population biology of the *S. suis* isolates from the clinically healthy pig herds of China and the UK. This *S. suis* population showed a higher diversity than the disease-associated isolates on serotypes and sequence types. A significant effect of temperature is identified on carriage of *S. suis* on the tonsils of healthy pigs (Zou *et al.*, 2018).

Genomics

Genome sequencing is extremely important for understanding the characteristics of *S. suis*. To date, more than 1000 genome sequences of *S. suis* strains are available in public databases. For example, whole genomes of the two virulent serotype 2 strains 98HAH12 and 05ZYH33 were isolated from the outbreaks in China in 1998 and 2005, respectively, and were sequenced in 2007. By comparing these genomes, a novel 89-Kbpathogenicity island (PAI) was present in both strains, but not in P1/7 (Chen *et al.*, 2007). Subsequent studies demonstrated that some genes in this 89K PAI are strongly associated with virulence and virulence regulation, such as the two two-component signal transduction systems (TCSTS) and six stand-alone transcriptional regulators (Li *et al.*, 2008; Xu *et al.*, 2018). Also, SalK/SalR and NisK/NisR of TCSTS have been confirmed to be related to virulence, resistance to PMN-mediated killing, and phagocytosis by macrophages (Li *et al.*, 2008; Xu *et al.*, 2014). Among the six stand-alone transcriptional regulators, TstS was obviously upregulated *in vivo* which promotes STSLS (Xu *et al.*, 2018). In addition, a Type-IVC secretion system consisting of four genes has been identified, which can transfer secretion of plasmid type 89K to other bacteria. It is also related to the virulence, and knockout studies of the genes impaired its ability to trigger host immune response (Zhang *et al.*, 2012b; Li *et al.*, 2011c; Zhao *et al.*, 2011; Yin *et al.*, 2016).

In 2009, Holden et al. analyzed the whole-genome sequences of three *S. suis* strains, European P1/7 from pigs, and two strains from human cases: SC84 from China and BM407 from Vietnam (Holden *et al.*, 2009). The P1/7 and SC84 chromosomes have a conserved structure, while the BM407 chromosome has a large inversion. Three 90-Kb regions that carry some drug resistance genes were found in the two isolates from humans. Upon comparing the P1/7 chromosome with those of other *Streptococcus* species, *S. suis* was phylogenetically distinct, although many housekeeping genes, such as fatty acid metabolism genes, nucleotide biosynthesis genes, and macromolecule biosynthesis genes display the highest levels of conservation. In contrast, some mobile genetic elements in the P1/7 genome have the lowest conservation compared with those in other *Streptococcus* species.

In addition to serotype 2, the genomes of several other serotype strains were also resolved, which indicated that there is significant genomic diversity among the different strains of *S. suis* (Zhang *et al.*, 2011b). What important information can we glean from the genomic differences among various types of *S. suis*? Weinert et al.studied the genetic differences among the systemic, respiratory, and non-clinical *S. suis* isolates from pigs and humans using high-quality clinical data and genome sequences, and found that the clinical *S. suis* isolates have a smaller genome size than the non-clinical ones, but are more likely to encode virulence factors. Human disease isolates are limited to a single-virulent population, originating in the 1920s when pig production was intensified. No consistent genomic differences are observed between pig and human isolates. High rates of recombination occur in the genomes of this bacterium, suggesting that virulence of *S. suis* may increase anywhere in the world (Weinert *et al.*, 2015).

Virulence factors and pathogenesis

Virulence-associated factors have been identified, and they provide a likely explanation as to why the strain is so highly virulent since the outbreak of *S. suis* serotype 2 in China in 2005 (Chen *et al.*, 2007). Although some comprehensive studies have been performed, some results were contradictory due to differences in strain backgrounds and different animal models of infectious disease in the evaluation of virulence (Auger *et al.*, 2017; Segura *et al.*, 2017).

The polysaccharide capsule (CPS) is considered to be essential for bacterial virulence by inhibiting the signaling pathways involved in phagocytosis (Smith *et al.*, 1999; Segura *et al.*, 2004; Chabot-Roy *et al.*, 2006; Lecours *et al.*, 2011) and by the evasion of neutrophil extracellular traps (NETs) (Zhao *et al.*, 2015). However, a few avirulent *S. suis* strains also contain CPS *(Berthelot-Herault et al.*, 2001; Berthelot-Herault *et al.*, 2005), indicating that the virulence is multifactorial and does not exclusively rely on CPS structure. Furthermore, CPS structure may interfere with some important virulence properties, such as the inhibition of bacterial adherence, invasion of host cells (Benga *et al.*, 2004; Tenenbaum *et al.*, 2009; Segura *et al.*, 2016), and biofilm formation (Tanabe *et al.*, 2010). Interestingly, the CPS structure could be retrieved after *in vivo* passage of a non-encapsulated *S. suis* (Auger *et al.*, 2016), indicating that CPS structure could be regulated *in vivo,* and that its contribution to virulence is complicated. In fact, CPS structure could be regulated by several factors, such as the availability of glucose or other carbohydrates, pH, and temperature (Smith *et al.*, 2001; Wu *et al.*, 2011). In particular, glucose availability could also regulate CPS gene expression through catabolite control protein A (CcpA) *(Willenborg et al.*, 2011), which is also why virulence-associated factor HP0197 contributed to the virulence through the regulation of CCPA activity (Zhang *et al.*, 2012a; Yuan *et al.*, 2013). Together, these studies indicate that CPS is an important virulence factor in the behavior of *S. suis.*

Suilysin (SLY) is a member of the pore-forming cholesterol-dependent cytolysin family of toxins (Palmer, 2001; Xu *et al.*, 2010). Examination of the crystal structure further confirmed the cytotoxic properties of SLY (Xu *et al.*, 2010). For a long time, SLY was thought to contribute to meningitis (Takeuchi *et al.*, 2014), supported by the fact that SLY can remodel the cytoskeleton of human brain microvascular endothelial cells by activating RhoA and Rac1 GTPase (Lv *et al.*, 2014) and can increase vascular permeability through the blood-brain barrier (Chen *et al.*, 2016; Liu *et al.*, 2017). The toxic effect of SLY benefits the adherence of the pathogen to airway cells, which subsequently causes loss and apoptosis of ciliated cells for invasion (Meng *et al.*, 2016). Therefore, SLY plays an important role in the invasion of host cells and induction of cell death (Allen *et al.*, 2001; Lun *et al.*, 2003; Vanier *et al.*, 2004; Ferrando *et al.*, 2015; Meng *et al.*, 2016). In addition, SLY is involved in inflammatory responses through TLR4 (Lecours *et al.*, 2011; Bi *et al.*, 2015; Zhang *et al.*, 2016a; Zhang *et al.*, 2016b), and SLY induced platelet aggregation and also platelet-neutrophil complexes formation through the pore-dependent Ca2+ influx (Zhang *et al.*, 2016a; Zhang *et al.*, 2016b). It has been reported that SLY is partially involved in cytokine release and also contributes to bacterial escape of opsonophagocytosis (Lecours *et al.*, 2011). Therefore, SLY is also an important virulence factor (Allen *et al.*, 2001), which has been the target of drugs in previous studies (Li *et al.*, 2017a; Zhang *et al.*, 2018b). Notably, the expression level of SLY seems to be associated with the virulence of *S. suis* strains (He *et al.*, 2014; Takeuchi *et al.*,

2014). Nonetheless, many virulent strains, especially those from North America, do not produce SLY, which suggests that SLY is not a requirement for the virulence of *S. suis* (Berthelot-Herault *et al.*, 2000; Fittipaldi *et al.*, 2011).

Muramidase-released protein (MRP) and extracellular factor protein were originally identified to be associated with the virulence of *S. suis* (Vecht *et al.*, 1991). Previous work has shown that MRP binds to fibrinogen and facilitates attachment to and traversal of human brain microvascular endothelial cells by increasing transendothelial cell permeability, thereby promoting the development of *Streptococcus suis* meningitis (Wang *et al.*, 2015). However, a subsequent study indicated that the virulence of MRP-deficient mutant was not decreased in comparison with that of the parental strain (Baums and Valentin-Weigand, 2009), suggesting that MRP is required for the virulence of some strains.

Factor H (FH) is an important negative regulator of the alternative complement pathway. *S. suis* can secrete FH-binding protein (FHBP) to bind the host complement component C3 and FH that reduces opsonophagocytosis. FHBP can also enhance the adherence to and invasion of host cells (Li *et al.*, 2016), thereby contributing to the virulence (Pian *et al.*, 2012). In addition, globotriaosylceramide (Gb3) has been identified as the receptor of FHBP, which contributes to *S. suis* infection-induced activation of myosin light chain 2 through Rho/ROCK signaling in hCMEC/D3 cells. This also contributes to the traversal of *S. suis* across the human blood-brain barrier. However, more comprehensive studies are required to determine the role of FHBP in unlocking the blood-brain barrier (Kong *et al.*, 2017) and accessing the central nervous system (Auger and

Gottschalk, 2017). Nonetheless, a double mutant-lacking FHBP and FHBP was similarly phagocytosed by human macrophages and killed by pig blood when compared to the wild-type strain. This suggests that the recruitment of factor H to the *S. suis* cell surface is multifactorial and redundant (Roy *et al.*, 2016). In fact, more factor H-binding proteins were identified to be required for the virulence of the bacteria (Li *et al.*, 2017c).

Oxidative stress is an ubiquitous challenge faced by pathogens, and stress response systems can play an important role in the virulence of pathogenic bacteria (Requena, 2012). *S. suis* has evolved certain strategies to tolerate oxidative stress, which in turn contributes to its virulence. For example, *S. suis* can express NADH oxidase (Zheng *et al.*, 2017) and superoxide dismutase (Fang *et al.*, 2015) to scavenge reactive oxygen species. The improved resistance to reactive oxygen species is also a factor through which hsdS *(Xu et al.*, 2017a), a serine/threonine phosphatase 1 (Fang *et al.*, 2017), heme-binding protein SntA (Wan *et al.*, 2017), FNR-like protein (Willenborg *et al.*, 2016), and PnuC (Li *et al.*, 2018a) contribute to *S. suis* virulence.

In addition to receptor-mediated capture and phagocytosis, neutrophils can also attack pathogens by an antimicrobial mechanism called NET-mediated bacterial killing, which plays an important role in the clearance of *S. suis in vivo (Zhao et al.*, 2016). *S. suis* can induce the formation of NETs both *in vitro* and *in vivo* (de Buhr *et al.*, 2014; de Buhr *et al.*, 2017). Interestingly, the bacterium has evolved certain strategies to evade this killing, and specifically can inhibit the formation of NETs through biofilms (Ma *et al.*, 2017), resist the NET-mediated trapping through CPS structure (Zhao *et al.*, 2015), and degrade the NET structure to resist killing through the secretion of DNase SsnA (de Buhr *et al.*, 2014; de Buhr *et al.*, 2015).

Interestingly, the *S. suis* ST7 strain contains an 89-Kb genomic island, which includes genes encoding the components of a type IV secretion system involved in the pathogenesis of *S. suis* (Li *et al.*, 2011c; Zhao *et al.*, 2011). Subtilisin-like protease-1 secreted through the type IV secretion system contributes to the high virulence of *Streptococcus suis* 2 (Yin *et al.*, 2016), and a novel PPIase molecule, SP1, has been shown to interact with a component of innate immunity, peptidoglycan recognition protein (PGLYRP-1), and to perturb the PGLYRP-1-mediated bacteriostatic effect by interacting with the protein PGLYRP-1 (Wang *et al.*, 2017b).

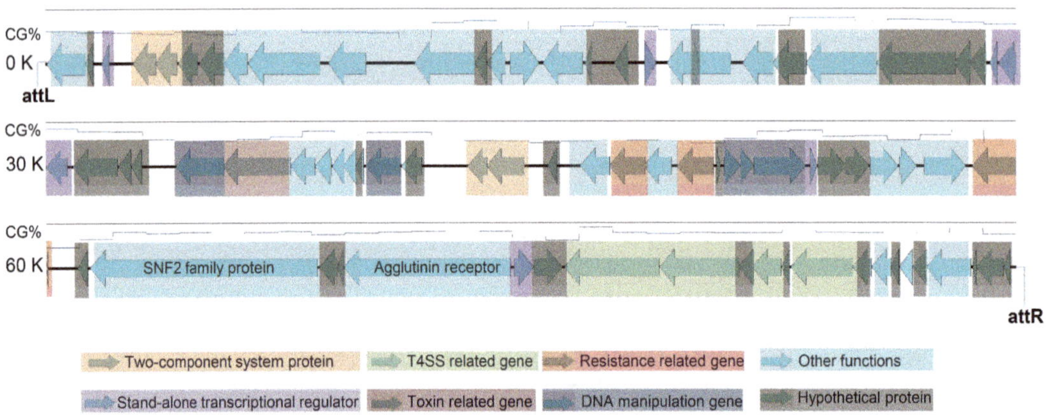

Figure 1. The structure schematic model of 89K PAI . Important elements are presented in different colors.

In addition to the identification of these novel virulence-associated factors, characterization of the pathogenic roles of virulence factor proteins of other pathogenic bacteria homologous to those in *S. suis* was also performed. Immunoglobulin A protease (IgAP) was identified and confirmed to be involved in the virulence of *S. suis* (Zhang *et al.*, 2010; Zhang *et al.*, 2011a), although the identified IgA protease is part of the zinc metalloprotease family and may have other functions besides this (Bek-Thomsen *et al.*, 2012). Interestingly, the findings also showed that Mac is an important marker of virulence for other *Streptococcus* strains but is not essential for *S. suis* virulence in strain P1/7 in natural, healthy hosts without specific IgM. Moreover, the immunogenicity of Mac does not appear to correlate with its significance for virulence (Xiao *et al.*, 2017a). The IgM degrading enzyme could reduce the amount of IgM bound to the bacterial surface, which is a novel complement evasion mechanism (Rungelrath *et al.*, 2018).

A few regulators were confirmed to be involved in the virulence of *S. suis* in addition to these virulence-associated factors. Fifteen groups of two-component systems including one orphan response regulator were predicted in two virulent *S. suis* 2 strains (98HAH12 and 05ZYH33), and nine of them have been confirmed to regulate the virulence (Li *et al.*, 2008; Li *et al.*, 2011a; Xu *et al.*, 2014; Yuan *et al.*, 2017; Chang *et al.*, 2018; Velikova, 2018; Zhong *et al.*, 2018; Zheng *et al.*, 2018a). In addition to the two-component system, there are other Stand-Alone Transcriptional Regulator that regulate the virulence of *S. suis*. For example, the Rgg regulator regulates genes associated with non-glucose carbohydrate metabolism, DNA recombination, protein biosynthesis, and others affecting *S. suis* 2 metabolism and virulence (Zheng *et al*., 2011). Two Spx ortholog regulators, SpxA1 and SpxA2, have been associated with multiple stress tolerance and virulence of *S. suis* 2 (Zheng *et al*., 2014). Also, TstS regulator located at 89K PAI has been identified to regulate the virulence of *S. suis* 2 and to stimulate the release of cytokines (Xu *et al*., 2018).

In addition to the regulation of CPS synthesis by CcpA, *S. suis* can also control manganese homeostasis by MntE, which explains why MntE is involved in the virulence of *S. suis* (Xu *et al*., 2017b). Also, (p)ppGpp synthetases can regulate the expression of virulence-related genes involved in morphology and virulence (Zhu *et al*., 2016). In addition, MsmK is an ATPase that contributes to the utilization of multiple carbohydrates and host colonization of *S. suis*. Additionally, STK and CodY have been confirmed to be central regulators of its virulence (Tan *et al.,* 2015; Tan *et al*., 2017), and GidA, a tRNA modification enzyme, was confirmed to contribute to its growth and virulence (Gao *et al*., 2016). In addition to these regulators, small RNA rss04 was also reported to regulate virulence by regulating capsule synthesis and inducing biofilm formation, as determined in a mouse infection model (Xiao *et al*., 2017b).

It has been recognized that the identification and characterization of novel virulence-associated factors are effective ways of understanding the pathogenesis of *S. suis*. Additional studies have also identified several other virulence-associated factors, such as Formate-tetrahydrofolate ligase (Zheng *et al*., 2016), HP1330 (Zhang *et al*., 2017b), SadP (Ferrando *et al*., 2017), SBP2 (Yu *et al*., 2016), vapE (Ji *et al*., 2016), oligopeptide-binding protein (OppA)

(Zheng *et al.*, 2018b), SssP1 (Zhang *et al.*, 2018a), lysM (Wu *et al.*, 2016), and Dnase (Haas *et al.*, 2014). However, in line with the concern of Segura et al. (YEAR) that the identified factors might be strain-specific (Segura *et al.*, 2017), the identification of critical virulent factors requires more comprehensive studies.

These identified novel factors strengthen our understanding about pathogenesis, but the studies that mainly focused on these virulent factors could not fully elucidate the primary pathogenic mechanism of *S. suis.* As any infectious disease is the result of the interaction of pathogen and host, more concern should be focused on the host response to infections in further studies.

Vaccine development

The development of vaccines against *S. suis* is an optimal approach to controlling its infection. The first vaccine developed for this purpose was based on all of its bacterins. An early experimental study reported that formalin-killed pathogenic SS2 could stimulate a complete protective response against homologous challenge in piglets (Holt *et al.*, 1990). However, a field study reported that the immunization of pigs with a commercial bacterin vaccine failed to protect against nursery mortality (Torremorell *et al.*, 1997). Aside from the use of whole *S. suis* inactivated vaccine, the inactive method (Holt *et al.*, 1990; Pallares *et al.*, 2004), very high doses of bacterin (Holt *et al.*, 1990), and the adjuvant used in the formulation (Pallares *et al.*, 2004) are important for the protective efficacy of the vaccine. In fact, Chinese swine farms began to use SS2 whole bacterins vaccine, which was prepared using high doses of bacterin in oil adjuvant and conferred a detectable level of CPS2 antibodies, as determined using ELISA (Jin *et al.*, 2006). Subsequently, the prevalence of serotype 2 has decreased significantly, although it is still quite prevalent in China. However, the protection conferred by all bacterins is either serotype or strain-dependent (Pallares *et al.*, 2004). There are several serotypes that are prevalent in swine farms in different countries, including China (Wei *et al.*, 2009), for which the development of novel vaccines is required.

The development of a live avirulent vaccine may provide better protective efficacy, as strong humoral immunity is induced following challenges with live bacteria (Buttner *et al.*, 2012). Temperature-sensitive mutants of *S. suis* were tested as vaccines and conferred protection only against homologous challenge in mice (Kebede *et al.*, 1990). Developed non-encapsulated isogenic mutants would be avirulent and may provide better cross-protection because CPS is serotype-specific. A live vaccine was developed based on a non-encapsulated serotype 2 mutant; it induced partial protection only against mortality and failed to prevent the development of clinical signs in pigs challenged with the wild-type strain (Wisselink *et al.*, 2002). However, another study indicated that pigs vaccinated with a non-encapsulated mutant exhibited a survival rate of 100% and presented only minor clinical signs after challenge in the wild-type strain (Fittipaldi *et al.*, 2007). The protection against different serotypes should be confirmed by further studies. Besides disruption of the CPS gene, disruption of other virulence-associated genes would also be promising for the development of an attenuated vaccine. For example, a double-deletion mutant (SsPep/SsPspC–/–) (Hu *et al.*, 2015), a five-deletion mutant (sly, scpA, ssnA, FHBP, and ssads) (Li *et al.*, 2018b), and a mutant with ssnA deletion (Li *et al.*,

2017b) have offered good protection. In contrast, the isogenic OFS mutant failed to induce opsonizing antibodies and protection (Kock *et al.*, 2009). In addition to the constructed isogenic mutants, isolated native avirulent strains could also be used as strains for vaccines (Quessy *et al.*, 1994; Quessy *et al.*, 1995; Busque *et al.*, 1997). For example, Yao et al used an avirulent strain from healthy pigs for a vaccine, which induced complete protection against SS2 infection (Yao *et al.*, 2015; Wang *et al.*, 2017a).

Another strategy to develop a universal vaccine is based on immunogenic proteins, which could confer cross-protection. For example, subunit vaccines using suilysin (Jacobs *et al.*, 1996; Du *et al.*, 2013) or MRP and extracellular protein factors (Wisselink *et al.*, 2001) have been shown to protect pigs against homologous and heterologous strains. However, in some geographical regions, their application is hindered by the presence of a substantial number of virulent strains that do not express these proteins.

SAO (Li *et al.*, 2006; Li *et al.*, 2007; Hsueh *et al.*, 2017) and the 38-kDa protein (Okwumabua and Chinnapapakkagari, 2005) were identified as vaccine candidate antigens. However, it is necessary to identify more immunogenic proteins based on the strategy of developing a universal vaccine. To date, several technologies have been applied to identify novel immunogenic proteins (Zhang and Lu, 2007a; b; Jing *et al.*, 2008; Zhang *et al.*, 2008; Gu *et al.*, 2009; Liu *et al.*, 2009; Mandanici *et al.*, 2010; Gomez-Gascon *et al.*, 2012) such as enolase (Esgleas *et al.*, 2008; Esgleas *et al.*, 2009; Feng *et al.*, 2009; Zhang *et al.*, 2009b), 6-phosphogluconatedehydrogenase (Tan *et al.*, 2008; Tan *et al.*,

2009), SsuiDRAFT 0103 (Aranda *et al.*, 2008), cation-regulated proteins (Aranda *et al.*, 2009), HP0197 (Zhang *et al.*, 2009a), RTX family exoprotein A, epidermal surface antigen, immunoglobulin G-binding protein (Liu *et al.*, 2009), SAT/HP0272 (Chen *et al.*, 2010; Mandanici *et al.*, 2010), pilus subunit (Garibaldi *et al.*, 2010), SsPepO (Li *et al.*, 2011b), HP0245 (Li *et al.*, 2011d), SsnA (Gomez-Gascon *et al.*, 2014; Gomez-Gascon *et al.*, 2016), Lmb (Zhang *et al.*, 2014), AbpB (Huang *et al.*, 2015), IdeSsuis (Seele *et al.*, 2015), Sbp (Zhou *et al.*, 2015), the type II histidine triad protein HtpsC (Li *et al.*, 2015), and IgA protease (Fu et *al.*, 2016). However, it should be noted that there are contradictory results about the protective efficacy of some antigens. For example, enolase was confirmed to be both protective (Feng *et al.*, 2009; Zhang *et al.*, 2009b) and nonprotective (Esgleas *et al.*, 2009). This may have been due to different adjuvants used in the formation of the vaccine. Therefore, the development of technology for evaluating the protection conferred by a subunit vaccine is important, and an assay based on porcine dendritic cells has been developed to assess the immunological behavior of vaccines and the polarizing effect of adjuvants (Martelet *et al.*, 2017).

CPS is the most external bacterial layer in contact with the host, and the antibodies are highly opsonizing and protective *(Charland et al., 1997; Calzas et al.*, 2017). However, free CPS is nonimmunogenic, and carbohydrate-based vaccines (glycoconjugate vaccines) were successfully discovered for many encapsulated pathogens, such as *Haemophilus influenzae* (Hiberix), *Neisseria meningitidis* (MenACWY), and *Streptococcus pneumoniae* (PCV13) (Roy and Shiao, 2011; Bottomley *et al.*, 2012). The structure and method of biosynthesis of capsular polysaccharides of serotypes 2 and 1/2 have now been characterized (Van Calsteren *et al.*, 2016), and a serotype 2 capsular polysaccharide glycoconjugate vaccine was confirmed to induce potent IgM and isotype-switched IgGs in mice and pigs, yielding functional activity in vitro and protection against lethal challenge in vivo. These are all features of a T-dependent response (Goyette-Desjardins *et al.*, 2016), which suggests the possibility of developing polyvalent glycoconjugate vaccines against *S. suis* infection.

Conclusions and future perspectives

With the rapid development of genome sequencing technology and the progress of molecular biology technology, important progress has been made in the understanding of the pathogenesis of *S. suis*. Despite these advances, there are still many limitations. Future studies on *S. suis* should include the following aspects: continue etiology work to elucidate changes in prevalent serotypes, focus on pathogen-environment-host interactions, and develop novel multivalent vaccines and diagnostic reagents.

Acknowledgements

This work was supported by grants from the National Key R&D Program of China (2017YFD0500200), the National Natural Science Foundation of China (31772714, 31572501), the Hubei Province Natural Science Foundation for Innovative Research Groups (2016CFA015), and the earmarked fund for China Agriculture Research System(CARS-35).

References

Allen, A.G., Bolitho, S., Lindsay, H., Khan, S., Bryant, C., Norton, P., Ward, P., Leigh, J., Morgan, J., Riches, H., Eastty, S., and Maskell, D. (2001). Generation and characterization of a defined mutant of *Streptococcus suis* lacking suilysin. Infect Immun *69*, 2732-2735.

Aranda, J., Garrido, M.E., Cortes, P., Llagostera, M., and Barbe, J. (2008). Analysis of the protective capacity of three *Streptococcus suis* proteins induced under divalent-cation-limited conditions. Infect Immun *76*, 1590-1598.

Aranda, J., Garrido, M.E., Fittipaldi, N., Cortes, P., Llagostera, M., Gottschalk, M., and Barbe, J. (2009). Protective capacities of cell surface-associated proteins of *Streptococcus suis* mutants deficient in divalent cation-uptake regulators. Microbiology *155*, 1580-1587.

Auger, J.P., Chuzeville, S., Roy, D., Mathieu-Denoncourt, A., Xu, J., Grenier, D., and Gottschalk, M. (2017). The bias of experimental design, including strain background, in the determination of critical *Streptococcus suis* serotype 2 virulence factors. PLoS One *12*, e0181920.

Auger, J.P., and Gottschalk, M. (2017). The *Streptococcus suis* factor H-binding protein: A key to unlocking the blood-brain barrier and access the central nervous system? Virulence*8*, 1081-1084.

Auger, J.P., Meekhanon, N., Okura, M., Osaki, M., Gottschalk, M., Sekizaki, T., and Takamatsu, D. (2016). *Streptococcus suis* Serotype 2 Capsule In Vivo. Emerg Infect Dis *22*, 1793-1796.

Baums, C.G., and Valentin-Weigand, P. (2009). Surface-associated and secreted factors of *Streptococcus suis* in epidemiology, pathogenesis and vaccine development. Anim Health Res Rev 10, 65-83.

Bek-Thomsen, M., Poulsen, K., and Kilian, M. (2012). Occurrence and evolution of the paralogous zinc metalloproteases IgA1 protease, ZmpB, ZmpC, and ZmpD in Streptococcus pneumoniae and related commensal species. MBio *3*, pii: e00303-12.

Benga, L., Goethe, R., Rohde, M., and Valentin-Weigand, P. (2004). Non-encapsulated strains reveal novel insights in invasion and survival of *Streptococcus suis* in epithelial cells. Cell Microbiol *6*, 867-881.

Berthelot-Herault, F., Cariolet, R., Labbe, A., Gottschalk, M., Cardinal, J.Y., and Kobisch, M. (2001). Experimental infection of specific pathogen free piglets with French strains of *Streptococcus suis* capsular type 2. Can J Vet Res *65*, 196-200.

Berthelot-Herault, F., Gottschalk, M., Morvan, H., and Kobisch, M. (2005). Dilemma of virulence of *Streptococcus suis*: Canadian isolate 89-1591 characterized as a virulent strain using a standardized experimental model in pigs. Can J Vet Res *69*, 236-240.

Berthelot-Herault, F., Morvan, H., Keribin, A.M., Gottschalk, M., and Kobisch, M. (2000). Production of muraminidase-released protein (MRP), extracellular factor (EF) and suilysin by field isolates of *Streptococcus suis* capsular types 2, 1/2, 9, 7 and 3 isolated from swine in France. Vet Res *31*, 473-479.

Bi, L., Pian, Y., Chen, S., Ren, Z., Liu, P., Lv, Q., Zheng, Y., Zhang, S., Hao, H., Yuan, Y., and Jiang, Y. (2015). Toll-like receptor 4 confers inflammatory response to Suilysin. Front Microbiol 6, 644.

Bottomley, M.J., Serruto, D., Safadi, M.A., and Klugman, K.P. (2012). Future challenges in the elimination of bacterial meningitis. Vaccine *30* Suppl 2, B78-86.

Busque, P., Higgins, R., Caya, F., and Quessy, S. (1997). Immunization of pigs against *Streptococcus suis* serotype 2 infection using a live avirulent strain. Can J Vet Res *61*, 275-279.

Buttner, N., Beineke, A., De Buhr, N., Lilienthal, S., Merkel, J., Waldmann, K.H., Valentin-Weigand, P., and Baums, C.G. (2012). *Streptococcus suis* serotype 9 bacterin immunogenicity and protective efficacy. Vet Immunol Immunopathol *146*, 191-200.

Calzas, C., Taillardet, M., Fourati, I.S., Roy, D., Gottschalk, M., Soudeyns, H., Defrance, T., and Segura, M. (2017). Evaluation of the Immunomodulatory Properties of *Streptococcus suis* and Group B Streptococcus Capsular Polysaccharides on the Humoral Response. Pathogens *6*, pii: E16.

Chabot-Roy, G., Willson, P., Segura, M., Lacouture, S., and Gottschalk, M. (2006). Phagocytosis and killing of *Streptococcus suis* by porcine neutrophils. Microb Pathog *41*, 21-32.

Chang, P., Li, W., Shi, G., Li, H., Yang, X., Xia, Z., Ren, Y., Li, Z., Chen, H., Bei, W. (2018). The VraSR regulatory system contributes to virulence in *Streptococcus suis* via resistance to innate immune defenses. Virulence *9*, 771-782.

Charland, N., Jacques, M., Lacouture, S., and Gottschalk, M. (1997). Characterization and protective activity of a monoclonal antibody against a capsular epitope shared by *Streptococcus suis* serotypes 1, 2 and 1/2. Microbiology *143* (Pt 11), 3607-3614.

Chen, B., Zhang, A., Li, R., Mu, X., He, H., Chen, H., and Jin, M. (2010). Evaluation of the protective efficacy of a newly identified immunogenic protein, HP0272, of *Streptococcus suis*. FEMS Microbiol Lett *307*, 12-18.

Chen, C., Tang, J., Dong, W., Wang, C., Feng, Y., Wang, J., Zheng, F., Pan, X.,

Liu, D., Li, M., Song, Y., Zhu, X., Sun, H., Feng, T., Guo, Z., Ju, A., Ge, J., Dong, Y., Sun, W., Jiang, Y., Wang, J., Yan, J., Yang, H., Wang, X., Gao, G.F., Yang, R., Wang, J., and Yu, J. (2007). A glimpse of streptococcal toxic shock syndrome from comparative genomics of S. suis 2 Chinese isolates. PLoS One*2,* e315.

Chen, S., Xie, W., Wu, K., Li, P., Ren, Z., Li, L., Yuan, Y., Zhang, C., Zheng, Y., Lv, Q., Jiang, H., and Jiang, Y. (2016). Suilysin Stimulates the Release of Heparin Binding Protein from Neutrophils and Increases Vascular Permeability in Mice. Front Microbiol *7*, 1338.

De Buhr, N., Neumann, A., Jerjomiceva, N., Von Kockritz-Blickwede, M., and Baums, C.G. (2014). *Streptococcus suis* DNase SsnA contributes to degradation of neutrophil extracellular traps (NETs) and evasion of NET-mediated antimicrobial activity. Microbiology *160*, 385-395.

De Buhr, N., Reuner, F., Neumann, A., Stump-Guthier, C., Tenenbaum, T., Schroten, H., Ishikawa, H., Muller, K., Beineke, A., Hennig-Pauka, I., Gutsmann, T., Valentin-Weigand, P., Baums, C.G., and Von Kockritz-Blickwede, M. (2017). Neutrophil extracellular trap formation in the *Streptococcus suis*-infected cerebrospinal fluid compartment. Cell Microbiol *19*, 10.1111/cmi.12649.

De Buhr, N., Stehr, M., Neumann, A., Naim, H.Y., Valentin-Weigand, P., Von Kockritz-Blickwede, M., and Baums, C.G. (2015). Identification of a novel DNase of *Streptococcus suis* (EndAsuis) important for neutrophil extracellular trap (NET) degradation during exponential growth. Microbiology *161*, 838-850.

Du, H., Huang, W., Xie, H., Ye, C., Jing, H., Ren, Z., and Xu, J. (2013). The genetically modified suilysin, rSLY(P353L), provides a candidate vaccine that suppresses proinflammatory response and reduces fatality following infection with *Streptococcus suis*. Vaccine *31*, 4209-4215.

Esgleas, M., Dominguez-Punaro, M.D., Li, Y., Harel, J., Dubreuil, J.D., and Gottschalk, M. (2009). Immunization with SsEno fails to protect mice against challenge with *Streptococcus suis* serotype 2. FEMS Microbiol Lett *294*, 82-88.

Esgleas, M., Li, Y., Hancock, M.A., Harel, J., Dubreuil, J.D., and Gottschalk, M. (2008). Isolation and characterization of alpha-enolase, a novel fibronectin-binding protein from *Streptococcus suis*. Microbiology *154*, 2668-2679.

Fang, L., Fan, P., Yang, Y., Zhou, J., Shen, H., and Fang, W. (2017). A serine/threonine phosphatase 1 of *Streptococcus suis* type 2 is an important virulence factor. J Vet Sci.

Fang, L., Shen, H., Tang, Y., and Fang, W. (2015). Superoxide dismutase of *Streptococcus suis* serotype 2 plays a role in anti-autophagic response by scavenging reactive oxygen species in infected macrophages. Vet Microbiol *176*, 328-336.

Feng, L., Zhu, J., Chang, H., Gao, X., Gao, C., Wei, X., Yuan, F., and Bei, W. (2016). The CodY regulator is essential for virulence in *Streptococcus suis* serotype 2. Sci Rep *6*, 21241.

Feng, Y., Pan, X., Sun, W., Wang, C., Zhang, H., Li, X., Ma, Y., Shao, Z., Ge, J., Zheng, F., Gao, G.F., and Tang, J. (2009). *Streptococcus suis* enolase functions as a protective antigen displayed on the bacterial cell surface. J Infect Dis *200*, 1583-1592.

Feng, Y., Zhang, H., Ma, Y., and Gao, G.F. (2010). Uncovering newly emerging variants of *Streptococcus suis*, an important zoonotic agent. Trends Microbiol *18*, 124-131.

Ferrando, M.L., De Greeff, A., Van Rooijen, W.J., Stockhofe-Zurwieden, N.,

Nielsen, J., Wichgers Schreur, P.J., Pannekoek, Y., Heuvelink, A., Van Der Ende, A., Smith, H., and Schultsz, C. (2015). Host-pathogen interaction at the intestinal mucosa correlates with zoonotic potential of *Streptococcus suis*. J Infect Dis*212*, 95-105.

Ferrando, M.L., Willemse, N., Zaccaria, E., Pannekoek, Y., Van Der Ende, A., and Schultsz, C. (2017). Streptococcal Adhesin P (SadP) contributes to *Streptococcus suis* adhesion to the human intestinal epithelium. PLoS One *12*,e0175639.

Fittipaldi, N., Harel, J., D'amours, B., Lacouture, S., Kobisch, M., and Gottschalk, M. (2007). Potential use of an unencapsulated and aromatic amino acid-auxotrophic *Streptococcus suis* mutant as a live attenuated vaccine in swine. Vaccine *25*, 3524-3535.

Fittipaldi, N., Xu, J., Lacouture, S., Tharavichitkul, P., Osaki, M., Sekizaki, T., Takamatsu, D., and Gottschalk, M. (2011). Lineage and Virulence of *Streptococcus suis* Serotype 2 Isolates from North America. Emerg Infect Dis *17*, 2239-2244.

Fongcom A, Pruksakorn S, Mongkol R, Tharavichitkul P, and Yunim N. (2001). *Streptococcus suis* infection in Northern Thailand. J Assoc Thail *84*, 1502-1508.

Fu, L., Zhao, J., Lin, L., Zhang, Q., Xu, Z., Han, L., Xie, C., Zhou, R., Jin, M., and Zhang, A. (2016). Characterization of IgA1 protease as a surface protective antigen of *Streptococcus suis* serotype 2. Microbes Infect *18*, 285-289.

Gao, T., Tan, M., Liu, W., Zhang, C., Zhang, T., Zheng, L., Zhu, J., Li, L., and Zhou, R. (2016). GidA, a tRNA Modification Enzyme, Contributes to the Growth, and Virulence of *Streptococcus suis* Serotype 2. Front Cell Infect Microbiol *6*, 44.

Garibaldi, M., Rodriguez-Ortega, M.J., Mandanici, F., Cardaci, A., Midiri, A.,

Papasergi, S., Gambadoro, O., Cavallari, V., Teti, G., and Beninati, C. (2010). Immunoprotective activities of a *Streptococcus suis* pilus subunit in murine models of infection. Vaccine *28*, 3609-3616.

Gomez-Gascon, L., Cardoso-Toset, F., Amarilla, P.S., Tarradas, C., Carrasco, L., Olaya-Abril, A., Jimenez-Munguia, I., Rodriguez-Ortega, M.J., and Luque, I. (2014). A new recombinant SsnA protein combined with aluminum hydroxide protects mouse against *Streptococcus suis*. Vaccine *32*, 6992-6999.

Gomez-Gascon, L., Cardoso-Toset, F., Tarradas, C., Gomez-Laguna, J., Maldonado, A., Nielsen, J., Olaya-Abril, A., Rodriguez-Ortega, M.J., and Luque, I. (2016). Characterization of the immune response and evaluation of the protective capacity of rSsnA against *Streptococcus suis* infection in pigs. Comp Immunol Microbiol Infect Dis *47*, 52-59.

Gomez-Gascon, L., Luque, I., Olaya-Abril, A., Jimenez-Munguia, I., Orbegozo-Medina, R.A., Peralbo, E., Tarradas, C., and Rodriguez-Ortega, M.J. (2012). Exploring the pan-surfome of *Streptococcus suis*: Looking for common protein antigens. J Proteomics*75*, 5654-5666.

Gottschalk, M., Lacouture, S., Bonifait, L., Roy, D., Fittipaldi, N., and Grenier, D. (2013). Characterization of *Streptococcus suis* isolates recovered between 2008 and 2011 from diseased pigs in Québec, Canada. Vet Microbiol *162*, 819-825.

Gottschalk, M., Xu, J., Calzas, C., and Segura M. (2010). *Streptococcus suis*: a new emerging or an old neglected zoonotic pathogen?Future Microbiol *5*, 371-391.

Goyette-Desjardins, G., Calzas, C., Shiao, T.C., Neubauer, A., Kempker, J., Roy, R., Gottschalk, M., and Segura, M. (2016). Protection Against *Streptococcus suis*

Serotype 2 Infection Using a Capsular Polysaccharide Glycoconjugate Vaccine. Infect Immun *84*, 2059 –2075.

Gu, H., Zhu, H., and Lu, C. (2009). Use of in vivo-induced antigen technology (IVIAT) for the identification of *Streptococcus suis* serotype 2 in vivo-induced bacterial protein antigens. BMC Microbiol *9*, 201.

Gurung, M., Tamang, M.D., Moon, D.C., Kim, S.R., Jeong, J.H., Jang, G.C., Jung, S.C., Park, Y.H., and Lim, S.K. (2015). Molecular basis of resistance to selected antimicrobial agents in the emerging zoonotic pathogen *Streptococcus suis*. J Clin Microbiol *53*, 2332-2336.

Haas, B., Bonifait, L., Vaillancourt, K., Charette, S.J., Gottschalk, M., and Grenier, D. (2014). Characterization of DNase activity and gene in *Streptococcus suis* and evidence for a role as virulence factor. BMC Res Notes *7*, 424.

He, Z., Pian, Y., Ren, Z., Bi, L., Yuan, Y., Zheng, Y., Jiang, Y., and Wang, F. (2014). Increased production of suilysin contributes to invasive infection of the *Streptococcus suis* strain 05ZYH33. Mol Med Rep *10*, 2819-2826.

Holden, M.T., Hauser, H., Sanders, M., Ngo, T.H., Cherevach, I., Cronin, A., Goodhead, I., Mungall, K., Quail, M.A., Price, C., Rabbinowitsch, E., Sharp, S., Croucher, N.J., Chieu, T.B., Mai, N.T., Diep, T.S., Chinh, N.T., Kehoe, M., Leigh, J.A., Ward, P.N., Dowson, C.G., Whatmore, A.M., Chanter, N., Iversen, P., Gottschalk, M., Slater, J.D., Smith, H.E., Spratt, B.G., Xu, J., Ye, C., Bentley, S., Barrell, B.G., Schultsz, C., Maskell, D.J., Parkhill, J. (2009). Rapid Evolution of Virulence and Drug Resistance in the Emerging Zoonotic Pathogen *Streptococcus suis*. PLoS One *4*, e6072.

Holt, M.E., Enright, M.R., and Alexander, T.J. (1990). Immunisation of pigs with killed cultures of *Streptococcus suis* type 2. Res Vet Sci *48*, 23-27.

Hsueh, K.J., Cheng, L.T., Lee, J.W., Chung, Y.C., Chung, W.B., and Chu, C.Y. (2017). Immunization with *Streptococcus suis* bacterin plus recombinant Sao protein in sows conveys passive immunity to their piglets. BMC Vet Res *13*, 15.

Hu, J., You, W., Wang, B., Hu, X., Tan, C., Liu, J., Chen, H., and Bei, W. (2015). Construction, characterization and evaluation of the protective efficacy of the *Streptococcus suis* double mutant strain DeltaSsPep/DeltaSsPspC as a live vaccine candidate in mice. Microbiol Res *170*, 87-94.

Huang, K., Yuan, Z., Li, J., Zhang, Q., Xu, Z., Yan, S., Zhang, A., and Jin, M. (2015). Identification and characterisation a surface-associated arginine peptidase in *Streptococcus suis* serotype 2. Microbiol Res *170*, 168-176.

Jacobs, A.A., Van Den Berg, A.J., and Loeffen, P.L. (1996). Protection of experimentally infected pigs by suilysin, the thiol-activated haemolysin of *Streptococcus suis*. Vet Rec *139*, 225-228.

Ji, X., Sun, Y., Liu, J., Zhu, L., Guo, X., Lang, X., and Feng, S. (2016). A novel virulence-associated protein, vapE, in *Streptococcus suis* serotype 2. Mol Med Rep *13*, 2871-2877.

Jin, M., A Zhang, Kang, C., Shi, J., Zheng, P., Tu, J., Wu, B., and Chen, H. (2006). DETECTION OF ANTIBODIES AGAINST *STREPTOCOCCUS SUIS* CAPSULAR TYPE 2 IN CHINA USING AN INDIRECT ELISA. Proceedings of the 19th IPVS Congress *2*, 263.

Jing, H.B., Yuan, J., Wang, J., Yuan, Y., Zhu, L., Liu, X.K., Zheng, Y.L., Wei, K.H., Zhang, X.M., Geng, H.R., Duan, Q., Feng, S.Z., Yang, R.F., Cao, W.C., Wang, H.L., and Jiang, Y.Q. (2008). Proteome analysis of *Streptococcus suis* serotype 2. Proteomics *8*, 333-349.

Kebede, M., Chengappa, M.M., and Stuart, J.G. (1990). Isolation and characterization of temperature-sensitive mutants of *Streptococcus suis*: efficacy trial of the mutant vaccine in mice. Vet Microbiol *22*,249-257.

Kerdsin, A., Dejsirilert, S., Puangpatra, P., Sripakdee, S., Chumla, K., Boonkerd, N., Polwichai, P., Tanimura, S., Takeuchi, D., Nakayama, T., Nakamura, S., Akeda, Y.,Gottschalk, M., Sawanpanyalert, P., and Oishi, K. (2011). Genotypic profile of *Streptococcus suis* serotype 2 and clinical features of infection in humans, Thailand. Emerg Infect Dis*17*, 835-42.

Kock, C., Beineke, A., Seitz, M., Ganter, M., Waldmann, K.H., Valentin-Weigand, P., and Baums, C.G. (2009). Intranasal immunization with a live *Streptococcus suis* isogenic ofs mutant elicited suilysin-neutralization titers but failed to induce opsonizing antibodies and protection. Vet Immunol Immunopathol*132*, 135-145.

Kong, D., Chen, Z., Wang, J., Lv, Q., Jiang, H., Zheng, Y., Xu, M., Zhou, X., Hao, H., and Jiang, Y. (2017). Interaction of factor H-binding protein of *Streptococcus suis* with globotriaosylceramide promotes the development of meningitis. Virulence *8*, 1290-1302.

Lecours, M.P., Gottschalk, M., Houde, M., Lemire, P., Fittipaldi, N., and Segura, M. (2011). Critical Role for *Streptococcus suis* Cell Wall Modifications and Suilysin in Resistance to Complement-Dependent Killing by Dendritic Cells. J Infect Dis *204*, 919-929.

Li, G., Lu, G.J., Qi, Z.M., Li, H.G., Wang, L., Wang, Y.H., Liu, B.W., Niu, X.D., Deng, X.M., and Wang, J.F. (2017a). Morin Attenuates *Streptococcus suis* Pathogenicity in Mice by Neutralizing Suilysin Activity. Frontiers in Microbiology *8*, 460.

Li, J., Tan, C., Zhou, Y., Fu, S., Hu, L., Hu, J., Chen, H., and Bei, W. (2011a). The two-component regulatory system CiaRH contributes to the virulence of

Streptococcus suis 2. Vet Microbiol *148*, 99-104.

Li, J., Xia, J., Tan, C., Zhou, Y., Wang, Y., Zheng, C., Chen, H., and Bei, W. (2011b). Evaluation of the immunogenicity and the protective efficacy of a novel identified immunogenic protein, SsPepO, of *Streptococcus suis* serotype 2. Vaccine *29*, 6514-6519.

Li, L.L., Liao, X.P., Sun, J., Yang, Y.R., Liu, B.T., Yang, S.S., Zhao, D.H. and Liu, Y.H. (2012). Antimicrobial resistance, serotypes, and virulence factors of *Streptococcus suis* isolates from diseased pigs. Foodborne Pathog Dis *9*, 583-588.

Li, M., Cai, R.J., Li, C.L., Song, S., Li, Y., Jiang, Z.Y., and Yang, D.X. (2017b). Deletion of ssnA Attenuates the Pathogenicity of *Streptococcus suis* and Confers Protection against Serovar 2 Strain Challenge. PLoS One *12*, e0169791.

Li, M., Shao, Z.Q., Guo, Y., Wang, L., Hou, T., Hu, D., Zheng, F., Tang, J., Wang, C., Feng, Y., Gao, J., and Pan, X. (2015). The type II histidine triad protein HtpsC is a novel adhesion with the involvement of *Streptococcus suis* virulence. Virulence, *6*, 631-641.

Li, M., Shen, X., Yan, J., Han, H., Zheng, B., Liu, D., Cheng, H., Zhao, Y., Rao, X., Wang, C., Tang, J., Hu, F., and Gao, G.F. (2011c). GI-type T4SS-mediated horizontal transfer of the 89K pathogenicity island in epidemic *Streptococcus suis* serotype 2. Mol Microbiol *79*, 1670-1683.

Li, M., Wang, C., Feng, Y., Pan, X., Cheng, G., Wang, J., Ge, J., Zheng, F., Cao, M., Dong, Y., Liu, D., Wang, J., Lin, Y., Du, H., Gao, G.F., Wang, X., Hu, F., and Tang, J. (2008). SalK/SalR, a two-component signal transduction system, is essential for full virulence of highly invasive *Streptococcus suis* serotype 2. PLoS One*3*, e2080.

Li, Q., Zhang, Y., Dechao, D., Yanfei, Y., and Zhang, W. (2018a). Characterization and functional analysis of PnuC that is involved in the oxidative stress tolerance and virulence of *Streptococcus suis* serotype 2. Vet Microbiol *216*, 198-206.

Li, Q., Ma, C.F., Fu, Y., He, Y.N., Yu, Y.F., Du, D.C., Yao, H.C., Lu, C.P., and Zhang, W. (2017c). Factor H specifically capture novel Factor H-binding proteins of *Streptococcus suis* and contribute to the virulence of the bacteria. MicrobiolRes*196*, 17-25.

Li, W., Hu, X., Liu, L., Chen, H., and Zhou, R. (2011d). Induction of protective immune response against *Streptococcus suis* serotype 2 infection by the surface antigen HP0245. FEMS Microbiol Lett *316*, 115-122.

Li, X., Liu, P., Gan, S., Zhang, C., Zheng, Y., Jiang, Y., and Yuan, Y. (2016). S. suis protein Fhb: the mechanisms of host-pathogen protein complexes formation and bacterial immune evasion. J Biol Chem *291*, 17122-17132.

Li, Y., Gottschalk, M., Esgleas, M., Lacouture, S., Dubreuil, J.D., Willson, P., and Harel, J. (2007). Immunization with recombinant Sao protein confers protection against *Streptococcus suis* infection. Clin Vaccine Immunol *14*, 937-943.

Li, Y., Martinez, G., Gottschalk, M., Lacouture, S., Willson, P., Dubreuil, J.D., Jacques, M., and Harel, J. (2006). Identification of a surface protein of *Streptococcus suis* and evaluation of its immunogenic and protective capacity in pigs. Infect Immun *74*, 305-312.

Li, Z., Chang, P., Xu, J., Tan, C., Wang, X., Bei, W., and Li, J. (2018b). A *Streptococcus suis* live vaccine suppresses streptococcal toxic shock-like syndrome and provides sequence type independent protectionTITLE. J Infect Dis doi: 10.1093/infdis/jiy512. [Epub ahead of print].

Liu, H.T., Zhu, S., Sun, Y.Y., Li, N., Gu, J.M., Sun, C.J., Feng, X., Han, W.Y., Jiang, J.X., and Lei, L.C. (2017). Selection of Potential Virulence Factors Contributing to *Streptococcus suis* Serotype 2 Penetration into the Blood-Brain Barrier in an In Vitro Co-Culture Model. J Microbiol Biotechnol *27*, 161-170.

Liu, L., Cheng, G., Wang, C., Pan, X., Cong, Y., Pan, Q., Wang, J., Zheng, F., Hu, F., and Tang, J. (2009). Identification and experimental verification of protective antigens against *Streptococcus suis* serotype 2 based on genome sequence analysis. Curr Microbiol *58*, 11-17.

Lun, S., Perez-Casal, J., Connor, W., and Willson, P.J. (2003). Role of suilysin in pathogenesis of *Streptococcus suis* capsular serotype 2. Microb Pathog *34*, 27-37.

Lun, Z.R., Wang, Q.P., Chen, X.G, Li, A.X., Zhu, X.Q. (2007). *Streptococcus suis*: an emerging zoonotic pathogen. Lancet Infect Dis *7*, 201-209.

Lv, Q., Hao, H., Bi, L., Zheng, Y., Zhou, X., and Jiang, Y. (2014). Suilysin remodels the cytoskeletons of human brain microvascular endothelial cells by activating RhoA and Rac1 GTPase. Protein Cell *5*, 261-264.

Ma, F., Yi, L., Yu, N.W., Wang, G.Y., Ma, Z., Lin, H.X., and Fan, H.J. (2017). *Streptococcus suis* Serotype 2 Biofilms Inhibit the Formation of Neutrophil Extracellular Traps. Front Cell Infect Microbiol *7*, 86.

Mai, N.T., Hoa, N.T., Nga, T.V., Linh le, D., Chau, T.T., Sinh, D.X., Phu, N.H., Chuong ,L.V., Diep, T.S., Campbell, J., Nghia, H.D., Minh, T.N., Chau, N.V., de Jong, M.D., Chinh, N.T.,Hien, T.T., Farrar, J., and Schultsz, C. (2008). *Streptococcus suis* meningitis in adults in Vietnam.Clin Infect Dis *46*, 659-667.

Mandanici, F., Gomez-Gascon, L., Garibaldi, M., Olaya-Abril, A., Luque, I., Tarradas, C., Mancuso, G., Papasergi, S., Barcena, J.A., Teti, G., Beninati, C., and Rodriguez-Ortega, M.J. (2010). A surface protein of *Streptococcus suis* serotype 2 identified by proteomics protects mice against infection. J Proteomics *73*, 2365-2369.

Martelet, L., Lacouture, S., Goyette-Desjardins, G., Beauchamp, G., Surprenant, C., Gottschalk, M., and Segura, M. (2017). Porcine Dendritic Cells as an In Vitro Model to Assess the Immunological Behaviour of *Streptococcus suis* Subunit Vaccine Formulations and the Polarizing Effect of Adjuvants. Pathogens *6*, pii: E13.

Meng, F., Wu, N.H., Seitz, M., Herrler, G., and Valentin-Weigand, P. (2016). Efficient suilysin-mediated invasion and apoptosis in porcine respiratory epithelial cells after streptococcal infection under air-liquid interface conditions. Sci Rep *6*, 26748.

Nghia, H.D, Tu le, T.P., Wolbers, M., Thai, C.Q., Hoang, N.V., Nga, T.V., Thao le, T.P., Phu, N.H., Chau, T.T., Sinh, D.X., Diep, T.S., Hang, H.T., Truong, H., Campbell, J., Chau, N.V., Chinh, N.T., Dung, N.V., Hoa, N.T., Spratt, B.G., Hien, T.T., Farrar, J., and Schultsz, C. (2011). Risk factors of *Streptococcus suis* infection in Vietnam. A case-control study. PLoS One *6*, e17604.

Nomoto, R., Maruyama, F., Ishida, S., Tohya, M., Sekizaki, T., Osawa, R. (2015). Reappraisal of the taxonomy of *Streptococcus suis* serotypes 20, 22 and 26: *Streptococcus parasuis* sp. nov. Int J Syst Evol Microbiol *65*, 438-43.

Nutravong, T., Angkititrakul, S., Jiwakanon, N., Wongchanthong, W., Dejsirilerts, S., Nawa, Y. (2014). Identification of major *Streptococcus suis* serotypes 2, 7, 8 and 9 isolated from p igs and humans in upper northeastern Thailand. Southeast Asian J Trop Med Public Health *45*, 1173-1181.

Okura, M., Osaki, M., Nomoto, R., Arai, S., Osawa, R., Sekizaki, T., and Takamatsu, D. (2016). Current taxonomical situation of *Streptococcus suis*. Pathogens *5*, 45.

Okwumabua, O., and Chinnapapakkagari, S. (2005). Identification of the gene encoding a 38-kilodalton immunogenic and protective antigen of *Streptococcus suis*. Clin Diagn Lab Immunol *12*, 484-490.

Oh, S.I., Jeon, A.B., Jung, B.Y., Byun, J.W., Gottschalk, M., Kim, A., Kim, J.W., and Kim, H.Y. (2017). Capsular serotypes, virulence-associated genes and antimicrobial susceptibility of *Streptococcus suis* isolates from pigs in Korea. J Vet Med Sci *79*, 780-787.

Pallares, F.J., Schmitt, C.S., Roth, J.A., Evans, R.B., Kinyon, J.M., and Halbur, P.G. (2004). Evaluation of a ceftiofur-washed whole cell *Streptococcus suis* bacterin in pigs. Can J Vet Res 68, 236-240.

Palmer, M. (2001). The family of thiol-activated, cholesterol-binding cytolysins. Toxicon *39*, 1681-1689.

Pian, Y., Gan, S., Wang, S., Guo, J., Wang, P., Zheng, Y., Cai, X., Jiang, Y., and Yuan, Y. (2012). Fhb, a novel factor H-binding surface protein, contributes to the antiphagocytic ability and virulence of *Streptococcus suis*. Infect Immun *80*, 2402-2413.

Quessy, S., Dubreuil, J.D., Caya, M., and Higgins, R. (1995). Discrimination of virulent and avirulent *Streptococcus suis* capsular type 2 isolates from different geographical origins. Infect Immun *63*, 1975-1979.

Quessy, S., Dubreuil, J.D., and Higgins, R. (1994). Immunization of mice against *Streptococcus suis* serotype 2 infections using a live avirulent strain. Can J Vet Res *58*, 299-301.

Requena, J.M. (2012). Stress Response in Microbiology. Caister Academic Press.

Roy, D., Grenier, D., Segura, M., Mathieu-Denoncourt, A., and Gottschalk, M. (2016). Recruitment of Factor H to the *Streptococcus suis* Cell Surface is Multifactorial. Pathogens *5*, pii: E47.

Roy, R., and Shiao, T.C. (2011). Organic Chemistry and Immunochemical Strategies in the Design of Potent Carbohydrate-based Vaccines. Chimia *65*, 24-29.

Rungelrath, V., Weisse, C., Schutze, N., Muller, U., Meurer, M., Rohde, M., Seele, J., Valentin-Weigand, P., Kirschfink, M., Beineke, A., Schrodl, W., Bergmann, R., and Baums, C.G. (2018). IgM cleavage by *Streptococcus suis* reduces IgM bound to the bacterial surface and is a novel complement evasion mechanism. Virulence *9*, 1314-1337.

Seele, J., Hillermann, L.M., Beineke, A., Seitz, M., Von Pawel-Rammingen, U., Valentin-Weigand, P., and Baums, C.G. (2015). The immunoglobulin M-degrading enzyme of *Streptococcus suis*, IdeSsuis, is a highly protective antigen against serotype 2. Vaccine *33*, 2207-2212.

Segura, M., Calzas, C., Grenier, D., and Gottschalk, M. (2016). Initial steps of the pathogenesis of the infection caused by *Streptococcus suis*: fighting against nonspecific defenses. FEBS Lett *590*, 3772-3799.

Segura, M., Fittipaldi, N., Calzas, C., and Gottschalk, M. (2017). Critical *Streptococcus suis* Virulence Factors: Are They All Really Critical? Trends in Microbiology *25*,586-U294.

Segura, M., Gottschalk, M., and Olivier, M. (2004). Encapsulated *Streptococcus suis* inhibits activation of signaling pathways involved in phagocytosis. Infect

Immun *72*, 5322-5330.

Segura, M., Zheng, Han., Greeff, A., Gao, G.F., Grenier, D., Jiang, Y., Lu, C., Maskell, D., Oishi, K., Okura, M.,, Osawa, Ro., Schultsz, C., Schwerk, C., Sekizaki, T., Smith, H., Srimanote, P., Takamatsu D., Tang, J., Tenenbaum, T., Tharavichitkul, P., Hoa, N.T., Valentin-Weigand, P., Wells, J.M., Wertheim, H., Zhu, B., Gottschalk, M., and Xu, J. (2014). Latest developments on *Streptococcus suis*: an emerging zoonotic pathogen: part 1.Future Microbiol *9*, 441-444.

Smith, H.E., Buijs, H., De Vries, R.R., Wisselink, H.J., Stockhofe-Zurwieden, N., and Smits, M.A. (2001). Environmentally regulated genes of *Streptococcus suis*: identification by the use of iron-restricted conditions in vitro and by experimental infection of piglets. Microbiology *147*, 271-280.

Smith, H.E., Damman, M., Van Der Velde, J., Wagenaar, F., Wisselink, H.J., Stockhofe-Zurwieden, N., and Smits, M.A. (1999). Identification and characterization of the cps locus of *Streptococcus suis* serotype 2: the capsule protects against phagocytosis and is an important virulence factor. Infect Immun *67*, 1750-1756.

Takeuchi, D., Akeda, Y., Nakayama, T., Kerdsin, A., Sano, Y., Kanda, T., Hamada, S., Dejsirilert, S., and Oishi, K. (2014). The contribution of suilysin to the pathogenesis of *Streptococcus suis* meningitis. J Infect Dis *209*, 1509-1519.

Tan, C., Fu, S., Liu, M., Jin, M., Liu, J., Bei, W., and Chen, H. (2008). Cloning, expression and characterization of a cell wall surface protein, 6-phosphogluconate-dehydrogenase, of *Streptococcus suis* serotype 2. Vet Microbiol *130,* 363-370.

Tan, C., Liu, M., Liu, J., Yuan, F., Fu, S., Liu, Y., Jin, M., Bei, W., and Chen, H.

(2009). Vaccination with *Streptococcus suis* serotype 2 recombinant 6PGD protein provides protection against S. suis infection in swine. FEMS Microbiol Lett *296*, 78-83.

Tan, M.F., Gao, T., Liu, W.Q., Zhang, C.Y., Yang, X., Zhu, J.W., Teng, M.Y., Li, L., and Zhou, R. (2015). MsmK, an ATPase, Contributes to Utilization of Multiple Carbohydrates and Host Colonization of *Streptococcus suis*. PLoS One *10*, e0130792.

Tan, M.F., Liu, W.Q., Zhang, C.Y., Gao, T., Zheng, L.L., Qiu, D.X., Li, L., and Zhou, R. (2017). The involvement of MsmK in pathogenesis of the *Streptococcus suis* serotype 2. Microbiologyopen *6*, e00433.

Tanabe, S., Bonifait, L., Fittipaldi, N., Grignon, L., Gottschalk, M., and Grenier, D. (2010). Pleiotropic effects of polysaccharide capsule loss on selected biological properties of *Streptococcus suis*. Can J Vet Res *74*, 65-70.

Tang, J., Wang, C., Feng, Y., Yang, W., Song, H., Chen, Z., Yu, H., Pan, X., Zhou, X., Wang, H., Wu, B., Wang ,H., Zhao, H., Lin, Y., Yue, J., Wu, Z., He, X., Gao, F., Khan, A.H., Wang, J., Zhao, G.P., Wang, Y., Wang, X., Chen, Z, and Gao, G.F. (2006). Streptococcal toxic shock syndrome caused by *Streptococcus suis* serotype 2. PLoS Med *3*, e151.

Taniyama, D., Sakurai, M., Sakai, T., Kikuchi, T., Takahashi, T. (2016). Human case of bacteremia due to *Streptococcus suis* serotype 5 in Japan: The first report and literature review. IDCases *6*,36-38.

Takeuchi, D., Akeda, Y., Nakayama, T., Kerdsin, A., Sano, Y., Kanda, T., Hamada, S., Dejsirilert, S., and Oishi, K. (2014). The contribution of suilysin to the pathogenesis of *Streptococcus suis* meningitis. J Infect Dis *209*, 1509-1519.

Tenenbaum, T., Papandreou, T., Gellrich, D., Friedrichs, U., Seibt, A., Adam, R.,

Wewer, C., Galla, H.J., Schwerk, C., and Schroten, H. (2009). Polar bacterial invasion and translocation of *Streptococcus suis* across the blood-cerebrospinal fluid barrier in vitro. Cell Microbiol *11*, 323-336.

Thongkamkoon, P., Kiatyingangsulee, T., and Gottschalk M. (2017). Serotypes of *Streptococcus suis* isolated from healthy pigs in Phayao Province, Thailand. BMC Res Notes *10*,53.

Tohya, M., Arai, S., Tomida, J., Watanabe, T., Kawamura, Y., Katsumi, M., Ushimizu, M., Ishida-Kuroki, K., Yoshizumi, M., Uzawa, Y., Iguchi, S., Yoshida, A., Kikuchi, K., Sekizaki, T. (2017). Defining the taxonomic status of *Streptococcus suis* serotype 33: the proposal for *Streptococcus ruminantium* sp. nov. Int J Syst Evol Microbiol *67*, 3660-3665.

Torremorell, M., Pijoan, C., and Trigo, E. (1997). Vaccination against *Streptococcus suis*: effect on nursery mortality. J Swine Health Prod*5*, 139-143.

Van Calsteren, M.R., Goyette-Desjardins, G., Gagnon, F., Okura, M., Takamatsu, D., Roy, R., Gottschalk, M., and Segura, M. (2016). Explaining the serological characteristics of *Streptococcus suis* serotypes 1 and 1/2 from their capsular polysaccharide structure and biosynthesis. J Biol Chem *291*, 8387-8398.

Vanier, G., Segura, M., Friedl, P., Lacouture, S., and Gottschalk, M. (2004). Invasion of porcine brain microvascular endothelial cells by *Streptococcus suis* serotype 2. Infect Immun 72, 1441-1449.

Vecht, U., Wisselink, H.J., Jellema, M.L., and Smith, H.E. (1991). Identification of two proteins associated with virulence of *Streptococcus suis* type 2. Infect Immun *59*, 3156-3162.

Velikova, N. (2018). The VraSR regulatory system contributes to virulence in *Streptococcus suis* via resistance to innate immune defenses. Virulence *9*,

681-682.

Wan, Y., Zhang, S., Li, L., Chen, H., and Zhou, R. (2017). Characterization of a novel streptococcal heme-binding protein SntA and its interaction with host antioxidant protein AOP2. Microb Pathog *111*, 145-155.

Wang, J., Feng, Y.J., Wang, C.J., Zheng, F., Hassan, B., Zhi, L.M., Li, W.J., Yao, Y., He, E., Jiang, S.B., and Tang, J.Q. (2017a). Genome-wide analysis of an avirulent strain that induces protective immunity against challenge with virulent *Streptococcus Suis* serotype 2. BMC Microbiol *17*, 152.

Wang, J., Feng, Y., Wang, C., Srinivas, S., Chen, C., Liao, H., He, E., Jiang, S., and Tang, J. (2017b). Pathogenic *Streptococcus strains* employ novel escape strategy to inhibit bacteriostatic effect mediated by mammalian peptidoglycan recognition protein. Cell Microbiol *19*, 10.1111/cmi.12724

Wang, J., Kong, D., Zhang, S., Jiang, H., Zheng, Y., Zang, Y., Hao, H., and Jiang, Y. (2015). Interaction of fibrinogen and muramidase-released protein promotes the development of *Streptococcus suis* meningitis. Front Microbiol *6*, 1001.

Wei, Z., Li, R., Zhang, A., He, H., Hua, Y., Xia, J., Cai, X., Chen, H., and Jin, M. (2009). Characterization of *Streptococcus suis* isolates from the diseased pigs in China between 2003 and 2007. Vet Microbiol *137*, 196-201.

Weinert, L.A., Chaudhuri, R.R., Wang, J., Peters, S.E., Corander, J., Jombart, T., Baig, A., Howell, K.J., Vehkala, M., Välimäki, N., Harris, D., Chieu, T.T., Van Vinh Chau, N., Campbell, J., Schultsz, C., Parkhill, J., Bentley, S.D., Langford, P.R., Rycroft, A.N., Wren, B.W., Farrar, J., Baker, S., Hoa, N.T., Holden, M.T., Tucker, A.W., Maskell, D.J., and BRaDP1T Consortium. (2015). Genomic signatures of human and animal diseasein the zoonotic pathogen *Streptococcus suis.* Nat Commun *6*, 6740.

Wertheim, H.F., Nguyen, H.N., Taylor, W., Lien, T.T., Ngo, H.T., Nguyen, T.Q., Nguyen, B.N., Nguyen, H.H., Nguyen, H.M., Nguyen, C.T., Dao, T.T., Nguyen, T.V., Fox, A., Farrar, J., Schultsz, C., Nguyen, HD, Nguyen, K.V., and Horby, P. (2009). *Streptococcus suis*, an important cause of adult bacterial meningitis in northen Vitnam. PLoS One *4*, e5973.

Willenborg, J., Fulde, M., De Greeff, A., Rohde, M., Smith, H.E., Valentin-Weigand, P., and Goethe, R. (2011). Role of glucose and CcpA in capsule expression and virulenceof *Streptococcus suis*. Microbiology *157*, 1823-1833.

Willenborg, J., Koczula, A., Fulde, M., De Greeff, A., Beineke, A., Eisenreich, W., Huber, C., Seitz, M., Valentin-Weigand, P., and Goethe, R. (2016). FlpS, the FNR-Like Protein of *Streptococcus suis* Is an Essential, Oxygen-Sensing Activator of the Arginine Deiminase System. Pathogens *5*, pii: E51.

Wisselink, H.J., Stockhofe-Zurwieden, N., Hilgers, L.A., and Smith, H.E. (2002). Assessment of protective efficacy of live and killed vaccines based on a non-encapsulated mutant of *Streptococcus suis* serotype 2. Vet Microbiol *84*, 155-168.

Wisselink, H.J., Vecht, U., Stockhofe-Zurwieden, N., and Smith, H.E. (2001). Protection of pigs against challenge with virulent *Streptococcus suis* serotype 2 strains by a muramidase-released protein and extracellular factor vaccine. Vet Rec *148*, 473-477.

Wu, T., Zhao, Z., Zhang, L., Ma, H., Lu, K., Ren, W., Liu, Z., Chang, H., Bei, W., Qiu, Y., and Chen, H. (2011). Trigger factor of *Streptococcus suis* is involved in stress tolerance and virulence. Microb Pathog *51*, 69-76.

Wu, Z., Shao, J., Ren, H., Tang, H., Zhou, M., Dai, J., Lai, L., Yao, H., Fan, H.,

Chen, D., Zong, J., and Lu, C. (2016). A *Streptococcus suis* LysM domain surface protein contributes to bacterial virulence. Vet Microbiol *187*, 64-69.

Xiao, G., Wu, Z., Zhang, S., Tang, H., Wang, F., and Lu, C. (2017a). Mac Protein is not an Essential Virulence Factor for the Virulent Reference Strain *Streptococcus suis* P1/7. Curr Microbiol *74*, 90-96.

Xiao, G.H., Tang, H.Y., Zhang, S.M., Ren, H.Y., Dai, J., Lai, L.Y., Lu, C.P., Yao, H.C., Fan, H.J., and Wu, Z.F. (2017b). *Streptococcus suis* small RNA rss04 contributes to the induction of meningitis by regulating capsule synthesis and by inducing biofilm formation in a mouse infection model. Vet Microbiol *199*, 111-119.

Xu, B., Zhang, P., Li, W., Liu, R., Tang, J., and Fan, H. (2017a). hsdS, Belonging to the Type I Restriction-Modification System, Contributes to the *Streptococcus suis* Serotype 2 Survival Ability in Phagocytes. Front Microbiol *8*, 1524.

Xu, J., Fu, S., Liu, M., Xu, Q., Bei, W., Chen, H., and Tan, C. (2014). The two-component system NisK/NisR contributes to the virulence of *Streptococcus suis* serotype 2. Microbiol Res *169*, 541-546.

Xu, J., Zheng, C., Cao, M., Zeng, T., Zhao, X., Shi, G., Chen, H., and Bei, W. (2017b). The manganese efflux system MntE contributes to the virulence of *Streptococcus suis* serotype 2. Microb Pathog *110*, 23-30.

Xu, L., Huang, B., Du, H., Zhang, X.C., Xu, J., Li, X., and Rao, Z. (2010). Crystal structure of cytotoxin protein suilysin from *Streptococcus suis*. Protein Cell *1*, 96-105.

Xu, Z., Chen, B., Zhang, Q., Liu, L., Zhang, A., Yang, Y., Huang, K., Yan, S., Yu, J., Sun, X., Jin, M. (2018). Streptococcus suis 2 Transcriptional Regulator TstS Stimulates Cytokine Production and Bacteremia to Promote Streptococcal Toxic

Shock-Like Syndrome. Front Microbiol 9, 1309.

Yao, X., Li, M., Wang, J., Wang, C., Hu, D., Zheng, F., Pan, X., Tan, Y., Zhao, Y., Hu, L., Tang, J., and Hu, F. (2015). Isolation and characterization of a native avirulent strain of *Streptococcus suis* serotype 2: a perspective for vaccine development. Sci Rep *5*,9835.

Ye, C., Bai, X., Zhang , J., Jing, H., Zheng, H., Du, H., Cui, Z., Zhang, S., Jin, D., Xu, Y., Xiong, Y., Zhao A., Luo, X., Sun, Q., Gottschalk, M., and Xu, J. (2008). Spread of *Streptococcus suis* sequence type 7, China. Emerg Infect Dis *14*, 787-791.

Yin, S., Li, M., Rao, X., Yao, X., Zhong, Q., Wang, M., Wang, J., Peng, Y., Tang, J., Hu, F., and Zhao, Y. (2016). Subtilisin-like protease-1 secreted through type IV secretion system contributes to high virulence of *Streptococcus suis* 2. Sci Rep *6*, 27369.

Yu, H., Jing, H., Chen, Z., Zheng, H., Zhu, X., Wang, H., Wang, S., Liu, L., Zu, R., Luo, L., Xiang, N., Liu, H., Liu, X., Shu, Y., Lee, S.S., Chuang , S.K., Wang, Y., Xu, J., and Yang, W. (2006).Human *Streptococcus suis* outbreak, Sichuan, China. Emerg Infect Dis *12*, 914-920.

Yu, Y., Qian, Y., Du, D., Xu, C., Dai, C., Li, Q., Liu, H., Shao, J., Wu, Z., and Zhang, W. (2016). SBP2 plays an important role in the virulence changes of different artificial mutants of *Streptococcus suis*. Mol Biosyst *12*, 1948-1962.

Yuan, F.Y., Tan, C., Liu, Z.W., Yang, K., Zhou, D., Liu, W., Duan, Z.Y., Guo, R., Chen, H.C., Tian, Y.X., and Bei, W.C. (2017). The 1910HK/RR two-component system is essential for the virulence of *Streptococcus suis* serotype 2. Microb Pathog *104*, 137-145.

Yuan, Z.Z., Yan, X.J., Zhang, A.D., Chen, B., Shen, Y.Q., and Jin, M.L. (2013). Molecular mechanism by which surface antigen HP0197 mediates host cell attachment in the pathogenic bacteria *Streptococcus suis*. J Biol Chem *288*, 956-963.

Zhang, A., Chen, B., Li, R., Mu, X., Han, L., Zhou, H., Chen, H., and Meilin, J. (2009a). Identification of a surface protective antigen, HP0197 of *Streptococcus suis* serotype 2. Vaccine *27*, 5209-5213.

Zhang, A., Chen, B., Mu, X., Li, R., Zheng, P., Zhao, Y., Chen, H., and Jin, M. (2009b). Identification and characterization of a novel protective antigen, Enolase of *Streptococcus suis* serotype 2. Vaccine *27*, 1348-1353.

Zhang, A., Chen, B., Yuan, Z., Li, R., Liu, C., Zhou, H., Chen, H., and Jin, M. (2012a). HP0197 contributes to CPS synthesis and the virulence of *Streptococcus suis* via CcpA. PLoS One *7*, e50987.

Zhang, A., Mu, X., Chen, B., Han, L., Chen, H., and Jin, M. (2011a). IgA1 protease contributes to the virulence of *Streptococcus suis*. Vet Microbiol *148*, 436-439.

Zhang, A., Mu, X., Chen, B., Liu, C., Han, L., Chen, H., and Jin, M. (2010). Identification and characterization of IgA1 protease from *Streptococcus suis*. Vet Microbiol *140*, 171-175.

Zhang, A., Xie, C., Chen, H., and Jin, M. (2008). Identification of immunogenic cell wall-associated proteins of *Streptococcus suis* serotype 2. Proteomics *8*, 3506-3515.

Zhang A., Yang M., Hu P., Wu J., Chen B., Hua Y., Yu J., Chen H., Xiao J., and Jin M. (2011b). Comparative genomic analysis of *Streptococcus suis* reveals significant genomic diversity among different serotypes. BMC Genomics *12*, 523.

Zhang, C.Y., Sun, W., Tan, M.F., Dong, M.M., Liu, W.Q., Gao, T., Li, L., Xu, Z.F., and Zhou, R. (2017a). The Eukaryote-Like Serine/Threonine Kinase STK Regulates the Growth and Metabolism of Zoonotic *Streptococcus suis*. Front Cell Infect Microbiol *7*, 66.

Zhang, Q., Huang, J., Yu, J., Xu, Z., Liu, L., Song, Y., Sun, X., Zhang, A., and Jin, M. (2017b). HP1330 Contributes to *Streptococcus suis* Virulence by Inducing Toll-Like Receptor 2- and ERK1/2-Dependent Pro-inflammatory Responses and Influencing In Vivo S. suis Loads. Front Immunol *8*, 869.

Zhang, S.W., Wang, J.P., Chen, S.L., Yin, J.Y., Pan, Z.Y., Liu, K.K., Li, L., Zheng, Y.L., Yuan, Y., and Jiang, Y.Q. (2016a). Effects of Suilysin on *Streptococcus suis*-Induced Platelet Aggregation. Front Cell Infect Microbiol *6*, 128.

Zhang, S.W., Zheng, Y.L., Chen, S.L., Huang, S.J., Liu, K.K., Lv, Q.Y., Jiang, Y.Q., and Yuan, Y. (2016b). Suilysin-induced Platelet-Neutrophil Complexes Formation is Triggered by Pore Formation-dependent Calcium Influx. SciRep*6*, 36787.

Zhang, W., and Lu, C.P. (2007a). Immunoproteomic assay of membrane-associated proteins of *Streptococcus suis* type 2 China vaccine strain HA9801. Zoonoses Public Health *54*, 253-259.

Zhang, W., and Lu, C.P. (2007b). Immunoproteomics of extracellular proteins of Chinese virulent strains of *Streptococcus suis* type 2. Proteomics *7*, 4468-4476.

Zhang, W., Rong, C., Chen, C., Gao, G. (2012b). Type-IVC secretion system: a novel subclass of type IV secretion system (T4SS) common existing in gram-positive genus *Streptococcus*. PLoS One *7*, e46390.

Zhang, Y., Lu, P., Pan, Z., Zhu, Y., Ma, J., Zhong, X., Dong, W., Lu, C., and Yao, H. (2018a). Sssp1, a *Streptococcus suis* fimbriae-like protein transported by SecY2/A2 system contributes to bacterial virulence. Appl Environ Microbiol doi:

10.1128/AEM.01385-18. [Epub ahead of print]

Zhang, Y., Zong, B., Wang, X., Zhu, Y., Hu, L., Li, P., Zhang, A., Chen, H., Liu, M., and Tan, C. (2018b). Fisetin Lowers *Streptococcus suis* serotype 2 Pathogenicity in Mice by Inhibiting the Hemolytic Activity of Suilysin. Front Microbiol *9***,** 1723.

Zhang, Y.M., Shao, Z.Q., Wang, J., Wang, L., Li, X., Wang, C., Tang, J., and Pan, X. (2014). Prevalent distribution and conservation of *Streptococcus suis* Lmb protein and its protective capacity against the Chinese highly virulent strain infection. Microbiol Res *169*, *395*-401.

Zhao, J., Lin, L., Fu, L., Han, L., and Zhang, A. (2016). Neutrophils extracellular Taps play an important role in clearance of *Streptococcus suis* in vivo. Microbiol Immunol *60*, 228-233.

Zhao, J., Pan, S., Lin, L., Fu, L., Yang, C., Xu, Z., Wei, Y., Jin, M., and Zhang, A. (2015). *Streptococcus suis* serotype 2 strains can induce the formation of neutrophil extracellular traps (NETs) and evade trapping. FEMS Microbiol Lett *362*, pii: fnv022.

Zhao, Y., Liu, G., Li, S., Wang, M., Song, J., Wang, J., Tang, J., Li, M., and Hu, F. (2011). Role of a type IV-like secretion system of *Streptococcus suis* 2 in the development of streptococcal toxic shock syndrome. J Infect Dis *204*, 274-281.

Zheng, C., Li, L., Ge, H., Meng, H., Li, Y., Bei, W., Zhou, X. (2018a). Role of two-component regulatory systems in the virulence of *Streptococcus suis*. Microbiol Res *214*, 123-128.

Zheng, C., Ren, S., Xu, J., Zhao, X., Shi, G., Wu, J., Li, J., Chen, H., and Bei, W. (2017). Contribution of NADH oxidase to oxidative stress tolerance and virulence of *Streptococcus suis* serotype 2. Virulence *8*, 53-65.

Zheng, C., Xu, J., Li, J., Hu, L., Xia, J., Fan, J., Guo, W., Chen, H., Bei, W.

(2014). Two Spx regulators modulate stress tolerance and virulence in *Streptococcus suis* serotype 2. PLoS One *9*, e108197.

Zheng, C., Xu, J., Shi, G., Zhao, X., Ren, S., Li, J., Chen, H., and Bei, W. (2016). Formate-tetrahydrofolate ligase is involved in the virulence of *Streptococcus suis* serotype 2. Microb Pathog *98*, 149-154.

Zheng, F., Ji, H., Cao, M., Wang, C., Feng, Y., Li, M., Pan, X., Wang, J., Qin, Y., Hu, F., Tang, J. (2011). Contribution of the Rgg transcription regulator to metabolism and virulence of *Streptococcus suis* serotype 2. Infect Immun *79*,1319-1328.

Zheng, F., Shao, Z.Q., Hao, X., Wu, Q., Li, C., Hou, H., Hu, D., Wang, C., Pan, X. (2018b). Identification of oligopeptide-binding protein (OppA) and its role in the virulence of *Streptococcus suis* serotype 2. Microb Pathog *118*, 322-329.

Zhong, X., Zhang, Y., Zhu, Y., Dong, W., Ma, J., Pan, Z., Roy, S., Lu, C., Yao, H. (2018). The Two-Component Signaling System VraSRss Is Critical for Multidrug Resistance and Full Virulence in *Streptococcus suis* Serotype 2. Infect Immun *86*, e00096-18.

Zhou, Y., Wang, Y., Deng, L., Zheng, C., Yuan, F., Chen, H., Bei, W., and Li, J. (2015). Evaluation of the protective efficacy of four novel identified membrane associated proteins of *Streptococcus suis* serotype 2. Vaccine *33*, 2254-2260.

Zhu, J., Tang, J., Guo, H. (2001). Identification of the pathogens responsible for toxic shock syndrome in the severe acute outbreak. Chin J Contag *19*,17-19.

Zhu, J., Zhang, T., Su, Z., Li, L., Wang, D., Xiao, R., Teng, M., Tan, M., and Zhou, R. (2016). (p)ppGpp synthetases regulate the pathogenesis of zoonotic *Streptococcus suis*. Microbiol Res *191*, 1-11.

Zou, G., Zhou, J., Xiao, R., Zhang, L., Cheng, Y., Jin, H., Li, L., Zhang, L., Wu, B.,

Qian, P., Li, S., Ren, L., Wang, J., Oshota, O., Hernandez-Garcia, J., Wileman, T.M., Bentley, S., Weinert, L., Maskell, D.J., Tucker, A.W.D., and Zhou, R. (2018). The effect of environmental and management associated factors on prevalence and diversity of *Streptococcus suis* in clinically healthy pig herds of China and UK.Appl Environ Microbiol *84*, e02590-17.

 https://doi.org/10.21775/9781912530229.12

The Adhesion and Invasion Mechanisms of Streptococci

Qingsong Jiang, Xuedong Zhou, Lei Cheng, Mingyun Li*

State Key Laboratory of Oral Diseases, West China school of Stomatology, Sichuan University, Chengdu, China

* Corresponding Author: limingyun@scu.edu.cn

Abstract

Streptococci are common human pathogens, colonizing multiple parts of the human body such as the upper respiratory tract, urethra, gastrointestinal tract, and oral cavity. Since they cause a variety of serious infections including heart diseases, meningitis, and oral diseases, streptococci are considered to play an important role in human diseases. Two critical steps in the pathogenesis of streptococcal infection are the adhesion to and invasion of host cells. This invasion is a strategy of streptococci to evade the host immune response and antibiotic therapy, as well as to penetrate to deeper tissues. To establish interaction between bacteria and host cells, adhesion is the initial step. To effectively adhere to host cells, streptococci express multiple adhesins, and the expression of different adhesins may lead to distinct mechanisms of subsequent invasion. The binding of streptococcal molecules to host proteins triggers downstream signal transduction in the host cells, leading to the uptake of bacteria. In this review, we present the adhesion and invasion mechanisms of different streptococci and the interaction with host cells leading to internalization.

Introduction

The *Streptococcus* genus is made up of common gram-positive bacteria, and includes two major pathogenic species, *Streptococcus pyogenes* (also known as group A streptococcus, GAS) and *Streptococcus pneumoniae*, and other invasive pathogens such as *Streptococcus agalactiae* (group B Streptococcus, GBS), *Streptococcus suis,* and *Streptococcus mitis.* GAS was reported to cause more than 500,000 deaths worldwide anually (Carapetis et al. 2005). GAS-induced disease involves the invasion of different types of eukaryotic cells via distinct mechanisms. M protein and streptococcal fibronectin binding protein I (SfbI) are two main virulence factors that are critical to invasion (Cunningham 2000; Molinari et al. 2000; Cywes and Wessels 2001; Rohde et al. 2003; Amelung et al. 2011; Siemens et al. 2011). Streptococci use these virulence factors to evade antibiotic treatment and host immune responses such as phagocytotic clearance (Kwinn and Nizet 2007; Nitsche-Schmitz et al. 2007; Nizet 2007; Smeesters et al. 2010; Barnett et al. 2013), and can cause a wide spectrum of human infections including superficial skin infections and life-threatening diseases (Cunningham 2000; Olsen and Musser 2010; Liang et al. 2012). Additionally, GAS infection can lead to post-infection sequelae, such as post-streptococcal glomerulonephritis, rheumatic fever, and rheumatic heart disease (Luca-Harari et al. 2009; Jackson et al. 2011; Steer et al. 2012; Walker et al. 2014). The human upper respiratory tract can be asymptomatically colonized by the Gram-positive and opportunistic pathogen *S. pneumoniae*. Although colonization by *S. pneumoniae* can occur at any stage in life, infants and elderly are the most vulnerable groups (van der Poll and Opal 2009; Shak et al. 2013; Valles et al. 2016). The morbidity from *S. pneumoniae* infections remains high worldwide, in spite of the available treatments

(Alonsodevelasco et al. 1995; van der Poll and Opal 2009). Many virulence factors contribute to its pathogenicity, including: (i) toxins, such as pneumolysin; (ii) adhesins, such as pneumococcal surface protein A (PspA), choline-binding protein A (CbpA; also known as PspC), and pneumococcal adherence and virulence protein A (PavA); (iii) extracellular enzymes, such as surface neuraminidase (Nan) A, NanB, NanC, and pneumococcal phospholipase A2 (PLA2); (iv) the pneumococcal capsule; and (v) pili. (Werdan et al. 2014; Iovino et al. 2016; Weiser et al. 2018; Yau et al. 2018). Using these virulence factors, *S. pneumoniae* is able to colonize host cells and translocate to deeper tissues, causing human diseases ranging from meningitis and sepsis to otitis media and pneumonia, or even invading the heart through the coronary circulation (Tonnaer et al. 2006; Brown et al. 2014; Wunderink and Waterer 2014; Gilley et al. 2016; Valles et al. 2016). Another Gram-positive and opportunistic pathogen, which can colonize the human genitourinary and gastrointestinal tract, is *S. agalactiae* (Verani et al. 2010). Numerous virulence factors of *S. agalactiae* have been identified to date, including pili, fibrinogen binding protein, Alpha C protein (ACP), serine rich repeat proteins (Srr), capsular polysaccharides (CPS), factor H-binding protein, superoxide dismutase (SodA), Streptococcal C5a peptidase of GBS (ScpB) and CAMP factor (LaPenta et al. 1997; Schubert et al. 2004; Rajagopal 2009; Landwehr-Kenzel and Henneke 2014). *S. agalactiae* plays an important role in the development of neonatal septicemia and meningitis, as well as invasive infections in immunocompromised adults and elderly (Verani et al. 2010; Pimentel et al. 2016). Maternal *S. agalactiae* infections can lead to neonatal diseases, by attaching to maternal vaginal epithelial cells via host extracellular matrix (ECM) molecules, then penetrating the uterine compartment and moving to the neonatal lung, followed by its entry into the blood system, and eventually causing pneumonia, sepsis, and meningitis in the newborn (Rajagopal 2009). Pneumonia and sepsis often occur within the first 7 days after birth while meningitis occurs during the period from the 7th day after birth to 2 or 3 months of age (Verani et al. 2010; Melin and Efstratiou 2013). Pneumonia developing in the first week after birth often leads to respiratory failure and worsens to septic shock rapidly. Neonatal meningitis will often result in severe sequelae such as deafness, visual impairment, cognitive impairment, and seizures (Schuchat 1998; Maisey et al. 2008; Libster et al. 2012; Melin and Efstratiou 2013). Based on its capsular polysaccharide immunologic reactivity, GBS has been classified into ten serotypes (Ia, Ib, II, III, IV, V, VI, VII, VIII, IX) (Rosa-Fraile et al. 2014). Serotypes Ia, Ib, III, and V are estimated to cause 85–90% of all GBS diseases (Hickman et al. 1999; Rajagopal 2009), while serotype III is responsible for causing meningitis (Edmond et al. 2012). Although it is known that GBS needs to penetrate the blood-brain barrier (BBB) to cause meningitis, the mechanism behind this is not clearly known. *S. suis* is a major swine pathogen colonizing the digestive and respiratory tracts of pigs. It can cause several diseases in pigs including septicemia, arthritis, endocarditis, pneumonia, and meningitis (Gottschalk et al. 2010; Goyette-Desjardins et al. 2014). It is also reported that *S. suis* can be transmitted from pigs to humans via contact, and is associated with streptococcal toxic shock syndrome (STSS), septicemia, arthritis, and meningitis (Tang et al. 2006; Feng et al. 2010). The virulence factors of *S. suis* include CPS, muraminidase-released protein, extracellular factor, suilysin (SLY), and hyaluronic acid lyase (Vecht et al. 1991; Okwumabua et al. 1999; Gottschalk and Segura 2000; Haas et al. 2015; Segura et al. 2017). At least 35 serotypes of *S. suis* have been identified based on different CPS antigens (Okura et al. 2016), among which serotype 2 is the major serotype isolated from humans infected by *S. suis* (Gottschalk et al. 2007; Mai et al. 2008). This pathogen can cross epithelial barriers and enter the bloodstream, causing localized and/or systemic infections, or causing meningitis by crossing the BBB (Haas and Grenier 2018; Zheng et al. 2018). *S. suis* is increasingly gaining attention for its prevalence worldwide as an emerging zoonotic pathogen. In 1998 and 2005, the epidemics occurring in China caused mortality up to 56% and 18.6%, respectively (Tang et al. 2006; Feng et al. 2010). In addition, *S. suis* infections in Northern Vietnam are thought to be a major cause of adult bacterial meningitis (Nguyen et al. 2008; Wertheim et al. 2009). In Thailand, *S. suis* is also thought to be an emerging human pathogen (Kerdsin et al. 2009), as by the end of 2013, the number of human infection cases of *S. suis* had risen to 1642 (Goyette-Desjardins et al. 2014). Although the high pathogenicity of *S. suis* has resulted in an increasing number of studies about the mechanisms by which *S. suis* causes infection, it is still not fully clear. *S. mitis* is one of the commensal and relatively benign bacteria colonizing the skin, genitourinary tract, and gastrointestinal tract, and it can be found on almost all surfaces of the oral cavity, such as teeth, tongue, and mucosal surfaces (Carrascosa et al. 1994; Pearce et al. 1995; Aas et al. 2005). Although the mechanism of pathogenesis is still not fully clear, *S. mitis* can cause multiple human diseases (Doern and Burnham 2010), including vasculitis and endocarditis, or even sacroiliitis (Mitchell 2011; Al-Farsi et al. 2018; Basaranoglu et al. 2018). In the Mitis group of streptococci, *S. mitis* and *S. pneumoniae* are closely related, and both have evolved to express a similar collection of virulence factors (Kilian et al. 2008). Johnston et al. showed that in five *S. mitis* isolates, 72-83% of the virulence genes of *S. pneumoniae* are present, among which *sodA* is present in all five strains (Johnston et al. 2010). Though originating from a common ancestor, compared to *S. pneumoniae, S. mitis* is more capable of adapting to a commensal lifestyle, which may be explained by a loss of virulence factors in *S. mitis* (Kilian et al. 2008; Rukke et al. 2014). In this review, we focus on the molecules involved in host invasion by different streptococcus bacteria strains, the interaction of bacterial adhesins or invasins with their respective receptors, and how they mediate streptococcal invasion. For host invasion by GAS, the most common streptococcus bacteria strain, we discuss the two main invasion pathways, which are mediated by SfbI and M protein.

GAS invasion

It has been demonstrated that highly pathogenic GAS strains can cause invasive disease in humans (Watanabe et al. 2016). GAS invasion is highly dependent on invasins exposed on the bacterium surface such as M protein and SfbI, which trigger two main uptake mechanisms: (i) The binding of M protein to α5β1 integrins facilitates host cytoskeleton rearrangement to cause membrane ruffling, and (ii) the binding of SfbI protein to host cells activates a caveolae-mediated internalization process, both of which eventually lead to the uptake of GAS (Dombek et al. 1999; Molinari et al. 2000; Rohde et al. 2003) (Figure 1). M protein is considered one of the most robust virulence factors of GAS strains (Cunningham 2000). It is present in the GAS cell wall, binds to host cell surface proteins and contributes to the interaction between the bacterium and host cells (Fischetti 2016). Subtyped by the M protein gene (*emm*), different M strains possess different pathogenicity, and M1 and M3 strains are most prevalent in invasive diseases (Bessen et al. 1996; Sellers et al. 1996; Kiska et al. 1997; Ikebe et al. 2015; Stevens and Bryant 2016; Watanabe et al. 2016). Another molecule

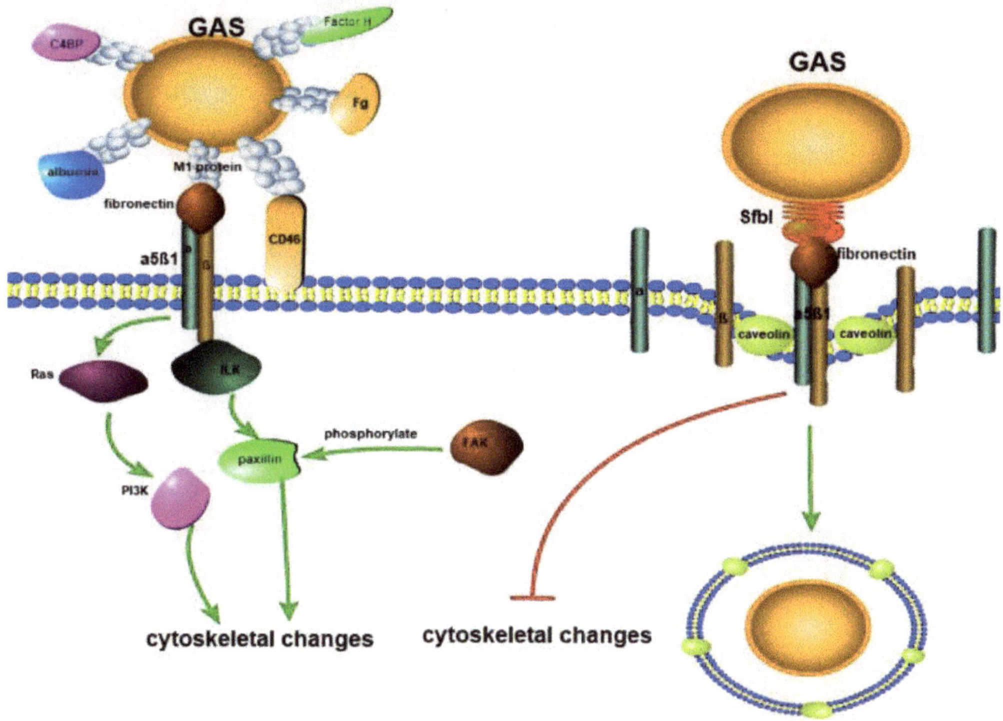

Figure 1. Two distinct mechanisms of GAS invasion. M protein and Sfbl protein are two major invasins of GAS. M1 protein, a subtype of M protein, can bind with fibronectin (Fn). Subsequently, Fn binds to α5β1 integrins, which triggers the intracellular signaling. Activated Ras is able to recruit and activate phosphatidylinositol 3-kinase (PI 3-K), leading to the cytoskeletal changes. Alternatively, integrin-linked kinase (ILK) can bind with β integrin, and transmit signals to paxillin, which provides docking sites for actin binding protein. Also, paxillin can be phosphorylated by focal adhesion kinase (FAK), leading to more docking sites. Additionally, M1 protein can bind to other molecules to facilitate invasion, such as albumin, CD46, C4BP, factor H and fibrinogen (Fg). During Sfbl-mediated invasion, the binding of Sfbl protein to fibronectin leads to the exposure of the RGD region of fibronectin, which facilitates the binding of fibronectin to α5β1 integrins. Then, the intracellular signaling leading to cytoskeleton rearrangement is inhibited, followed by integrin clustering and caveolae aggregation, which contributes to caveolae-mediated invasion.

that plays an important role in the adhesion to epithelial cells is Sfbl, which is known to promote the uptake of streptococci by eukaryotic cells (Hanski and Caparon 1992; Molinari et al. 1997). Rohde et al. (Rohde et al. 2003) demonstrated that invasion by some GAS strains was dependent on Sfbl, a fibronectin-binding protein, which was essential in the caveolae-mediated endocytic process. Here we review the two invasins involved in the process of GAS invasion.

M protein

M protein is considered the major virulence molecule on the surface of *S. pyogenes* (Maxted 1956). It is of great importance to *S. pyogenes* as M protein-knockout mutants with all surface and secretory molecules but lacking M protein can not survive in human blood containing phagocytes (Perezcasal et al. 1992). The M protein forms a dimeric coiled-coil α-helix, consisting of a highly variable terminal region (Brandt et al. 1997; McNamara et al. 2008). It possesses a highly conserved C-terminal domain and a hypervariable N-terminal domain. The N-terminal region extends into the extracellular space and the C-terminal region is anchored in the cell surface (Fischetti 1989, 1991). More than 200 different serotypes of M protein such as M6, M12, and M18 have been identified, classified by distinct N-terminal regions determined by their *emm* gene. Each host individual is likely to be infected by different group A streptococcal types in different stages of their lifetime (Lancefield 1962; Beall et al. 1996). Among all serotypes of M protein, M1 protein is considered as the primary invasin of the highly invasive strain 90-226, as it contributes to about 90% of measured cell invasion by streptococci (Cue et al. 1998). Another experiment that supports this view confirmed that human endothelial cells can phagocytose M1 wild-type GAS but not an M1 knockout mutant (Ochel et al. 2014). M1 protein is involved in cellular invasion by *S. pyogenes*, and a highly invasive M1 strain was reported to rely heavily on the expression of M1 (Cue and Cleary 1997; Jadoun et al. 1997). M1 protein is able to interact with several cellular receptors, including membrane cofactor protein (CD46), a molecule on the surface of most human cells (Liszewski et al. 1991). CD46 was demonstrated to effectively promote epithelial cell invasion by M1 and M3 strains (Rezcallah et al. 2005). In their study, deletion of the CD46 cytoplasmic domain significantly reduced the ability to invade epithelial cells by streptococci (Rezcallah et al. 2005). A recent study showed that CD46 was able to allow GGS_124 penetrate into deep tissues of mice, causing a higher mortality rate (Yoshida et al. 2016). These results suggest that CD46 plays an important role in streptococcal invasion. Other types of cell surface receptors involved in bacterial invasion, are integrins. This was shown by the fact that uptake of the M1 90-226 strain by epithelial cells can be blocked by antibodies against α5β1 integrin (Cue et al. 1998; Dombek et al. 1999). M1 protein is unable to bind to integrins directly, but can bind to the ECM protein fibronectin (Fn), and Fn possesses the ability to bind to integrins (Hynes 1992; Cue et al. 1998). After binding to integrins, M1 protein interacts with Ras, a small G protein, to activate and recruit phosphatidylinositol 3-kinase (PI3K), which in turn induces cytoskeletal changes that are required for GAS invasion (Cantrell 2001; Chan et al. 2002; Purushothaman et al. 2003). This pathway also involves integrin-linked kinase (ILK), paxillin, and focal adhesion kinase (FAK). By providing several docking sites for actin binding protein and signaling molecules, paxillin forms an essential adaptor (Turner 2000). ILK is able to interact with paxillin via a paxillin binding subdomain and can bind to the cytoplasmic region of β integrins by its C-terminus (Nikolopoulos and Turner 2001). FAK can phosphorylate paxillin and creates additional docking sites for cytoskeleton rearrangement (Vindis et al. 2004). The binding of M1 protein to Fn initiates FAK-induced phosphorylation (Wang et al. 2007) and eventually triggered the formation of membrane protrusions involved in phagocytosis of the streptococcal chain, leading to a zipper-like invasion by GAS (Ochel et al. 2014). M protein also binds to other host proteins to mediate the interaction between GAS and host cells, including C4b binding protein (C4BP), factor H, fibrinogen (Fg), and albumin (Ryc et al. 1989; Akesson et al. 1994; Johnsson et al. 1996). Thus, M protein significantly enhances the virulence of GAS by binding or recruiting host proteins.

M3 strains of GAS have been isolated from invasive streptococcal disease patients and also contribute to invasive streptococcal disease (Terao et al. 2002; Sumby et al. 2006). M3 strains have a weak ability to invade HEp-2 cells, which could be explained by their lack M1 protein and SfbI. In contrast to this, M3 strains express the Fn-binding protein of group A streptococci type B (*fbaB*) gene (Terao et al. 2002), which encodes FbaB protein that was demonstrated to facilitate invasion of endothelial cells (Nerlich et al. 2009). The FbaB-mediated invasion consists of four steps: (i) membrane protrusions form at the invasion site; (ii) F-actin accumulates in the vicinity of invading streptococci and Ras-related C3 botulinum toxin substrate 1 (Rac1), a small GTPase, accumulates and becomes activated; (iii) the phagosomal membrane acquires endosomal marker proteins; and (iv) the bacteria are transported to the terminal lysosomal compartment (Nerlich et al. 2009; Amelung et al. 2011).

Sfbl invasion

Fibronectin (Fn) possesses a string of five Fn type I (F1) modules within the N-terminal domain (Desimone et al. 1992), which are functionally recognized by Sfbl protein. Sfbl consists of three major regions: (i) a C-terminal region that contains several fibronectin-binding repeats (FBR region), (ii) a region containing a number of proline-rich repeats (PRR region), and (iii) an N-terminal region with several aromatic amino acid residues (AroD region) (Talay et al. 1994; Towers et al. 2003; Towers et al. 2004). Molinari et al. (Molinari et al. 1997) demonstrated that the fibronectin-binding domains of Sfbl protein play a role in mediation of eukaryotic cell invasion by *S. pyogenes*. Sfbl-dependent invasion involves the use of host cell caveolae (Rohde et al. 2003). With shapes ranging from almost flat to cup-like depressed, to flask-like in some cases (Schlormann et al. 2010), caveolae are invaginations that are present in eukaryotic plasma membranes. They are defined as 60-80nm wide pits that exert several biological functions, including mediating the uptake of bacteria (Parton and del Pozo 2013). One caveola consists of many caveolin 1 (CAV1) proteins, which form the primary structure of caveolae (Pelkmans and Zerial 2005; Parton and del Pozo 2013). Sfbl protein binds to a fibrin-binding fragment of fibronectin by two distinct domains, and eukaryotic cell invasion by streptococci is effectively increased by the cooperation of these two regions (Talay et al. 2000). The Sfbl protein binds to fibronectin via a tandem β-zipper cooperation, in which the fibronectin-binding region (FnBR) forms a sequence that is anti-parallel to the C-terminal sequence of each F1 module (Schwarz-Linek et al. 2003; Bingham et al. 2008). The binding of Sfbl protein to fibronectin involves opening of the RGD region via changes in the fibronectin quaternary structure. This facilitates the binding of the RGD region to α5β1 integrins on the surface of host cells and inhibits intracellular signaling by rearranging the cytoskeleton. Both competitive RGD peptides and antibodies against the β-subunit of the integrin can block the interaction between the RGD-region of fibronectin and integrins on host cell surfaces, inhibiting the invasion of streptococci significantly (Jadoun et al. 1998; Ozeri et al. 1998; Molinari et al. 2000; Ozeri et al. 2001). The tripeptide RGD region in fibronectin and other extracellular ligands, which enables recognition by integins, is not found in SfbI, so the binding of SfbI protein to fibronectin plays an important role in the process of streptococcal invasion. Binding of SfbI to fibronectin leads to subsequent integrin clustering and caveolae aggregation, formation of large encapsulating invaginations, and eventually promotes the uptake of streptococci. Once inside the host cells, the streptococci stay in a membrane-bound compartment in the cytoplasm, termed caveosomes. Contributing to its survival in cytoplasm, these SfbI-expressing streptococci possess a unique ability to bypass the lysosomal degradation machinery of the host cells (Rohde et al. 2003). It was also demonstrated that the uptake efficiency is affected by the amount of bound fibronectin on the bacterial surface (Ozeri et al. 1998). This led to the hypothesis that a threshold of fibronectin bound to integrins is necessary to trigger signaling. Schwarz-Linek et al. discovered that one SfbI protein can bind up to five fibronectin molecules (Schwarz-Linek et al. 2003), which means one streptococcus can bind to a great number of integrins on a host cell, which benefits its invasion. Ezrin, another host protein involved in SfbI-mediated invasion (Tsukita et al. 1997; Rox et al. 2017), consists of a C-terminal domain, an N-terminal domain, a poly-proline region, and an α-helical domain (Tsukita et al. 1997). During the uptake process of bacteria, Ezrin plays an effective role in the signal transduction of host protein kinases and cell receptors, such as PI3K, which acts as an important factor in the invasion pathway (Hirao et al. 1996; Schoenwaelder and Burridge 1999; Bretscher et al. 2002).

S. pneumoniae invasion

In most cases, *S. pneumoniae* primarily adheres to and invades upper respiratory tract cells and subsequently spreads to other tissues through the bloodstream, which can be promoted by sialic acid (Kadioglu et al. 2008; Henriques-Normark and Tuomanen 2013; Iovino et al. 2013; Hatcher et al. 2016). Murine lung invasion by *S. pneumoniae* has been shown to be enhanced by administering sialic acid (Trappetti et al. 2009). After initial adhesion, *S. pneumoniae* can invade cells of the respiratory tract via at least two pathways: (i) *S. pneumoniae* binds to platelet-activating factor receptor (PAFR) on the surface of respiratory epithelial activated by cytokines through phosphorylcholine (ChoP) moieties, and is subsequently engulfed by host cells via the PAFR recycling pathway (Cundell et al. 1995). (ii) Another pathway similar to this is triggered by the binding of the polymeric immunoglobulin receptor (PIGR) to CbpA, followed by pneumococcal translocation through epithelium (Zhang et al. 2000). Binding of platelet-activating factor (PAF) to PAFR facilitates the signal transduction of host cells (Shukla 1992), and binding of *S. pneumoniae* to PAFR promotes the adhesion to and internalization by host endothelial cells (Cundell et al. 1995; Ring et al. 1998; Radin et al. 2005). *S. pneumoniae* has been shown bind to PIGR present on host nasopharyngeal epithelial cells, which enhanced the translocation through epithelium. As antibody against PIGR effectively inhibits the adherence, PIGR is thought to participate in the adhesion process, (Zhang et al. 2000). The two pathways both lead to the release of pneumococci into the interstitium. Consistent with the experiments of Iovino et al., neither of these two pathways involves the disruption of tight junctions between host cells (Iovino et al. 2013). PIGR and CbpA are also involved in brain invasion. A recent study demonstrated that brain invasion by *S. pneumoniae* can be mediated by a pilus-related adhesin (RrgA) that is able to interact with PIGR and platelet endothelial cell adhesion molecule (PECAM-1). Pneumococcal entry into the brain of mice was significantly blocked by antibodies against these two proteins (Iovino et al. 2017). In addition, CbpA can bind to the laminin receptor (LR), which facilitates streptococci crossing the BBB, eventually leading to meningitis (Orihuela et al. 2009). Agarwal et al. (Agarwal et al. 2013) identified another invasion pathway mediated by the host complement protein C1q. C1q acts as a bridge molecule between pneumococci and host cells, by binding to pneumococcal exposed surface proteins via its globular heads and by binding host cell receptors via its N-terminal stalk region, thereby significantly promoting the uptake of *S. pneumoniae* by host epithelial and endothelial cells. Additionally, van Ginkel et al. (van Ginkel et al. 2003) showed that *S. pneumoniae* is able to spread to olfactory nerves from nasopharyngeal epithelium, providing a novel mechanism for *S. pneumonia* invasion of the brain via a non-hematologic route. Streptococcal invasion of the heart also involves the binding of LR and PAFR with their ligands, as demonstrated by the reduction of microlesion formation in $PAFR^{-/-}$ mice or by using antibody against LR. (Brown et al. 2014) (Figure 2)

S. agalactiae invasion

To activate downstream signaling and eventually invade host cells, *S. agalactiae* interacts with proteins on the host cell surface (Beckmann et al. 2002; Cheng et al. 2002; Bolduc and Madoff 2007; De Gaetano et al. 2018). *S. agalactiae* can bind to vitronectin (Vtn), an ECM protein, via its plasminogen binding surface Protein (PbsP) present in the cell wall, and Vtn acts as a bridge molecule between *S. agalactiae* and host cells (De Gaetano et al. 2018). Vtn plays a role in the invasion process of invasive pathogens, and Vtn-mediated binding can initiate integrin-dependent internalization of bacteria (Grashoff et al. 2004; Singh et al. 2010). Vtn is present in extracellular fluids and endogenous secretions, is expressed on the surface of different cell types, and can be upregulated upon injury and inflammation (Hallstrom et al. 2016; Aulakh 2018). PbsP, which is anchored in the cell wall, possesses a plasminogen-binding region and one or two streptococcal surface repeat regions, which contribute to GBS virulence and

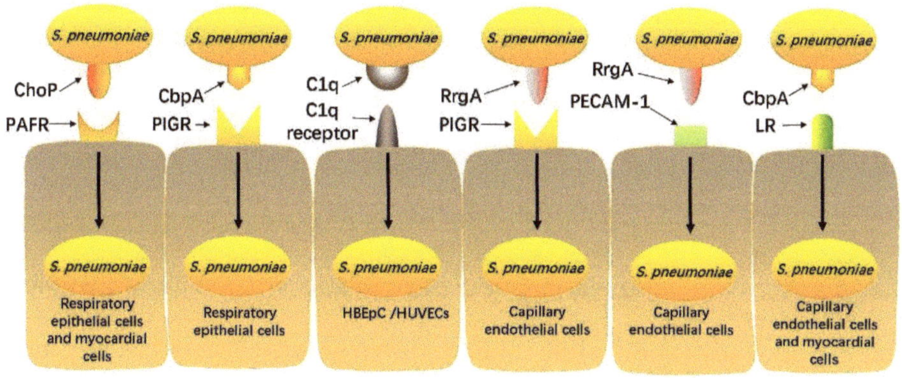

Figure 2. Different molecular mechanisms by which *S. pneumoniae* invades host cells. The invasion of *S. pneumoniae* is dependent on the binding of bacterial surface proteins with host receptors. Different ligand-receptor interactions are involved in different host cell types.

are essential in plasmin-dependent invasion (Bumbaca et al. 2004; Buscetta et al. 2016). Binding of ligands to host-cell receptors also activates Ras, which subsequently triggers the recruitment of PI3K to the membrane of host cells (Reuther and Der 2000). Activated PI3K interacts with GTPases, both of which can be activated by the binding of *S. agalactiae* directly, and are responsible for actin rearrangement (Burnham et al. 2007a; Burnham et al. 2007b). Akt, a serine kinase that can modulate downstream effector molecules is also modulated by PI3K, which eventually affects actin rearrangement (Stokoe 2005; Burnham et al. 2007a); an indispensable step in cell invasion by GBS (Boone and Tyrrell 2012).

ACP, a cell wall anchored protein, is also involved in GBS invasion. The N-terminal domain of ACP has been classified into two structural regions (D1 and D2). D1 includes a β sandwich while D2 consists of 3 antiparallel α helix coils (Auperin et al. 2005). An important function in cell invasion is exerted by the N-terminal domain of ACP, since the soluble form of the N-terminal domain significantly inhibits GBS internalization (Bolduc et al. 2002). Subsequent studies revealed that both D1 and D2 are involved in the invasion process. GBS affinity to α1β1 integrin and its ability to invade cervical epithelial cells was decreased by converting a KTD motif within D1 to KTA (Bolduc and Madoff 2007). Moreover, when a mutation was induced in a glycosaminoglycan-binding site within D2, invasion was blocked (Baron et al. 2007). These results suggest both D1 and D2 can mediate GBS entry, but the mechanisms may be distinct.

Additional studies have shown that fibrinogen-binding proteins (Fbs) were relevant to cell invasion by GBS, including FbsB (Schubert et al. 2004), the Srr that can mediate the uptake of GAS by microvascular endothelial cells (Seo et al. 2012; Seo et al. 2013), and FbsC (Buscetta et al. 2014). A novel molecule named fibronectin-binding protein A (SfbA) has been identified as well, which contributes to GBS invasion of brain endothelial cells and meningitis (Mu et al. 2014). This view is supported by a subsequent study that demonstrated that the ability to breach the BBB and to cause meningitis was reduced when *sfbA* mutant strains were used to infect mice (Stoner et al. 2015).

A significant number of studies report that β-hemolysin/cytolysin (β-H/C) triggers cytolysis of host cells, thereby promoting GBS invasion of epithelial and endothelial barriers, including the BBB, as well as endothelial and epithelial cells of the host lung (Nizet et al. 1996; Nizet et al. 1997; Gibson et al. 1999; Doran et al. 2002; Doran et al. 2003; Liu et al. 2004). β-H/C-deficient GBS mutants showed attenuated virulence to cause infections such as meningitis, sepsis, and pneumonia (Doran et al. 2002; Doran et al. 2003; Hensler et al. 2005). Additionally, several other life-threatening diseases can be induced by β-H/C, including cardiac impairment and liver failure (Ring et al. 2002; Hensler et al. 2008), underscoring the importance of β-H/C in GBS-mediated disease

S. suis invasion

S. suis is able to interact with epidermal cells in wounded skin or intestinal epithelial cells (Gottschalk et al. 2010; Segura et al. 2016), and the gastrointestinal tract is a common entry site both in humans and pigs (Ferrando et al. 2015). Ferrando et al. (Ferrando et al. 2015) demonstrated that *S. suis* was able to cause the disruption of intercellular tight junctions and subsequently cross the intestinal epithelial barrier in humans through a paracellular route. *S. suis* can eventually cause meningitis, after reaching the bloodstream and crossing the BBB, (Fulde and Valentin-Weigand 2013; Haas and Grenier 2018). To cause meningitis, crossing the BBB is a critical step for bacteria, but the mechanism is not clearly known. Several mechanisms by which *S. suis* penetrates the BBB have been shown, such as invasion of the cells making up the BBB via the endocytic pathway and disruption of intercellular tight junctions (Gottschalk et al. 2010; Fittipaldi et al. 2012; Fulde and Valentin-Weigand 2013). Kong et al. (Kong et al. 2017) recently demonstrated that *S. suis* translocation across the BBB was significantly promoted by the interaction of factor H-binding protein of *S. suis* with globotriaosylceramide of host cells (Kong et al. 2017). Another virulence factor of *S. suis* named SLY can trigger pore-formation in host cell membranes and also contributes to the invasion of *S. suis* (Seitz et al. 2013). Norton et al. showed that the invasion of epithelial cells by SLY-negative strains was inhibited (Norton et al. 1999). SLY is thought to contribute to the formation of membrane ruffles, thereby facilitating the invasion into epithelial cells by *S. suis* (Benga et al. 2004; Seitz et al. 2013). In addition to epithelial cells, SLY also plays a role in the invasion of human brain microvascular endothelial cells (HBMEC). Lv et al. (Lv et al. 2014) demonstrated that *S. suis* invasion into HBMEC can be facilitated by SLY, which can activate GTPase Ras homolog gene family member A (RhoA) and Rac1 to remodel the actin cytoskeleton. Nevertheless, since a SLY knockout mutant also disseminates in the host, SLY is not indispensable in *S. suis* infections (Lun et al. 2003). Additionally, by translocation through the blood-cerebrospinal fluid barrier, *S. suis* can invade the host brain, and this process is associated with its adherence to and entry into the choroid plexus epithelial cells and transportation by endocytic vacuoles (Haas and Grenier 2018).

S. mitis invasion

S. mitis can interact with human gingival fibroblasts (HGFs), and its entry into HGFs is associated with both bacterial virulence factors and host proteins (Rukke et al. 2012; Di Giacomo et al. 2013; Cataldi et al. 2016; Di Giulio et al. 2018). The invasion of HGFs by *S. mitis* is mediated by Focal Adhesion Kinase (FAK), integrin β1, and the two cytoskeleton proteins vinculin and F-actin (Cataldi et al. 2016; Di Giulio et al. 2018). Chitlac-nAg silver nanoparticles can enhance the internalization of *S. mitis*, which is thought to be a strategy of bacteria to evade the toxicity of silver. Nevertheless, the effects of Chitlac-nAg on the invasion of *S. mitis* can be inhibited by saliva, and it is hypothesized that proteins within the saliva play a role in inhibiting the uptake of *S. mitis* (Di Giulio et al. 2018). Similar to *S. pneumoniae*, to which it is closely related, *S. mitis* was shown to possess a capsule, which is an important virulence factor (Kilian et al. 2008). The capsule of *S. pneumoniae* has been well studied, and is associated with numerous functions such as evasion of mucus-mediated clearance, and modulation of inflammatory responses (Bootsma et al. 2007; Nelson et al. 2007). Although it was shown that the capsule of *S. mitis* inhibits its adherence to oral epithelial cells (Rukke et al. 2016), the underlying mechanism remains unknown.

Conclusion

Streptococci can form a severe threat to human health, and novel members of streptococci are constantly isolated, such as *Streptococcus tigurinus*, causing infective endocarditis (Zbinden et al. 2012). Recent years, great progress has been made in uncovering the molecular mechanisms of streptococcal adhesion and invasion. Streptococci are difficult to eradicate due to their ability of adherence to and invasion of eukaryotic cells. By expressing distinct adhesins and invasins in different environments, streptococcal infections can be difficult to treat. Both extracellular and intracellular molecules are involved in the adhesion and invasion, and there is a demand for a better understanding of these associated molecules. In spite of the advances in antimicrobial therapy based on the current molecular research, streptococcal infections still cause a high morbidity and mortality from diseases such as meningitis. Thus, identification and targeting of molecular patterns involved in invasive bacterial infections, and pathways by which streptococci adhere to and invade host cells, are crucial in developing novel effective therapeutic strategies.

References

Aas J.A., Paster B.J., Stokes L.N., Olsen I. and Dewhirst F.E. (2005). Defining the normal bacterial flora of the oral cavity. J. Clin. Microbiol. *43*, 5721-5732. https://dx.doi.org/10.1128/Jcm.43.11.5721-5732.2005

Agarwal V., Ahl J., Riesbeck K. and Blom A.M. (2013). An Alternative Role of C1q in Bacterial Infections: Facilitating *Streptococcus pneumoniae* Adherence and Invasion of Host Cells. J. Immunol. *191*, 4235-4245. https://dx.doi.org/10.4049/jimmunol.1300279

Akesson P., Schmidt K.H., Cooney J. and Bjorck L. (1994). M1 protein and protein H: IgGFc- and albumin-binding streptococcal surface proteins encoded by adjacent genes. Biochem. J. *300 (Pt 3)*, 877-886

Al-Farsi F., Al-Busaidi I. and Al-Zeedi K. (2018). Acute *Streptococcus mitis* Sacroiliitis in a Teenager with Unclear Source of Bacteremia: A Case Report and Literature Review. Case. Rep. Infect. Dis. *2018*, 2616787. https://dx.doi.org/10.1155/2018/2616787

Alonsodevelasco E., Verheul A.F.M., Verhoef J. and Snippe H. (1995). *Streptococcus-Pneumoniae* - Virulence Factors, Pathogenesis, and Vaccines. Microbiol. Rev. *59*, 591-602

Amelung S., Nerlich A., Rohde M., Spellerberg B., Cole J.N., Nizet V., Chhatwal G.S. and Talay S.R. (2011). The FbaB-type fibronectin-binding protein of *Streptococcus pyogenes* promotes specific invasion into endothelial cells. Cell. Microbiol. *13*, 1200-1211. https://dx.doi.org/10.1111/j.1462-5822.2011.01610.x

Aulakh G.K. (2018). Neutrophils in the lung: "the first responders". Cell Tissue Res. *371*, 577-588. https://dx.doi.org/10.1007/s00441-017-2748-z

Auperin T.C., Bolduc G.R., Baron M.J., Heroux A., Filman D.J., Madoff L.C. and Hogle J.M. (2005). Crystal structure of the n-terminal domain of the group B Streptococcus alpha C protein. J. Biol. Chem. *280*, 18245-18252. https://dx.doi.org/10.1074/jbc.M412391200

Barnett T.C., Liebl D., Seymour L.M., Gillen C.M., Lim J.Y., Larock C.N., Davies M.R., Schulz B.L., Nizet V., Teasdale R.D., *et al.* (2013). The globally disseminated M1T1 clone of group A Streptococcus evades autophagy for intracellular replication. Cell Host Microbe *14*, 675-682. https://dx.doi.org/10.1016/j.chom.2013.11.003

Baron M.J., Filman D.J., Prophete G.A., Hogle J.M. and Madoff L.C. (2007). Identification of a glycosaminoglycan binding region of the alpha c protein that mediates entry of group B streptococci into host cells. J. Biol. Chem. *282*, 10526-10536. https://dx.doi.org/10.1074/jbc.M608279200

Basaranoglu S.T., Ozsurekci Y., Aykac K., Aycan A.E., Bicakcigil A., Altun B., Sancak B., Cengiz A.B., Kara A. and Ceyhan M. (2018). *Streptococcus mitis/oralis* Causing Blood Stream Infections in Pediatric Patients. Jpn. J. Infect. Dis. https://dx.doi.org/10.7883/yoken.JJID.2018.074

Beall B., Facklam R. and Thompson T. (1996). Sequencing emm-specific PCR products for routine and accurate typing of group A streptococci. J. Clin. Microbiol. *34*, 953-958

Beckmann C., Waggoner J.D., Harris T.O., Tamura G.S. and Rubens C.E. (2002). Identification of novel adhesins from group B streptococci by use of phage display reveals that c5a peptidase mediates fibronectin binding. Infect. Immun. *70*, 3309-3309

Benga L., Goethe R., Rohde M. and Valentin-Weigand P. (2004). Non-encapsulated strains reveal novel insights in invasion and survival of *Streptococcus suis* in epithelial cells. Cell. Microbiol. *6*, 867-881. https://dx.doi.org/10.1111/j.1462-5822.2004.00409.x

Bessen D.E., Sotir C.M., Readdy T.L. and Hollingshead S.K. (1996). Genetic correlates of throat and skin isolates of group A streptococci. J. Infect. Dis. *173*, 896-900

Bingham R.J., Rudino-Pinera E., Meenan N.A.G., Schwarz-Linek U., Turkenburg J.P., Hook M., Garman E.F. and Potts J.R. (2008). Crystal structures of fibronectin-binding sites from *Staphylococcus aureus* FnBPA in complex with fibronectin domains. Proc. Natl. Acad. Sci. U. S. A. *105*, 12254-12258. https://dx.doi.org/10.1073/pnas.0803556105

Bolduc G.R., Baron M.J., Gravekamp C., Lachenauer C.S. and Madoff L.C. (2002). The alpha C protein mediates internalization of group B Streptococcus within human cervical epithelial cells. Cell. Microbiol. *4*, 751-758. https://dx.doi.org/10.1046/j.1462-5822.2002.00227.x

Bolduc G.R. and Madoff L.C. (2007). The group B streptococcal alpha C protein binds alpha(1)beta(1)-integrin through a novel KTD motif that promotes internalization of GBS within human epithelial cells. Microbiology-Sgm *153*, 4039-4049. https://dx.doi.org/10.1099/mic.0.2007/009134-0

Boone T.J. and Tyrrell G.J. (2012). Identification of the actin and plasminogen binding regions of group B streptococcal phosphoglycerate kinase. J. Biol. Chem. *287*, 29035-29044. https://dx.doi.org/10.1074/jbc.M112.361261

Bootsma H.J., Egmont-Petersen M. and Hermans P.W. (2007). Analysis of the in vitro transcriptional response of human pharyngeal epithelial cells to adherent *Streptococcus pneumoniae*: evidence for a distinct response to encapsulated strains. Infect. Immun. *75*, 5489-5499.

https://dx.doi.org/10.1128/IAI.01823-06

Brandt E.R., Hayman W.A., Currie B., Pruksakorn S. and Good M.F. (1997). Human antibodies to the conserved region of the M protein: opsonization of heterologous strains of group A streptococci. Vaccine *15*, 1805-1812. https://dx.doi.org/10.1016/S0264-410x(97)00178-3

Bretscher A., Edwards K. and Fehon R.G. (2002). ERM proteins and merlin: Integrators at the cell cortex. Nat. Rev. Mol. Cell. Bio. *3*, 586-599. https://dx.doi.org/10.1038/nrm882

Brown A.O., Mann B., Gao G.L., Hankins J.S., Humann J., Giardina J., Faverio P., Restrepo M.I., Halade G.V., Mortensen E.M., *et al.* (2014). *Streptococcus pneumoniae* Translocates into the Myocardium and Forms Unique Microlesions That Disrupt Cardiac Function. Plos Pathogens *10*, e1004383. https://dx.doi.org/10.1371/journal.ppat.1004383

Bumbaca D., Littlejohn J.E., Nayakanti H., Rigden D.J., Galperin M.Y. and Jedrzejas M.J. (2004). Sequence analysis and characterization of a novel fibronectin-binding repeat domain from the surface of *Streptococcus pneumoniae*. Omics *8*, 341-356. https://dx.doi.org/10.1089/omi.2004.8.341

Burnham C.A., Shokoples S.E. and Tyrrell G.J. (2007a). Invasion of HeLa cells by group B streptococcus requires the phosphoinositide-3-kinase signalling pathway and modulates phosphorylation of host-cell Akt and glycogen synthase kinase-3. Microbiol. *153*, 4240-4252. https://dx.doi.org/10.1099/mic.0.2007/008417-0

Burnham C.A.D., Shokoples S.E. and Tyrrell G.J. (2007b). Rac1, RhoA, and Cdc42 participate in HeLa cell invasion by group B streptococcus. Fems. Microbiol. Lett. *272*, 8-14. https://dx.doi.org/10.1111/j.1574-6968.2007.00768.x

Buscetta M., Firon A., Pietrocola G., Biondo C., Mancuso G., Midiri A., Romeo L., Galbo R., Venza M., Venza I., *et al.* (2016). PbsP, a cell wall-anchored protein that binds plasminogen to promote hematogenous dissemination of group B Streptococcus. Mol. Microbiol. *101*, 27-41. https://dx.doi.org/10.1111/mmi.13357

Buscetta M., Papasergi S., Firon A., Pietrocola G., Biondo C., Mancuso G., Midiri A., Romeo L., Teti G., Speziale P., *et al.* (2014). FbsC, a Novel Fibrinogen-binding Protein, Promotes *Streptococcus agalactiae*-Host Cell Interactions. J. Biol. Chem. *289*, 21003-21015. https://dx.doi.org/10.1074/jbc.M114.553073

Cantrell D.A. (2001). Phosphoinositide 3-kinase signalling pathways. J. Cell. Sci

114, 1439-1445

Carapetis J.R., Steer A.C., Mulholland E.K. and Weber M. (2005). The global burden of group A streptococcal diseases. Lancet. Infect. Dis. *5*, 685-694. https://dx.doi.org/10.1016/S1473-3099(05)70267-X

Carrascosa M., Perezcastrillon J.L., Sampedro I., Valle R., Cillero L. and Mendez M.A. (1994). Lung Abscess Due to *Streptococcus-Mitis* - Case-Report and Review. Clin. Infect. Dis. *19*, 781-783. https://dx.doi.org/10.1093/clinids/19.4.781

Cataldi A., Gallorini M., Di Giulio M., Guarnieri S., Mariggio M.A., Traini T., Di Pietro R., Cellini L., Marsich E. and Sancilio S. (2016). Adhesion of human gingival fibroblasts/*Streptococcus mitis* co-culture on the nanocomposite system Chitlac-nAg. J. Mater. Sci.-Mater. Med. *27*, 88. https://dx.doi.org/https://dx.doi.org/10.1007/s10856-016-5701-x

Chan T.O., Rodeck U., Chan A.M., Kimmelman A.C., Rittenhouse S.E., Panayotou G. and Tsichlis P.N. (2002). Small GTPases and tyrosine kinases coregulate a molecular switch in the phosphoinositide 3-kinase regulatory subunit. Cancer Cell *1*, 181-191. https://dx.doi.org/10.1016/s1535-6108(02)00033-8

Cheng Q., Stafslien D., Purushothaman S.S. and Cleary P. (2002). The group B streptococcal C5a peptidase is both a specific protease and an invasin. Infec. Immun. *70*, 3309-3309

Cue D., Dombek P.E., Lam H. and Cleary P.P. (1998). *Streptococcus pyogenes* serotype M1 encodes multiple pathways for entry into human epithelial cells. Infec. Immunity *66*, 4593-4601

Cue D.R. and Cleary P.P. (1997). High-frequency invasion of epithelial cells by *Streptococcus pyogenes* can be activated by fibrinogen and peptides containing the sequence RGD. Infec. Immun. *65*, 2759-2764

Cundell D.R., Gerard N.P., Gerard C., Idanpaan-Heikkila I. and Tuomanen E.I. (1995). *Streptococcus pneumoniae* anchor to activated human cells by the receptor for platelet-activating factor. Nature *377*, 435-438. https://dx.doi.org/10.1038/377435a0

Cunningham M.W. (2000). Pathogenesis of group A streptococcal infections. Clin. Microbiol. Rev. *13*, 470-511.

Cywes C. and Wessels M.R. (2001). Group A Streptococcus tissue invasion by CD44-mediated cell signalling. Nature *414*, 648-652. https://dx.doi.org/10.1038/414648a

De Gaetano G.V., Pietrocola G., Romeo L., Galbo R., Lentini G., Giardina M., Biondo C., Midiri A., Mancuso G., Venza M., *et al.* (2018). The *Streptococcus agalactiae* cell wall-anchored protein PbsP mediates adhesion to and invasion of epithelial cells by exploiting the host vitronectin/alphav integrin axis. Mol. Microbiol. *110*, 82-94. https://dx.doi.org/10.1111/mmi.14084

Desimone D.W., Norton P.A. and Hynes R.O. (1992). Identification and Characterization of Alternatively Spliced Fibronectin Messenger-Rnas Expressed in Early Xenopus Embryos. Dev. Biol. *149*, 357-369. https://dx.doi.org/10.1016/0012-1606(92)90291-N

Di G.V., Pacella S., Rapino M., Di Giulio M., Zara S., Pasquantonio G., Cellini L. and Cataldi A. (2013). pPKC alpha regulates through integrin beta 1 human gingival fibroblasts/*Streptococcus mitis* adhesion in response to HEMA. Int. End. J. *46*, 1164-1172. https://dx.doi.org/https://dx.doi.org/10.1111/iej.12113

Di Giulio M., Di Valerio V., Bosco D., Marsich E., Cataldi A., Cellini L. and Sancilio S. (2018). Molecular mechanisms driving *Streptococcus mitis* entry into human gingival fibroblasts in presence of chitlac-nAg and saliva. J. Mater. Sci. Mater. Med. *29*, 36. https://dx.doi.org/10.1007/s10856-018-6040-x

Doern C.D. and Burnham C.A.D. (2010). It's Not Easy Being Green: the Viridans Group Streptococci, with a Focus on Pediatric Clinical Manifestations. J. Clin. Microbiol. *48*, 3829-3835. https://dx.doi.org/10.1128/Jcm.01563-10

Dombek P.E., Cue D., Sedgewick J., Lam H., Ruschkowski S., Finlay B.B. and Cleary P.P. (1999). High-frequency intracellular invasion of epithelial cells by serotype M1 group A streptococci: M1 protein-mediated invasion and cytoskeletal rearrangements. Mol. Microbiol. *31*, 859-870. https://dx.doi.org/10.1128/Jcm.01563-10

Doran K.S., Chang J.C., Benoit V.M., Eckmann L. and Nizet V. (2002). Group B streptococcal beta-hemolysin/cytolysin promotes invasion of human lung epithelial cells and the release of interleukin-8. J. Infect. Dis. *185*, 196-203. https://dx.doi.org/10.1086/338475

Doran K.S., Liu G.Y. and Nizet V. (2003). Group B streptococcal beta-hemolysin/cytolysin activates neutrophil signaling pathways in brain endothelium and contributes to development of meningitis. J. Clin. Invest. *112*, 736-744. https://dx.doi.org/10.1172/JCI17335

Edmond K.M., Kortsalioudaki C., Scott S., Schrag S.J., Zaidi A.K.M., Cousens S. and Heath P.T. (2012). Group B streptococcal disease in infants aged younger than 3 months: systematic review and meta-analysis. Lancet *379*,

547-556. https://dx.doi.org/10.1016/S0140-6736(11)61651-6

Feng Y., Zhang H., Ma Y. and Gao G.F. (2010). Uncovering newly emerging variants of *Streptococcus suis*, an important zoonotic agent. Trends Microbiol. *18*, 124-131. https://dx.doi.org/10.1016/j.tim.2009.12.003

Ferrando M.L., de Greeff A., van Rooijen W.J.M., Stockhofe-Zurwieden N., Nielsen J., Schreur P.J.W., Pannekoek Y., Heuvelink A., van der Ende A., Smith H., *et al.* (2015). Host-pathogen Interaction at the Intestinal Mucosa Correlates With Zoonotic Potential of *Streptococcus suis*. J. Infect. Dis. *212*, 95-105. https://dx.doi.org/10.1093/infdis/jiu813

Fischetti V.A. (1989). Streptococcal M-Protein - Molecular Design and Biological Behavior. Clin. Microbiol. Rev. 2, 285-314

Fischetti V.A. (1991). Streptococcal M-Protein. Sci Am *264*, 58-65. https://dx.doi.org/10.1038/scientificamerican0691-58

Fischetti V.A. (2016) M Protein and Other Surface Proteins on Streptococci. In: Ferretti JJ, Stevens DL, Fischetti VA (eds) Streptococcus pyogenes : Basic Biology to Clinical Manifestations. Oklahoma City (OK),

Fittipaldi N., Segura M., Grenier D. and Gottschalk M. (2012). Virulence factors involved in the pathogenesis of the infection caused by the swine pathogen and zoonotic agent *Streptococcus suis*. Future Microbiol. *7*, 259-279. https://dx.doi.org/10.2217/Fmb.11.149

Fulde M. and Valentin-Weigand P. (2013). Epidemiology and Pathogenicity of Zoonotic Streptococci. Host-Pathogen Interactions in Streptococcal Diseases *368*, 49-81. https://dx.doi.org/10.1007/82_2012_277

Gibson R.L., Nizet V. and Rubens C.E. (1999). Group B streptococcal beta-hemolysin promotes injury of lung microvascular endothelial cells. Pediatr. Res. *45*, 626-634. https://dx.doi.org/10.1203/00006450-199905010-00003

Gilley R.P., Gonzalez-Juarbe N., Shenoy A.T., Reyes L.F., Dube P.H., Restrepo M.I. and Orihuela C.J. (2016). Infiltrated Macrophages Die of Pneumolysin-Mediated Necroptosis following Pneumococcal Myocardial Invasion. Infect. Immun. *84*, 1457-1469. https://dx.doi.org/10.1128/IAI.00007-16

Gottschalk M. and Segura M. (2000). The pathogenesis of the meningitis caused by *Streptococcus suis*: the unresolved questions. Vet. Microbiol. *76*, 259-272

Gottschalk M., Segura M. and Xu J. (2007). *Streptococcus suis* infections in humans: the Chinese experience and the situation in North America. Anim. Health. Res. Rev. *8*, 29-45.

https://dx.doi.org/10.1017/S1466252307001247

Gottschalk M., Xu J., Calzas C. and Segura M. (2010). *Streptococcus suis*: a new emerging or an old neglected zoonotic pathogen? Future Microbiol. *5*, 371-391. https://dx.doi.org/10.2217/fmb.10.2

Goyette-Desjardins G., Auger J.P., Xu J., Segura M. and Gottschalk M. (2014). *Streptococcus suis*, an important pig pathogen and emerging zoonotic agent-an update on the worldwide distribution based on serotyping and sequence typing. Emerg. Microbes Infec. *3*. https://dx.doi.org/10.1038/emi.2014.45

Grashoff C., Thievessen I., Lorenz K., Ussar S. and Fassler R. (2004). Integrin-linked kinase: integrin's mysterious partner. Curr. Opin. Cell Biol. *16*, 565-571. https://dx.doi.org/10.1016/j.ceb.2004.07.004

Haas B. and Grenier D. (2018). Understanding the virulence of *Streptococcus suis*: A veterinary, medical, and economic challenge. Med. Mal. Infect. *48*, 159-166. https://dx.doi.org/10.1016/j.medmal.2017.10.001

Haas B., Vaillancourt K., Bonifait L., Gottschalk M. and Grenier D. (2015). Hyaluronate lyase activity of *Streptococcus suis* serotype 2 and modulatory effects of hyaluronic acid on the bacterium's virulence properties. BMC Res. Notes *8*, 722. https://dx.doi.org/10.1186/s13104-015-1692-9

Hallstrom T., Singh B., Kraiczy P., Hammerschmidt S., Skerka C., Zipfel P.F. and Riesbeck K. (2016). Conserved Patterns of Microbial Immune Escape: Pathogenic Microbes of Diverse Origin Target the Human Terminal Complement Inhibitor Vitronectin via a Single Common Motif. Plos One *11*, e0147709. https://dx.doi.org/ARTN 10.1371/journal.pone.0147709

Hanski E. and Caparon M. (1992). Protein F, a fibronectin-binding protein, is an adhesin of the group A streptococcus *Streptococcus pyogenes*. Proc. Natl. Acad. Sci. U. S. A. *89*, 6172-6176

Hatcher B.L., Hale J.Y. and Briles D.E. (2016). Free Sialic Acid Acts as a Signal That Promotes *Streptococcus pneumoniae* Invasion of Nasal Tissue and Nonhematogenous Invasion of the Central Nervous System. Infec. Immun. *84*, 2607-2615. https://dx.doi.org/10.1128/Iai.01514-15

Henriques-Normark B. and Tuomanen E.I. (2013). The Pneumococcus: Epidemiology, Microbiology, and Pathogenesis. Csh. Perspect. Med. *3* ,a010215.

Hensler M.E., Liu G.Y., Sobczak S., Benirschke K., Nizet V. and Heldt G.P. (2005). Virulence role of group B Streptococcus beta-hemolysin/cytolysin in a neonatal rabbit model of early-onset pulmonary infection. J. Infect. Dis.

191, 1287-1291. https://dx.doi.org/10.1086/428946

Hensler M.E., Miyamoto S. and Nizet V. (2008). Group B Streptococcal beta-Hemolysin/Cytolysin Directly Impairs Cardiomyocyte Viability and Function. Plos One *3*, e2446.

Hickman M.E., Rench M.A., Ferrieri P. and Baker C.J. (1999). Changing epidemiology of group B streptococcal colonization. Pediatrics *104*, 203-209. https://dx.doi.org/10.1542/peds.104.2.203

Hirao M., Sato N., Kondo T., Yonemura S., Monden M., Sasaki T., Takai Y., Tsukita S. and Tsukita S. (1996). Regulation mechanism of ERM (ezrin/radixin/moesin) protein/plasma membrane association: Possible involvement of phosphatidylinositol turnover and Rho-dependent signaling pathway. J. Cell Biol. *135*, 37-51. https://dx.doi.org/10.1083/jcb.135.1.37

Hynes R.O. (1992). Integrins - Versatility, Modulation, and Signaling in Cell-Adhesion. Cell *69*, 11-25. https://dx.doi.org/10.1016/0092-8674(92)90115-S

Ikebe T., Tominaga K., Shima T., Okuno R., Kubota H., Ogata K., Chiba K., Katsukawa C., Ohya H., Tada Y.*, et al.* (2015). Increased prevalence of group A streptococcus isolates in streptococcal toxic shock syndrome cases in Japan from 2010 to 2012. Epidemiol. Infect. *143*, 864-872. https://dx.doi.org/10.1017/S0950268814001265

Iovino F., Engelen-Lee J.Y., Brouwer M., van de Beek D., van der Ende A., Seron M.V., Mellroth P., Muschiol S., Bergstrand J., Widengren J.*, et al.* (2017). pIgR and PEC AM-1 bind to pneumococcal adhesins RrgA and PspC mediating bacterial brain invasion. J. Exp. Med. *214*, 1619-1630. https://dx.doi.org/10.1084/jem.20161668

Iovino F., Orihuela C.J., Moorlag H.E., Molema G. and Bijlsma J.J.E. (2013). Interactions between Blood-Borne *Streptococcus pneumoniae* and the Blood-Brain Barrier Preceding Meningitis. Plos One *8*, e68408.

Iovino F., Seinen J., Henriques-Normark B. and van Dijl J.M. (2016). How Does *Streptococcus pneumoniae* Invade the Brain? Trends Microbiol. *24*, 307-315. https://dx.doi.org/10.1016/j.tim.2015.12.012

Jackson S.J., Steer A.C. and Campbell H. (2011). Systematic Review: Estimation of global burden of non-suppurative sequelae of upper respiratory tract infection: rheumatic fever and post-streptococcal glomerulonephritis. Trop. Med. Int. Health *16*, 2-11. https://dx.doi.org/10.1111/j.1365-3156.2010.02670.x

Jadoun J., Burstein E., Hanski E. and Sela S. (1997). Proteins M6 and F1 are required for efficient invasion of group A streptococci into cultured epithelial cells. Adv. Exp. Med. Biol. *418*, 511-515

Jadoun J., Ozeri V., Burstein E., Skutelsky E., Hanski E. and Sela S. (1998). Protein F1 is required for efficient entry of *Streptococcus pyogenes* into epithelial cells. Journal of Infectious Diseases *178*, 147-158. https://dx.doi.org/Doi 10.1086/515589

Johnsson E., Thern A., Dahlback B., Heden L.O., Wikstrom M. and Lindahl G. (1996). A highly variable region in members of the streptococcal M protein family binds the human complement regulator C4BP. J. Immunol. *157*, 3021-3029

Johnston C., Hinds J., Smith A., van der Linden M., Van Eldere J. and Mitchell T.J. (2010). Detection of Large Numbers of Pneumococcal Virulence Genes in Streptococci of the Mitis Group. J. Clin. Microbiol. *48*, 2762-2769. https://dx.doi.org/10.1128/Jcm.01746-09

Kadioglu A., Weiser J.N., Paton J.C. and Andrew P.W. (2008). The role of *Streptococcus pneumoniae* virulence factors in host respiratory colonization and disease. Nature Rev. Microbiol. *6*, 288-301. https://dx.doi.org/10.1038/nrmicro1871

Kerdsin A., Oishi K., Sripakdee S., Boonkerd N., Polwichai P., Nakamura S., Uchida R., Sawanpanyalert P. and Dejsirilert S. (2009). Clonal dissemination of human isolates of *Streptococcus suis* serotype 14 in Thailand. J. Med. Microbiol. *58*, 1508-1513. https://dx.doi.org/10.1099/jmm.0.013656-0

Kilian M., Poulsen K., Blomqvist T., Havarstein L.S., Bek-Thomsen M., Tettelin H. and Sorensen U.B. (2008). Evolution of *Streptococcus pneumoniae* and its close commensal relatives. PLoS One *3*, e2683. https://dx.doi.org/10.1371/journal.pone.0002683

Kiska D.L., Thiede B., Caracciolo J., Jordan M., Johnson D., Kaplan E.L., Gruninger R.P., Lohr J.A., Gilligan P.H. and Denny F.W., Jr. (1997). Invasive group A streptococcal infections in North Carolina: epidemiology, clinical features, and genetic and serotype analysis of causative organisms. J. Infect. Dis. *176*, 992-1000

Kong D.C., Chen Z., Wang J.P., Lv Q.Y., Jiang H., Zheng Y.L., Xu M.K., Zhou X.Y., Hao H.J. and Jiang Y.Q. (2017). Interaction of factor H-binding protein of *Streptococcus suis* with globotriaosylceramide promotes the development of meningitis. Virulence *8*, 1290-1302. https://dx.doi.org/10.1080/21505594.2017.1317426

Kwinn L.A. and Nizet V. (2007). How group A Streptococcus circumvents host phagocyte defenses. Future Microbiol. *2*, 75-84. https://dx.doi.org/10.2217/17460913.2.1.75

Lancefield R.C. (1962). CURRENT KNOWLEDGE OF TYPE-SPECIFIC M ANTIGENS OF GROUP A STREPTOCOCCI. J. Immunol. *89*, 307

Landwehr-Kenzel S. and Henneke P. (2014). Interaction of *Streptococcus agalactiae* and Cellular Innate Immunity in Colonization and Disease. Front. Immunol. *5*, 519. https://dx.doi.org/10.3389/fimmu.2014.00519

LaPenta D., Framson P., Nizet V. and Rubens C. (1997) Epithelial cell invasion by group B streptococci is important to virulence. In: Horaud T, Bouvet A, Leclercq R, deMontclos H, Sicard M (eds) Streptococci and the Host, vol 418. Adv. Exp. Med. Biol. pp 631-634

Liang Y.M., Liu X.R., Chang H.S., Ji L.L., Huang G.Y., Fu Z., Zheng Y.J., Wang L.B., Li C.R., Shen Y., *et al.* (2012). Epidemiological and molecular characteristics of clinical isolates of *Streptococcus pyogenes* collected between 2005 and 2008 from Chinese children. J. Med. Microbiol. *61*, 975-983. https://dx.doi.org/10.1099/jmm.0.042309-0

Libster R., Edwards K.M., Levent F., Edwards M.S., Rench M.A., Castagnini L.A., Cooper T., Sparks R.C., Baker C.J. and Shah P.E. (2012). Long-term outcomes of group B streptococcal meningitis. Pediatrics *130*, e8-15. https://dx.doi.org/10.1542/peds.2011-3453

Liszewski M.K., Post T.W. and Atkinson J.P. (1991). Membrane Cofactor Protein (Mcp or Cd46) - Newest Member of the Regulators of Complement Activation Gene-Cluster. Annu. Rev. Immunol. *9*, 431-455. https://dx.doi.org/DOI 10.1146/annurev.iy.09.040191.002243

Liu G.Y., Doran K.S., Lawrence T., Turkson N., Puliti M., Tissi L. and Nizet V. (2004). Sword and shield: linked group B streptococcal beta-hemolysin/cytolysin and carotenoid pigment function to subvert host phagocyte defense. Proc. Natl. Acad. Sci. U. S. A. *101*, 14491-14496. https://dx.doi.org/10.1073/pnas.0406143101

Luca-Harari B., Darenberg J., Neal S., Siljander T., Strakova L., Tanna A., Creti R., Ekelund K., Koliou M., Tassios P.T., *et al.* (2009). Clinical and microbiological characteristics of severe *Streptococcus pyogenes* disease in Europe. J. Clin. Microbiol. *47*, 1155-1165. https://dx.doi.org/10.1128/JCM.02155-08

Lun S.C., Perez-Casal J., Connor W. and Willson P.J. (2003). Role of suilysin in pathogenesis of *Streptococcus suis* capsular serotype 2. Microb. Pathogenes. *34*, 27-37. https://dx.doi.org/10.1016/S0882-4010(02)00192-4

Lv Q.Y., Hao H.J., Bi L.L., Zheng Y.L., Zhou X.Y. and Jiang Y.Q. (2014). Suilysin remodels the cytoskeletons of human brain microvascular endothelial cells by activating RhoA and Rac1 GTPase. Protein & Cell *5*, 261-264. https://dx.doi.org/10.1007/s13238-014-0037-0

Mai N.T., Hoa N.T., Nga T.V., Linh le D., Chau T.T., Sinh D.X., Phu N.H., Chuong L.V., Diep T.S., Campbell J., *et al.* (2008). *Streptococcus suis* meningitis in adults in Vietnam. Clin. Infect. Dis. *46*, 659-667. https://dx.doi.org/10.1086/527385

Maisey H.C., Doran K.S. and Nizet V. (2008). Recent advances in understanding the molecular basis of group B Streptococcus virulence. Expert. Rev. Mol. Med. *10*, e27. https://dx.doi.org/10.1017/S1462399408000811

Maxted W.R. (1956). The indirect bactericidal test as a means of identifying antibody to the M antigen of *Streptococcus pyogenes*. Br. J. Exp .Pathol. *37*, 415-422

McNamara C., Zinkernagel A.S., Macheboeuf P., Cunningham M.W., Nizet V. and Ghosh P. (2008). Coiled-coil irregularities and instabilities in group A Streptococcus M1 are required for virulence. Science *319*, 1405-1408. https://dx.doi.org/10.1126/science.1154470

Melin P. and Efstratiou A. (2013). Group B streptococcal epidemiology and vaccine needs in developed countries. Vaccine *31 Suppl 4*, D31-42. https://dx.doi.org/10.1016/j.vaccine.2013.05.012

Mitchell J. (2011). *Streptococcus mitis*: walking the line between commensalism and pathogenesis. Mol. Oral Microbiol. *26*, 89-98. https://dx.doi.org/10.1111/j.2041-1014.2010.00601.x

Molinari G., Rohde M., Guzman C.A. and Chhatwal G.S. (2000). Two distinct pathways for the invasion of *Streptococcus pyogenes* in non-phagocytic cells. Cell. Microbiol. *2*, 145-154. https://dx.doi.org/DOI 10.1046/j.1462-5822.2000.00040.x

Molinari G., Talay S.R., Valentin-Weigand P., Rohde M. and Chhatwal G.S. (1997). The fibronectin-binding protein of Streptococcus pyogenes, SfbI, is involved in the internalization of group A streptococci by epithelial cells. Infect. Immun. *65*, 1357-1363

Mu R., Kim B.J., Paco C., Del Rosario Y., Courtney H.S. and Doran K.S. (2014). Identification of a Group B Streptococcal Fibronectin Binding Protein, SfbA, That Contributes to Invasion of Brain Endothelium and Development of Meningitis. Infect. Immun. *82*, 2276-2286. https://dx.doi.org/10.1128/Iai.01559-13

Nelson A.L., Roche A.M., Gould J.M., Chim K., Ratner A.J. and Weiser J.N. (2007).

Capsule enhances pneumococcal colonization by limiting mucus-mediated clearance. Infect. Immun. *75*, 83-90. https://dx.doi.org/10.1128/IAI.01475-06

Nerlich A., Rohde M., Talay S.R., Genth H., Just I. and Chhatwal G.S. (2009). Invasion of Endothelial Cells by Tissue-invasive M3 Type Group A Streptococci Requires Src Kinase and Activation of Rac1 by a Phosphatidylinositol 3-Kinase-independent Mechanism. J. Biol. Chem. *284*, 20319-20328. https://dx.doi.org/10.1074/jbc.M109.016501

Nguyen T.H.M., Ngo T.H., Tran V.T.N., Le D.L., Tran T.H.C., Dinh X.S., Nguyen H.P., Ly V.C., To S.D., James C.*, et al.* (2008). *Streptococcus suis* meningitis in adults in Vietnam. Clin. Infect. Dis. *46*, 659-667. https://dx.doi.org/10.1086/527385

Nikolopoulos S.N. and Turner C.E. (2001). Integrin-linked kinase (ILK) binding to paxillin LD1 motif regulates ILK localization to focal adhesions. J. Biol. Chem. *276*, 23499-23505. https://dx.doi.org/.10.1074/jbc.M102163200

Nitsche-Schmitz D.P., Rohde M. and Chhatwal G.S. (2007). Invasion mechanisms of Gram-positive pathogenic cocci. Thromb Haemost *98*, 488-496

Nizet V. (2007). Understanding how leading bacterial pathogens subvert innate immunity to reveal novel therapeutic targets. J. Allergy Clin. Immun. *120*, 13-22. https://dx.doi.org/10.1016/j.jaci.2007.06.005

Nizet V., Gibson R.L., Chi E.Y., Framson P.E., Hulse M. and Rubens C.E. (1996). Group B streptococcal beta-hemolysin expression is associated with injury of lung epithelial cells. Infect. Immun. *64*, 3818-3826

Nizet V., Gibson R.L. and Rubens C.E. (1997). The role of group B streptococci beta-hemolysin expression in newborn lung injury. Adv. Exp. Med. Biol. *418*, 627-630

Norton P.M., Rolph C., Ward P.N., Bentley R.W. and Leigh J.A. (1999). Epithelial invasion and cell lysis by virulent strains of *Streptococcus suis* is enhanced by the presence of suilysin. Fems Immunol. Med. Microbiol. *26*, 25-35. https://dx.doi.org/10.1111/j.1574-695X.1999.tb01369.x

Ochel A., Rohde M., Chhatwal G.S. and Talay S.R. (2014). The M1 Protein of Streptococcus pyogenes Triggers an Innate Uptake Mechanism into Polarized Human Endothelial Cells. J. Innate Immun. *6*, 585-596. https://dx.doi.org/10.1159/000358085

Okura M., Osaki M., Nomoto R., Arai S., Osawa R., Sekizaki T. and Takamatsu D. (2016). Current Taxonomical Situation of *Streptococcus suis*. Pathogens *5,* 45. https://dx.doi.org/10.3390/pathogens5030045

Okwumabua O., Abdelmagid O. and Chengappa M.M. (1999). Hybridization analysis of the gene encoding a hemolysin (suilysin) of *Streptococcus suis* type 2: evidence for the absence of the gene in some isolates. FEMS Microbiol. Lett. *181*, 113-121. https://dx.doi.org/10.1111/j.1574-6968.1999.tb08833.x

Olsen R.J. and Musser J.M. (2010). Molecular Pathogenesis of Necrotizing Fasciitis. Annu. Rev. Pathol-Mech. *5*, 1-31. https://dx.doi.org/10.1146/annurev-pathol-121808-102135

Orihuela C.J., Mahdavi J., Thornton J., Mann B., Wooldridge K.G., Abouseada N., Oldfield N.J., Self T., Ala'Aldeen D.A.A. and Tuomaneni E.I. (2009). Laminin receptor initiates bacterial contact with the blood brain barrier in experimental meningitis models. J. Clin. Invest. *119*, 1638-1646. https://dx.doi.org/10.1172/Jci36759

Ozeri V., Rosenshine I., Ben-Ze'ev A., Bokoch G.M., Jou T.S. and Hanski E. (2001). De novo formation of focal complex-like structures in host cells by invading Streptococci. Molecul. Microbiol. *41*, 561-573. https://dx.doi.org/10.1046/j.1365-2958.2001.02535.x

Ozeri V., Rosenshine I., Mosher D.F., Fassler R. and Hanski E. (1998). Roles of integrins and fibronectin in the entry of *Streptococcus pyogenes* into cells via protein F1. Molecul. Microbiol. *30*, 625-637. https://dx.doi.org/10.1046/j.1365-2958.1998.01097.x

Parton R.G. and del Pozo M.A. (2013). Caveolae as plasma membrane sensors, protectors and organizers. Nat. Rev. Mol. Cell Bio. *14*, 98-112. https://dx.doi.org/10.1038/nrm3512

Pearce C., Bowden G.H., Evans M., Fitzsimmons S.P., Johnson J., Sheridan M.J., Wientzen R. and Cole M.F. (1995). Identification of Pioneer Viridans Streptococci in the Oral Cavity of Human Neonates. J. Med. Microbiol. *42*, 67-72. https://dx.doi.org/10.1099/00222615-42-1-67

Pelkmans L. and Zerial M. (2005). Kinase-regulated quantal assemblies and kiss-and-run recycling of caveolae. Nature *436*, 128-133. https://dx.doi.org/10.1038/nature03866

Perezcasal J., Caparon M.G. and Scott J.R. (1992). INTRODUCTION OF THE EMM6 GENE INTO AN EMM-DELETED STRAIN OF STREPTOCOCCUS-PYOGENES RESTORES ITS ABILITY TO RESIST PHAGOCYTOSIS. Res. Microbiol. *143*, 549-558. https://dx.doi.org/10.1016/0923-2508(92)90112-2

Pimentel B.A., Martins C.A., Mendonca J.C., Miranda P.S., Sanches G.F., Mattos-Guaraldi A.L. and Nagao P.E. (2016). *Streptococcus agalactiae* infection in cancer patients: a five-year study. Eur. J. Clin. Microbiol. Infect. Dis. *35*, 927-933. https://dx.doi.org/10.1007/s10096-016-2617-9

Purushothaman S.S., Wang B.N. and Cleary P.P. (2003). M1 protein triggers a phosphoinositide cascade for group A Streptococcus invasion of epithelial cells. Infec. Immun. *71*, 5823-5830. https://dx.doi.org/10.1128/Iai.71.10.5823-5830.2003

Radin J.N., Orihuela C.J., Murti G., Guglielmo C., Murray P.J. and Tuomanen E.I. (2005). beta-Arrestin 1 participates in platelet-activating factor receptor-mediated endocytosis of *Streptococcus pneumoniae*. Infect. Immun. *73*, 7827-7835. https://dx.doi.org/10.1128/IAI.73.12.7827-7835.2005

Rajagopal L. (2009). Understanding the regulation of Group B Streptococcal virulence factors. Future Microbiol. *4*, 201-221. https://dx.doi.org/10.2217/17460913.4.2.201

Reuther G.W. and Der C.J. (2000). The Ras branch of small GTPases: Ras family members don't fall far from the tree. Curr. Opin. Cell Biol. *12*, 157-165

Rezcallah M.S., Hodges K., Gill D.B., Atkinson J.P., Wang B.N. and Cleary P.P. (2005). Engagement of CD46 and alpha 5 beta 1 integrin by group A streptococci is required for efficient invasion of epithelial cells. Cell. Microbiol. *7*, 645-653. https://dx.doi.org/10.1111/j.1462-5822.2005.00497.x

Ring A., Braun J.S., Pohl J., Nizet V., Stremmel W. and Shenep J.L. (2002). Group B streptococcal beta-hemolysin induces mortality and liver injury in experimental sepsis. J. Infect. Dis. *185*, 1745-1753. https://dx.doi.org/10.1086/340818

Ring A., Weiser J.N. and Tuomanen E.I. (1998). Pneumococcal trafficking across the blood-brain barrier. Molecular analysis of a novel bidirectional pathway. J. Clin. Invest. *102*, 347-360. https://dx.doi.org/10.1172/JCI2406

Rohde M., Muller E., Chhatwal G.S. and Talay S.R. (2003). Host cell caveolae act as an entry-port for group A streptococci. Cell. Microbiol. *5*, 323-342. https://dx.doi.org/10.1046/j.1462-5822.2003.00279.x

Rosa-Fraile M., Dramsi S. and Spellerberg B. (2014). Group B streptococcal haemolysin and pigment, a tale of twins. Fems Microbiol. Rev. *38*, 932-946. https://dx.doi.org/10.1111/1574-6976.12071

Rox K., Rohde M., Chhatwal G.S. and Muller R. (2017). Disorazoles Block Group A Streptococcal Invasion into Epithelial Cells Via Interference with the Host Factor Ezrin. Cell Chem. Biol. *24*, 159-170

Rukke H.V., Engen S.A., Schenck K. and Petersen F.C. (2016). Capsule expression in *Streptococcus mitis* modulates interaction with oral keratinocytes and alters susceptibility to human antimicrobial peptides. Mol. Oral Microbiol. *31*, 302-313. https://dx.doi.org/10.1111/omi.12123

Rukke H.V., Hegna I.K. and Petersen F.C. (2012). Identification of a functional capsule locus in *Streptococcus mitis*. Mol. Oral Microbiol. *27*, 95-108. https://dx.doi.org/10.1111/j.2041-1014.2011.00635.x

Rukke H.V., Kalluru R.S., Repnik U., Gerlini A., Jose R.J., Periselneris J., Marshall H., Griffiths G., Oggioni M.R., Brown J.S., *et al.* (2014). Protective role of the capsule and impact of serotype 4 switching on *Streptococcus mitis*. Infect. Immun. *82*, 3790-3801. https://dx.doi.org/10.1128/IAI.01840-14

Ryc M., Beachey E.H. and Whitnack E. (1989). Ultrastructural localization of the fibrinogen-binding domain of streptococcal M protein. Infect. Immun. *57*, 2397-2404

Schlormann W., Steiniger F., Richter W., Kaufmann R., Hause G., Lemke C. and Westermann M. (2010). The shape of caveolae is omega-like after glutaraldehyde fixation and cup-like after cryofixation. Histochem. Cell Biol. *133*, 223-228. https://dx.doi.org/10.1007/s00418-009-0651-8

Schoenwaelder S.M. and Burridge K. (1999). Bidirectional signaling between the cytoskeleton and integrins. Curr. Opin. Cell Biol. *11*, 274-286. https://dx.doi.org/10.1016/S0955-0674(99)80037-4

Schubert A., Zakikhany K., Pietrocola G., Meinke A., Speziale P., Eikmanns B.J. and Reinscheid D.J. (2004). The fibrinogen receptor FbsA promotes adherence of *Streptococcus agalactiae* to human epithelial cells. Infect. Immun. *72*, 6197-6205. https://dx.doi.org/10.1128/IAI.72.11.6197-6205.2004

Schuchat A. (1998). Epidemiology of group B streptococcal disease in the United States: shifting paradigms. Clin. Microbiol. Rev. *11*, 497-513

Schwarz-Linek U., Werner J.M., Pickford A.R., Gurusiddappa S., Kim J.H., Pilka E.S., Briggs J.A.G., Gough T.S., Hook M., Campbell I.D., *et al.* (2003). Pathogenic bacteria attach to human fibronectin through a tandem beta-zipper. Nature *423*, 177-181. https://dx.doi.org/10.1038/nature0189

Segura M., Calzas C., Grenier D. and Gottschalk M. (2016). Initial steps of the pathogenesis of the infection caused by *Streptococcus suis*: fighting against nonspecific defenses. FEBS Lett. *590*, 3772-3799. https://dx.doi.org/10.1002/1873-3468.12364

Segura M., Fittipaldi N., Calzas C. and Gottschalk M. (2017). Critical *Streptococcus suis* Virulence Factors: Are They All Really Critical? Trends. Microbiol. *25*, 585-599. https://dx.doi.org/10.1016/j.tim.2017.02.005

Seitz M., Baums C.G., Neis C., Benga L., Fulde M., Rohde M., Goethe R. and

Valentin-Weigand P. (2013). Subcytolytic effects of suilysin on interaction of *Streptococcus suis* with epithelial cells. Vet. Microbiol. *167*, 584-591. https://dx.doi.org/10.1016/j.vetmic.2013.09.010

Sellers B.J., Woods M.L., Morris S.E. and Saffle J.R. (1996). Necrotizing group A streptococcal infections associated with streptococcal toxic shock syndrome. Am. J. Surg. *172*, 523-527; discussion 527-528. https://dx.doi.org/10.1016/S0002-9610(96)00248-6

Seo H.S., Minasov G., Seepersaud R., Doran K.S., Dubrovska I., Shuvalova L., Anderson W.F., Iverson T.M. and Sullam P.M. (2013). Characterization of fibrinogen binding by glycoproteins Srr1 and Srr2 of *Streptococcus agalactiae*. J. Biol. Chem. *288*, 35982-35996. https://dx.doi.org/10.1074/jbc.M113.513358

Seo H.S., Mu R., Kim B.J., Doran K.S. and Sullam P.M. (2012). Binding of glycoprotein Srr1 of *Streptococcus agalactiae* to fibrinogen promotes attachment to brain endothelium and the development of meningitis. PLoS Pathog. *8*, e1002947. https://dx.doi.org/10.1371/journal.ppat.1002947

Shak J.R., Vidal J.E. and Klugman K.P. (2013). Influence of bacterial interactions on pneumococcal colonization of the nasopharynx. Trends. Microbiol. *21*, 129-135. https://dx.doi.org/10.1016/j.tim.2012.11.005

Shukla S.D. (1992). Platelet-Activating-Factor Receptor and Signal Transduction Mechanisms. Faseb J. *6*, 2296-2301

Siemens N., Patenge N., Otto J., Fiedler T. and Kreikemeyer B. (2011). *Streptococcus pyogenes* M49 Plasminogen/Plasmin Binding Facilitates Keratinocyte Invasion via Integrin-Integrin-linked Kinase (ILK) Pathways and Protects from Macrophage Killing. J. Biol. Chem. *286*, 21612-21622. https://dx.doi.org/10.1074/jbc.M110.202671

Singh B., Su Y.C. and Riesbeck K. (2010). Vitronectin in bacterial pathogenesis: a host protein used in complement escape and cellular invasion. Mol. Microbiol. *78*, 545-560. https://dx.doi.org/10.1111/j.1365-2958.2010.07373.x

Smeesters P.R., McMillan D.J. and Sriprakash K.S. (2010). The streptococcal M protein: a highly versatile molecule. Trends. Microbiol. *18*, 275-282. https://dx.doi.org/10.1016/j.tim.2010.02.007

Steer A.C., Lamagni T., Curtis N. and Carapetis J.R. (2012). Invasive group a streptococcal disease: epidemiology, pathogenesis and management. Drugs *72*, 1213-1227. https://dx.doi.org/10.2165/11634180-000000000-00000

Stevens D.L. and Bryant A.E. (2016) Severe Group A Streptococcal Infections. In:

Ferretti JJ, Stevens DL, Fischetti VA (eds) *Streptococcus pyogenes* : Basic Biology to Clinical Manifestations. Oklahoma City (OK),

Stokoe D. (2005). The phosphoinositide 3-kinase pathway and cancer. Expert Rev. Mol. Med. *7*, 1-22. https://dx.doi.org/10.1017/S1462399405009361

Stoner T.D., Weston T.A., Trejo J. and Doran K.S. (2015). Group B Streptococcal Infection and Activation of Human Astrocytes. Plos One *10*, e0128431. https://dx.doi.org/10.1371/journal.pone.0128431

Sumby P., Whitney A.R., Graviss E.A., Deleo F.R. and Musser J.M. (2006). Genome-wide analysis of group A streptococci reveals a mutation that modulates global phenotype and disease specificity. Plos Pathog. *2*, 41-49. https://dx.doi.org/10.1371/journal.ppat.0020005

Talay S.R., Valentinweigand P., Timmis K.N. and Chhatwal G.S. (1994). Domain-Structure and Conserved Epitopes of Sfb Protein, the Fibronectin-Binding Adhesin of *Streptococcus-Pyogenes*. Mol. Microbiol. *13*, 531-539. https://dx.doi.org/10.1111/j.1365-2958.1994.tb00448.x

Talay S.R., Zock A., Rohde M., Molinari G., Oggioni M., Pozzi G., Guzman C.A. and Chhatwal G.S. (2000). Co-operative binding of human fibronectin to Sfbl protein triggers streptococcal invasion into respiratory epithelial cells. Cell. Microbiol. *2*, 521-535. https://dx.doi.org/10.1046/j.1462-5822.2000.00076.x

Tang J., Wang C., Feng Y., Yang W., Song H., Chen Z., Yu H., Pan X., Zhou X., Wang H., *et al.* (2006). Streptococcal toxic shock syndrome caused by *Streptococcus suis* serotype 2. Plos Med. *3*, e151. https://dx.doi.org/10.1371/journal.pmed.0030151

Terao Y., Kawabata S., Nakata M., Nakagawa I. and Hamada S. (2002). Molecular characterization of a novel fibronectin-binding protein of *Streptococcus pyogenes* strains isolated from toxic shock-like syndrome patients. J. Biol. Chem. *277*, 47428-47435. https://dx.doi.org/10.1074/jbc.M209133200

Tonnaer E.L., Graamans K., Sanders E.A. and Curfs J.H. (2006). Advances in understanding the pathogenesis of pneumococcal otitis media. Pediatr. Infect. Dis. J. *25*, 546-552. https://dx.doi.org/10.1097/01.inf.0000222402.47887.09

Towers R.J., Fagan P.K., Talay S.R., Currie B.J., Sriprakash K.S., Walker M.J. and Chhatwal G.S. (2003). Evolution of sfbl encoding streptococcal fibronectin-binding protein I: Horizontal genetic transfer and gene mosaic structure. J. Clin. Microbiol. *41*, 5398-5406. https://dx.doi.org/10.1128/Jcm.41.12.5398-5406.2003

Towers R.J., Gal D., McMillan D., Sriprakash K.S., Currie B.J., Walker M.J.,

Chhatwal G.S. and Fagan P.K. (2004). Fibronectin-binding protein gene recombination and horizontal transfer between group A and G streptococci. J. Clin. Microbiol. *42*, 5357-5361. https://dx.doi.org/10.1128/Jcm.42.11.5357-5361.2004

Trappetti C., Kadioglu A., Carter M., Hayre J., Iannelli F., Pozzi G., Andrew P.W. and Oggioni M.R. (2009). Sialic acid: a preventable signal for pneumococcal biofilm formation, colonization, and invasion of the host. J. Infect. Dis. *199*, 1497-1505. https://dx.doi.org/10.1086/598483

Tsukita S., Yonemura S. and Tsukita S. (1997). ERM proteins: Head-to-tail regulation of actin-plasma membrane interaction. Trends. Biochem. Sci. *22*, 53-58. https://dx.doi.org/0.1016/S0968-0004(96)10071-2

Turner C.E. (2000). Paxillin interactions. J. Cell Sci. *113*, 4139-4140

Valles J., Diaz E., Martin-Loeches I., Bacelar N., Saludes P., Lema J., Gallego M., Fontanals D. and Artigas A. (2016). Evolution over a 15-year period of the clinical characteristics and outcomes of critically ill patients with severe community-acquired pneumonia. Med. Intensiva. *40*, 238-245. https://dx.doi.org/10.1016/j.medin.2015.07.005

van der Poll T. and Opal S.M. (2009). Pathogenesis, treatment, and prevention of pneumococcal pneumonia. Lancet *374*, 1543-1556. https://dx.doi.org/10.1016/S0140-6736(09)61114-4

van Ginkel F.W., McGhee J.R., Watt J.M., Campos-Torres A., Parish L.A. and Briles D.E. (2003). Pneumococcal carriage results in ganglioside-mediated olfactory tissue infection. Proc. Natl. Acad. Sci. U. S. A. *100*, 14363-14367. https://dx.doi.org/10.1073/pnas.2235844100

Vecht U., Wisselink H.J., Jellema M.L. and Smith H.E. (1991). Identification of two proteins associated with virulence of *Streptococcus suis* type 2. Infect. Immun. *59*, 3156-3162

Verani J.R., McGee L., Schrag S.J., Division of Bacterial Diseases N.C.f.I., Respiratory Diseases C.f.D.C. and Prevention. (2010). Prevention of perinatal group B streptococcal disease--revised guidelines from CDC, 2010. MMWR Recomm. Rep. *59*, 1-36

Vindis C., Teli T., Cerretti D.P., Turner C.E. and Huynh-Do U. (2004). EphB1-mediated cell migration requires the phosphorylation of paxillin at Tyr-31/Tyr-118. J. Biol. Chem. *279*, 27965-27970. https://dx.doi.org/10.1074/jbc.M401295200

Walker M.J., Barnett T.C., McArthur J.D., Cole J.N., Gillen C.M., Henningham A., Sriprakash K.S., Sanderson-Smith M.L. and Nizet V. (2014). Disease manifestations and pathogenic mechanisms of Group A Streptococcus.

Clin. Microbiol. Rev. *27*, 264-301. https://dx.doi.org/10.1128/CMR.00101-13

Wang B.N., Li S.Y., Dedhar S. and Cleary P.P. (2007). Paxillin phosphorylation: bifurcation point downstream of integrin-linked kinase (ILK) in streptococcal invasion. Cell. Microbiol. *9*, 1519-1528. https://dx.doi.org/10.1111/j.1462-5822.2007.00889.x

Watanabe S., Takemoto N., Ogura K. and Miyoshi-Akiyama T. (2016). Severe invasive streptococcal infection by *Streptococcus pyogenes* and *Streptococcus dysgalactiae subsp. equisimilis*. Microbiol. Immunol. *60*, 1-9. https://dx.doi.org/10.1111/1348-0421.12334

Weiser J.N., Ferreira D.M. and Paton J.C. (2018). *Streptococcus pneumoniae*: transmission, colonization and invasion. Nature Rev. Microbiol. *16*, 354-367. https://dx.doi.org/10.1038/s41579-018-0001-8

Werdan K., Dietz S., Loffler B., Niemann S., Bushnaq H., Silber R.E., Peters G. and Muller-Werdan U. (2014). Mechanisms of infective endocarditis: pathogen-host interaction and risk states. Nat. Rev. Cardiol. *11*, 35-50. https://dx.doi.org/10.1038/nrcardio.2013.174

Wertheim H.F.L., Nguyen H.N., Taylor W., Lien T.T.M., Ngo H.T., Nguyen T.Q., Nguyen B.N.T., Nguyen H.H., Nguyen H.M., Nguyen C.T., *et al.* (2009). *Streptococcus suis*, an Important Cause of Adult Bacterial Meningitis in Northern Vietnam. Plos One *4,* e5973. https://dx.doi.org/10.1371/journal.pone.0005973

Wunderink R.G. and Waterer G.W. (2014). Community-Acquired Pneumonia. New Engl. J. Med. *370*, 543-551. https://dx.doi.org/10.1056/NEJMcp1214869

Yau B., Hunt N.H., Mitchell A.J. and Too L.K. (2018). BloodBrain Barrier Pathology and CNS Outcomes in *Streptococcus pneumoniae* Meningitis. Int. J. Mol. Sci. *19*. https://dx.doi.org/10.3390/ijms19113555

Yoshida H., Takahashi T., Nakamura M., Overby A., Takahashi T., Ubukata K. and Matsui H. (2016). A highly susceptible CD46 transgenic mouse model of subcutaneous infection with *Streptococcus dysgalactiae subspecies equisimilis*. J. Infect. Chemother. *22*, 229-234. https://dx.doi.org/10.1016/j.jiac.2016.01.001

Zbinden A., Mueller N.J., Tarr P.E., Eich G., Schulthess B., Bahlmann A.S., Keller P.M. and Bloemberg G.V. (2012). *Streptococcus tigurinus*, a Novel Member of the *Streptococcus mitis* Group, Causes Invasive Infections. J. Clin. Microbiol. *50*, 2969-2973. https://dx.doi.org/10.1128/Jcm.00849-12

Zhang J.R., Mostov K.E., Lamm M.E., Nanno M., Shimida S., Ohwaki M. and

Tuomanen E. (2000). The polymeric immunoglobulin receptor translocates pneumococci across human nasopharyngeal epithelial cells. Cell *102*, 827-837. https://dx.doi.org/10.1016/S0092-8674(00)00071-4

Zheng C., Li L., Ge H., Meng H., Li Y., Bei W. and Zhou X. (2018). Role of two-component regulatory systems in the virulence of *Streptococcus suis*. Microbiol. Res. *214*, 123-128. https://dx.doi.org/10.1016/j.micres.2018.07.002

Chapter 13

https://doi.org/10.21775/9781912530229.13

Molecular Evolution of Clinical Pathogenic Streptococci

Dongdong Li[1] and Chuanmin Tao[1*]

[1]Division of Clinical microbiology, Department of Laboratory Medicine, West China Hospital, Sichuan University, Chengdu, China.
*Corresponding Author: taocm@scu.edu.cn

Abstract

The genus *Streptococcus* comprises a wide variety of pathogenic and commensal gram-positive bacteria, many of which the pathogenic species cause severe, invasive infections that account for a high burden of morbidity and mortality. Here, we reviewed the evolution of representative virulence factors, capsule in *Streptococcus pneumoniae*, M protein in *Streptococcus pyogenes* (GAS), biofilm in *Streptococcus agalactiae* (GBS) and some oral *Streptococcus*, as well as the effect caused by evolution, antibiotic resistance and vaccine escape. Thanks to the rapid development of whole genome sequence (WGS) data, the impact of genetic recombination to the *Streptococcus* evolution has been proved. As to adaptive evolution caused by antibiotics, vaccine and so on, continuous surveillance is an essential to monitor evolution of *Streptococcus* causing disease. This knowledge is invaluable to the development of preventative and control strategies against this important pathogen.

Introduction

The genus *Streptococcus* comprises a wide variety of pathogenic and commensal gram-positive bacteria, many of which the pathogenic species cause severe, invasive infections that account for a high burden of morbidity and mortality. Among them, *S. pyogenes* (commonly as an alternative to group A streptococci, GAS), *S. agalactiae* (group B streptococci, GBS) and *S. pneumoniae* are the three most important streptococcal human pathogens, opportunistic pathogens which colonizes the skin, digestive and genitourinary tract (Donati et al., 2015; Edwards, 2010). Apart from these bacteria, the streptococci also include the viridans group streptococci (VGS = anginosus, mitis, mutans and salivarius groups), *Streptococcus suis* (Kohler, 2007; Olson et al., 2013) and so on. These streptococci exist oral unique environment, forming a characteristic microbiome. The streptococcal commensal flora intimately interacts and communicates with other bacterial species and the human host to support colonization. Being generally commensal colonizers of humans and other animals, streptococci express a wealth of colonization factors that allow them to adhere to epithelium, including capsule, pili and matrix-interacting proteins, and of virulence-related proteins (Nobbs et al., 2015). Streptococcal virulence factors usually participate in multiple events such as colonization, internalization, invasion of the host tissues and dissemination, systemic toxicity, and, evasion of the host defense (Sitkiewicz, 2017). The understanding of recent advances in descriptions of molecular mechanisms of streptococcal virulence, coupled with massive sequencing efforts to isolate genomes, have allowed the field to better understand the molecular and evolutionary changes leading to pandemic strains (Brown et al., 2012; Wilkening and Federle, 2017).

Clinical interventions-vaccine and/or antibiotics generated a selection pressure that drove the evolution of vaccine-escape mutants and strains that were highly resistant against antibiotics (Straume et al., 2015), as a result of gene transfer intra- and inter-species gene transfer involving *Streptococcus* (Donati et al.,

2015). For *Streptococcus* to survive, rapid adaptation to clinical interventions and its human immune responses is essential (Croucher et al., 2011). The plasticity of bacterial genomes ensures rapid adaptation to changing environmental conditions and enables bacteria to colonize new niches successfully. Molecular mechanisms contributing to genome plasticity include point mutations, genome rearrangements, mobile genetic elements, homologous or non-homologous recombination (Darmon and Leach, 2014). The rapid influx of whole genome sequence (WGS) data and the advent of novel analysis methods and powerful computational tools for population genetics and evolution studies has transformed our understanding of how genetic recombination drives streptococci adaptation and evolution. As we all know, evolutionary studies have highlighted the mechanisms that arises phenotypes changements. In this paper we present an overview of the latest results on the population structure, genomics and mechanisms of genome evolution of a selection of species from the genus *Streptococcus*. The species discussed include three major pathogens, *S. pneumonia*, *S. agalactiae* and *S. pyogenes.* Oral *Streptococcus* is also considered.

Genome of *Streptococcus* and evolution

Bacteria are haploid in naturally, possessing a single copy of most genes. The *Streptococcus* genomes can be best described as consisting of core and accessory genomes. The core genome in pneumococcus represents those genes that are common to all human pneumococcal isolates. These include key colonization (adhesions, matrix-interacting proteins and pili) and virulence genes (M protein, proteases and so on) as well as genes that enable adaptation and regulation of metabolism in response to environment alterations, and anaerobic conditions (Nobbs et al., 2015). The core genome size is dependent on the number of isolates used during the analysis, and this gradually decreases with the addition of new genomes until a plateau is reached (Tettelin et al., 2005). The accessory genome represents the set of genes specific to a isolate, which are not shared in their entirety across the whole pneumococcal population. In other words, the accessory genome does not contain genes essential to cell survival however it plays an important role in bacterial pathogen evolution (Jackson et al., 2011). For example, antibiotic resistance is largely due to the acquisition of mobile genetic elements that harbor antibiotic resistance determinants (Jackson et al., 2011). From the complete genomes, the basic information of GAS genome is as follows: the genome is composed of a single chromosome that is approximately 1.8-Mbp long with a GC content of 38.5 ± 0.1% on average, and includes 1,826 ± 92 protein-coding regions, 5 or 6 rRNA operons, and from 57 to 67 tRNA encoding genes (Maruyama et al., 2016). Pneumococci possess a 2.1 megabases (Mb) pair circular genome that consists of over 2000 predicted protein coding regions and approximately 5% insertion elements. On average the core genome of *S. pneumoniae* consists of 1647 predicted coding sequences. Variation in the pneumococcal core genome is predominantly introduced by random mutations and homologous recombination that involves both short and long stretches of DNA, whilst recombination involving unrelated loci is more restricted to the accessory genome. The accessory genome does not contain genes essential to cell survival however it plays an important role in bacterial pathogen evolution. This is largely due to the acquisition of mobile genetic elements that harbor antibiotic resistance determinants and virulence factors. Comparative genome analysis among organisms of the same species not only shows a high degree of similarity in gene content and organization, but also a high degree of sequence heterogeneity as evidenced by the large number of single nucleotide polymorphisms present. Comparative genome analysis between the species showed that *S. pyogenes* was more closely related to *S. agalactiae* than with *S. pneumoniae* or *Streptococcus mutans* (Ferretti et al., 2004).

Molecular Mechanism of Streptococcus Evolution

Molecular mechanisms contributing to genome plasticity include point mutations, genome rearrangements, mobile genetic elements and horizontal gene transfer (Darmon and Leach, 2014). Genetic recombination is the exchange of genetic material between different organisms which leads to production of offspring with combinations of traits that differ from those found in either parent recombination occurs between similar molecules of DNA. Horizontal gene transfer (HGT) is the sharing of genetic material among organisms other than by the DNA transmission from parent to offspring, which enable organisms to acquire new genes and phenotypes. HGT, a fundamental process in the genome evolution of bacteria (Gogarten and Townsend, 2005), can accelerate genome innovation and evolution (Jain et al., 2003) and spreads genetic diversity by moving genes across species boundaries. Genes are preferentially exchanged among organisms sharing similar genome size, genome G+C composition, carbon utilization, and oxygen tolerance. It enables bacteria to evolve rapidly through the acquisition of novel genetic determinants, or genetic determinants that are homologous to existing DNA, which were not previously resident within the recipient's genome (Nobbs et al., 2015). In bacteria, uptake of exogenous DNA is achieved via three main processes: transformation, conjugation and transduction. Of the three known HGT mechanism operating in bacteria, natural genetic transformation has been studied in most detail in *S. pneumonia*. Bacterial conjugation occurs when there is direct cell-to-cell contact of the bacteria exchanging DNA via a sex-pili that protrudes from one cell into the other. During transduction, DNA is transferred from one bacterium to another by viruses that infect bacteria called bacteriophages. Bacterial transformation involves the acquisition of exogenous DNA from the bacterial surroundings followed by the integration of the acquired DNA into the host cell genome.

Recombination events vary in size, which identified in the pneumococci range from very small fragments to thousands of base pairs (bp) (Croucher et al., 2013; Mostowy et al., 2014; Wyres et al., 2013). Two classes of recombination have been proposed based on the sizes of the recombination events: 1) micro-recombination which involve single and short stretches of DNA occur more frequently. 2) macro-recombination, which are usually rare and consists of multi-fragment replacements of DNA (Andam and Hanage, 2015; Mostowy et al., 2014). Mostowy et al proposed several mathematical models to describe the rate and size of recombination in the evolutionary history of two very distinct pneumococcal lineages and concluded that macro-recombination is comparatively rarely observed in nature (Mostowy et al., 2014). But Cowley et al proved that macro-recombination is likely is more likely to occur in environments that permit cell-to-cell contact (Cowley et al., 2018). Macro-recombination was associated with major phenotypic changes, including serotype-switching events, and thus was a major driver of the diversification of the pathogen (Mostowy et al., 2014). A single nucleotide change in a Streptococcal housekeeping gene is at least 50-fold more likely to change as a result of recombination than as a result of mutation (Feil et al., 2000; Feil et al., 1999). Thanks to the whole genome sequencing of a large panel of isolates and comparative genomics, it has recently become clear that HGT both within and between species is the major driver of genome evolution of these organisms, allowing them to quickly evolve their genomic content in response to environmental stimuli and share a large number of genes important for virulence (Chewapreecha et al., 2014). For example, large imports of genes from *Streptococcus mitis* and other *anginosus*, and s*alivarius* group streptococci to *S. pneumoniae* have been reported to contribute to the evolution and structural diversity of the pneumococcal capsule (Kilian et al., 2014). Phylogenomics of GAS and closely related species revealed that interspecific HGT events among species of the pyogenic division are well-documented. Its closest genetic relatives are *Streptococcus canis* and *Streptococcus dysgalactiae* subspecies *equisimilis*. Richards et al (Richards et al., 2011) provided strong evidence for two cases of interspecies HGT within the shared bovine environment: bovine *S. agalactiae* with *Streptococcus uberis* (nisin U operon) and *Streptococcus dysgalactiae* subsp *dysgalactiae* (lactose operon). They also found evidence for HGT, involving the salivaricin operon, between the bovine *S. agalactiae* strain and either *S. pyogenes* or *Streptococcus salivarius*. Also, we know that frequency of recombination varies by niche for example higher levels of recombination have been observed in carriage than during septic infection.

Natural genetic transformation

Natural genetic transformation has been studied in most detail in *S. pneumoniae* (Straume et al., 2015). *S. pneumoniae* is competent for natural genetic transformation, a property that enables the pneumococcus to acquire new traits by taking up naked DNA from the environment and incorporating it into its genome through homologous recombination. Four distinct steps (Straume et al., 2015) in *S. pneumoniae*, the transformation processes were described: regulation of competence induction, DNA-uptake and homologous recombination, the fratricide mechanism and termination of the competence period. This process involved many genes. The proteins required for this process are conserved between species and are produced during a specific physiological state known as competence. The review of competence regulation has revealed several cues regarding its physiological function in a species (Fontaine et al., 2015). Recent findings regarding competence regulation by the ComCDE and ComRS cell–cell signaling pathways and the Clp proteolytic system are specifically highlighted (Fontaine et al., 2015). The comX-inducing peptide (XIP) of *Streptococcus mutans* is a key regulatory element in the activation of genetic competence, which allows cells to take up extracellular DNA (Kaspar et al., 2017; Shanker et al., 2016). Flechard et al investigated on a large scale if loci coding for early competence factors (ComX and the two pheromone-dependent signalling systems ComCDE and ComRS) of streptococci are especially targeted by transposable elements. Transformation is even more common among the streptococci than has been recognized (Morrison et al., 2013). At the other extreme, no species of the *S. pyogenes* has been reported to transform in laboratory culture (Mashburn-Warren et al., 2012).

When transformation involves DNA exchange from closely related loci, it is known as homologous recombination. When transformation occurs between unrelated loci, it is known as non-homologous recombination. Homologous recombination also occurs between mobile genetic elements (MGE) such as insertion sequences (IS), integrons, bacteriophages, plasmids and transposons, considered being part of the accessory genome (non-core genome). The evolutionary function of natural competence remains controversial, as imported DNA can act as a source of substrates or can be integrated into the genome. Exogenous homologous DNA can also be used for genome repair. Veening and Blokesch (Veening and Blokesch, 2017) proposed that predation of non-related neighboring bacteria coupled with competence regulation might function as an active strategy for DNA acquisition, and argued that the forced release of DNA from killed bacteria and the transfer of non-clonal genetic material have important roles in bacterial evolution. Clinical interventions as drivers of genome evolution in *S. pneumoniae* demonstrate that natural genetic transformation plays a key role in these processes. Johnston et al (Johnston et al., 2013) showed that transformation produces a population containing many different merodiploid cells. Merodiploidy provides opportunities for evolution of new genetic traits via alteration of duplicated genes, unrestricted by functional selective pressure such as antibiotic treatment in *S. pneumoniae*, reinforces the plasticity potential of this bacterium and transformable species generally.

The role of mobile genetic elements (MGEs)

Mobile genetic elements (transposable elements, integrons, plasmids and bacteriophages) play an important role in pneumococcal evolution, both as an additional means of mobilizing genetic material. Integrons are genetic elements that can help bacteria to enhance their ability of adaptation and evolution through the acquisition, stockpiling, excision and differential expression of new genes. The genes are usually contained in a gene structure called cassette. In Streptococci, *S. pneumoniae* was proved to carry class 1 integron. There are two key advantages owning by the integron system for the genomic innovation. First, the new cassette will not disturb the existing genes. Second, the newly integrated cassettes is ready for the natural selection (Gillings, 2014). The Tn5253 and Tn916 ICE families are commonly found in pneumococci and are important vehicles for the dissemination of antibiotic resistance in pneumococci and between *Streptococcus* species (Roberts and Mullany, 2011). These transposons are particularly known for carrying and transferring the tetracycline resistance determinant *tet* (M) and chloramphenicol acetyltransferase (*cat*) resistance gene. Some Tn916-like elements also carry other resistance genes, such as *erm* (B)-mediated erythromycin resistance, and genes associated with other accessory functions (Cochetti et al., 2007). The contributions of ICEs and phages to pneumococcal biology and pathogenicity further highlight the remarkable ability of these bacteria to exchange genes through different transfer mechanisms (transformation, transduction and conjugation). However, the diversity of these pneumococcal mobile elements and phages has not been fully explored and remain under-appreciated. Croucher et al (Croucher et al., 2014) showed that prophage content is highly variable even within pneumococcal lineages, suggesting frequent horizontal transmission that would necessitate rapidly diversifying antiphage mechanisms to prevent these viruses sweeping through populations. But recently, Brueggemman et al (Brueggemann et al., 2017) proved that some prophages persisted over long time periods. Pneumococcal prophages are likely to play a more important role in pneumococcal biology and evolution than previously recognized, but the vast array of prophage DNA within these pneumococcal genomes warrants a detailed exploration. We anticipate that genome sequencing of pneumococcal samples from diverse geographical origins will provide additional insight into the extent of the genetic diversity of ICEs and phages, and how these combines with other loci to make up the accessory genome of pneumococci.

HGT is a pervasive mechanism of diversification in many microbial species, but its primary evolutionary role remains controversial. Much recent research has emphasized the adaptive benefit of acquiring novel DNA. But Croucher et al (Croucher et al., 2016) argued intragenomic conflict provides a coherent framework for understanding the evolutionary origins of HGT. The competence for transformation was found to provide an effective defense against parasitic MGEs. MGEs inhibit transformation through integrative disruption of genes encoding the competence machinery across many species and through secretion of DNases to reduce the concentration of extracellular DNA. Therefore, this framework is able to explain both common properties of MGEs, and the seemingly paradoxical bacterial behaviors of transformation and cell-cell killing within clonally related populations, as the consequences of intragenomic conflict between self-replicating chromosomes and parasitic MGEs. The antagonistic nature of the different mechanisms of HGT over short timescales means their contribution to bacterial evolution is likely to be substantially greater than previously appreciated. Regarded as a common genetic element responsible for horizontal gene transfer, integrons are widely distributed in various pathogens considered as a determinant in the acquisition and evolution of antibiotic resistance. Li et al (Li et al., 2017) summarized the occurrence, pathogenicity and virulence mediated by integrons in typical Gram-positive microorganisms (*Staphylococcus*, *Enterococcus*, *Corynebacterium* and *Streptococcus*) and the role of integrons in antibiotic resistance.

Mosaic genes

Interspecies recombination results in mosaicism, i.e. genes that are composed of alternate blocks of nucleotides derived from a donor and its recipient. These blocks may be the result of small fragments that have recombined within a gene or the result of almost the whole gene being replaced by incoming genetic material (Dowson et al., 1989). Whichever outcome, the new mosaic usually encodes a protein with a different activity to that of the original recipient and possibly to that of the donor. Following integration of advantageous mosaics, such as penicillin resistance conferred by alterations within genes encoding penicillin-binding proteins, these become fixed within the population (Coffey et al., 1999). Once fixed, these successful mosaics can then readily spread laterally from species to species. For example, the DNA segments encoding the transpeptidase domains of PBP1a, PBP2x and PBP2b in highly resistant clinical isolates have a mosaic structure that have been generated by several successive recombination events. It is difficult to trace the sources of the various mosaic blocks. However, at least in some cases it has been shown that they originate from other species, namely *Streptococcus mitis* and *Streptococcus oralis* (Chi et al., 2007; Dowson et al., 1993; Sibold et al., 1994).

Crosstalk between vertical and horizontal gene transfer

Both intrinsic (vertical) and acquired mechanisms (HGT) affect susceptibility to a large variety of antibiotics (El Moujaber et al., 2017). Horizontal gene transfer is a key process in the evolution of bacteria and also represents a source of genetic variation in eukaryotes. Meanwhile, the vertical gene transfer was not talked much here. Among elements participating in gene transfer, thousands of small (<10 kb) mobile bacterial plasmids that replicate by the rolling circle mechanism represent a driving force in the spread of antibiotic resistances. In general, these plasmids are built as genetic modules that encode a replicase, an antibiotic-resistance determinant, and a relaxase that participates in their conjugative mobilization. Lorenzo-Diaz et al (Lorenzo-Diaz et al., 2017) reported that the MobM conjugative relaxase encoded by the promiscuous plasmid pMV158 participates in regulation of the plasmid copy number by transcriptional repression of the antisense RNA, thus increasing the number of plasmid molecules ready to be horizontally transferred (mobilization) and/or vertically inherited (replication), which is the first report of a type of crosstalk between genetic modules involved in vertical and horizontal gene flow. Much more work involved in vertical and horizontal gene transfer is need in future.

Virulence factors and evolution

Pathogenicity implies that a bacterium can be harmful to the host and virulence indicates the degree to which this pathogenicity is expressed. Pathogenicity and virulence depend on multiple factors such as capsule, M protein, biofilm, and resistance to immune responses, vaccine, antibiotics and other defensive mechanisms of the host, competitiveness, inflammation, invasion, translocation, toxin production and cytotoxicity.

Serotype replacement and capsular switching

In pneumococci, genetic exchange at the capsular locus is greatly facilitated by the syntenic organization and homology of genes in the capsular locus. This process of substituting the genes encoding one type of capsule with genes encoding for another is referred to as serotype or capsular switching , as a strategy to increase pneumococcal virulence (Sabharwal et al., 2014). This population level change in the serotype distribution, most often involving pre-existing clones and serotypes that were already in circulation before vaccine implementation, is known as serotype replacement. In Casablanca, the leading serotypes of penicillin non-susceptible *S. pneumoniae* (PNSP) were 14 (33 vs. 57%) and 19A (18 vs. 14%) before and after vaccination among children. For adults, serotypes 19A (53%) and 23F (24%) were the dominant serotypes in the pre-vaccination period, while serotype 14 (22%) was the most prevalent after vaccination (Diawara et al., 2017). Although there was evidence of natural capsule switching in the absence of vaccine induced selection pressure, most of the highly prevalent capsule-switched isolates were associated with acquisition of vaccine-targeted capsules (Chaguza et al., 2017). Changes in serotype prevalence among pneumococcal populations result from both serotype replacement and serotype (capsular) switching. This can be attributed to the phenomena of “serotype replacement”, the expansion of preexisting NVT pneumococci, and/or “serotype switching”, and a change of serotype of a single clone by alteration or exchange of its cps locus. The capsular polysaccharide

(CPS) the main virulence factor of *S. pneumonia*, of which there are more than 90 known serotypes protects against phagocytosis during invasive pneumococcal disease and may also prevent clearance during nasopharyngeal colonization. A historical perspective study (Wyres et al., 2013) of seven decades showed that capsular switching events were presumably the result of nucleotide substitution and/or deletion, and recombination-imports of various lengths, inserted at different points around the cps locus, with or without the adjacent *pbp* sequences. Sabharwal et al (Sabharwal et al., 2014) proved that capsular switch events can result in *S. pneumoniae* strains of enhanced virulence for respiratory tract infection. Mostowy et al (Mostowy et al., 2017) found that found cps to be an evolutionary hotspot with elevated substitution and recombination rates to generate novel serotypes by recombination. Chiba et al (Chiba et al., 2017) reported a penicillin-resistant *S. pneumoniae* (PRSP) isolate from an adult patient, characterized as a recombinant strain derived from PRSP of serotype 23F with the cps locus (20.3 kb) replaced by that of a penicillin-susceptible strain of serotype 3. Much of serotype replacement is due to clonal expansion of serotype variants extant prior to vaccination and not those that arise through capsule switching. This should be clear because it may be erroneously interested that capsule switching is a major driver for serotype replacement. Moreover, switching was much more likely to exchange one serotype for another within the same serogroup than expected by chance (Croucher et al., 2015). Nontypeable pneumococci (NTPn) fall into several categories and reviewed (Andam and Hanage, 2015): those that do not possess capsule genes, pneumococci that are phenotypically non-encapsulated but possess down-regulated or defective capsule genes, and a population of atypical non-encapsulated isolates that are genetically divergent from most pneumococci. Sporadic or random switching between encapsulated and nonencapsulated states may play an important role in pneumococcal population dynamics and antibiotic resistance. Andam et al (Andam and Hanage, 2015) illustrated a possible model that there is a switching between encapsulated and non-encapsulated states in a closely related group of pneumococci. One strain loses its ability to produce a capsule through gene loss, deletion or insertion of a stop codon within the capsule locus. The strain may eventually regain its original capsule type or another capsule of a different serotype, which may also correlate with increased resistance due to recombination.

M protein in GAS

The M protein, first described in 1927 by Rebecca Lancefield, who demonstrated it to be a primary source of strain-specific immunity (Metzgar and Zampolli, 2011) coats the surface of GAS. M protein inhibits phagocytosis, promotes adhesion to epithelial cells, and helps the bacteria evade innate immune defenses (Smeesters et al., 2010). There are >200 different serotypes of M protein (i.e., M6, M12, M18, M24, and so on), an individual may become infected by more than one group A streptococcal type during a lifetime (Pinto et al., 2016). The M protein is hyper variable and has long served as the primary target for epidemiological typing of GAS. More than 80 distinct M types were identified, and the M type-specific determinants map to the fibril tips, encoded by the 5' end of *emm* genes. More recently, a sequence-based *emm* typing scheme was implemented, based on extensive nt sequence differences at the 5' end of the *emm* gene, whereby a unique *emm* type is defined as having <92% sequence identity over the nt sequence corresponding to the first 30 codons of the mature M protein. Among the 234 *emm* types recognized to date are >1200 distinct allelic forms of the emm type-specific regions of *emm* genes, known as *emm* subtypes (Bessen et al., 2015). All GAS isolates harbor an *emm* gene. In addition, many GAS strains have paralogous *emm*-like genes lying immediately upstream and downstream of *emm*, and a few strains have only the downstream *emm*-like locus. Thus, a given GAS strain can have one or two *emm*-like genes, in addition to *emm*; the upstream *emm*-like gene is often referred to as mrp, and the downstream *emm*-like gene is often referred to as enn (Bessen et al., 2015). Comparative analyses of the genomes of GAS clearly demonstrated the importance of MGEs, LGT, and recombination in generating diversity between *emm*-types. Most GAS genomes possess multiple bacteriophages, which in turn often carry virulence genes predicted to change the virulence of a lineage (McMillan et al., 2013). Integrative conjugative elements (ICE) and remnant MGEs are also common. The *S. pyogenes* genome is remarkable for its content of prophages, streptococcal phage-like chromosomal islands (SpyCIs), and other mobile genetic elements (MGEs), such as integrative and conjugative elements (ICEs) (Maruyama et al., 2016). In GAS, prophage regions have been shown to encode virulence factors such as exotoxins. In addition, it has been reported that genome rearrangements can generate a new prophage in an M3 strain (Maruyama et al., 2016). Generalized transduction occurs in GAS and is mediated by bacteriophages (McShan and Nguyen, 2016). Generally, the susceptibility of cells to phage-mediated transduction probably varies by growth state and genetic background, both of which could influence horizontal transfer (McShan and Nguyen, 2016). The majority of the genome prophages (72%) are found to be integrated into genes encoded on the lagging strand. No prophages have been found to target genes in the hypervariable regions, which include the M-protein (*emm*) or the streptococcal pilus. Lysogenic bacteriophages of GAS are capable of mediating transfer of antibiotic resistance by transduction. Strains with bacteriophage T12-like prophages can produce transducing lysates that are capable of transferring resistance to tetracycline, chloramphenicol, macrolides, lincomycin, and clindamycin, following lysogen induction. Generalized transduction transfer of erythromycin and streptomycin resistance, following mitomycin C treatment of endogenous prophages, has also been observed (McShan and Nguyen, 2016). Horizontal transfer of serotype-related genes and the emergence of new strains/clones may be a result of selective pressures conferred by the host immune response (Rouge et al., 2016). Studies suggest that the acquisition of new MGEs, resulting in the elaboration of novel virulence gene repertoire is an important facet underlying the changes in virulence (McMillan et al., 2013). GAS has a rich evolutionary history of horizontal transfer among its core genes (Bessen et al., 2011). In countries where streptococcal disease is endemic, and *emm* type diversity is high, the conditions for LGT of the *emm*-gene is much greater than non-endemic countries (McMillan et al., 2013).

Streptococcus population evolution

Establishing the contribution of co-adaptation and competition in the maintenance of discrete lineages is important since the outcome of certain interventions, such as vaccination depends crucially on these underlying determinants of population structure. Populations of *S. pneumoniae* are typically structured into groups of closely related organisms or lineages, but it is not clear whether they are maintained by selection or neutral processes. Camilli et al (Camilli et al., 2017) showed the impact of pneumococcal conjugate vaccine (PCV7 and PCV13) on pneumococcal invasive diseases in Italian children and insight into evolution of pneumococcal population structure. Recombination events (Duvvuri et al., 2016) were observed by whole genome sequencing among non-PCV 13 serotypes (22F, 15A and 8) populations after the in introduction of pneumococcal conjugate vaccines. Serotype 22F (ST433) has emerged into two sub-populations, with 28% (7/25) exhibiting recombination events, and five also acquiring macrolide resistance as a result of recombination. Lourenco et al (Lourenco et al., 2017) indicated that lineages evolved through immune selection on the groEL chaperone protein. The groEL protein is part of the groESL operon and enables a large range of proteins to fold correctly within the physical environment of the nasopharynx, thereby explaining why lineage structure is so stable within *S. pneumoniae* despite high levels of genetic transfer (Lourenco et al., 2017). Hence, vaccine strategies based on groESL variants would target entire lineages instead, including all uncommon serotypes within and thereby preventing their expansion. Both pre- and post-genomic studies have revealed that HGT and recombination occurs between these two organisms and plays a major role in shaping the population structure (McNeilly and McMillan, 2014). *S. pneumoniae* from invasive (IPD) and noninvasive pneumococcal disease (NIPD) were studied by serotype and multilocus sequence typing (MLST) for population structure characteristics (Zhou et al., 2017). The seven predominant STs were ST271, ST320, ST876, ST3173, ST236, ST81 and ST342, which were mainly associated with serotypes 19F, 19A, 14, 6A, 19F, 1, and 1/23F, respectively. The colonization rate of GBS in postmenopausal women was 17.8%. Capsular type III was predominant (34.6%), followed by type V (22.4%). The most frequent sequence type (ST) was 19 (23.3%), followed by 23 (18.7%), 1 (16.8%) and 17 (12.1%). Isolates were assembled into three phylogenetic groups from ST-19, ST-23 and ST-17 founders. All isolates were susceptible to penicillin, whereas resistance to erythromycin and clindamycin was recorded in 23.4% and 20.6% of isolates, respectively. The population structure of GBS is highly diverse and contains different STs (Molto-Garcia et al., 2016). Athey et al, hypothesized that *Streptococcus suis* serotype 2 strains of the ST25 lineage are genetically heterogeneous, and proved that the importance of lateral gene transfer and recombination as drivers of diversity in *Streptococcus suis* (Athey et al., 2016). More attention should pay to MLST genotyping for *Streptococcus* and evolution relationship between different ST types, as well as the relationship of ST types to virulence and antibiotic resistance profile. The sequences of seven housekeeping loci used for *Streptococcus* MLST genotyping are stored in the databases hosted on PubMLST (https://pubmlst.org/).

Evolution in *Streptococcus* biofilms

Biofilms are defined as structured communities of microorganisms that are attached to a surface and enmeshed in an extracellular polymeric matrix (Flemming et al., 2016; Hobley et al., 2015). Dental caries (tooth decay) is a polymicrobial biofilm disease driven by the diet and microbiota–matrix interactions that occur on a solid surface. Early studies prove that microorganisms residing within biofilms are embedded in a matrix containing extracellular polymeric substances (EPS) (Hobley et al., 2015; Lebeaux et al.,

2014). Obviously, polymicrobial interactions and the local biofilm microenvironment play instrumental roles in modulating health and disease conditions (Dewhirst, 2016; Koo et al., 2013; Mira et al., 2017). Coevolution of *S. mutans* with the increased sugar consumption largely explains how well adapted *S. mutans* is to colonize and thrive on teeth when its human host ingests sugars. *S. mutans* can utilize a wide variety of carbohydrates to produce EPS and acids for adaptation of stress resistance and bacterial competence (Bowen and Koo, 2011). Furthermore, pathological process of the dental caries relies on the milieu within which the organisms interact, and acids accumulate. Thus, *S. mutans* can contribute to the biofilm matrix assembly as the main producer of insoluble glucans among oral bacteria and reset the microenvironment for other aciduric-cariogenic bacteria to thrive and become established (Johansson et al., 2016; Simon-Soro and Mira, 2015). If the commensal environment is supplied with sugar, *S. mutans* can work synergistically with other aciduric species and cause ecological changes to shape the biofilm community, structure, and metabolism conducive to caries development. Carriage of the pneumococcus in the nasopharynx is thought to be mediated by biofilm formation, an environment where isogenic populations frequently give rise to morphological colony variants, including small colony variant (SCV) phenotypes. Churton et al (Churton et al., 2016) revealed that SCVs exhibit reduced growth rates, reduced capsule expression, altered metabolic profiles, and increased biofilm formation, and all SCVs studied had mutations within the DNA-directed RNA polymerase delta subunit (RpoE). Furthermore, recurrent mutation of the pneumococcal *rpoE* gene presents an unprecedented level of parallel evolution in pneumococcal biofilm development. The formation of GBS biofilm can be strongly influenced by environmental conditions like PH and glucose concentration (D'Urzo et al., 2014), as well as endogenous factors including pili and bacterial surface adhesin of GBS (BsaB/ FbsC) (Buscetta et al., 2014; Jiang and Wessels, 2014). No matter what the specific GBS or host factor is, the evolutionarily conserved interaction between GBS factor and host surface molecular always allow GBS to survival and multiply to an enough community, and then cause harm to the host with virulence element. Obviously, biofilm formation enhances fomite survival of *S. pneumoniae* and *S. pyogenes* (Marks et al., 2014).

Evolution in immune escape

The host has developed multiple immune responses to fight the pathogens. Simultaneously, these bacterial also evolved complex mechanisms to survive in host internal environment. To avoid immune clearance, GBS can degrade host-derived intermediate immunological molecules to inactive form which cannot stimulate host immune response pathway. Simultaneously, these bacterial also evolved complex mechanisms to survive in host internal environment. To avoid immune clearance, GBS can degrade host-derived intermediate immunological molecules to inactive form which cannot stimulate host immune response pathway. Hyaluronan (HA), known to have an important role in immune surveillance, can be synthesized by cervical fibroblasts, epithelial cells and immune cells (Akgul et al., 2014). In the condition of infectious or non-infectious tissue injury, HA will be degraded by host hyaluronidases into HA fragments which can interact with TLR2/4 (Toll-like receptor) to elicit inflammatory responses. It was believed that most clinical GBS strains can secrete hyaluronidase. With the help of hyaluronidase, GBS can degrade HA fragments produced by host to non-stimulatory disaccharide subunits and then facilitate immune evasion. Furthermore, the HA disaccharides can inhibit TLR2/4 pathways activated by HA fragments or other TLR2/4 ligands. Then the immune recognition and production of pro-inflammatory cytokine will be blocked, and the GBS pathogens can escape the host immune detection (Kolar et al., 2015). Induction of type I interferon (IFN) is an important host immune response to GBS infection. The production of type I IFN can be activated by the sensing of bacterial nucleic acids present in host cytoplasm (Monroe et al., 2010). The cyclic GMP-AMP synthase (cGAS), as a host DNA receptor, can be activated by GBS double-stranded DNA and produce a cyclic dinucleotide (cdN). These GBS derived cdNs can bind stimulator of IFN genes (STING) to activate type I IFN production and promote inflammation. Andrade et al reported an ectonucleotidase CdnP expressed by GBS can hydrolyze extracellular cyclic dinucleotides, block the stimulation of STING and inhibit the production of type I IFN, ultimately leading to immune escape (Andrade et al., 2016). The histidine triad proteins (HTPs), also known as Pht proteins in *S. pneumoniae*, constitute a family of surface-exposed proteins that exist in many pathogenic streptococcal species. Although many studies have revealed the importance of HTPs in streptococcal physiology and pathogenicity, little is known about their origin and evolution. In previous study (Shao et al., 2013), several major findings were made. First, *htp* genes originated earlier than the *Streptococcu*s genus and gene-loss events have occurred among three streptococcal groups, resulting in the absence of the *htp* gene in the bovis, mutans and salivarius group streptococci. Second, the copy number of htp genes in other groups of *Streptococcus* is variable, ranging from one to four functional copies. Third, both phylogenetic evidence and domain structure analyses support the division of two htp subfamilies, designated as htp I and htp II. Although present mainly in pyogenic group and in *Streptococcus suis*, htp II members are distinct from htp I due to the presence of an additional leucine-rich-repeat domain at the C-terminus. Finally, htp genes exhibit a faster nucleotide substitution rate than do housekeeping genes. Specifically, the regions outside the HTP domains are under strong positive selection. This distinct evolutionary pattern likely helped *Streptococcus* to easily escape from recognition by host immunity.

Evolution associated with antibiotics and vaccine

Evolution drive drug resistance

Genetic recombination plays an important role in the development of antibiotic resistance in pneumococci. Antibiotic induced stress is known to induce competence in pneumococci; during the competence phase, the pneumococci acquire exogenous DNA, which may include genes that confer resistance to antibiotics (Slager et al., 2014). It has been reported that recombination replacements are responsible for the mosaic structure typically observed in penicillin binding proteins (PBP) genes in *S. pneumoniae* (Dowson et al., 1989). Mutations in PBP genes confer resistance to β-lactam antibiotics including penicillin, amoxicillin and cefotaxime (Hakenbeck et al., 1999; Li et al., 2016). Hakenbeck et al (Hakenbeck et al., 2012) showed that clinical isolates display a mosaic structure of the affected PBP genes, the result of interspecies gene transfer and recombination events. Fan et al (Fan et al., 2009) proved that both the ComD and ComC are involved in the drug resistance of *S. pneumoniae* to cefotaxime. Recombination also mediates the dissemination of transposon and integrative conjugative elements (ICEs) that carry an array of antibiotic resistance determinants, throughout the pneumococcal population. Such mobile genetic elements include the Tn916-likeMGEs, Tn5251 (a Tn916-like element) (Provvedi et al., 1996), Tn5252 (Provvedi et al., 1996), Tn5253 (a composite of Tn5251 and Tn5252 transposons) (Provvedi et al., 1996) and many others. The Tn916, Tn5251, Tn5252, Tn5253 and Tn1545 transposons carry tetM gene, which confers resistance to tetracycline (Chaguza et al., 2015).

The ability to effectively treat pneumococcal infection has been compromised due to the acquisition of antibiotic resistance, particularly to β-lactam drugs (Appelbaum, 2002). Resistance to β-lactam antibiotics in pneumococci is due to alterations in PBPs, especially PBPs 2X, 1A and 2B (Coffey et al., 1991). Recently, two studies have used whole-genome sequencing to analyze isolates from the PMEN1 genetic lineage. The first study (Croucher et al., 2011) focused solely on the PMEN1 lineage and described considerable genomic diversity believed to have originated by horizontal gene transfer in response to antimicrobial and vaccine selective pressures. The second study (Kong et al., 2013) focused on understanding the evolution of penicillin resistance among pneumococci and revealed a surprising directional transmission of penicillin-resistance genes and other genes associated with virulence and antibiotic resistance from the PMEN1 clone to several genetically unrelated clones. All the isolates were resistant to penicillin, cotrimoxazole and chloramphenicol, and shared the same pbp1a allele, whereas multiple alleles of *pbp2b, pbp2x, pspA* and *pspC* were detected. Of the isolates, 89.7% were tetracycline resistant and 60.3% were macrolide resistant, and resistance was associated with different Tn916-like transposons. Genetic variability was observed among PMEN1 isolates collected in our area over the past 20 years (Domenech et al., 2014).

Generally, the proportions of PNSP in Casablanca, Morocco between pre- and post-vaccination periods (PCV-13 introduced in 2010 and replaced by PCV-10) were 31 and 13%. There were 21 *pbp* genotypes in the pre-vaccination period vs. 12 for post-vaccination period. PFGE clustering showed six clusters of PNSP grouped into three clusters specific to pre-vaccination period (clusters I, II and III), two clusters specific to post-period (clusters V and VI) and a cluster (IV) that contained clones belonging to the two periods of vaccination. It was demonstrated a high degree of genetic diversity among PNSP. PFGE clustering combined with *pbp* genotyping revealed that vaccination can change the population structure of PNSP (Diawara et al., 2017).

The prevalence of fluoroquinolone-resistant *S. pneumoniae* (FQRSP) decreased from the initial survey to the second survey (PPV23). Prevalence increased in the third survey (fluoroquinolone restriction) (Ben-David et al., 2014). Fluoroquinolone nonsusceptibility among pneumococci results mainly from point mutations in the QRDR topoisomerase genes (Ben-David et al., 2014; Richter et al., 2005). In the current study, identical mutations were found in different serotypes among the strains in adults. This finding suggests that the mutation occurred before capsular switch. Olivieri et al (Olivieri et al., 2015) evaluated temporal fluctuations in macrolide resistance rates, analyzing genetic determinants of resistance and clonal evolution in a population of 2744 *S.*

pyogenes isolates collected in an area of central Italy, which suggested that changes in bacterial population structure, rather than horizontal transfer of resistance determinants, plays a major epidemiological role in *S. pyogenes*. Zhou et al (Zhou et al., 2014) held a study to identify the mechanism responsible for the horizontal transfer of transposon Tn2010 in *S. pneumoniae*, and the genomic alterations introduced by the transfer process Tn2010 tended to be transferred by transformation rather than conjugation in *S. pneumoniae*, and the spread of Tn2010 could have a profound effect on the evolution of the genome. Many other recombinations were scattered throughout the genome of the transformants in addition to transposon Tn2010. The acquisition of Tn2010 with negligible fitness cost may facilitate spread of the transposon.

Understanding how changes in antibiotic consumption affect the prevalence of antibiotic resistance in bacterial pathogens is important for public health. In a number of bacterial species, including *S. pneumoniae*, the prevalence of resistance has remained relatively stable despite prolonged selection pressure from antibiotics (Lehtinen et al., 2017). The evolutionary processes allowing the robust coexistence of antibiotic sensitive and resistant strains are not fully understood. Lehtinen et al (Lehtinen et al., 2017) suggested that heterogeneity in duration of carriage is a partial explanation for the coexistence of sensitive and resistant strains and that factors determining bacterial duration of carriage will also affect the prevalence of resistance.

Evolution drive vaccine escape

Capsule switching is a natural process that occurs when different pneumococcal serotypes 'swap' their capsular polysaccharide locus through alteration of the capsule biosynthesis locus via mutations (single base changes, insertions or deletions) or genetic recombination. PCVs directly work against a set of pneumococcal serotypes whose capsular polysaccharides have been targeted in the vaccine formulation. The introduction of PCVs has led to the emergence of the capsule-switch variants arising due to the vaccine derived selective pressures. Current vaccine strategies that target a selection of capsular serotypes can lead to the expansion of non-vaccine serotypes (Chaguza et al., 2017; Di Pasquale et al., 2017; Latasa Zamalloa et al., 2017). An increase in the prevalence of these capsule-switched variants can result in serotype replacement, which is the increase of NVT associated pneumococcal clones that follows the decrease in VT associated clones (Wyres et al., 2013). Apart from capsule switch, serotype replacement that occurs after introduction of PCVs is mainly caused by 'serotype unmasking' (Lipsitch, 1999). Unmasking is the process where less prevalent or 'masked' NVT serotypes rise in prevalence to occupy the ecological niche vacated by the 'more' competitive VT serotypes after vaccination (Lipsitch, 1999). However, genetic variation may not always result in phenotypic differences which produce novel serotypes. With the introduction of high throughput whole genome sequencing, discovery of novel genotypic variants is not unexpected. Kapatai et al (Kapatai et al., 2017) found a novel variant of the serotype 23B CPS operon, characterized as a novel genotypic subtype (23B1) with 70% homology to the published 23B CPS sequence. Although the 23B1 variant appears to have no phenotypic impact and cannot be considered as novel serotype, it appears to have led to a genetic restructuring of the UK serotype 23B population. Another work using whole genome sequencing finds a robust and unexpected pattern of serotype switching in a sample of bacteria collected following the introduction of routine anti-pneumococcal vaccination: switching was much more likely to exchange one serotype for another within the same serogroup than expected by chance (Croucher et al., 2015). However, higher rates of within-serogroup switching could not be fully explained by either more frequent, shorter recombinations, or by genetic linkage to genes involved in β–lactam resistance. This suggested the observed pattern was a consequence of selection for preserving serogroup. Several hypotheses are presented and tested to explain this pattern, including limitations of genetic recombination, interactions between the genes that determine serotype and the rest of the genome, and the constraints imposed by bacterial metabolism.

Conclusion and future directions

The impact of genetic recombination to the *Streptococcus* evolution has been proved. Horizontal gene transfer is a key process in the evolution of bacteria and also represents a source of genetic variation in Streptococci. Thanks to the rapid influx of whole genome sequence (WGS) data and the advent of novel analysis methods, future studies should aim to identify the functional roles of certain recombination events in order to provide further insights into pathogenesis, carriage dynamics and strain transmission. Long term evolution-population genomics and tissue tropism, as well as short term evolution focusing comparative genomics in epidemics and outbreaks should be gain much more attention. Furthermore, crosstalks between different recombination mechanisms also should be focused. Here, we reviewed the evolution of representative virulence factors, capsule in *S. pneumoniae*, M protein in GAS, biofilm in GBS and some oral *Streptococcus*, as well as the effect caused by evolution, antibiotic resistance and vaccine escape. Multiply phenotype change, for example capsular switch and multiple antimicrobial nonsusceptibility mutations, can occur in some clone. As to adaptive evolution caused by antibiotics, vaccine and so on, continuous surveillance is an essential to monitor evolution of *S. pneumoniae* causing disease. In order to better survey, control and prevent the emergence of drug-resistant or vaccine escape strains, antimicrobial stewardship, national surveillance and public awareness programs should be developed.

Acknowledgements

We thank Doctor Yanhua Zhao, Shu Feng and Zhonghao Wang for help with references annotation and management, and page proof in this work.

References

Akgul, Y., Word, R.A., Ensign, L.M., Yamaguchi, Y., Lydon, J., Hanes, J., and Mahendroo, M. (2014). Hyaluronan in cervical epithelia protects against infection-mediated preterm birth. The Journal of Clinical Investigation. 124(12), 5481-5489. DOI: 10.1172/JCI78765.

Andam, C.P., and Hanage, W.P. (2015). Mechanisms of genome evolution of Streptococcus. Infection, genetics and evolution. 33, 334-342. DOI: 10.1016/j.meegid.2014.11.007.

Andrade, W.A., Firon, A., Schmidt, T., Hornung, V., Fitzgerald, K.A., Kurt-Jones, E.A., Trieu-Cuot, P., Golenbock, D.T., and Kaminski, P.A. (2016). Group B Streptococcus degrades cyclic-di-AMP to modulate STING-dependent type I interferon production. Cell Host Microbe. 20(1), 49-59. Published online 2016/07/16 DOI: 10.1016/j.chom.2016.06.003.

Appelbaum, P.C. (2002). Resistance among Streptococcus pneumoniae: implications for drug selection. Clinical infectious diseases. 34(12), 1613-1620. DOI: 10.1086/340400.

Athey, T.B., Teatero, S., Takamatsu, D., Wasserscheid, J., Dewar, K., Gottschalk, M., and Fittipaldi, N. (2016). Population structure and antimicrobial resistance profiles of Streptococcus suis serotype 2 sequence type 25 strains. PLoS One. 11(3), e0150908. DOI: 10.1371/journal.pone.0150908.

Ben-David, D., Schwaber, M.J., Adler, A., Masarwa, S., Edgar, R., Navon-Venezia, S., Schwartz, D., Porat, N., Kotlovsky, T., Polivkin, N., et al. (2014). Persistence and complex evolution of fluoroquinolone-resistant Streptococcus pneumoniae clone. Emerging infectious diseases. 20(5), 799-805. DOI: 10.3201/eid2005.130142.

Bessen, D.E., Kumar, N., Hall, G.S., Riley, D.R., Luo, F., Lizano, S., Ford, C.N., McShan, W.M., Nguyen, S.V., Dunning Hotopp, J.C., et al. (2011). Whole-genome association study on tissue tropism phenotypes in group A Streptococcus. Journal of bacteriology. 193(23), 6651-6663. DOI: 10.1128/JB.05263-11.

Bessen, D.E., McShan, W.M., Nguyen, S.V., Shetty, A., Agrawal, S., and Tettelin, H. (2015). Molecular epidemiology and genomics of group A Streptococcus. Infection, genetics and evolution. 33, 393-418. DOI: 10.1016/j.meegid.2014.10.011.

Bowen, W.H., and Koo, H. (2011). Biology of Streptococcus mutans-derived glucosyltransferases: role in extracellular matrix formation of cariogenic biofilms. Caries research. 45(1), 69-86. DOI: 10.1159/000324598.

Brown, S.P., Cornforth, D.M., and Mideo, N. (2012). Evolution of virulence in opportunistic pathogens: generalism, plasticity, and control. Trends in microbiology. 20(7), 336-342. DOI: 10.1016/j.tim.2012.04.005.

Brueggemann, A.B., Harrold, C.L., Rezaei Javan, R., van Tonder, A.J., McDonnell, A.J., and Edwards, B.A. (2017). Pneumococcal prophages are diverse, but not without structure or history. Scientific reports. 7, 42976. DOI: 10.1038/srep42976.

Buscetta, M., Papasergi, S., Firon, A., Pietrocola, G., Biondo, C., Mancuso, G., Midiri, A., Romeo, L., Teti, G., Speziale, P., et al. (2014). FbsC, a novel fibrinogen-binding protein, promotes Streptococcus agalactiae-host cell interactions. Journal of Biological Chemistry. 289(30), 21003-21015. DOI: 10.1074/jbc.M114.553073.

Camilli, R., D'Ambrosio, F., Del Grosso, M., Pimentel de Araujo, F., Caporali, M.G., Del Manso, M., Gherardi, G., D'Ancona, F., Pantosti, A., and Pneumococcal Surveillance, G. (2017). Impact of pneumococcal conjugate vaccine (PCV7 and PCV13) on pneumococcal invasive diseases in Italian children and insight into evolution of pneumococcal population structure. Vaccine. 35(35 Pt B), 4587-4593. DOI: 10.1016/j.vaccine.2017.07.010.

Chaguza, C., Cornick, J.E., Andam, C.P., Gladstone, R.A., Alaerts, M., Musicha, P., Peno, C., Bar-Zeev, N., Kamng'ona, A.W., Kiran, A.M., et al. (2017). Population genetic structure, antibiotic resistance, capsule switching and evolution of invasive pneumococci before conjugate vaccination in Malawi. Vaccine. 35(35 Pt B), 4594-4602. DOI: 10.1016/j.vaccine.2017.07.009.

Chaguza, C., Cornick, J.E., and Everett, D.B. (2015). Mechanisms and impact of genetic recombination in the evolution of Streptococcus pneumoniae. Computational and structural biotechnology journal. 13, 241-247. DOI: 10.1016/j.csbj.2015.03.007.

Chewapreecha, C., Harris, S.R., Croucher, N.J., Turner, C., Marttinen, P., Cheng, L., Pessia, A., Aanensen, D.M., Mather, A.E., Page, A.J., et al. (2014). Dense genomic sampling identifies highways of pneumococcal recombination. Nature genetics. 46(3), 305-309. DOI: 10.1038/ng.2895.

Chi, F., Nolte, O., Bergmann, C., Ip, M., and Hakenbeck, R. (2007). Crossing the barrier: evolution and spread of a major class of mosaic pbp2x in Streptococcus pneumoniae, S. mitis and S. oralis. International journal of medical microbiology. 297(7-8), 503-512. DOI: 10.1016/j.ijmm.2007.02.009.

Chiba, N., Murayama, S.Y., Morozumi, M., Iwata, S., and Ubukata, K. (2017). Genome evolution to penicillin resistance in serotype 3 Streptococcus pneumoniae by capsular switching. Antimicrobial agents and chemotherapy. 61(9), e00478-17. DOI: 10.1128/AAC.00478-17.

Churton, N.W., Misra, R.V., Howlin, R.P., Allan, R.N., Jefferies, J., Faust, S.N., Gharbia, S.E., Edwards, R.J., Clarke, S.C., and Webb, J.S. (2016). Parallel evolution in Streptococcus pneumoniae biofilms. Genome biology and evolution. 8(5), 1316-1326. DOI: 10.1093/gbe/evw072.

Cochetti, I., Tili, E., Vecchi, M., Manzin, A., Mingoia, M., Varaldo, P.E., and Montanari, M.P. (2007). New Tn916-related elements causing erm(B)-mediated erythromycin resistance in tetracycline- susceptible pneumococci. The Journal of antimicrobial chemotherapy. 60(1), 127-131. DOI: 10.1093/jac/dkm120.

Coffey, T.J., Daniels, M., Enright, M.C., and Spratt, B.G. (1999). Serotype 14 variants of the Spanish penicillin-resistant serotype 9V clone of Streptococcus pneumoniae arose by large recombinational replacements of the cpsA-pbp1a region. Microbiology. 145 (Pt8), 2023-2031. DOI: 10.1099/13500872-145-8-2023.

Coffey, T.J., Dowson, C.G., Daniels, M., Zhou, J., Martin, C., Spratt, B.G., and

Musser, J.M. (1991). Horizontal transfer of multiple penicillin-binding protein genes, and capsular biosynthetic genes, in natural populations of Streptococcus pneumoniae. Molecular microbiology. 5(9), 2255-2260.

Cowley, L.A., Petersen, F.C., Junges, R., Jimson, D.J.M., Morrison, D.A., and Hanage, W.P. (2018). Evolution via recombination: cell-to-cell contact facilitates larger recombination events in Streptococcus pneumoniae. PLoS genetics. 14(6), e1007410. DOI: 10.1371/journal.pgen.1007410.

Croucher, N.J., Coupland, P.G., Stevenson, A.E., Callendrello, A., Bentley, S.D., and Hanage, W.P. (2014). Diversification of bacterial genome content through distinct mechanisms over different timescales. Nature communications. 5, 5471. DOI: 10.1038/ncomms6471.

Croucher, N.J., Harris, S.R., Fraser, C., Quail, M.A., Burton, J., van der Linden, M., McGee, L., von Gottberg, A., Song, J.H., Ko, K.S., et al. (2011). Rapid pneumococcal evolution in response to clinical interventions. Science. 331(6016), 430-434. DOI: 10.1126/science.1198545.

Croucher, N.J., Kagedan, L., Thompson, C.M., Parkhill, J., Bentley, S.D., Finkelstein, J.A., Lipsitch, M., and Hanage, W.P. (2015). Selective and genetic constraints on pneumococcal serotype switching. PLoS genetics. 11(3), e1005095. DOI: 10.1371/journal.pgen.1005095.

Croucher, N.J., Mitchell, A.M., Gould, K.A., Inverarity, D., Barquist, L., Feltwell, T., Fookes, M.C., Harris, S.R., Dordel, J., Salter, S.J., et al. (2013). Dominant role of nucleotide substitution in the diversification of serotype 3 pneumococci over decades and during a single infection. PLoS genetics. 9(10), e1003868. DOI: 10.1371/journal.pgen.1003868.

Croucher, N.J., Mostowy, R., Wymant, C., Turner, P., Bentley, S.D., and Fraser, C. (2016). Horizontal DNA transfer mechanisms of bacteria as weapons of intragenomic conflict. PLoS biology. 14(3), e1002394. DOI: 10.1371/journal.pbio.1002394.

D'Urzo, N., Martinelli, M., Pezzicoli, A., De Cesare, V., Pinto, V., Margarit, I., Telford, J.L., and Maione, D. (2014). Acidic pH strongly enhances In vitro biofilm formation by a subset of hypervirulent ST-17 Streptococcus agalactiae strains. Applied and Environmental Microbiology. 80(7), 2176-2185. DOI: 10.1128/aem.03627-13.

Darmon, E., and Leach, D.R. (2014). Bacterial genome instability. Microbiology and molecular biology reviews. 78(1), 1-39. DOI: 10.1128/MMBR.00035-13.

Dewhirst, F.E. (2016). The oral microbiome: critical for understanding oral health and disease. Journal of the California Dental Association. 44(7), 409-410.

Di Pasquale, M., Aliberti, S., Azzari, C., Moriondo, M., Nieddu, F., Blasi, F., and Mantero, M. (2017). Serotypes and antibiotic susceptibility of Streptococcus pneumoniae isolated from hospitalized patients with community-acquired pneumonia in Italy. SAGE open medicine. 5, 2050312117720058. DOI: 10.1177/2050312117720058.

Diawara, I., Barguigua, A., Katfy, K., Nayme, K., Belabbes, H., Timinouni, M., Zerouali, K., and Elmdaghri, N. (2017). Molecular characterization of penicillin non-susceptible Streptococcus pneumoniae isolated before and after pneumococcal conjugate vaccine implementation in Casablanca, Morocco. Annals of clinical microbiology and antimicrobials. 16(1), 23. DOI: 10.1186/s12941-017-0200-6.

Domenech, A., Ardanuy, C., Grau, I., Calatayud, L., Pallares, R., Fenoll, A., Brueggemann, A.B., and Linares, J. (2014). Evolution and genetic diversity of the Spain 23F-ST81 clone causing adult invasive pneumococcal disease in Barcelona (1990-2012). The Journal of antimicrobial chemotherapy. 69(4), 924-931. DOI: 10.1093/jac/dkt473.

Donati, C., Glaser, P., and Mitchell, T. (2015). Evolution and molecular epidemiology of streptococci. Infection, genetics and evolution. 33, 333. DOI: 10.1016/j.meegid.2015.05.028.

Dowson, C.G., Coffey, T.J., Kell, C., and Whiley, R.A. (1993). Evolution of penicillin resistance in Streptococcus pneumoniae; the role of Streptococcus mitis in the formation of a low affinity PBP2B in S. pneumoniae. Molecular microbiology. 9(3), 635-643.

Dowson, C.G., Hutchison, A., Brannigan, J.A., George, R.C., Hansman, D., Linares, J., Tomasz, A., Smith, J.M., and Spratt, B.G. (1989). Horizontal transfer of penicillin-binding protein genes in penicillin-resistant clinical isolates of Streptococcus pneumoniae. Proceedings of the National Academy of Sciences of the United States of America. 86(22), 8842-8846.

Duvvuri, V.R., Deng, X., Teatero, S., Memari, N., Athey, T., Fittipaldi, N., and Gubbay, J.B. (2016). Population structure and drug resistance patterns of emerging non-PCV-13 Streptococcus pneumoniae serotypes 22F, 15A, and 8 isolated from adults in Ontario, Canada. Infection, genetics and evolution. 42, 1-8. DOI: 10.1016/j.meegid.2016.04.007.

Edwards, M.S. (2010). Streptococcus agalactiae (Group B Streptococcus). Mandell, Douglas, and Benett's Principles and Practice of Infectious Diseases. 2655-2666.

El Moujaber, G., Osman, M., Rafei, R., Dabboussi, F., and Hamze, M. (2017). Molecular mechanisms and epidemiology of resistance in Streptococcus pneumoniae in the Middle East region. Journal of medical microbiology. 66(7), 847-858. DOI: 10.1099/jmm.0.000503.

Fan, H., Sun, A.H., Xia, X.P., Sun, Q., and Yan, J. (2009). [Recombinant expression of Streptococcus pneumoniae comD/E/C genes and correlation of ComD/C with beta-lactam antibiotic resistance]. Zhejiang da xue xue bao. Yi xue ban. 38(3), 276-282.

Feil, E.J., Enright, M.C., and Spratt, B.G. (2000). Estimating the relative contributions of mutation and recombination to clonal diversification: a comparison between Neisseria meningitidis and Streptococcus pneumoniae. Res Microbiol. 151(6), 465-469.

Feil, E.J., Maiden, M.C., Achtman, M., and Spratt, B.G. (1999). The relative contributions of recombination and mutation to the divergence of clones of Neisseria meningitidis. Molecular biology and evolution. 16(11), 1496-1502. DOI: 10.1093/oxfordjournals.molbev.a026061.

Ferretti, J.J., Ajdic, D., and McShan, W.M. (2004). Comparative genomics of streptococcal species. Indian J Med Res. 119 Suppl, 1-6.

Flemming, H.C., Wingender, J., Szewzyk, U., Steinberg, P., Rice, S.A., and Kjelleberg, S. (2016). Biofilms: an emergent form of bacterial life. Nature Reviews Microbiology. 14(9), 563-575. DOI: 10.1038/nrmicro.2016.94.

Fontaine, L., Wahl, A., Flechard, M., Mignolet, J., and Hols, P. (2015). Regulation of competence for natural transformation in streptococci. Infection, genetics and evolution. 33, 343-360. DOI: 10.1016/j.meegid.2014.09.010.

Gillings, M.R. (2014). Integrons: past, present, and future. Microbiology and molecular biology reviews. 78(2), 257-277. DOI: 10.1128/MMBR.00056-13.

Gogarten, J.P., and Townsend, J.P. (2005). Horizontal gene transfer, genome innovation and evolution. Nature Reviews Microbiology. 3(9), 679-687. DOI: 10.1038/nrmicro1204.

Hakenbeck, R., Bruckner, R., Denapaite, D., and Maurer, P. (2012). Molecular mechanisms of beta-lactam resistance in Streptococcus pneumoniae. Future microbiology. 7(3), 395-410. DOI: 10.2217/fmb.12.2.

Hakenbeck, R., Grebe, T., Zahner, D., and Stock, J.B. (1999). Beta-lactam resistance in Streptococcus pneumoniae: penicillin-binding proteins and non-penicillin-binding proteins. Molecular microbiology. 33(4), 673-678.

Hobley, L., Harkins, C., MacPhee, C.E., and Stanley-Wall, N.R. (2015). Giving structure to the biofilm matrix: an overview of individual strategies and emerging common themes. FEMS microbiology reviews. 39(5), 649-669. DOI: 10.1093/femsre/fuv015.

Jackson, R.W., Vinatzer, B., Arnold, D.L., Dorus, S., and Murillo, J. (2011). The influence of the accessory genome on bacterial pathogen evolution. Mobile genetic elements. 1(1), 55-65. DOI: 10.4161/mge.1.1.16432.

Jain, R., Rivera, M.C., Moore, J.E., and Lake, J.A. (2003). Horizontal gene transfer accelerates genome innovation and evolution. Molecular biology and evolution. 20(10), 1598-1602. DOI: 10.1093/molbev/msg154.

Jiang, S., and Wessels, M.R. (2014). BsaB, a novel adherence factor of group B Streptococcus. Infection and Immunity. 82(3), 1007-1016. DOI:

10.1128/iai.01014-13.

Johansson, I., Witkowska, E., Kaveh, B., Lif Holgerson, P., and Tanner, A.C. (2016). The microbiome in populations with a low andhigh prevalence of caries. Journal of dental research. 95(1), 80-86. DOI: 10.1177/0022034515609554.

Johnston, C., Caymaris, S., Zomer, A., Bootsma, H.J., Prudhomme, M., Granadel, C., Hermans, P.W., Polard, P., Martin, B., and Claverys, J.P. (2013). Natural genetic transformation generates a population of merodiploids in Streptococcus pneumoniae. PLoS genetics. 9(9), e1003819. DOI: 10.1371/journal.pgen.1003819.

Kapatai, G., Sheppard, C.L., Troxler, L.J., Litt, D.J., Furrer, J., Hilty, M., and Fry, N.K. (2017). Pneumococcal 23B molecular subtype identified using whole genome sequencing. Genome biology and evolution. 9(8), 2122-2135. DOI: 10.1093/gbe/evx092.

Kaspar, J., Underhill, S.A.M., Shields, R.C., Reyes, A., Rosenzweig, S., Hagen, S.J., and Burne, R.A. (2017). Intercellular communication via the comX-Inducing Peptide (XIP) of Streptococcus mutans. Journal of bacteriology. DOI: 10.1128/JB.00404-17.

Kilian, M., Riley, D.R., Jensen, A., Bruggemann, H., and Tettelin, H. (2014). Parallel evolution of Streptococcus pneumoniae and Streptococcus mitis to pathogenic and mutualistic lifestyles. mBio. 5(4), e01490-01414. DOI: 10.1128/mBio.01490-14.

Kohler, W. (2007). The present state of species within the genera Streptococcus and Enterococcus. International journal of medical microbiology. 297(3), 133-150. DOI: 10.1016/j.ijmm.2006.11.008.

Kolar, S.L., Kyme, P., Tseng, C.W., Soliman, A., Kaplan, A., Liang, J., Nizet, V., Jiang, D., Murali, R., Arditi, M., et al. (2015). Group B Streptococcus evades host immunity by degrading hyaluronan. Cell Host Microbe. 18(6), 694-704. DOI: 10.1016/j.chom.2015.11.001.

Kong, I.G., Sato, A., Yuki, Y., Nochi, T., Takahashi, H., Sawada, S., Mejima, M., Kurokawa, S., Okada, K., Sato, S., et al. (2013). Nanogel-based PspA intranasal vaccine prevents invasive disease and nasal colonization by Streptococcus pneumoniae. Infection and immunity. 81(5), 1625-1634. DOI: 10.1128/IAI.00240-13.

Koo, H., Falsetta, M.L., and Klein, M.I. (2013). The exopolysaccharide matrix: a virulence determinant of cariogenic biofilm. Journal of dental research. 92(12), 1065-1073. DOI: 10.1177/0022034513504218.

Latasa Zamalloa, P., Sanz Moreno, J.C., Ordobas Gavin, M., Barranco Ordonez, M.D., Insua Marisquerena, E., Gil de Miguel, A., Fernandez Chavez, A.C., and Garcia-Comas, L. (2018). Trends of invasive pneumococcal disease and its serotypes in the Autonomous Community of Madrid. Enferm Infecc Microbiol Clin. 36(10) ,612-620.DOI: 10.1016/j.eimc.2017.10.026.

Lebeaux, D., Ghigo, J.M., and Beloin, C. (2014). Biofilm-related infections: bridging the gap between clinical management and fundamental aspects of recalcitrance toward antibiotics. Microbiology and molecular biology reviews. 78(3), 510-543. DOI: 10.1128/MMBR.00013-14.

Lehtinen, S., Blanquart, F., Croucher, N.J., Turner, P., Lipsitch, M., and Fraser, C. (2017). Evolution of antibiotic resistance is linked to any genetic mechanism affecting bacterial duration of carriage. Proceedings of the National Academy of Sciences of the United States of America. 114(5), 1075-1080. DOI: 10.1073/pnas.1617849114.

Li, Y., Metcalf, B.J., Chochua, S., Li, Z., Gertz, R.E., Jr., Walker, H., Hawkins, P.A., Tran, T., Whitney, C.G., McGee, L., et al. (2016). Penicillin-binding protein transpeptidase signatures for tracking and predicting beta-lactam resistance levels in Streptococcus pneumoniae. mBio. 7(3). DOI: 10.1128/mBio.00756-16.

Li, Y., Yang, L., Fu, J., Yan, M., Chen, D., and Zhang, L. (2017). Microbial pathogenicity and virulence mediated by integrons on Gram-positive microorganisms. Microbial pathogenesis. 111, 481-486. DOI: 10.1016/j.micpath.2017.09.035.

Lipsitch, M. (1999). Bacterial vaccines and serotype replacement: lessons from Haemophilus influenzae and prospects for Streptococcus pneumoniae.

Emerging infectious diseases. 5(3), 336-345. DOI: 10.3201/eid0503.990304.

Lorenzo-Diaz, F., Fernandez-Lopez, C., Lurz, R., Bravo, A., and Espinosa, M. (2017). Crosstalk between vertical and horizontal gene transfer: plasmid replication control by a conjugative relaxase. Nucleic acids research. 45(13), 7774-7785. DOI: 10.1093/nar/gkx450.

Lourenco, J., Watkins, E.R., Obolski, U., Peacock, S.J., Morris, C., Maiden, M.C.J., and Gupta, S. (2017). Lineage structure of Streptococcus pneumoniae may be driven by immune selection on the groEL heat-shock protein. Scientific reports. 7(1), 9023. DOI: 10.1038/s41598-017-08990-z.

Marks, L.R., Reddinger, R.M., and Hakansson, A.P. (2014). Biofilm formation enhances fomite survival of Streptococcus pneumoniae and Streptococcus pyogenes. Infection and immunity. 82(3), 1141-1146. DOI: 10.1128/IAI.01310-13.

Maruyama, F., Watanabe, T., and Nakagawa, I. (2016). Streptococcus pyogenes genomics. In: Streptococcus pyogenes: Basic Biology to Clinical Manifestations, J.J. Ferretti, D.L. Stevens, and V.A. Fischetti eds. (Oklahoma City (OK)).

Mashburn-Warren, L., Morrison, D.A., and Federle, M.J. (2012). The cryptic competence pathway in Streptococcus pyogenes is controlled by a peptide pheromone. Journal of bacteriology. 194(17), 4589-4600. DOI: 10.1128/JB.00830-12.

McMillan, D.J., Sanderson-Smith, M.L., Smeesters, P.R., and Sriprakash, K.S. (2013). Molecular markers for the study of streptococcal epidemiology. Current topics in microbiology and immunology. 368, 29-48. Published online 2012/11/28 DOI: 10.1007/82_2012_278.

McNeilly, C.L., and McMillan, D.J. (2014). Horizontal gene transfer and recombination in Streptococcus dysgalactiae subsp. equisimilis. Frontiers in microbiology. 5, 676. DOI: 10.3389/fmicb.2014.00676.

McShan, W.M., and Nguyen, S.V. (2016). The Bacteriophages of Streptococcus pyogenes. In: Streptococcus pyogenes: Basic Biology to Clinical Manifestations, J.J. Ferretti, D.L. Stevens, and V.A. Fischetti eds. (Oklahoma City (OK)).

Metzgar, D., and Zampolli, A. (2011). The M protein of group A Streptococcus is a key virulence factor and a clinically relevant strain identification marker. Virulence. 2(5), 402-412. DOI: 10.4161/viru.2.5.16342.

Mira, A., Simon-Soro, A., and Curtis, M.A. (2017). Role of microbial communities in the pathogenesis of periodontal diseases and caries. Journal of clinical periodontology. 44(Suppl 18), S23-S38. DOI: 10.1111/jcpe.12671.

Molto-Garcia, B., Liebana-Martos Mdel, C., Cuadros-Moronta, E., Rodriguez-Granger, J., Sampedro-Martinez, A., Rosa-Fraile, M., Gutierrez-Fernandez, J., Puertas-Priet, A., and Navarro-Mari, J.M. (2016). Molecular characterization and antimicrobial susceptibility of hemolytic Streptococcus agalactiae from post-menopausal women. Maturitas. 85, 5-10. DOI: 10.1016/j.maturitas.2015.11.007.

Monroe, K.M., McWhirter, S.M., and Vance, R.E. (2010). Induction of type I interferons by bacteria. Cell Microbiol. 12(7), 881-890. Published online 2010/05/21 DOI: 10.1111/j.1462-5822.2010.01478.x.

Morrison, D.A., Guedon, E., and Renault, P. (2013). Competence for natural genetic transformation in the Streptococcus bovis group streptococci S. infantarius and S. macedonicus. Journal of bacteriology. 195(11), 2612-2620. DOI: 10.1128/JB.00230-13.

Mostowy, R., Croucher, N.J., Hanage, W.P., Harris, S.R., Bentley, S., and Fraser, C. (2014). Heterogeneity in the frequency and characteristics of homologous recombination in pneumococcal evolution. PLoS genetics. 10(5), e1004300. DOI: 10.1371/journal.pgen.1004300.

Mostowy, R.J., Croucher, N.J., De Maio, N., Chewapreecha, C., Salter, S.J., Turner, P., Aanensen, D.M., Bentley, S.D., Didelot, X., and Fraser, C. (2017). Pneumococcal capsule synthesis locus cps as evolutionary hotspot with potential to generate novel serotypes by recombination. Molecular biology and evolution. 34(10), 2537-2554. DOI: 10.1093/molbev/msx173.

Nobbs, A.H., Jenkinson, H.F., and Everett, D.B. (2015). Generic determinants of Streptococcus colonization and infection. Infection, genetics and evolution. 33, 361-370. DOI: 10.1016/j.meegid.2014.09.018.

Olivieri, R., Morandi, M., Zanchi, A., Tordini, G., Pozzi, G., De Luca, A., and Montagnani, F. (2015). Evolution of macrolide resistance in Streptococcus pyogenes over 14 years in an area of central Italy. Journal of medical microbiology. 64(10), 1186-1195. DOI: 10.1099/jmm.0.000146.

Olson, A.B., Kent, H., Sibley, C.D., Grinwis, M.E., Mabon, P., Ouellette, C., Tyson, S., Graham, M., Tyler, S.D., Van Domselaar, G., et al. (2013). Phylogenetic relationship and virulence inference of Streptococcus Anginosus Group: curated annotation and whole-genome comparative analysis support distinct species designation. BMC genomics. 14, 895. DOI: 10.1186/1471-2164-14-895.

Pinto, T.C., Kegele, F.C., Dias, C.A., Barros, R.R., Peralta, J.M., Merquior, V.L., Carvalho, M.D., Chochua, S., Hawkins, P., McGee, L., et al. (2016). Streptococcus pneumoniae serotypes 9 and 14 circulating in Brazil over a 23-year period prior to introduction of the 10-valent pneumococcal conjugate vaccine: role of international clones in the evolution of antimicrobial resistance and description of a novel genotype. Antimicrobial agents and chemotherapy. 60(11), 6664-6672. DOI: 10.1128/AAC.00673-16.

Provvedi, R., Manganelli, R., and Pozzi, G. (1996). Characterization of conjugative transposon Tn5251 of Streptococcus pneumoniae. FEMS microbiology letters. 135(2-3), 231-236.

Richards, V.P., Lang, P., Bitar, P.D., Lefebure, T., Schukken, Y.H., Zadoks, R.N., and Stanhope, M.J. (2011). Comparative genomics and the role of lateral gene transfer in the evolution of bovine adapted Streptococcus agalactiae. Infection, genetics and evolution. 11(6), 1263-1275. DOI: 10.1016/j.meegid.2011.04.019.

Richter, S.S., Heilmann, K.P., Beekmann, S.E., Miller, N.J., Rice, C.L., and Doern, G.V. (2005). The molecular epidemiology of Streptococcus pneumoniae with quinolone resistance mutations. Clinical infectious diseases. 40(2), 225-235. DOI: 10.1086/426817.

Roberts, A.P., and Mullany, P. (2011). Tn916-like genetic elements: a diverse group of modular mobile elements conferring antibiotic resistance. FEMS microbiology reviews. 35(5), 856-871. DOI: 10.1111/j.1574-6976.2011.00283.x.

Rouge, A., Wintzer-Wehekind, J., Demailly, B., Abdellaoui, M., Faurie, B., and Monsegu, J. (2016). [Purulent pericarditis in a patient with diabetes mellitus treated by percutaneous pericardiocentesis]. Annales de cardiologie et d'angeiologie. 65(5), 370-372. DOI: 10.1016/j.ancard.2016.09.007.

Sabharwal, V., Stevenson, A., Figueira, M., Orthopoulos, G., Trzcinski, K., and Pelton, S.I. (2014). Capsular switching as a strategy to increase pneumococcal virulence in experimental otitis media model. Microbes and infection. 16(4), 292-299. DOI: 10.1016/j.micinf.2013.12.002.

Shanker, E., Morrison, D.A., Talagas, A., Nessler, S., Federle, M.J., and Prehna, G. (2016). Pheromone recognition and selectivity by comR proteins among Streptococcus species. PLoS pathogens. 12(12), e1005979. DOI: 10.1371/journal.ppat.1005979.

Shao, Z.Q., Zhang, Y.M., Pan, X.Z., Wang, B., and Chen, J.Q. (2013). Insight into the evolution of the histidine triad protein (HTP) family in Streptococcus. PloS One. 8(3), e60116. DOI: 10.1371/journal.pone.0060116.

Sibold, C., Henrichsen, J., Konig, A., Martin, C., Chalkley, L., and Hakenbeck, R. (1994). Mosaic pbpX genes of major clones of penicillin-resistant Streptococcus pneumoniae have evolved from pbpX genes of a penicillin-sensitive Streptococcus oralis. Molecular microbiology. 12(6), 1013-1023.

Simon-Soro, A., and Mira, A. (2015). Solving the etiology of dental caries. Trends in microbiology. 23(2), 76-82. DOI: 10.1016/j.tim.2014.10.010.

Sitkiewicz, I. (2017). How to become a killer, or is it all accidental? Virulence strategies in oral streptococci. Molecular oral microbiology. 33(1):1-12. DOI: 10.1111/omi.12192.

Slager, J., Kjos, M., Attaiech, L., and Veening, J.W. (2014). Antibiotic-induced replication stress triggers bacterial competence by increasing gene dosage near the origin. Cell. 157(2), 395-406. DOI: 10.1016/j.cell.2014.01.068.

Smeesters, P.R., McMillan, D.J., and Sriprakash, K.S. (2010). The streptococcal M protein: a highly versatile molecule. Trends in microbiology. 18(6), 275-282.

DOI: 10.1016/j.tim.2010.02.007.

Straume, D., Stamsas, G.A., and Havarstein, L.S. (2015). Natural transformation and genome evolution in Streptococcus pneumoniae. Infection, genetics and evolution. 33, 371-380. DOI: 10.1016/j.meegid.2014.10.020.

Tettelin, H., Masignani, V., Cieslewicz, M.J., Donati, C., Medini, D., Ward, N.L., Angiuoli, S.V., Crabtree, J., Jones, A.L., Durkin, A.S., et al. (2005). Genome analysis of multiple pathogenic isolates of Streptococcus agalactiae: implications for the microbial "pan-genome". Proceedings of the National Academy of Sciences of the United States of America. 102(39), 13950-13955. DOI: 10.1073/pnas.0506758102.

Veening, J.W., and Blokesch, M. (2017). Interbacterial predation as a strategy for DNA acquisition in naturally competent bacteria. Nature Reviews Microbiology. 15(10), 621-629. DOI: 10.1038/nrmicro.2017.66.

Wilkening, R.V., and Federle, M.J. (2017). Evolutionary constraints shaping Streptococcus pyogenes-host interactions. Trends in microbiology. 25(7), 562-572. DOI: 10.1016/j.tim.2017.01.007.

Wyres, K.L., Lambertsen, L.M., Croucher, N.J., McGee, L., von Gottberg, A., Linares, J., Jacobs, M.R., Kristinsson, K.G., Beall, B.W., Klugman, K.P., et al. (2013). Pneumococcal capsular switching: a historical perspective. The Journal of infectious diseases. 207(3), 439-449. DOI: 10.1093/infdis/jis703.

Zhou, H., Guo, J., Qin, T., Ren, H., Xu, Y., Wang, C., and Xu, X. (2017). Serotype and MLST-based inference of population structure of clinical Streptococcus pneumoniae from invasive and noninvasive pneumococcal disease. Infection, genetics and evolution. 55, 104-111. DOI: 10.1016/j.meegid.2017.08.034.

Zhou, W., Yao, K., Zhang, G., Yang, Y., Li, Y., Lv, Y., and Feng, J. (2014). Mechanism for transfer of transposon Tn2010 carrying macrolide resistance genes in Streptococcus pneumoniae and its effects on genome evolution. The Journal of antimicrobial chemotherapy. 69(6), 1470-1473. DOI: 10.1093/jac/dku019.

 https://doi.org/10.21775/9781912530229.14

Progress of Antimicrobial Discovery Against the Major Cariogenic Pathogen *Streptococcus mutans*

Tao Cui[1,2*], Wenfu Luo[1], Letong Xu[1], Baoqiang Yang[1], Wen Zhao[1], Huaixing Cang[2*]

[1]Research & Development Institute of Northwestern Polytechnical University in Shenzhen, Shenzhen 518057, Guangdong, P. R. China.

[2]School of Life Sciences, Northwestern Polytechnical University, Xi'an 710072, Shaanxi, P. R. China.

*Corresponding authors: cuitao@nwpu.edu.cn, hxcang@nwpu.edu.cn

Abstract

Dental caries, also known as tooth decay, or cavities, remains a global public health issue. *Streptococcus mutans* is considered the most significant contributor to dental caries. Its cariogenicity typically depends on its unique metabolic activity and lifestyle, including acid production, acid tolerance and biofilm formation. Currently used anti-caries therapies, such as fluoride and chlorhexidine, are characterized by side-effects and drug resistance. Therefore, the development of alternative inhibitors against *S. mutans* growth is urgently needed. In the last decade, a larger number of natural products and their derivatives from plants, marine organisms and microorganisms were studied to evaluate their antibacterial activity against *S. mutans*. In addition, drug-repositioning base screening and target based high-throughput screening were employed, resulting in inspiring progresses in recent years. In this review, we summarized the available evidences regarding the inhibition of *S. mutans* growth. We focus on the sources, structures and potential mechanism of action of these inhibitors. Beside small molecular compounds, we also considered antibacterial peptides and protein inhibitors developed in this field.

Introduction

Tooth decay, also known as dental caries or cavities, is a significant public health problem worldwide (Pitts *et al.*, 2017). Approximately 2.43 billion people (36% of the population) have dental caries. Tooth decay is a result of demineralization of the tooth by acids made by oral microorganisms (Figure 1). Food debris and sugars on the surface of the tooth are metabolized by microorganisms, which produce acids and lead to tooth decay. The oral cavity is inhabited by over 600 different bacterial species (Dewhirst *et al.*, 2010; Kilian *et al.*, 2016), among which *S. mutans* is the most prosperous colonizer causing dental caries. During carbohydrate metabolism, *S. mutans* produce lactic acid, leading to a drop of pH in the environment, which subsequently inhibits the growth of other microbial competitors, making *S. mutans* a predominant species in the oral cavity.

Unlike other pathogens that live in the blood, cells or tissues that are of difficult access, *S. mutans* lives in the tooth surface, which is easily and mechanically accessible, thus, mechanical approaches for removing *S. mutans* can be considered. However, *S. mutans* can tightly adhere to the tooth surface and its efficient removal is difficult by mechanical method due to the formation of a biofilm. Dental plaque biofilm is made by a community of oral microorganisms, which adhere to the tooth surface, and *S. mutans* is the most predominant species (Ajdić *et al.*, 2002; Mitchell, 2003; Nobbs *et al.*, 2009; Flemming *et al.*, 2010; Kolenbrander *et al.*, 2010; Avilés-Reyes *et al.*, 2017).

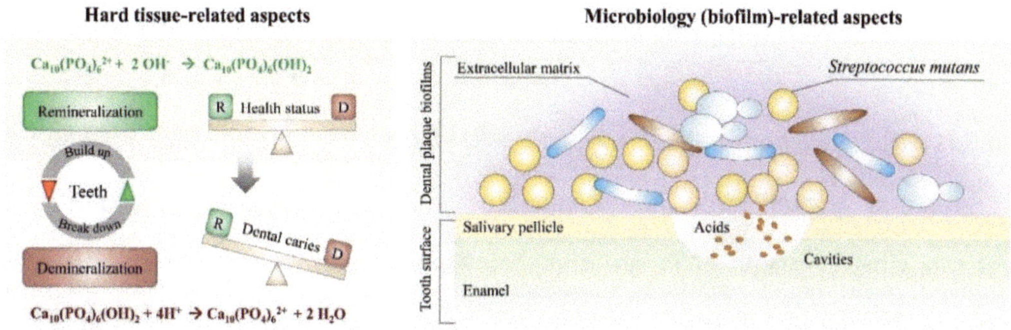

Figure 1. Dental caries and *S. mutans*. Dental caries is a result of tooth demineralization by acids produced by microorganisms in the dental plaque. Dental plaque is a multi-species and cross-kingdom biofilm in which *S. mutants* is the predominant species. $Ca_{10}(PO_4)_6(OH)_2$ is the formula of hydroxyapatite, which is the primary mineral of tooth enamel.

Taken together, at least three indispensable factors play essential roles in the cariogenicity of *S. mutans*. The first one is the acid production. Lactic acid produced by *S. mutans* metabolic activity breakdowns the tooth enamel and causes a pH decline of the surrounding environment, inhibiting the growth of other species and making *S. mutans* the dominant species in the oral cavity. The second one is the acid tolerance (Matsui *et al.*, 2010; Smith *et al.*, 2012; Baker *et al.*, 2017). Acid production results in a drop of pH in the local environment. Therefore, *S. mutans* should be (and it actually is) able to respond and adapt to this acid environment to survive. The third one is the biofilm formation. The formation of a biofilm results in multiple benefits for *S. mutans*.

Indeed, biofilm formation allows *S. mutans* to tightly adhere to the tooth surface, thus avoiding being removed. Moreover, its formation prevents the dilution or/and neutralization of acids produced in the biofilm by water or saliva. Furthermore, biofilm is helpful for adsorbing nutrition. Finally, biofilm formation reduces the access of antibiotics to *S. mutans,* thus increasing antibiotic resistance.

In this review, we summarized the available evidences regarding the inhibition of *S. mutans* growth. We focus on the sources, structures and potential mechanisms of action of these inhibitors. Beside small molecular compounds, we also considered antibacterial peptides and protein inhibitors developed in this field.

1. Small molecules currently used in dental cavity treatment

Currently, widely used anti-caries compounds include fluoride, chlorhexidine, and xylitol (Figure 2). Most toothpastes contain between 0.22% and 0.312% fluoride, usually in the form of sodium fluoride or sodium monofluorophosphate (Buzalaf *et al.*, 2011; Pandit *et al.*, 2011; Takahashi *et al.*, 2011; Liu *et al.*, 2012; Nóbrega *et al.*, 2016; Demonte *et al.*, 2017; Liao *et al.*, 2017). Sodium fluoride came into use to prevent tooth decay in the 1940s. The effects of fluoride on dental caries can be divided into the following two: 1. Fluoride reduces tooth enamel decay by the formation of fluorapatite and its incorporation into it; 2. Fluoride exhibits an inhibitory action on *S. mutans* growth. The glycolytic pathway contributes to the acid production by *S. mutans*. Fluoride inhibits the glycolytic pathway by directly targeting the enolase enzyme (Takahashi *et al.*, 2011). The accumulation of internal polysaccharide when sugars are in excess helps *S. mutans* to respond to starvation stress, thus contributing to cariogenicity by *S. mutans.* A recent study showed that sodium monofluorophosphate inhibits internal polysaccharide biosynthesis in *S. mutans* by targeting the *S. mutans* ADP-glucose pyrophosphorylase (Demonte *et al.*, 2017). Chlorhexidine is a broad-spectrum antibiotic, came into medical use in the 1950s and it is the most widely used antibacterial agent against *S. mutans.* Chlorhexidine is positively charged in solution and can bind to the bacterial cell wall, damaging its cell membrane (Liu *et al.*, 2012; Coelho *et al.*, 2017). Xylitol is a sugar alcohol used as a sweetener and it attracted popular attention because evidence suggests that it reduces the incidence of cavities. Several potential mechanisms of action are proposed for its anti-caries activity. Xylitol is non-fermentable and cannot be metabolized into lactic acid by *S. mutans*, as well as it cannot be used as an energy source although being taken up into the cell to interfere with bacterial growth and reproduction (Takahashi *et al.*, 2011). Xylitol can promote the remineralization of the tooth enamel by coordinating with and transporting polyvalent cations, such as Ca^{2+}.

Current anti-caries compounds mainly contribute to the prevention and reduction of the incidence of dental caries. However, they have very limited success in the treatment of already formed dental caries. In addition, they also face the challenges of side-effects and drug resistance. These compounds possess low selectivity, which may destroy the homeostasis of oral microbiome. Due to the extensive use of fluoride, many microorganisms developed a resistance against its effects. For example, fluoride antiporter protein mediates microbial fluoride resistance by exporting fluoride ions (F^-) to maintain a low F^- level in the cell (Liao *et al.*, 2017). Therefore, the development of alternative inhibitors against *S. mutans* growth is expected. In this review, we summarized current efforts on the development of inhibitors against *S. mutans* growth.

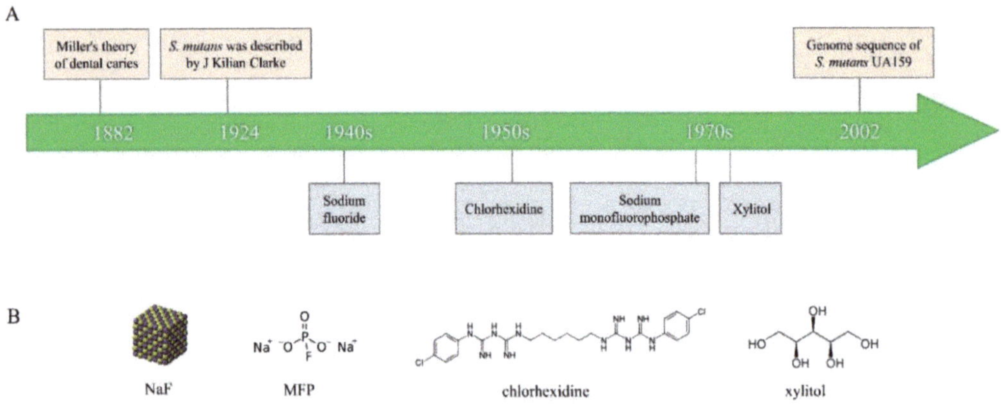

Figure 2. Timeline of our understanding on the pathology of dental caries and development of anti-caries agents. Current treatments for dental caries have limited success and face the problems of having side-effects and developing drug resistance. Alternative antimicrobials against *S. mutans* are required.

2. Natural products from plants

Natural products are rich sources of compounds with potential powers against several diseases, including caries. Plants are the important source of natural products (Koehn *et al.*, 2005; Harvey *et al.*, 2015). Numerous plant natural products exhibit inhibitory activities against *S. mutans* growth. According to their chemical structures, these compounds mainly include polyphenolic compounds, especially flavonoids (Figure 3). However, their mechanism of action against *S. mutans* survival is not very clear. The potential mechanisms include a specific action on cell membrane and a modification of its permeability, leading to cell death. At the transcriptional level, some compounds can induce the change in the expression of genes involved in biofilm formation. Some compounds can inhibit enzyme activity involved in acid production pathways. Interestingly, many of these compounds can be found in tea, thus suggesting its potential anti-caries effect (Jeon *et al.*, 2011; Cheng *et al.*, 2012; Abachi *et al.*, 2016).

Earlier studies suggested that flavonoids exhibit inhibitory activities against *S. mutans* growth. Flavonoids have the general structure of a 15-carbon skeleton, which consists of two phenyl rings and heterocyclic ring. This carbon structure can be abbreviated C6-C3-C6. Compounds in this category include apigenin (Koo *et al.*, 2003; Koo *et al.*, 2005; Jeon *et al.*, 2009), myricetin (Falsetta *et al.*, 2012), quercetin, kaempferol (Guan *et al.*, 2012) and epigallocatechin gallate (EGCG) (Xu *et al.*, 2011; Bai *et al.*, 2016). Apigenin affects the accumulation and polysaccharide content of *S. mutans* biofilms without major impact on bacterial viability (Koo *et al.*, 2003; Koo *et al.*, 2005; Jeon *et al.*, 2009). Myricetin inhibits the ability of *S. mutans* to assemble biofilms without altering cell viability. It inhibits glucosyltransferase (Gtf) activity and reduces the expression of the *gtfBC* gene cluster, which are involved in the synthesis of glucans (Falsetta *et al.*, 2012). Quercetin and kaempferol inhibit the growth of *S. mutans* and several other oral bacteria. In addition, They also significantly inhibit acidogenicity, acidurity and F-ATPase activity of *S. mutans* (Guan *et al.*, 2012). EGCG inhibits various cariogenic virulent factors of *S. mutans* at the transcriptional and enzymatic levels, leading to reduced acidogenicity and compromised stress tolerance (especially acid tolerance). Real-time PCR results demonstrated that EGCG significantly suppresses the *ldh*, *eno*, *atpD*, and *aguD* genes, which encode virulence factors that are involved in acid production and stress response of *S. mutans*. Inhibition of the enzymatic activity of F_1F_o-ATPase and lactate dehydrogenase is also observed (Xu et al., 2011).

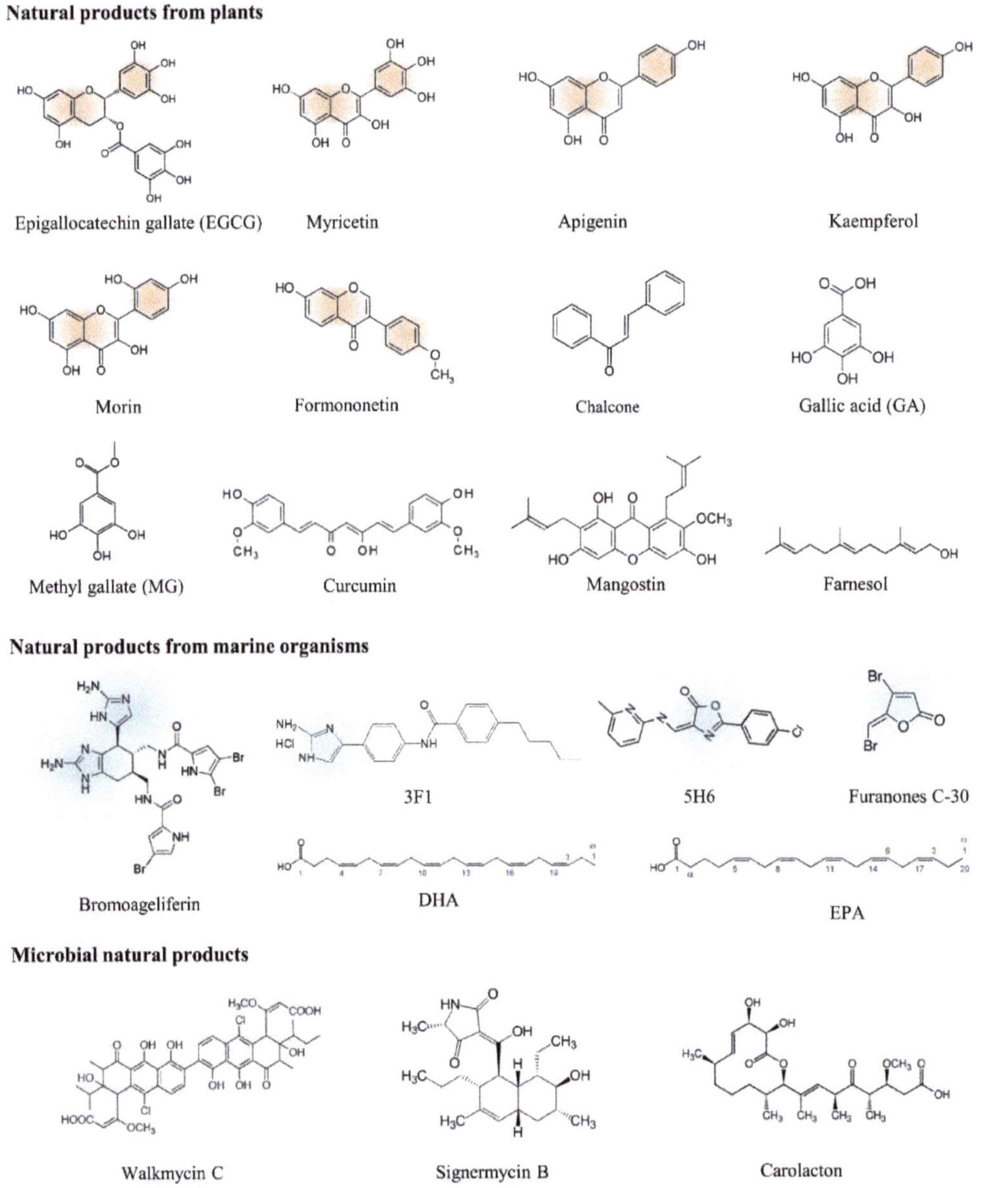

Figure 3. *S. mutans* growth inhibitors from plant natural products, marine natural products and microbial natural products. Natural products from plants mainly include polyphenolic compounds, especially flavonoids (EGCG, myricetin, apigenin, kaempferol, morin, and formononetin). Marine natural products mainly include heterocyclic compounds (imidazole ring in bromoageliferin and 3F1; oxazole ring in 5H6; furanone ring in furanones C-30) and fatty acids (DHA and EPA).

Several flavonoids, such as morin (Huang *et al.*, 2014) and formononetin (Park *et al.*, 2017) inhibit the activity of Sortase A (SrtA). SrtA is a membrane-associated transpeptidase responsible for the anchoring of surface-exposed proteins to the cell wall envelope of gram-positive bacteria (Hendrickx *et al.*, 2011), including *S. mutans.* A mutant *S. mutans* strain lacking *srtA* exhibits a decreased ability to colonize oral mucosa and teeth, consequently reducing caries formation. Therefore, this enzyme represents a potential target in the treatment and/or prevention of dental caries (Park *et al.*, 2017). Morin is another natural plant extract that inhibits *S. aureus* and *S. mutans* SrtA activity, the latter inhibited by an IC_{50} of 27.2 ± 2.6 μM. *S. mutans* adheres to the tooth surface through the sortase A (SrtA)-mediated cell wall anchored protein Pac. Western blotting demonstrated that 30 μM morin induces Pac protein partial release into the supernatant (Huang *et al.*, 2014). *trans*-chalcone is the precursor molecule of many flavonoids, and it inhibits *S. mutans* SrtA *in vitro* and biofilm formation. Mass spectrometry revealed that *trans*-chalcone forms a Michael addition adduct with the active site on SrtA cysteine (Wallock-Richards *et al.*, 2015). Recently, two new flavonoids, 7-hydroxy-6-methoxyflavanone and formononetin, were found as strong inhibitors against *S. mutans* SrtA activity, with IC_{50} values of 46.1 and 41.8 μM, respectively (Park *et al.*, 2017).

Other polyphenols with inhibitory activity against *S. mutans* survival include tannic acid (Sendamangalam *et al.*, 2011), gallic acid (GA) (Kang *et al.*, 2008, Sendamangalam *et al.*, 2011; Shao *et al.*, 2015) and methyl gallate (MG) (Kang *et al.*, 2008; Kacergius *et al.*, 2017). Tannic acid inhibits *S. mutans* biofilm formation. Enzyme activity assay indicated that tannic acid exhibits an inhibitory effect on two enzymes, glucosyltransferase and fructosyltransferase. These two enzymes are involved in the synthesis of glucans and fructans, the important components of extracellular polymers (Sendamangalam *et al.*, 2011). Tannic acid can be hydrolyzed into GA. GA can be found in many plants including tea leaves. It has inhibitory effects on the growth of cariogenic and periodontopathic bacteria. Moreover, GA inhibits the *in vitro* formation of *S. mutans* biofilms (Kang *et al.*, 2008; Sendamangalam *et al.*, 2011; Shao *et al.*, 2015). Methyl gallate is the methyl ester of gallic acid. It has antibacterial and antibiofilm effects on *S. mutans*. (Kang *et al.*, 2008; Kacergius *et al.*, 2017). In addition, MG inhibits the acidogenicity of *S. mutans* (Kacergius *et al.*, 2017).

Carvacrol and thymol are natural monoterpene phenols and can be found in the plant *Origanum vulgare* L. Carvacrol and thymol IC_{50} values against *S. mutans* were 65 and 54 µg/ml, respectively. The mechanism of action may be their involvement in the permeabilization and depolarization of the cytoplasmic membrane. Deformed and lysed cells as observed under scanning electron microscope further confirm that thymol and carvacrol treatment results in the lysis of the cells (Khan *et al.*, 2017).

Other natural products from plants include curcumin (Hu *et al.*, 2013a; Hu *et al.*, 2013b), α-mangostin (Nguyen *et al.*, 2011; Nguyen *et al.*, 2014) and *trans-trans* farnesol (Koo *et al.*, 2003; Jeon *et al.*, 2011). Curcumin is an inhibitor of *S. aureus* and *S. mutans* SrtA activity (Hu *et al.*, 2013a) and it inhibits *S. mutans* biofilm formation (Hu *et al.*, 2013b). Curcumin exerts a strong inhibitory activity against SrtA with a half maximal inhibitory concentration (IC_{50}) of 10.2±0.7 μM. α-mangostin is a xanthone purified from ethanolic extracts of the tropical plant *Garcinia mangostana* L. α-Mangostin is a potent inhibitor of acid production by *S. mutans* UA159. Studies on the mechanism of action suggest that α-mangostin is a multitarget inhibitor. α-Mangostin is active against membrane enzymes, including the F(H+)-ATPase and the phosphoenolpyruvate - sugar phosphotransferase system. It also inhibits the glycolytic enzymes aldolase, glyceraldehyde-3-phosphate dehydrogenase, and lactic dehydrogenase. Glycolysis in intact cells in suspensions or biofilm formation is inhibited by α-mangostin at concentrations of 12 and 120 μM, respectively. Other α-mangostin inhibitory effects include (i) malolactic fermentation, involved in alkali production from malate, and (ii) NADH oxidase, the major respiratory enzyme for *S. mutans* (Nguyen *et al.*, 2011; Nguyen *et al.*, 2014). *tt*-farnesol is a bioactive sesquiterpene alcohol commonly found in propolis (a beehive product) and citrus fruits. It disrupts the ability of *S. mutans* to form biofilms. In general, *tt*-farnesol significantly increases the membrane proton permeability and reduces the glycolytic activity of *S. mutans* in the planktonic state and in biofilms. It may also affect the competitiveness of *S. mutans* in a mixed-species environment by primarily disrupting the membrane function and physiology of this bacterium (Koo *et al.*, 2003; Jeon *et al.*, 2011). In addition, *tt*-farnesol inhibits acid production, acid tolerance, and polysaccharide synthesis in *S. mutans* (Falsetta *et al.*, 2012).

Crude extracts and essential oils

Plants from traditional medicine, folk medicine, and herbal medicine represent valuable resources for identifying bioactive compounds. The discovery of artemisinin as anti-malaria drug from traditional Chinese medicine is one of the most successful examples (Tu, 2011). Beside pure compounds, crude extracts and the essential oils from plants are also investigated to identify alternative antimicrobial agents against *S. mutans* (Morimoto-Yamashita *et al.*, 2011; Swadas *et al.*, 2016; Gulube *et al.*, 2016; Yadav *et al.*, 2016; Tofiño-Rivera *et al.*, 2016; Lall *et al.*, 2017; Janardhanan *et al.*, 2017; Liu *et al.*, 2017; Tardugno *et al.*, 2018). The constituents are unknown for most extracts and require further activity-based fractionation to identify the active constituents. It is worthy of attention that extracts from fruit, seed, fruit peel, plant leaves for tea beverages and plants used as condiments in cooking exhibit an inhibitory action against *S. mutans* growth.

3. Natural products from marine organisms

Marine organisms are another important source of natural products (Molinski *et al.*, 2009). Many marine natural products possess antibacterial and antibiofilm activity on a wide range of bacteria. For example, marine sponge-derived natural product bromoageliferin is a potent and versatile antibiofilm agent. Therefore, it attracts the interest of researchers in their aim of discovering potential inhibitors against *S. mutants* growth from marine natural products (Figure 3). Current reported marine natural products with antibacterial activity against *S. mutans* include oxazole derivatives (Chen *et al.*, 2016), 2-Aminoimidazole (2-AI) derivatives (Rogers *et al.*, 2008; Liu *et al.*, 2011; Pan *et al.*, 2015; Garcia *et al.*, 2017), free fatty acids (Huang *et al.*, 2010; Sun *et al.*, 2017) and furanones (He *et al.*, 2012).

Oxazole derivatives

Oxazole and its natural product-related derivatives, especially those from marine environment, possess significant biological activities, including anti-tumor, antibacterial, anti-viral, antimalaria, and anthelmintic activities. Recently, an oxazole derivative, named 5H6, was identified as a new inhibitor against *S. mutans* survival. 5H6 inhibits *S. mutans* biofilm formation and adhesion without affecting bacterial growth. *In vivo* study using a rat caries model showed that 5H6 disrupts the development of smooth-surface caries and sulcal-surface carious lesions to the same extent as treatment with fluoride at 250 µg/mL. A study on the mechanism of action suggested that 5H6 prevents the synthesis of extracellular polysaccharides by antagonizing Gtfs (Chen *et al.*, 2016).

2-AI derivatives

As mentioned above, marine sponge–derived bromoageliferin is a broad spectrum antibiofilm agent. The 2-AI functional group plays key roles in its antibiofilm activity. Therefore, many 2-AI derivatives were synthesized to identify more effective and specific biofilm inhibitors. Encouragingly, numerous 2-AI derivatives exhibit inhibitory activity against *S. mutans* biofilm formation (Liu *et al.*, 2011; Pan *et al.*, 2015; Garcia *et al.*, 2017). For example, 2-AI derivative 3F1 disperses mature *S. mutans* biofilm. However, 3F1 do not disperse biofilms formed by the commensal species *Streptococcus sanguinis* or *Streptococcus gordonii.* In addition, 3F1 treatment effectively prevents dental caries by controlling *S. mutans* in a rat caries model without perturbing the oral microbiota. These results suggest the specificity of 3F1 against *S. mutans* survival. The mechanism of 3F1 action is not dependent on well-characterized biofilm factors, including AgI/II, Gtfs, glucans, and eDNA, since they are often associated with the initial development of *S. mutans* biofilm, while proteins and factors involved in biofilm maintenance and maturation are largely unknown. Thus, 3F1 may represent a target as a novel biofilm-related protein potentially inhibiting a crucial interaction between *S. mutans* and the biofilm matrix *in vitro* (Garcia *et al.*, 2017).

Free fatty acids

Unsaturated fatty acids such as docosahexaenoic acid (DHA) and eicosapentaenoic acid (EPA) display antibacterial effects on mature *S. mutans* biofilm (Huang *et al.*, 2010; Sun *et al.*, 2017). Free fatty acids are found naturally at high levels in many marine organisms. DHA and EPA possess highly inhibitory effects on various Gram-positive and Gram-negative bacteria survival. DHA and EPA mainly affect the integrity of the bacterial plasma membrane, thereby leading to bacteria damage and death. In addition, DHA or EPA may target extracellular polysaccharides (EPS) synthesis, since an expression profile showed the downregulation of *gtfB* and *ftf* gene, which encode enzymes that catalyze the cleavage of sucrose to synthesize extracellular glucan and fructan polysaccharides, facilitating bacteria adherence and aggregation.

Furanones

Furanones are natural inhibitors of quorum sensing (QS) signaling system, which regulate biofilm formation in many species. The red marine alga *D. pulchra* developed a defense mechanism to protect itself from extensive bacterial colonization by producing brominated furanones as inhibitors. Synthetic furanone C-30 inhibits *S. mutans* biofilm formation. The mechanism of action of furanones is that these molecules, whose structures are similar to autoinducers of QS signaling system, can bind to QS response regulators but fail to activate them, thus inhibiting the QS system (He *et al.*, 2012).

4. Microbial natural products

Microorganisms usually secrete molecules to antagonize competitors. Therefore, many antibiotics were and are currently isolated from microbial natural products. Several microbial natural products are inhibitors of *S. mutans* survival (Figure 3). These compounds include walkmycin C (Eguchi *et al.*, 2011), signermycin B (Watanabe *et al.*, 2012), carolacton (Kunze *et al.*, 2010; Reck *et al.*, 2011; Li *et al.*, 2013; Sudhakar *et al.*, 2014) and vizantin (Takenaka *et al.*, 2016).

The distinguishing feature of these compounds is that they target signaling system of *S. mutans*. Walkmycin C inhibited the *in vitro* autophosphorylation activity of three purified *S. mutans* histidine kinases, including VicK, CiaH, and LiaS. Walkmycin C treatment represses biofilm formation, acid tolerance, and competence of *S. mutans* (Eguchi *et al.*, 2011). Signermycin B is identified as an inhibitor of histidine kinase WalK. It targets the conserved dimerization domain of WalK to inhibit autophosphorylation (Watanabe *et al.*, 2012).

Carolacton is a secondary metabolite isolated from *Sorangium cellulosum*. It has high activity against *S. mutans* biofilm formation. At a concentration of 10 nM carolacton, biofilm damage is already at 35% under anaerobic conditions. Planktonic growth of bacteria is only slightly impaired and no acute cytotoxicity against mouse fibroblasts is observed (Kunze *et al.*, 2010). A knock-out mutant for *comD*, encoding a histidine kinase specific for the competence stimulating peptide (CSP) is slightly less sensitive to carolacton than the wildtype (Kunze *et al.*, 2010). Competence related alternate sigma factor ComX expression is strongly reduced by carolacton. A comparative time series microarray analysis identified VicKRX and ComDE as two-component signal transduction system and genes involved in cell wall metabolism, playing essential roles in the response to carolacton treatment. A sensitivity testing of mutants with deletion of all the 13 viable histidine kinases and the serine/threonine protein kinase PknB of *S. mutans* identified only the ΔpknB deletion mutant as being insensitive to carolacton treatment. A strong overlap between the regulon of PknB in *S. mutans* and the genes affected by carolacton treatment was found. The results suggested that carolacton acts by interfering with PknB-mediated signaling in growing cells. The resulting altered cell wall morphology causes membrane damage and cell death at low pH (Reck *et al.*, 2011). Comparative proteome analysis identified 192 proteins whose expression is affected by carolacton. Function analysis of these proteins indicated that carolacton exerts its inhibitory effects by disturbing peptidoglycan biosynthesis and degradation, thereby causing damages to the integrity of the cell envelope, leading ultimately to cell death (Li *et al.*, 2013). A regulatory network analysis identified several regulators involved in the response of *S. mutans* to carolacton (Sudhakar *et al.*, 2014). It was recently found that carolacton targets FolD, a key enzyme from the folate-dependent C1 metabolism in *S. pneumonia* (Fu *et al.*, 2017).

Vizantin is a molecule derived from trehalose-6,6′-dimycolate, a major constituent of the outer surface membrane of the human pathogen *Mycobacterium tuberculosis*. Vizantin was originally developed as a safe immune-stimulating compound (Oda *et al.*, 2014). Recently, vizantin has also been found to have antibiofilm activity against *S. mutans*. Notably, supplementation with sulfated vizantin up to 50 μM do not affect either bacterial growth or biofilm formation, whereas 50 μM sulfated vizantin causes the detachment of the biofilm from the surface. Sulfated vizantin at the concentration of 50 μM upregulates *gtfB* and *gtfC* gene expression, but downregulates *gtfD* gene expression, suggesting an altered architecture in the biofilm (Takenaka *et al.*, 2016).

5. Drug repositioning

Drug repositioning is a drug development strategy that tries to find new applications of existing drugs (Ashburn and Thor, 2004). For example, a recent study found the anti-cancer activity of the alcohol-abuse drug disulfiram (Skrott *et al.*, 2017). Drug repositioning strategy has been used to investigate inhibitors against *S. mutans* growth in recent years (Figure 4). Several small-scale drug-repositioning studies identified trimetrexate analogues (Zhang *et al.*, 2015), saxagliptin (De *et al.*, 2016), zafirlukast (Gerits *et al.*, 2017), clotrimazole, econazole (Qiu *et al.*, 2017) and reserpine (Zeng *et al.*, 2017) as inhibitors against *S. mutans* growth. Particularly, a latest high-throughput drug-repositioning screening of 853 FDA-approved drugs identified 126 drugs with inhibitory activity against planktonic *S. mutans* growth. These drugs were further tested for activity against *S. mutans* biofilm and the results showed that 24 drugs inhibit biofilm formation, 6 drugs kill preexisting biofilms, 84 exhibit both biofilm inhibition and killing activity, and 12 has no activity against biofilms (Saputo *et al.*, 2017a).

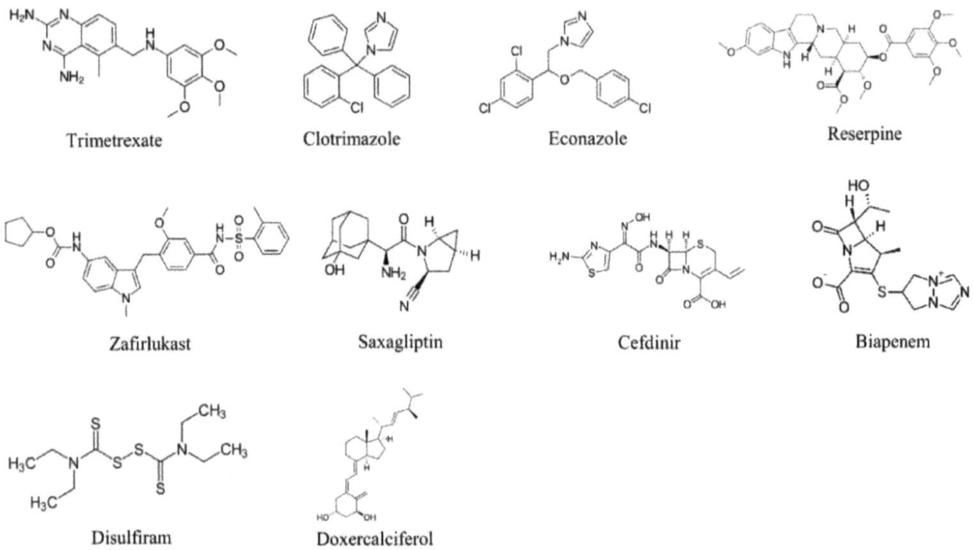

Figure 4. Examples of *S. mutans* growth inhibitors from drug-repositioning studies.

Trimetrexate analogs

Trimetrexate is a quinazoline derivative and targets human dihydrofolate reductase (DHFR). A library of analogs based on trimetrexate was screened and three new analogs that selectively inhibits *S. mutans* survival were identified. The most potent inhibitor has an IC_{50} of 454.0±10.2 nM related to the biofilm formation and 8.7±1.9 nM associated to DHFR activity in *S. mutans*. In contrast, the IC_{50} of this compound for human DHFR is approximately 1000 nM, a >100-fold decreased potency, demonstrating the high selectivity of the analog. An analog that exhibited the least potency against *S. mutans* biofilm formation has also the lowest activity towards *S. mutans* DHFR inhibition, further indicating that inhibition of biofilms is related to a reduced DHFR activity (Zhang *et al.*, 2005).

Saxagliptin

Saxagliptin is used in the treatment of type 2 diabetes. It targets human dipeptidyl peptidase-4 (DPP-4). *S. mutans* encodes an intracellular Xaa-Pro dipeptidyl aminopeptidase (XPDAP) which is similar to the human DPP-4 enzyme. XPDAP is a narrow-range serine protease, which cleaves oligopeptides with a proline as the penultimate residue from the N-terminus. Several studies have suggested importance of XPDAP in utilization of proline-rich salivary peptides as well as virulence in *S. mutans* (De *et al.*, 2016). The inhibition constants (K_i) of anti-human DPP-4 drugs sitagliptin, vildagliptin, and saxagliptin against *S. mutans* XPDAP activity were determined, showing that saxagliptin inhibits the activity of *S. mutans* XPDAP (K_i = 129 ± 16 μM) (De *et al.*, 2016).

Zafirlukast

Zafirlukast is used for the chronic treatment of asthma. It targets human leukotriene receptor antagonist (LTRA). Zafirlukast inhibits *S. mutans* growth (MIC: 25 μM), biofilm formation (MBIC: 50 μM) and reduces preformed biofilm (MBRC: 12.5 μM). In addition, zafirlukast exerts no toxicity on human osteoblasts (Gerits *et al.*, 2017).

Clotrimazole and Econazole

Dental plaque is not only a multi-species biofilm but also a cross-kingdom biofilm. For example, *C. albicans*, a fungal microorganism, is frequently detected with heavy infection by *S. mutans* in plaque-biofilms from children with early-childhood caries. Furthermore, several recent studies suggested the synergistic effects between *S. mutans* and *C. albicans* on biofilm formation (Hwang *et al.*, 2017; Ellepola *et al.*, 2017). Therefore, compounds with inhibitory activity against both *S. mutans* and *C. albicans* survival may have more effective anti-caries activity. Based on this hypothesis, inhibitory effects of eight antifungal drugs on *S. mutans* survival were determined in a recent study. Two antifungal drugs, clotrimazole and econazole show an antibacterial activity on *S. mutans* UA159 and *S. mutans* clinical isolates at 12.5 and 25 mg/L, respectively. Clotrimazole and econazole can also inhibit *S. mutans* biofilm formation and reduce the viability of preformed biofilm. In addition, they inhibit pH drop, lactate acid production and acid tolerance in *S. mutans*. The antifungal effects of clotrimazole and econazole are similar. They kill individual fungal cells by altering the permeability of their cell wall. They bind to phospholipids in the cell membrane and inhibit the biosynthesis of ergosterol and other sterols required for cell membrane formation, leading to cell death via loss of intracellular elements. As regard *S. mutans*, clotrimazole and econazole inhibit the expression of *S. mutans ldh* gene, which is involved in the acid production. However, their specific targets in *S. mutans* and their mechanism of action underlining multiple inhibitory activities on *S. mutans* are not very clear and require further analysis. Anyhow, this study provided a starting point in the development of inhibitors that target two relevant species that contribute in the formation of dental plaque biofilm (Qiu *et al.*, 2017).

Reserpine

Bacterial efflux pumps are membrane proteins that can extrude substrates out of the cells, including drugs, antibiotics, chemicals and signal molecules. Efflux pumps play important roles in bacterial pathogenicity, antibiotic resistance and biofilm formation. Two predicted ABC efflux pumps encoded by *rcrP* and *rcrQ* are associated with cellular physiology, genetic competence, and acid and oxidative stress tolerance of *S. mutans*. Therefore, inhibiting the efflux pump with its inhibitors may suppress the virulence properties of *S. mutans*. Inhibitory activities of several efflux pump inhibitors, including reserpine, verapamil hydrochloride, thioridazine hydrochloride and sodium orthovanadate against *S. mutans* growth were tested. Results indicated that reserpine suppresses acid tolerance, mutacin production and transformation efficiency in *S. mutans*, and modifies biofilm architecture and extracellular polysaccharide distribution. Studies on the mechanism of action suggested that reserpine inhibits glycosyltransferase activity in *S. mutans*. Results from quantitative real-time PCR demonstrated that reserpine significantly alters QS expression profile and virulence-associated genes (Zeng *et al.*, 2017)

Doxercalciferol

Bacitracin is a polypeptide-type drug that interferes with cell wall synthesis. Innate resistance to bacitracin is attributed to efflux by a conserved ABC-type transporter, which is encoded by the mbrABCD operon in *S. mutans*. Doxercalciferol is a vitamin D drug derivative that exhibits direct bactericidal activity against the cariogenic *S. mutans*. In addition, through the use of a combinatorial approach, doxercalciferol was shown to act synergistically with bacitracin, suggesting that doxercalciferol may act in a bacitracin-related pathway (Saputo *et al.*, 2017b).

6. Target-based screening

Target-based screening strategy firstly identifies essential proteins involved in the pathogenicity of bacteria. Secondly, the inhibitory activity of compounds against target proteins was tested by an enzyme activity experiment or another protein function experiment. If the three-dimensional structure of the target protein is available, a structure-based screening is usually performed before the experimental test. Afterwards, compounds with activity on target proteins are evaluated by phenotype assays (Terstappen *et al.*, 2007; Swinney and Anthony, 2011; Eder *et al.*, 2014). One advantage of target-based screening is that the screening can be easily performed in a high-throughput manner. In addition, target proteins selection provides the opportunity for selecting unique proteins of pathogens as candidate targets for screening, which is important for the discovery of inhibitors with high specificity. This is particularly important for inhibitory screening against *S. mutans*. The expected ideal inhibitor needs to be antibacterial to *S. mutans* but without destroying of oral microbiome. Two proteins, GtfC (Ren *et al.*, 2015; Ren *et al.*, 2016; Zhang *et al.*, 2017) and QS signaling protein ComA (Ishii *et al.*, 2017) have been used in target-based inhibitor screening against *S. mutans*.

Glucosyltransferase

Biofilm formation in *S. mutans* is promoted by major virulence factors known as Gtfs, which synthesize adhesive EPS. A structure-based virtual screening of approximately 150,000 commercially available compounds was performed against the crystal structure of the glucosyltransferase domain of the GtfC protein from *S. mutans*. Then, 51 chemically diverse lead compounds were further screened for their inhibitory activities on EPS synthesis *in vitro*, resulting in the identification of a quinoxaline derivative as a potential Gtf inhibitor. *In vitro* assays showed that this compound is capable of inhibiting EPS synthesis and biofilm formation in *S. mutans* by selectively antagonizing Gtfs instead of killing the bacteria directly. Moreover, the *in vivo* anti-caries efficacy of the compound was evaluated in a rat model. The results showed that the compound significantly reduces the incidence and severity of smooth and sulcal surface caries *in vivo* with a concomitant reduction in the percentage of *S. mutans* in the animals' dental plaque (Ren *et al.*, 2016). In another study, a small molecule library of 500,000 small molecule compounds was screened *in silico* against the GtfC catalytic domain. A total of 90 compounds with diverse scaffolds which vary in their functional groups, hydrophobicity, and H-bond accepting/donating capacity were then purchased and subjected to *in vitro* biofilm assays using *S. mutans*. Seven potent low micromolar inhibitors were identified. Two of these compounds were the most potent, as they inhibit more than 85% of *S. mutans* biofilms at 12.5 μM. Compounds #G16 and #G43 share several functional groups including a nitro group, heterocyclic rings, and a polar carbonyl functional property (Zhang *et al.*, 2017).

S. mutans expresses 3 genetically distinct Gtfs, including GtfB, GtfC, and GtfD. Recently, it was found that GtfB plays key roles in cross-kingdom biofilm formation of *S. mutans* and *C. albicans* (Hwang *et al.*, 2017; Ellepola *et al.*, 2017). This property makes GtfB a potential candidate target for a future inhibitor screening.

QS signaling system protein ComA

The QS system is a bacterial cell-cell signal communication system mediated by an inherent signal molecule called autoinducer. The ComABCDE pathway is a QS system in *S. mutans*. Autoinducer peptides are processed from the precursor ComC and concomitantly exported to the extracellular space by ComA and ComB. The accumulated autoinducers bind to the membrane-bound receptor kinase ComD, which subsequently phosphorylates the response regulator ComE to activate the transcription of a specific set of genes, such as those essential for the competence development and biofilm formation in *S. mutans*. ComA is a bi-functional ATP-binding cassette (ABC) transporter that comprises three domains: a N-terminal peptidase domain (PEP), a transmembrane domain, and a C-terminal nucleotide-binding domain. A high-throughput screening system using a fluorescence-labeled substrate was performed to identify inhibitors of PEP in *S. mutans*. The first screening of 164,514 compounds yielded 951 hits (0.58%) that inhibits PEP in *S. mutans* activity by >50% at a compound concentration of 20 μM. These compounds were further subjected to screening that examined inhibition against *S. mutans* biofilm formation. Finally, six compounds were found to inhibit biofilm formation without inhibiting cell growth. Two of the six compounds are quinuclidine derivatives, and the other four compounds have no primary chemical structure in common (Ishii *et al.*, 2017).

7. Antibacterial peptides and proteins

Besides small molecule inhibitors, antibacterial peptides were also investigated for its activity against *S. mutans* survival. Compared to small molecule antibiotics, the advantages of antibacterial peptides include broad-spectrum antimicrobial activity, low toxicity to mammalian cell and less selection stress. Less selection stress decreases the risk of antibiotic-resistant bacteria development, which is one of the biggest challenges in the treatment of bacterial infection. In addition, antibacterial peptides are encoded by genes and thus they can be easily modified with genetic engineering techniques to obtain more derivatives compared to small molecule derivatives obtained by chemical synthesis. Several antibacterial peptides, such as nisin (Tong *et al*, 2010; Yamakami *et al.*, 2013), IMB-2 (Mai *et al.*, 2011), decapeptide KSL (Liu *et al.*, 2011), cyclic lipopeptide 4 (CLP-4) (Min *et al.*, 2017) and lipopeptide C14KKc12K (Meir *et al.*, 2017), possess antimicrobial activity against *S. mutans.*

As we mentioned above, antimicrobial peptides are usually broad-spectrum. The use of antimicrobial peptides in the oral cavity may disturb the oral microbiome and cause unexpected problems. Therefore, it is better to improve the specificity of antimicrobial peptides against *S. mutans*. The development of the antimicrobial peptide IMB-2 is an enlightening work for this purpose. IMB-2 is modified from a marine-derived, broad-spectrum antimicrobial peptide pleurocidin. ComC signaling peptide pheromone (CSP) can bind to *S. mutans* ComC. The foundation for creating IMB-2 is based on the addition of a targeting domain of *S. mutans* ComC CSP to a killing domain of pleurocidin to generate a target-specific antimicrobial peptide. The results showed that nearly 95% of *S. mutans* bacteria lost their viability following exposure to fusion peptide IMB-2 (5.65 μM) for 15 min. In contrast, only 20% of *S. sanguinis* or *S. gordonii* bacteria were killed following the same exposure. A *S. mutans* mutant defective in the

CSP receptor resulted in 60% survival following exposure to IMB-2, suggesting that the targeted peptide predominantly bound to the CSP receptor to mediate the killing of the wild-type strain (Mai *et al.*, 2011).

The stability of the antimicrobial peptide in oral environment is another potential problem for antimicrobial peptide inhibitors against *S. mutans*. Nisin, produced by *Lactococcus lactis*, is an antibiotic peptide to effectively antagonize a broad spectrum of Gram-positive bacteria. Nisin targets the membrane-bound peptidoglycan precursor lipid II to inhibit peptidoglycan biosynthesis, leading to membrane pore formation. Nisin displays an antimicrobial activity against multiple oral microorganisms including *S. mutans* (Tong *et al.* 2010). The naked form of nisin is sensitive to proteolytic degradation and oxidation. To improve its stability, an encapsulated form of nisin within liposomes has been developed. The concentration of nisin-liposome required for an efficacious inhibition of glucan-biofilm synthesis is four times lower than that of naked nisin (Yamakami *et al.*, 2013).

Recently, two studies also tried to develop protein inhibitors against *S. mutans* (Yang *et al.*, 2016; Ito *et al.*, 2017). Bacteriophage-encoded lysins are peptidoglycan hydrolases that digest the bacterial cell wall, leading to the rapid lysis and death of bacteria. ClyR, an engineered lysin, is active against all clinical *S. mutans* isolates tested and *S. mutans* biofilms (Yang *et al.*, 2016). Another protein inhibitor against *S. mutans* is agaricus bisporus agglutinin (ABA), which belongs to lectin. The mechanism of action of ABA is largely different from the one of ClyR. Oral bacteria initiate biofilm formation by adhering to the tooth surface through an interaction of a lectin-like bacterial protein with carbohydrate chains on the pellicle. ABA binds to the carbohydrate chains and thus competitively inhibits the attachment of *S. mutans* to carbohydrate chains on the pellicle (Ito *et al.*, 2017).

Conclusion and future perspectives

Dental caries remains a worldwide public health problem. *S. mutans* is the primary pathogen in dental caries. Current anti-caries agents face the challenge of side-effects and drug resistance. Therefore, alternative therapies are expected. Recently years, many inhibitors against *S. mutans* were identified by phenotypic or target-based screening. Most of these inhibitors come from natural products and their derivatives. These compounds provide an important basis for the development of new anti-caries therapies in the future. However, several works also need to be addressed in the future. First, the inhibitory effects of many compounds were only tested by *in vitro* assay. Their anti-caries ability in animal model, influence on the oral microbiome and cytotoxicity need to be further evaluated. Second, inhibitory mechanisms at the molecular level are largely unknown for many of these compounds. Target identification of these inhibitors is required for further understanding their mechanism of action. Last, many of the currently proposed inhibitors are broad spectrum antimicrobials. Identifying essential and unique proteins in *S. mutans* is required for uncovering more candidate targets for a future target-base screening of inhibitors with high specificity.

Demonte, A.M., Asencion, Diez, M.D., Naleway, C., Iglesias, A.A., Ballicora, M.A. (2017). Monofluorophosphate Blocks Internal Polysaccharide Synthesis in *Streptococcus mutans*. PLoS. One. 12, e0170483 DOI: 10.1371/journal.pone.0170483

Dewhirst, F.E., Chen, T., Izard, J., Paster, B.J., Tanner, A.C., Yu, W.H., Lakshmanan, A., Wade, W.G. (2010). The human oral microbiome. J. Bacteriol. 192, 5002-5017 DOI: 10.1128/JB.00542-10

Eder, J., Sedrani, R., Wiesmann, C. (2014). The discovery of first-in-class drugs: origins and evolution. Nat. Rev. Drug. Discov. 13, 577-587 DOI: 10.1038/nrd4336

Eguchi, Y., Kubo, N., Matsunaga, H., Igarashi, M., Utsumi, R. (2011). Development of an antivirulence drug against *Streptococcus mutans*: repression of biofilm formation, acid tolerance, and competence by a histidine kinase inhibitor, walkmycin C. Antimicrob. Agents Chemother. 55, 1475-1484 DOI: 10.1128/AAC.01646-10

Ellepola, K., Liu, Y., Cao, T., Koo, H., Seneviratne, C.J. (2017). Bacterial GtfB Augments *Candida albicans* Accumulation in Cross-Kingdom Biofilms. J. Dent. Res. 96, 1129-1135 DOI: 10.1177/0022034517714414

Falsetta, M.L., Klein, M.I., Lemos, J.A., Silva, B.B., Agidi, S., Scott-Anne, K.K., Koo, H. (2012). Novel antibiofilm chemotherapy targets exopolysaccharide synthesis and stress tolerance in *Streptococcus mutans* to modulate virulence expression *in vivo*. Antimicrob. Agents Chemother. 56, 6201-6211 DOI: 10.1128/AAC.01381-12

Flemming, H.C., Wingender, J. (2010). The biofilm matrix. Nat. Rev. Microbiol. 8, 623-633 DOI: 10.1038/nrmicro2415

Freires, I.A., Denny, C., Benso, B., de, Alencar, S.M., Rosalen, P.L. (2015).

Antibacterial Activity of Essential Oils and Their Isolated Constituents against Cariogenic Bacteria: A Systematic Review. Molecules. 20, 7329-7358 DOI: 10.3390/molecules20047329

Fu, C., Sikandar, A., Donner, J., Zaburannyi, N., Herrmann, J., Reck, M., Wagner-Döbler, I., Koehnke, J., Müller, R. (2017). The natural product carolacton inhibits folate-dependent C1 metabolism by targeting FolD/MTHFD. Nat. Commun. 8, 1529 DOI: 10.1038/s41467-017-01671-5

Garcia, S.S., Blackledge, M.S., Michalek, S., Su, L., Ptacek, T., Eipers, P., Morrow, C., Lefkowitz, E.J., Melander, C., Wu, H. (2017). Targeting of *Streptococcus mutans* Biofilms by a Novel Small Molecule Prevents Dental Caries and Preserves the Oral Microbiome. J. Dent. Res. 96, 807-814 DOI: 10.1177/0022034517698096

Gerits, E., Van der Massen, I., Vandamme, K., De Cremer, K., De Brucker, K., Thevissen, K., Cammue, B.P.A., Beullens, S., Fauvart, M., Verstraeten, N., Michiels, J., Roberts, M. (2017). *In vitro* activity of the antiasthmatic drug zafirlukast against the oral pathogens *Porphyromonas gingivalis* and *Streptococcus mutans*. FEMS. Microbiol. Lett. 364 DOI: 10.1093/femsle/fnx005

Guan, X., Zhou, Y., Liang, X., Xiao, J., He, L., Li, J. (2012). Effects of compounds found in *Nidus Vespae* on the growth and cariogenic virulence factors of *Streptococcus mutans*. Microbiol. Res. 167, 61-68 DOI: 10.1016/j.micres.2011.03.002

Gulube, Z., Patel, M. (2016). Effect of *Punica granatum* on the virulence factors of cariogenic bacteria *Streptococcus mutans*. Microb. Pathog. 98, 45-49 DOI: 10.1016/j.micpath.2016.06.027

Harvey, A.L., Edrada-Ebel, R., Quinn, R.J. (2015). The re-emergence of natural capacity of *Rhus coriaria* L. extract and its major component methyl gallate on *Streptococcus mutans* biofilm formation by optical profilometry: Potential applications for oral health. Mol. Med. Rep. 16, 949-956 DOI: 10.3892/mmr.2017.6674

Kang, M.S., Oh, J.S., Kang, I.C., Hong, S.J., Choi, C.H. (2008). Inhibitory effect of methyl gallate and gallic acid on oral bacteria. J. Microbiol. 46, 744-750 DOI: 10.1007/s12275-008-0235-7

Khan, S.T., Khan, M., Ahmad, J., Wahab, R., Abd-Elkader, O.H., Musarrat, J., Alkhathlan, H.Z., Al-Kedhairy, A.A. (2017). Thymol and carvacrol induce autolysis, stress, growth inhibition and reduce the biofilm formation by *Streptococcus mutans*. AMB. Express. 7, 49 DOI: 10.1186/s13568-017-0344-y

Kilian, M., Chapple, I.L., Hannig, M., Marsh, P.D, Meuric, V., Pedersen, A.M., Tonetti, M.S., Wade, W.G., Zaura, E. (2016). The oral microbiome - an update for oral healthcare professionals. Br. Dent. J. 221, 657-666 DOI: 10.1038/sj.bdj.2016.865

Koehn, F.E., Carter, G.T. (2005). The evolving role of natural products in drug discovery. Nat. Rev. Drug Discov. 4, 206-220 DOI: 10.1038/nrd1657

Kolenbrander, P.E., Palmer, R.J. Jr., Periasamy, S., Jakubovics, N.S. (2010). Oral multispecies biofilm development and the key role of cell-cell distance. Nat. Rev. Microbiol. 8, 471-480 DOI: 10.1038/nrmicro2381

Koo, H., Allan, R.N., Howlin, R.P., Stoodley, P., Hall-Stoodley, L. (2017). Targeting microbial biofilms: current and prospective therapeutic strategies. Nat. Rev. Microbiol. 15, 740-755 DOI: 10.1038/nrmicro.2017.99

Koo, H., Hayacibara, M.F., Schobel, B.D., Cury, J.A, Rosalen, P.L, Park YK, Vacca-Smith AM, Bowen WH. (2003). Inhibition of *Streptococcus mutans* biofilm accumulation and polysaccharide production by apigenin and tt-farnesol. J Antimicrob Chemother. 52, 782-789 DOI: 10.1093/jac/dkg449

Koo, H., Schobel, B., Scott-Anne, K., Watson, G., Bowen, W.H., Cury, J.A., Rosalen, P.L., Park, Y.K. (2005). Apigenin and tt-farnesol with fluoride effects on *S. mutans* biofilms and dental caries. J. Dent. Res. 84, 1016-1020 DOI: 10.1177/154405910508401109

Kunze, B., Reck, M., Dötsch, A., Lemme, A., Schummer, D., Irschik, H., Steinmetz, H., Wagner-Döbler, I. (2010). Damage of *Streptococcus mutans* biofilms by carolacton, a secondary metabolite from the myxobacterium *Sorangium cellulosum*. BMC. Microbiol. 26, 10:199 DOI: 10.1186/1471-2180-10-199

Lall, N., Kishore, N., Bodiba, D., More, G., Tshikalange, E., Kikuchi, H., Oshima, Y. (2017). Alkaloids from aerial parts of *Annona senegalensis* against *Streptococcus mutans*. Nat. Prod. Res. 31, 1944-1947 DOI: 10.1080/14786419.2016.1263847

Li, J., Wang, W., Wang, Y., Zeng, A.P. (2013). Two-dimensional gel-based proteomic of the caries causative bacterium *Streptococcus mutans* UA159 and insight into the inhibitory effect of carolacton. Proteomics. 13, 3470-3477 DOI: 10.1002/pmic.201300077

Liao, Y., Brandt, B.W., Li., J, Crielaard, W., Van, Loveren, C., Deng, D.M. (2017). Fluoride resistance in *Streptococcus mutans*: a mini review. J. Oral Microbiol. 9, 1344509 DOI: 10.1080/20002297.2017.1344509

Liu, C., Worthington, R.J., Melander, C., Wu, H. (2011). A new small molecule specifically inhibits the cariogenic bacterium *Streptococcus mutans* in multispecies biofilms. Antimicrob. Agents Chemother. 55, 2679-2687 DOI:

10.1128/AAC.01496-10

Liu, J., Ling, J.Q., Zhang, K., Huo, L.J., Ning, Y. (2012). Effect of sodium fluoride, ampicillin, and chlorhexidine on *Streptococcus mutans* biofilm detachment. Antimicrob. Agents Chemother. 56, 4532-4535 DOI: 10.1128/AAC.00885-12

Liu, Y., Wang, L., Zhou, X., Hu, S., Zhang, S., Wu, H. (2011). Effect of the antimicrobial decapeptide KSL on the growth of oral pathogens and *Streptococcus mutans* biofilm. Int. J. Antimicrob. Agents. 37, 33-38 DOI: 10.1016/j.ijantimicag.2010.08.014

Liu, Y., Xu, Y., Song, Q., Wang, F., Sun, L., Liu, L., Yang, X., Yi, J., Bao, Y., Ma, H., Huang, H., Yu, C., Huang, Y., Wu, Y., Li, Y. (2017). Anti-biofilm Activities from *Bergenia crassifolia* Leaves against *Streptococcus mutans*. Front Microbiol. 8, 1738 DOI: 10.3389/fmicb.2017.01738

Mai, J., Tian, X.L., Gallant, J.W., Merkley, N., Biswas, Z., Syvitski, R., Douglas, S.E., Ling, J., Li, Y.H. (2011). A novel target-specific, salt-resistant antimicrobial peptide against the cariogenic pathogen *Streptococcus mutans*. Antimicrob. Agents Chemother. 55, 5205-5213 DOI: 10.1128/AAC.05175-11

Matsui, R., Cvitkovitch, D. (2010). Acid tolerance mechanisms utilized by *Streptococcus mutans*. Future Microbiol. 5, 403-417 DOI: 10.2217/fmb.09.129

Meir, O., Zaknoon, F., Cogan, U., Mor, A. (2017). A broad-spectrum bactericidal lipopeptide with anti-biofilm properties. 7, 2198 DOI: 10.1038/s41598-017-02373-0

Min, K.R., Galvis, A., Williams, B., Rayala, R., Cudic, P., Ajdic, D. (2017). Antibacterial and Antibiofilm Activities of a Novel Synthetic Cyclic Lipopeptide Against Cariogenic *Streptococcus mutans* UA159. Antimicrob. Agents Chemother. 61 DOI: 10.1128/AAC.00776-17

Mitchell, T.J. (2003). The pathogenesis of streptococcal infections: from tooth decay to meningitis. Nat. Rev. Microbiol. 1, 219-230 DOI: 10.1038/nrmicro771

Molinski, T.F., Dalisay, D.S., Lievens, S.L., Saludes, J.P. (2009). Drug development from marine natural products. Nat. Rev. Drug Discov. 8, 69-85 DOI: 10.1038/nrd2487

Morimoto-Yamashita, Y., Matsuo, M., Komatsuzawa, H., Kawahara, K., Kikuchi, K., Torii, M., Tokuda, M. (2011). MK615: a new therapeutic approach for the treatment of oral disease. Med. Hypotheses. 77, 258-260 DOI: 10.1016/j.mehy.2011.04.027

Nguyen, P.T., Falsetta, M.L., Hwang, G., Gonzalez-Begne, M., Koo, H. (2014). α-Mangostin disrupts the development of *Streptococcus mutans* biofilms and facilitates its mechanical removal. PLoS. One. 9, e111312 DOI: 10.1371/journal.pone.0111312

Nguyen, P.T., Marquis, R.E. (2011). Antimicrobial actions of α-mangostin against oral streptococci. Can. J. Microbiol. 57, 217-225 DOI: 10.1139/W10-122

Nobbs, A.H., Lamont, R.J., Jenkinson, H.F. (2009). *Streptococcus* adherence and colonization. Microbiol. Mol. Biol. Rev. 73, 407-450, DOI: 10.1128/MMBR.00014-09

Nóbrega, D.F., Fernández, C.E., Del, Bel, Cury, A.A., Tenuta, L.M., Cury, J.A. (2016). Frequency of Fluoride Dentifrice Use and Caries Lesions Inhibition and Repair. Caries Res. 50, 133-140 DOI: 10.1159/000444223

Oda, M., Yamamoto, H., Shibutani, M., Nakano, M., Yabiku, K., Tarui, T., Kameyama, N., Shirakawa, D., Obayashi, S., Watanabe, N., Nakase, H., Suenaga, M., Matsunaga, Y., Nagahama, M., Takahashi, H., Imagawa, H., Kurosawa, M., Terao, Y., Nishizawa, M., Sakurai, J. (2014). Vizantin inhibits endotoxin-mediated immune responses via the TLR 4/MD-2 complex. J. Immunol. 193, 4507-4514 DOI: 10.4049/jimmunol.1401796

Pan, W., Fan, M., Wu, H., Melander, C., Liu, C. (2015). A new small molecule inhibits *Streptococcus mutans* biofilms *in vitro* and *in vivo*. J. Appl. Microbiol. 119, 1403-1411 DOI: 10.1111/jam.12940

Pandit, S., Kim, J.E., Jung, K.H., Chang, K.W., Jeon, J.G. (2011). Effect of sodium fluoride on the virulence factors and composition of *Streptococcus mutans* biofilms. Arch. Oral Biol. 56, 643-649 DOI: 10.1016/j.archoralbio.2010.12.012

Park, W., Ahn, C.H., Cho, H., Kim, C.K., Shin, J., Oh, K.B. (2017). Inhibitory effects of flavonoids from *Spatholobus suberectus* on sortase A and sortase A-mediated aggregation of *Streptococcus mutans*. J. Microbiol. Biotechnol. 27, 1457-1460 DOI: 10.4014/jmb.1704.04001

Pitts, N.B., Zero, D.T., Marsh, P.D., Ekstrand, K., Weintraub, J.A., Ramos-Gomez, F., Tagami, J., Twetman, S., Tsakos, G., Ismail, A. (2017). Dental caries. Nat. Rev. Dis. Primers. 3, 17030 DOI: 10.1038/nrdp.2017.30

Qiu, W., Ren, B., Dai, H., Zhang, L., Zhang, Q., Zhou, X., Li, Y. Clotrimazole and econazole inhibit *Streptococcus mutans* biofilm and virulence *in vitro*. 73, 113-120 DOI: 10.1016/j.archoralbio.2016.10.011

Reck, M., Rutz, K., Kunze, B., Tomasch, J., Surapaneni, S.K., Schulz, S., Wagner-Döbler, I. (2011). The biofilm inhibitor carolacton disturbs membrane integrity and cell division of *Streptococcus mutans* through the serine/threonine protein kinase PknB. J. Bacteriol. 193, 5692-5706 DOI: 10.1128/JB.05424-11

Ren, Z., Chen, L., Li, J., Li, Y. (2016). Inhibition of *Streptococcus mutans* polysaccharide synthesis by molecules targeting glycosyltransferase activity. J.

Oral Microbiol. 8, 31095 DOI: 10.3402/jom.v8.31095

Ren, Z., Cui, T., Zeng, J., Chen, L., Zhang, W., Xu, X., Cheng, L., Li, M., Li, J., Zhou, X., Li, Y. (2015). Molecule Targeting Glucosyltransferase Inhibits *Streptococcus mutans* Biofilm Formation and Virulence. Antimicrob. Agents Chemother. 60, 126-135 DOI: 10.1128/AAC.00919-15

Rogers, S.A., Melander, C. (2008). Construction and screening of a 2-aminoimidazole library identifies a small molecule capable of inhibiting and dispersing bacterial biofilms across order, class, and phylum. Angew. Chem. Int. Ed. Engl. 47, 5229-5231 DOI: 10.1002/anie.200800862

Saputo, S., Faustoferri, R.C., Quivey, R.G. Jr. (2017a). A drug repositioning approach reveals *Streptococcus mutans* is susceptible to a diverse range of established antimicrobials and non-antibiotics. Antimicrob. Agents Chemother. 62 DOI: 10.1128/AAC.01674-17

Saputo, S., Faustoferri, R.C., Quivey, R.G. Jr. (2017b). Vitamin D compounds are bactericidal against *Streptococcus mutans* and target the bacitracin-associated efflux system. Antimicrob. Agents Chemother. 62 DOI: 10.1128/AAC.01675-17

Sendamangalam, V., Choi, O.K., Kim, D., Seo, Y. (2011). The anti-biofouling effect of polyphenols against *Streptococcus mutans*. Biofouling. 27, 13-19 DOI: 10.1080/08927014.2010.535897

Shao, D., Li, J., Li, J., Tang, R., Liu, L., Shi, J., Huang, Q., Yang, H. (2015). Inhibition of Gallic Acid on the Growth and Biofilm Formation of *Escherichia coli* and *Streptococcus mutans*. J. Food. Sci. 80, M1299-M1305 DOI: 10.1111/1750-3841.12902

Skrott, Z., Mistrik, M., Andersen, K.K., Friis, S., Majera, D., Gursky, J., Ozdian, T., Bartkova, J., Turi, Z., Moudry, P., Kraus, M., Michalova, M., Vaclavkova, J.,

Dzubak, P., Vrobel, I., Pouckova, P., Sedlacek, J., Miklovicova, A., Kutt, A., Li, J., Mattova, J., Driessen, C., Dou, Q.P., Olsen, J., Hajduch, M., Cvek, B., Deshaies, R.J., Bartek, J. (2017). Alcohol-abuse drug disulfiram targets cancer via p97 segregase adaptor NPL4. Nature. 552, 194-199 DOI: 10.1038/nature25016

Smith, E.G., Spatafora, G.A. (2012). Gene regulation in *S. mutans*: complex control in a complex environment. J. Dent. Res. 91, 133-141 DOI: 10.1177/0022034511415415

Sudhakar, P., Reck, M., Wang, W., He, F.Q., Wagner-Döbler, I., Zeng, A.P. (2014). Construction and verification of the transcriptional regulatory response network of *Streptococcus mutans* upon treatment with the biofilm inhibitor carolacton. BMC. Genomics. 15, 362 DOI: 10.1186/1471-2164-15-362

Sun, M., Dong, J., Xia, Y., Shu, R. (2017). Antibacterial activities of docosahexaenoic acid (DHA) and eicosapentaenoic acid (EPA) against planktonic and biofilm growing *Streptococcus mutans*. Microb. Pathog. 107, 212-218 DOI: 10.1016/j.micpath.2017.03.040

Swadas, M., Dave, B., Vyas, S.M., Shah, N. (2016). Evaluation and Comparison of the Antibacterial Activity against *Streptococcus mutans* of Grape Seed Extract at Different Concentrations with Chlorhexidine Gluconate: An *in vitro* Study. Int. J. Clin. Pediatr. Dent. 9, 181-185 DOI: 10.5005/jp-journals-10005-1360

Swinney, D.C., Anthony, J. How were new medicines discovered? (2011). Nat. Rev. Drug. Discov. 10, 507-519 DOI: 10.1038/nrd3480

Takahashi, N., Washio, J. (2011). Metabolomic effects of xylitol and fluoride on plaque biofilm *in vivo*. J. Dent. Res. 90, 1463-1468 DOI: 10.1177/0022034511423395

Takenaka, S., Oda, M., Domon, H., Ohsumi, T., Suzuki, Y., Ohshima, H.,

Yamamoto, H., Terao, Y., Noiri, Y. (2016). Vizantin inhibits bacterial adhesion without affecting bacterial growth and causes *Streptococcus mutans* biofilm to detach by altering its internal architecture. Biochem. Biophys. Res. Commun. 480, 173-179 DOI: 10.1016/j.bbrc.2016.10.021

Tardugno, R., Pellati, F., Iseppi, R., Bondi, M., Bruzzesi, G., Benvenuti, S. (2018). Phytochemical composition and *in vitro* screening of the antimicrobial activity of essential oils on oral pathogenic bacteria. Nat. Prod. Res. 32, 544-551 DOI: 10.1080/14786419.2017.1329730

Terstappen, G.C., Schlüpen, C., Raggiaschi, R., Gaviraghi, G. (2007). Target deconvolution strategies in drug discovery. Nat. Rev. Drug. Discov. 6, 891-903 DOI: 10.1038/nrd2410

Tofiño-Rivera, A., Ortega-Cuadros, M., Galvis-Pareja, D., Jiménez-Rios, H., Merini, L.J., Martínez-Pabón, M.C. (2016). Effect of *Lippia alba* and *Cymbopogon citratus* essential oils on biofilms of *Streptococcus mutans* and cytotoxicity in CHO cells. J. Ethnopharmacol. 194, 749-754 DOI: 10.1016/j.jep.2016.10.044

Tong, Z., Dong, L., Zhou, L., Tao, R., Ni, L. (2010). Nisin inhibits dental caries-associated microorganism *in vitro*. Peptides. 31, 2003-2008 DOI: 10.1016/j.peptides.2010.07.016

Tu, Y. (2011). The discovery of artemisinin (qinghaosu) and gifts from Chinese medicine. Nat. Med. 17, 1217-1220 DOI: 10.1038/nm.2471

Wallock-Richards, D.J., Marles-Wright, J., Clarke, D.J., Maitra, A., Dodds, M., Hanley, B., Campopiano, D.J. (2015). Molecular basis of *Streptococcus mutans* sortase A inhibition by the flavonoid natural product *trans*-chalcone. Chem. Commun. (Camb.). 51, 10483-10485 DOI: 10.1039/c5cc01816a

Watanabe, T., Igarashi, M., Okajima, T., Ishii, E., Kino, H., Hatano, M., Sawa, R., Umekita, M., Kimura, T., Okamoto, S., Eguchi, Y., Akamatsu, Y., Utsumi, R. (2012). Isolation and characterization of signermycin B, an antibiotic that targets the dimerization domain of histidine kinase WalK. Antimicrob. Agents Chemother. 56, 3657-63 DOI: 10.1128/AAC.06467-11

Xu, X., Zhou, X.D., Wu, C.D. The tea catechin epigallocatechin gallate suppresses cariogenic virulence factors of *Streptococcus mutans*. (2011). Antimicrob. Agents Chemother. 55, 1229-1236 DOI: 10.1016/j.archoralbio.2011.10.021

Yadav, R., Rai, R., Yadav, A., Pahuja, M., Solanki, S., Yadav, H. (2016). Evaluation of antibacterial activity of *Achyranthes aspera* extract against *Streptococcus mutans*: An *in vitro* study. J. Adv. Pharm. Technol. Res. 7, 149-152 DOI: 10.4103/2231-4040.191426

Yamakami, K., Tsumori, H., Sakurai, Y., Shimizu, Y., Nagatoshi, K., Sonomoto, K. (2013). Sustainable inhibition efficacy of liposome-encapsulated nisin on insoluble glucan-biofilm synthesis by *Streptococcus mutans*. Pharm. Biol. 51, 267-270 DOI: 10.3109/13880209.2012.717227

Yang, H., Bi, Y., Shang, X., Wang, M., Linden, S.B., Li, Y., Li, Y., Nelson, D.C., Wei, H. (2016). Antibiofilm Activities of a Novel Chimeolysin against *Streptococcus mutans* under Physiological and Cariogenic Conditions. Antimicrob. Agents Chemother. 60, 7436-7443 DOI: 10.1128/AAC.01872-16

Zeng, H., Liu, J., Ling, J. (2017). Efflux inhibitor suppresses *Streptococcus mutans* virulence properties. FEMS. Microbiol. Lett. 364 DOI: 10.1093/femsle/fnx033

Zhang, Q., Nguyen, T., McMichael, M., Velu, S.E., Zou, J., Zhou, X., Wu, H.

(2015). New small-molecule inhibitors of dihydrofolate reductase inhibit *Streptococcus mutans*. Int. J. Antimicrob. Agents. 46, 174-182 DOI: 10.1016/j.ijantimicag.2015.03.015

Zhang, Q., Nijampatnam, B., Hua, Z., Nguyen, T., Zou, J., Cai, X., Michalek, SM., Velu, SE., Wu, H. (2017). Structure-Based Discovery of Small Molecule Inhibitors of Cariogenic Virulence. Sci. Rep. 7:5974 DOI: 10.1038/s41598-017-06168-1

Chapter 15

https://doi.org/10.21775/9781912530229.15

Recent Advances in Pathogenic *Streptococcus* Vaccine Development

Hongren Wang[1▲], Zhen Qin[1▲], Mingyuan Li[1,2*]

[1]Department of Pathogenic Biology, West China School of Basic Medical Sciences & Forensic Medicine, Sichuan University.

[2]State Key Laboratory of Oral Diseases, Sichuan University

▲ Hongren Wang and Zhen Qin contributed equally to this work.

*Correspondence: lmy3985@sina.com

Abstract

Streptococcus pneumoniae (Spn) and *Streptococcus pyogenes* (Spy) cause many invasive and noninvasive diseases responsible for high morbidity and mortality worldwide. Safe, efficacious and affordable vaccines could have a significant, positive impact on the global infectious disease burden. Since the implementation of pneumococcal vaccine in the 1980s, the incidence of Spn infection has decreased significantly. Still so, these currently used multivalent polysaccharides and conjugated pneumococcal vaccines have some limitations. For Spy, there are even no vaccines available yet. There is an urgent need of new vaccines against Spn and Spy. Encouragingly, with the hard work of many investigators worldwide, a number of new vaccines candidates are developed with promising results. Of them, many have already entered the clinical trial stage. This review will describe the current status of Spn and Spy vaccine development, with particular focus on protein-based strategy.

Introduction

Human beings can be infected by many streptococcal species that are called pathogenic streptococci. Among them, *Streptococcus pneumoniae* (Spn) and *Streptococcus pyogenes* (also known as Group A streptococci, GAS) are major causes of many invasive and noninvasive diseases responsible for high morbidity and mortality worldwide. For these infectious diseases, safe and effective vaccines are the best way to prevent infections. In this review, first we would like to briefly summarize the weaknesses of current licensed serotype-specific Spn vaccines, then discuss recent progress in the development of pneumococcal and GAS vaccines, and share our opinions on vaccinal strategies to overcome Spn- and GAS-caused diseases in different situations.

1. Spn Vaccine

As a Gram-positive diplococcus and a part of commensal flora in the upper respiratory tract in human, Spn opportunistically results in various invasive and non-invasive diseases including meningitis, bacteremia, pneumonia, otitis media, sinusitis, etc(Centers for Disease Control and Prevention, 2015). Known to be the leading cause of bacterial pneumonia in children under the age of 5 (O'Brien et al., 2009; WHO, 2016), Spn is also a huge disease burden of the elderly mainly by causing community-acquired pneumonia (Drijkoningen and Rohde, 2014), posing a great threat on the health of humans (Bridy-Pappas et al., 2005) and raising a global economic concern (Boccalini et al., 2017; De Graeve and Beutels, 2004; Huang et al., 2011; Porchia et al., 2017). A World Health Organization (WHO) report showed that pneumonia accounts for 16% of all deaths of children under 5 years old, killing 920,136 children in 2015(WHO, 2016). Moreover, drug-resistance has been observed in Spn over decades worldwide (Cherazard et al.,

2017). Recently, WHO released a list of bacteria for which new antibiotics are urgently needed, and Spn made its mark (WHO, 2017). Since vaccines (including pneumococcal conjugate vaccines, PCVs) have a good reputation for solving the antimicrobial resistance problem (Laxminarayan et al., 2013; Laxminarayan et al., 2016; Lipsitch and Siber, 2016), the needs of Spn vaccines are strongly addressed.

Given that the encapsulated, instead of non-capsulated, pneumococcus accounts for Spn-related diseases (Centers for Disease Control and Prevention, 2015), scientists focused on pneumococcal capsule polysaccharides (CPs), with which Spn is typed due to their immunological distinction, providing the basis of existing pneumococcal vaccines (Bogaert et al., 2004b; Smit et al., 1977)(Figure 1).

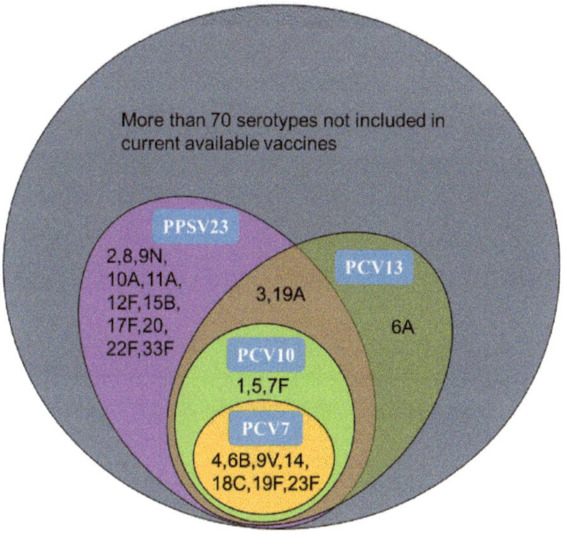

Figure 1. Serotypes contained in existing Spn vaccines. Among the over 90 serotypes of Spn, 24 were chosen to develop pneumococcal vaccines including one polysaccharide type (PPSV23) and three polysaccharide conjugate types (PCV7, PCV10 and PCV13) (WHO Publication, 2012).

Albeit with their great success in preventing serotype covered Spn diseases, serotype-based vaccines have several limitations, addressing the needs of improvement on their designing, research and development of alternative strategies such as protein-based vaccine.

1.1 Limitations of existing Spn vaccines

The limitations of current available pneumococcal vaccines were summarized in Figure 2, and each was briefly introduced as following.

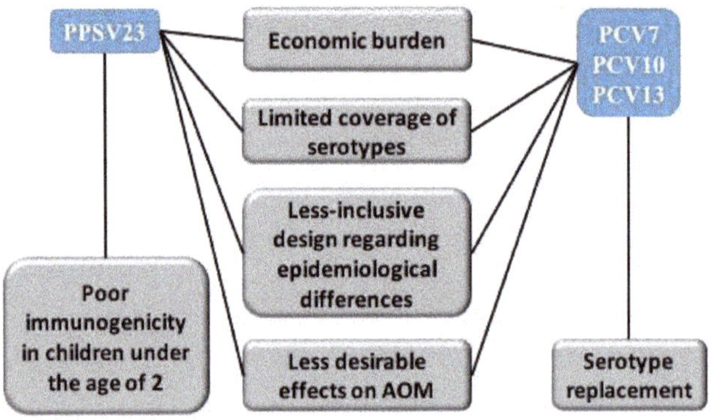

Figure 2. Limitations of current available pneumococcal vaccines.

1.1.1 Poor immunogenicity in children younger than 2

Licensed in 1983, PPSV23, containing 23 serotypes responsible for 85-90% invasive pneumococcal diseases (IPD) in the U.S, served well as being effective to protect against 56% diseases caused by serotype-covered Spn (Shapiro et al., 1991). However, this vaccine failed to elicit an immunogenic response in children under the age of 2, as it requires a T-cell-independent immunity of B cells (Heilmann, 1990), which is lacking in those younger than 2. As a result, vaccines with capsular polysaccharide conjugated to a protein carrier, diphtheria toxin (CRM197), were developed.

1.1.2 Limited coverage of serotypes, less inclusive design in terms of epidemiological differences and less economically friendly

This is an inherent disadvantage of a CPs-based vaccine, as its protection against pneumococcal infection is serotype-specific, which means that among the almost 100 types CPs of Spn, those included in current vaccines covered less than 1/3 out of all serotypes potentially leading to diseases (Figure 1). In addition, the distribution of prevalent strains varied epidemiologically in region, age and medical condition (Hausdorff et al., 2005) while it is technically irrational to wrap all CPs antigens in a single dose. Combined with the intrinsic need of increasing numbers of serotypes by PCVs, current vaccine manufacturing production made cost an issue of concern (Hanage, 2008; Josefsberg and Buckland, 2012; Ray, 2002), especially for developing countries.

1.1.3 Serotype replacement and poor performance on AOM-controlling

Serotype replacement, a phenomenon that vaccine-excluded serotypes (including drug-resistant strains) become more dominant than the pre-vaccination period, emerged after the administration of PCVs (Kaur et al., 2016; Miller et al., 2011), and is primarily attributed by reducing the carriage of strains which the serotypes were included in the vaccines. The interference on the asymptomatic carriage of respiratory tract commensal could also influence human health by breaking the equilibrium of the microbiota (Man et al., 2017). Moreover, the administration of CPs vaccines shows an undesirable effect on Spn-causing acute otitis media (AOM) (Jambo et al., 2010), which is partially thought to be the outcome of the replacement by other organisms such as *Staphylococcus aureus* (SA) (Fortanier et al., 2014).

1.2 Alternative strategies for the research and development of pneumococcal vaccine

Designing an effective pneumococcal vaccine largely depends on a better understanding of structures of Spn and how each component works as a whole during the interplay within the host (Kadioglu et al., 2008). In particular, protein-based pneumococcal vaccines have long been receiving attention (Daniels et al., 2016; Pichichero et al., 2016; Principi and Esposito, 2018) and are thought to be the most promising tactics with making up for the shortfalls of existing vaccines. In addition, some adjuvants such as pFL (a dendritic cell-targeting DNA-based adjuvant) showed potency in mucosal immunity enhancement, providing a new strategy to boost vaccinal function (Kataoka et al., 2017). Moreover, whole cell vaccines (WCVs), live attenuated or inactivated, have also been used as an alternative strategy by virtue of manufacturing convenience (Minor, 2015) and presenting a great number of antigens at once (Moffitt and Malley, 2011). Finally, modifications or adaptions of current CPs conjugate vaccines have been made by using conserved pneumococcal proteins as carriers (Table 2) to achieve WHO/GAVI target product profile (Moffitt and Malley, 2011).

1.2.1 Pneumococcal protein-based vaccine

1.2.1.1 Promising pneumococcal proteins utilized in vaccinal research and development

Due to their pathogenesis and antigenic characteristics, over 10 proteins (Table 1) of pneumococcus have been employed in many alternative strategies on vaccinal research and development (Rigden et al., 2003).

(1) Choline-binding proteins (CBPs)

Phosphorylcholine (ChoP), a sort of bacterial adhesin, constitutes lipoteichoic acids and cell wall teichoic acids of Spn (Cundell et al., 1995), and is strongly associated with colonization in the upper respiratory tract (Hammerschmidt, 2006). A group of proteins(Table 1), such as PspA, PspC, PcpA and LytA, being non-covalently anchored to ChoP, can bind to receptors on the epithelial cells of the host (Bergmann and Hammerschmidt, 2006).

PspA is expressed by all clinical pneumococcal isolates (Khan and Jan, 2017). By interfering the binding of C-reactive protein to phosphocholine, PspA inhibits complement-mediated opsonization (Mukerji et al., 2012), which in turn assists the immune evasion of Spn. Moreover, as a lactoferrin-binding protein, PspA can also protect Spn from killing by apolactoferrin (Shaper et al., 2004). PspC is another well-studied protein, also known as choline binding protein A (CbpA) as was found to be the major component of CBPs (Rosenow et al., 1997); or referred to as secretory pneumococcal surface protein A (SpsA) as binding to the ectodomain of polymeric immunoglobulin (Ig) receptor transporting secretory IgA on host respiratory epithelial cells, aiding to bacterial colonization and invasion (Dave et al., 2004; Zhang et al., 2000). PspC can also promote pneumococcal colonization by binding to C4b-binding protein, an inhibitor of the classical pathway (Dieudonne-Vatran et al., 2009). Moreover, PspC has been reported to inhibit C3b formation via binding to the heparin-binding FH domain of complement factor H (Cao et al., 2011; Dave et al., 2004), protecting Spn from complement-mediated opsonization (Quin et al., 2005). Existing on nearly all pathogenic pneumococci, PcpA is considered one of the potential candidate antigens for pneumococcal vaccine. To date, the ligand of PcpA is uncertain, while a study showed that CodY, a global nutritional repressor in bacteria, can activate PcpA, favoring bacterial adaptation to nutrients, and regulating the adherence to nasopharyngeal and lung epithelial cells (Hendriksen et al., 2008; Khan et al., 2012). Additionally, the expression of this protein is found to be inversely correlated to the concentration of manganese (Manzoor et al., 2015). LytA, also referred to as autolysin, is an enzyme cleaving the pneumococcal cell wall, promoting the release of pneumolysin (Ply) and other bacterial components such as peptidoglycan and teichoic acids (Mellroth et al., 2012; van der Poll and Opal, 2009). Recently, it has also been reported to inhibit C3 convertase formation (Andre et al., 2017).

(2) ATP-binding cassette transporter

The ATP-binding cassette (ABC) transport system provides energy for binding and transporting the solute (mainly metal) through the cell membrane by ATP hydrolysis (Jedrzejas, 2001), and includes a group of proteins such as PsaA, PiaA and PiuA(Table 1).

PsaA was thought to be a kind of adhesin due to its sequence homology to other adhesins of Spn(Sampson et al., 1994), but was shown by other studies that it is a lipoprotein of an ABC transport system, with the property of divalent metal-ion-binding(Lawrence et al., 1998; McAllister et al., 2004) and manganese-transporting(Dintilhac et al., 1997), preventing from oxidative stress(Kadioglu et al., 2008; Tseng et al., 2002). PiaA (pneumococcal ion acquisition A) and PiuA (pneumococcal ion uptake A) are two lipoprotein components of Spn for ion-acquisition and uptake, which is associated with bacterial growth and virulence (Brown et al., 2001a; van der Poll and Opal, 2009). Immunization with PiaA and PiuA protected mice from systemic infection of Spn (Brown et al., 2001b). SP2108 and SP0148 are solute-binding components of ABC transport systems, specifically binding to aromatic amino acids and maltose (Paton and Ogunniyi, 2011), respectively, enhancing bacterial growth and colonization (Shelburne et al., 2008). In recent years, they have been considered as potential vaccinal candidates due to their high conservation and mucosal immunity-triggering (Paton and Ogunniyi, 2011).

(3) LPXTG motif-binding proteins

The most universal mechanism for Gram-positive bacteria to anchor a set of surface proteins is that sortase transpeptidase recognizes an amino-acid sequence, LPXTG, harbored on the surface of these proteins (including NanA, RrgA, RrgB and PrtA), and cleaves the precursor between threonine and glycine of the LPXTG motif (Kadioglu et al., 2008; Navarre and Schneewind, 1994). Most LPXTG-containing proteins are anchored by StrA (Table 1), a pneumococcal virulence factor, which has been found to play an important role in aiding bacterial colonization and pathogenesis (Paterson and Mitchell, 2006).

(4) Neuraminidase

Known for its pre-invasive property (Brittan et al., 2012), neuraminidase (NA), an enzyme expressed on all pneumococci, shows potential to be an ideal vaccine target. There are at least 3 forms of NA: NanA, NanB and NanC(Table 1), out of which NanA is the most essential and well-studied.

NanA is encoded by all strains of Spn, and contains the LPXTG motif, covalently binding to the peptidoglycan (Camara et al., 1994). By cleaving the terminal sialic acid of glycoproteins of the epithelial cells in the host respiratory tract, NanA and

NanB expose more decoy receptors for Spn to bind, enhancing the pneumococcal attachment and invasion (Kadioglu et al., 2008). Recently, NanA was found to help pneumococcus escape the complement pathway in host (Andre et al., 2017). Research on the role of NanC in the pneumococcal pathogenesis is still lacking, though the finding that NanC is more commonly seen in isolates from cerebrospinal fluid than those from carriage suggested its tissue-specificity characteristic (Pettigrew et al., 2006).

(5) Pili proteins

Pili are strongly associated with pneumococcal adherence. There are 7 pilus-encoding genes (*rrgA*, *rrgB*, *rrgC, srtB, srtC, srtD, rlrA*) on the pathogenic island. Of them, *rrgA*, *rrgB* and *rrgC* respectively encode the subunits of pneumococcal pili 1, RrgA, RrgB and RrgC (Telford et al., 2006), which are anchored on the cell surface with the help of sortase (LeMieux et al., 2006).

RrgA has been found to act as the dominant adhesive element other than RrgB (Nelson et al., 2007), also known as pilus backbone (Gentile et al., 2011). RrgB is the most abundant subunit of pili 1 (Gentile et al., 2011), and has been reported to elicit a host inflammatory response (Gianfaldoni et al., 2007). Both RrgA and RrgB have been studied for vaccinal purposes for over a decade (LeMieux et al., 2006), while the structure and functional information of RrgC was rarely known until recent years after RrgC had been structurally described (Gianfaldoni et al., 2007), in which RrgC was found to bind the preformed pilus to the peptidoglycan with the catalytic activity of SrtA.

(6) Polyhistidine triad family

Polyhistidine triad (Pht) family includes a group of pneumococcal surface-expressed proteins, PhtA, PhtB, PhtD and PhtE, which are well-conserved and contain a histidine triad motif, playing an important role of the attachment of Spn to respiratory epithelial cells (Plumptre et al., 2013). Moreover, Pht proteins are considered to regulate metal homeostasis, especially the zinc storage, providing ion for pneumococcal invasion during early stages (Godfroid et al., 2011). Of the four proteins, PhtD and PhtE are mostly studied due to their prevalence on 97~100% Spn strains (Khan and Pichichero, 2012; Rioux et al., 2011).

(7) Cholesterol-dependent cytolysin family

Cholesterol-dependent cytolysin (CDC) is a family consisting of a group of proteins, such as pneumolysin (Ply), that lead to cell death by forming pores on cell membrane (Tilley et al., 2005). Ply, a transmembrane pore-forming oligomer, is a crucial pneumococcal virulence factor and expressed by all clinical isolates of Spn (Vernatter and Pirofski, 2013). Pneumococcal Ply can damage the host respiratory cells by binding to cholesterol-containing membranes, forming ring-shaped pores (van Pee et al., 2016), reinforcing the invasion of Spn. Within the hosts, Ply is able to trigger an inflammatory response through interaction with Toll-like receptors (TLRs) and activating NLRP3 inflammasomes (Vernatter and Pirofski, 2013), contributing to pneumococcal pathogenesis, including host-to-host transmission (Zafar et al., 2017). As the original form of Ply is highly toxic, it has been genetically modified to keep the property of immunogenicity instead of antigenicity (dPly). The immunization with dPly on mice successfully induced serotype-independent protection against pneumococcal diseases and colonization (Alexander et al., 1994).

Table 1. Pathogenesis-related proteins of Spn-potential targets for protein-based vaccine

Location	Protein family	Representative(s)
Cell-surface	Choline-binding	PspA PspC PcpA LytA LytB/C
	ATP-binding cassette (ABC) transporter	PsaA PiaA PiuA SP2108 SP0148
	LPXTG motif-binding	StrA
	Neuraminidases	NanA NanB NanC
	Pili	RrgA RrgB RrgC
	Polyhistidine triad family	PhtA PhtB PhtD PhtE
From cell matrix to outside cell	Cholesterol-dependent cytolysin (CDC) family	dPly (Genetically modified)

1.2.1.2 The forefront of Spn protein-based vaccines

Monovalent vaccine mainly focused on PspA. As one of the most abundant surface proteins expressing on all pneumococci, PspA becomes one of the forefront antigens of candidate vaccine since its immunization can elicit broadly protective serotype-independent serum antibodies in both humans and mice, regardless of its immunogenic variants (McCool et al., 2003; Nabors et al., 2000). Due to the low sequence homology between PspA and human cardiac myosin,

PspA is presumed to cause an autoimmune response (Ginsburg et al., 2012), whereas no evident proof has yet been found to confirm this possibility. Hence, PspA is still undergoing a series of research into animal experiments and clinical trials (Converso et al., 2017b; Entwisle et al., 2017; Goulart et al., 2017) with an avoidance of the involvement of the homology sequence to human cardiac myosin. A recombinant PspA has been tested in adults in a phase I clinical trial and has proved to be both safe and highly immunogenic (Briles et al., 2000). Thus, cross-reactivity with human should be tested and filtered for a rationale vaccine development.

In order to increase antibody-accessibility, vaccines involving multiple components have been springing out for decades (Moffitt and Malley, 2011) (Table 2) and have been well-reviewed elsewhere (Darrieux et al., 2015; Kim et al., 2017b; Odutola et al., 2017; Pichichero, 2017; Pichichero et al., 2016; Singh and Dutta, 2017). Among them, dPly, PhtD, PspA and PsaA are highly employed. Here we introduce two protein-based vaccines lately passed the clinical trial this year. ImmunoBiology Ltd. completed the phase I clinical trial on adults with PnuBioVax, a multi-valent recombinant vaccine, containing PspA, RrgA, RrgB, PsaA and dPly (Entwisle et al., 2017) (Table 2). This randomized clinical trial proved the safety and immunogenicity of PnuBioVax in all dosage levels, and more well-tolerant than its first investigation in humans where severe injection site reactions occurred. Presently, PnuBioVax is awaiting further tests on the pediatric population of whom the natural immunity is immature. In the same year, GSK tested the efficacy of PhiD-CV/dPly/PhtD-30, based on its well-tolerance showed in the first part study, on infants for a phase II clinical trial (Odutola et al., 2017) (Table 2). In this randomized, controlled, observer-blind trial, the nasopharyngeal carriage of Spn was particularly investigated, and the results showed that antibodies against PhtD and dPly were elicited by PhiD-CV/dPly/PhtD-30 without interfering the prevalence of pneumococcal carriage in the nasopharynx.

There are numerous recently developed protein-based vaccines undergoing a series of preclinical research. Of them, quite a few focused on mucosal immunity, which we particularly discussed later. Here we introduce research on adjuvant for pneumococcal vaccine development conducted by *Cibelly Goulart* et al. earlier this year, who used BCG as an adjuvant to deliver an infusion protein, rPspA-PdT, containing a fragment of PspA and a genetically detoxified Ply. In this prime-boost mice study, the administration of rBCG PspA-PdT followed by a boost of rPspA-PdT provided favorable effects on anti-lethal pneumococcal disease, reinforcement of complement deposition to Spn and inhibition of cytolysis by Ply (Goulart et al., 2017).

There are at least two facts that prompt scientists to look for a vaccine that boosting a strong mucosal immune response, i) Spn vaccines on the market are less effective against AOM which is the most common Spn-caused disease in children under the age of 5; ii) colonization on the mucosal surface is the primary step for Spn invasion. Since mucosal defense is critical in preventing AOM (Bergenfelz and Hakansson, 2017), recent years of experimental studies on pneumococcal vaccines illustrated a mucosal immunity-oriented trend, with the employment of adjuvant-assistance, pneumococcal surface proteins, whole cell mutants, etc. PspA, PhtD, PcpA, PlyD1 and PotD have been selected to develop vaccines focusing on mucosal immunity (Converso et al., 2017a; Kuipers et al., 2017; Xu et al., 2017). PPrV, a trivalent protein vaccine containing PhtD, PcpA and PlyD1, was shown to successfully elicit both serum and mucosal antibodies (IgG) that protect infant mice from pneumococcal bacteremia and AOM with the challenge of heterotypic strains (Xu et al., 2017). In another mice study, immunization with recombinant PotD can decrease the nasopharyngeal colonization by inducing immune response including production of IL-17 (Converso et al., 2017a). Additionally, mucosal immunity can be employed by other strategies. Similar to other adjuvants such as BCG, as mentioned above, dendritic cell-targeting DNA-based nasal adjuvants have shown their potential in immune response enhancement, especially for mucosal immunity promotion (Kataoka et al., 2017). Furthermore, innate and adaptive IL-17 responses in mice can be achieved by using γ-irradiated RX1 LytA/PdT and attenuated SPY1 strain of Spn, respectively (Babb et al., 2016; Gao et al., 2016).

1.2.2 WCVs

There are at least two prominent advantages of using whole cells of the specific organism as vaccine: cost-effectiveness and broad-coverage of antigens (Minor, 2015). WCVs are classified as the live attenuated and the inactivated. To date, two types of WCVs for Spn have been under clinical trial stage. An avirulent *Salmonella Typhi* strain was used as a vector to delivery PspA for enhancing immune responses in mice with high efficiency (Wang et al., 2010), promising results from this study drove further investigations of its potential use in humans. A few years later, 09-RASV-Sp-01, attenuated recombinant *Salmonella Typhi* vectors (RASV) expressing PspA, was used for a phase I dose-escalation trial to test its safety and immunogenicity (Frey et al., 2013) (Table 2). In this study, three RASV vaccines synthesizing PspA-RX1 antigen were employed to immunize adults, and the results proved that all three RASV pneumococcal vaccines were safe and well-tolerated, but with limited immunogenicity which could be a result from pre-existing immunity. Another pneumococcal WCV is an inactivated non-encapsulated whole cell vaccine, RX1 LytA/PdT, which was employed in a phase I clinical trial on adults, and showed safety, well-tolerance, and immunogenicity with high level of IgG antibodies against Ply and PspA (Alderson et al., 2014) (Table 2).

Preclinical efforts have also been made these years. *Eun-Hye Kim* et al. successfully constructed Spn *pep27* mutant (EHpep27), a live attenuated vaccine with deletion of pneumococcal virulent gene, *pep27* (Kim et al., 2012). Mice were intranasally immunized with EHpep27 at various dose levels, in which no virulent effect was detected, IgG antibody was elicited without the help of adjuvant, and serotype-independent protection against lethal pneumococcal challenge was achieved. EHpep27 was then tested for its immunogenicity and protection against non-typeable strains (Kim et al., 2016). In this study, EHpep27 showed a wide range of protection and long-lasting immunity, making itself a very potent vaccine candidate.

Table 2. Alternative strategies for research and development of pneumococcal vaccines under clinical trial

Vaccines		**Institute**	**Status**	**References**
Type	**Name (Components)**			
Protein	PspA	Sanofi-Pasteur	phase I complete	(Darrieux et al., 2015; Pichichero et al., 2016)
	PspA and PsaA	Sanofi-Pasteur/CDC	phase I complete	(Darrieux et al., 2015; Pichichero et al., 2016)

	PhpA and PhtB	GSK	phase II complete	(Pichichero, 2017; Pichichero et al., 2016)
	IC47 (PsaA, PcsB and StkP)	Valneva Austria GmbH/PATH	phase I complete	(Darrieux et al., 2015; Kim et al., 2017a; Pichichero et al., 2016)
	PhtD/PcpA and PlyD1	International Centre for Diarrhoeal Disease Research, Bangladesh/Sanofi-Pasteur	phase I complete	(Darrieux et al., 2015; Kim et al., 2017a; Pichichero, 2017; Pichichero et al., 2016)
	dPly/PhtD w/PHiD-CV	GSK	phase II complete	(Darrieux et al., 2015)
	PhiD-CV/dPly/PhtD-30	GSK	phase II complete	(Odutola et al., 2017; Pichichero, 2017)
	w/DTPa-HBV-IPV/Hib	GSK	phase II complete	(Darrieux et al., 2015; Pichichero et al., 2016)
	SP0148,1912,2108	Genocea	phase IIa complete	(Pichichero, 2017; Pichichero et al., 2016)
	PnuBioVax (PspA, RrgB, RrgA, PsaA and dPly)	ImmunoBiology Ltd.	phase I complete	(Entwisle et al., 2017)
CPs-protein conjugate	V114 (1, 3, 4, 5, 6A, 6B, 7F, 9V, 14, 18C, 19A, 19F, 22F, 23F and 33F and CRM197 protein)	Merck Sharp & Dohme Corp	phase II complete	(Kim et al., 2017a)
	V114 w/Alum (1, 3, 4, 5, 6A, 6B, 7F, 9V, 14, 18C, 19A, 19F, 22F, 23F and 33F and CRM197 protein)	Merck Sharp & Dohme Corp	phase II complete	(Kim et al., 2017a)
	PhtD/dPly w/or w/o PCV10	GSK	phase I complete	(Darrieux et al., 2015; Pichichero, 2017; Pichichero et al., 2016)
	PhtD/dPly/w/PCV10	GSK	phase II complete	(Darrieux et al., 2015; Pichichero, 2017; Pichichero et

				al., 2016)
	PHiD-CV w/PPV 23 (1, 4, 5, 6B, 7F, 9V, 14, 18C, 19F, and 23F and non-typeable H.influenzae protein D conjugate vaccine)	GSK	phase III complete	(Pichichero et al., 2016)
	PHiD-CV (1, 4, 5, 6B, 7F, 9V, 14, 18C, 19F, and 23F and non-typeable H.influenzae protein D conjugate vaccine)	GSK	phase III complete	(Darrieux et al., 2015)
Whole-cell	RX1 LytA/ PdT (killed non-encapsulated whole cell vaccine)	Boston Children's Hospital Instituto Butantan/PATH	phase I complete	(Darrieux et al., 2015; Kim et al., 2017a; Pichichero, 2017; Pichichero et al., 2016)
	09-RASV-Sp-01 (Attenuated Salmonella typhi expressing PspA)	Arizona State University/Biodesign Instititе	phase I complete	(Darrieux et al., 2015; Kim et al., 2017a)

2. GAS Vaccine Development

Gram positive bacteria GAS, which is responsible for substantial worldwide morbidity and mortality, can cause a wide range of invasive and noninvasive diseases, including pharyngitis, impetigo, necrotizing fasciitis (NF), and immune mediated diseases such as acute rheumatic fever (ARF), rheumatic heart disease (RHD) and acute post-streptococcal glomerulonephritis (APSGN). GAS can also cause toxin-mediated diseases including scarlet fever and streptococcal toxic shock syndrome (STSS) (Sims Sanyahumbi et al., 2016). Although GAS infections can be treated by one of the cheapest and oldest antibiotics known—penicillin, the global burden of GAS disease is substantial, and most infections, especially those in severe cases, are in low- and middle-income countries (LMIC). There is an urgent need for a safe, efficacious and affordable GAS vaccine to reduce the prevalence and the sequelae caused by this organism.

2.1 Feasibility for GAS vaccine development

A clear understanding of the mechanisms of protective immunity to GAS infection and identification of immune correlates of protection are essential for the rational design of a GAS vaccine. Although accurate protection mechanisms are needed to figure out with extra efforts, a large amount of serologic data from natural history studies and pre-clinical studies suggests good feasibility for GAS vaccine development and encourages investigators throwing themselves into this field. The phenomenon that peak incidence of GAS infections occurs in childhood and declines in adulthood indicates that natural infection with GAS would lead to a protective immunity. This is because of accumulation of protective bactericidal antibodies in serum following multiple GAS infections during childhood. According to the research of Lancefield, immunity against GAS infections has been believed to be type specific (Dale et al., 2016; Lancefield, 1959). Recently, a body of evidences indicate that mucosal immunity plays important roles in the protection against GAS infection, especially in the pharyngitis cases (Chen et al., 2016; Ghaffar et al., 2016; Loh et al., 2017; Marasini et al., 2017; Marasini et al., 2016; Mortensen et al., 2017; Schulze et al., 2017). These findings suggest that systemic IgG and local IgA are involved in the protection against GAS infection and provide implications in vaccine strategies.

2.2 The challenges of GAS vaccine development

Many investigators have been working on GAS vaccine development for more than 90 years, nevertheless, no licensed vaccines are currently available to prevent GAS infection (Dale et al., 2016; Excler and Kim, 2016; Gandhi et al., 2017; Schodel et al., 2017). Vaccine development is impeded for several reasons.

2.2.1 The epidemiology of GAS is extremely complex.

As mentioned previously, the presence of type-specific antibodies were responsible for protection against the homologous serotype of GAS infection. This made serotyping to be the first method to categorized different GAS strains, also known as "M typing". This time-consuming method has been progressively replaced by "*emm* typing" which is based on PCR and sequencing of the hypervariable region of the M gene (Beall et al., 1996). This *emm* typing method is widely used in molecular epidemiology investigation of GAS currently and through this method, 223 *emm* types have been identified to date. The *emm* type distribution represents considerable variation at both the country and global regional level. Unfavourably for vaccine development, limited data on *emm* type distribution in LIMC is available and the distribution patterns are completely different between LIMC and industrialized countries. In industrialized countries, the most common circulating types are *emm* 1, 4, 6 and 12, but in LIMC the diversity of *emm* types is much greater and no particular *emm* type is highly dominant (Sims Sanyahumbi et al., 2016). This complexity makes vaccine development very difficult because it is hard to cover all the GAS strains, at least, most of them.

2.2.2 There are some autoimmune epitopes in GAS and this makes the vaccine safety a major concern.

GAS infection can cause autoimmune diseases such as acute rheumatic fever (ARF) (Karthikeyan and Guilherme, 2018), this fact implies GAS vaccines must be designed very meticulously and it does make great obstacles for investigators.

Lots of studies indicate both cross-reactive antibodies and T cells have roles in ARF pathogenesis (Cunningham, 2000; Tandon et al., 2013). Fortunately, the autoimmune epitopes were found mainly existing in the M protein, precisely speaking, in the B-repeats region of M protein. This led to the strategy using the full length of M protein as antigen in the vaccine design being abandoned.

2.2.3 The socio-economy factor can influence GAS vaccine development.

Most GAS infections occur in developing countries and some specific populations in industrialized countries. (Chang, 2012; Gandhi et al., 2017; Zuhlke et al., 2017). From the perspective of investment-return, this is very hard to attract biological companies to invest GAS vaccine development.

2.2.4 The correlation of protection for GAS vaccine needs to be researched in-depth.

Establishing a standardized immunoassay that correlates with immune protection is essential in vaccine development. However, despite decades of research, currently used immunoassays to assess antibodies in GAS research are ELISA and functional bactericidal assays. ELISA gives a purely quantitative measurement of antibodies which include non-functional antibodies. The latter protocol can measure functional antibodies, but whole human blood must be used and this makes it labor-intensive and inter-assay variability. Recently, to overcome these limitations, Lorenz and colleagues developed a high-throughput opsonophagocytic assay for the determination of functional antibody activity against GAS using bioluminescence (Lorenz et al., 2017). This assay is believed to replace the two methods currently used and to make clinical trials easier.

2.3 Current status of GAS Vaccine Candidates

GAS vaccine development started in the early 1920s with whole heat-killed GAS, cell wall or M proteins (Dale et al., 2016). Although protective effectivity was observed, these relatively crude preparations had severe problems including reactogenicity caused by contaminating antigens, and risk of leading to ARF because of the autoimmune epitopes. With the advance in purification technology and understanding of the GAS related autoimmune mechanism, great progress has been made in GAS vaccine development. Many vaccine candidates have been developed to preclinical and clinical stages. Generally, these vaccine candidates can be classified into M protein based vaccines and other antigen based vaccines and summarized in Table 3.

2.3.1 M protein based vaccines

As a major surface-associated virulence factor and protective antigen, the M protein has been extensively researched for many years. The M protein is a coiled-coil protein consisting of A-C 3 domains. A-repeat/N-terminal domain is highly variable and used for *emm* typing; B-repeat domain contains autoimmune but not protective epitopes; C-repeat domain is highly conserved. Up to now, most notably promising vaccine candidates are based on the M protein. Early studies indicated antibodies against M protein could opsonize the homologous type strains and promote C3-mediated phagocytosis (Lancefield, 1962). The protective epitopes were found mainly existing in the N-terminal hypervariable region of M protein. These findings resulted in a vaccine design strategy that use only the N-terminal M peptides from different *emm* type strains in multivalent vaccines to devoid of potential autoepitopes (Dale et al., 2011).

2.3.1.1 Multivalent M protein based vaccines

Just like multivalent pneumococcal vaccines, theoretically, multivalent GAS vaccines could also be developed using the N-terminal M peptides linked in tandem based on the *emm* typing epidemiological data. The first multivalent GAS vaccine using this strategy was reported by Dale et al. in 1999 (Dale, 1999), in which protective M protein peptides from serotypes 1,3,5,6,19, and 24 of GAS were selected to form a hexavalent vaccine. This vaccine evoked high titers of antibodies against all six serotypes of GAS included in the vaccine. In the subsequent phase I trial, the hexavalent vaccine showed good tolerance and immunogenicity in humans, which represents a critical step in the development of a multivalent vaccine using hybrid fusion protein (Kotloff et al., 2004). Shortly afterwards, the multivalent vaccine was expanded to a 26-valent vaccine (Hu et al., 2002; McNeil et al., 2005) and later to a 30-valent vaccine (Dale et al., 2011; Dale et al., 2013). Both vaccines underwent phase I clinical trials in human adult volunteers and were shown to be safe and immunogenic. Compared to the 26-valent vaccine, the 30-valent vaccine has an increased coverage of circulating *emm* types worldwide, mainly in North America and Europe (Dale et al., 2011; Dale et al., 2013). Imperfectly, these two vaccines provide good coverage of circulating strains of GAS in industrialized countries but poor coverage in LIMC. A recent epidemiological study in Western Australia shows the new 30-valent M-protein GAS vaccine's potential coverage is 70% (Speers et al., 2017). Interestingly, cross-reactive immune responses were observed and raised new opportunities to optimize multivalent vaccines with broader coverage. This cross-protection can be partly explained by the *emm*-cluster typing system which classifies the 223 *emm* types into 48 functional *emm* clusters (McMillan et al., 2013). Given the cross-protection mechanism, the combination of selected *emm* types included in multivalent GAS vaccines could be further optimized to remove some redundant types within the same *emm* cluster and add some new cluster representative types which are absent in current multivalent vaccine candidates.

2.3.1.2 Conserved region of M-protein based vaccines

To overcome the imperfect coverage of multivalent vaccines, vaccines were designed using the highly conserved C-repeat portion of the M protein and believed to protect against all GAS strains. Bessen and colleagues used the entire C-terminal region of M protein conjugated to the mucosal adjuvant cholera toxin B subunit to construct a recombinant vaccine (Bessen and Fischetti, 1988). The results demonstrated that immune response evoked by conserved portions of M protein reduced GAS infection at the nasopharyngeal mucosa in the mouse model and highlighted the role of conserved region peptide-specific local mucosal Ig in controlling GAS colonization of the throat.

However, low levels of antibodies to selected peptides were observed to bind to cleaved or denatured myosin (Vashishtha and Fischetti, 1993), thus raising safety concerns on this strategy. This led to define minimal epitopes for GAS vaccine design to eliminate the hidden danger.

According to the minimal epitope strategy, P145 was identified in the C-repeat region of the M-protein exhibiting immunogenicity and no cross-reactive to human tissue (Pruksakorn et al., 1992). Furthermore, J8 and J14 vaccines were derived from P145 containing shorter single minimal B cell epitopes. J8 was a 28-mer synthetic peptide which contained a central 12 amino acids B cell epitope (J8i) with the flanking sequences derived from the yeast DNA binding protein GCN4 to fold J8i as a helix (Relf et al., 1996). To enhance the immunogenicity of this poor immunogenic one-epitope peptide, J8 was conjugated to the diphtheria toxoid (DT) to form J8-DT (Batzloff et al., 2003). This vaccine was highly immunogenic and the induced opsonic IgG had significant protection following intraperitoneal GAS challenge (Batzloff et al., 2003; Pandey et al., 2009; Pandey et al., 2013). Virus-like particles (VLPs) was used to carry J8 peptide as a GAS vaccine J8-VLPs. Results of the sublingual immunization indicated effective immune responses in both systemic and mucosal compartments. The saliva isolated from mice immunized with J8-VLPs demonstrated opsonizing activity against GAS in vitro. This suggests the vaccine's potential prevention of pharyngitis (Seth et al., 2016). Furthermore, J8-DT has completed a double-blinded Phase I pilot trial with no adverse events and a good antibody response (Dale et al., 2016). Furthermore, J8 was also conjugated to CRM197 to form the J8-CRM197 vaccine and shown immunogenic in mice and non-human primates (Caro-Aguilar et al., 2013). Subsequently, vaccine efficacy was further improved by incorporation of SpyCEP (Pandey et al., 2017). An alternative strategy using rational sequence modification to improve immunogenicity of p145 was recently reported (Nordstrom et al., 2017). After modification, the peptide can elicit higher titer antibodies with significantly higher affinity.

Another promising conserved M-protein based vaccine candidate is StreptInCor, which composed of 55 amino acid residues of the C-terminal region of the M5 protein containing both T and B cell protective epitopes (Guilherme et al., 2011; Guilherme et al., 2006; Guilherme et al., 2009). A molecular epidemiology study in Sao Paulo, Brazil and in silico analysis of vaccine coverage capacity indicated that StreptInCor could provide 71.0 % coverage in this area with high diversity of

GAS strains (Freschi de Barros et al., 2015). Studies demonstrated that StreptInCor is effective against several GAS strains and can prevent infection without causing autoimmune reactions (De Amicis et al., 2014; Freschi de Barros et al., 2015; Guerino et al., 2011; Guilherme et al., 2011; Guilherme et al., 2013; Guilherme et al., 2009; Postol et al., 2013). Moving forward on the basis of these immunogenicity and safety data, this vaccine has entered into phase I of clinical trials (Steer et al., 2016).

Table 3. Current status of promising GAS M protein-based vaccines

Vaccine candidates	Current status	Significant findings	References
6-valent M protein-based vaccine	completed phase I clinical trials	Safe, immunogenic, elicits antibodies to all homologous serotypes.	(Dale, 1999; Hall et al., 2004; Kotloff et al., 2004)
26-valent M protein-based vaccine	phase II clinical trials	Safe, immunogenic, elicits antibodies to all homologous serotypes; good coverage in industrialized countries but poor in LIMC.	(McNeil et al., 2005)
30-valent M protein-based vaccine	phase I clinical trials	Safe, immunogenic, elicits antibodies to all homologous and some heterologous serotypes; better coverage than the 26-valent vaccine.	(Dale et al., 2011; Dale et al., 2013)
J8/J14/P145	phase I clinical trials	Safe, immunogenic, elicits antibodies to different serotypes; no cross-reactive antibodies; vaccines were optimized based on this candidate to improve immunogenicity significantly.	(Pandey et al., 2017) (Nordstrom et al., 2017)
StreptInCor	phase I clinical trials	Safe, immunogenic, elicits antibodies to different serotypes; no cross-reactive	(Freschi de Barros et al., 2015; Guerino et al., 2011; Guilherme et al.,

		antibodies.	2011; Guilherme et al., 2013; Postol et al., 2013)

2.2.2 Non-M protein based vaccines

Since multivalent vaccines cannot provide full coverage against all GAS strains, and that there are lessons of serotype replacement from the multivalent pneumococcal vaccines implementation, highly conserved non-M protein GAS antigens received more and more attention. Several promising candidates were discovered with some encouraging results, such as GAS carbohydrate, C5a peptidase (Park and Cleary, 2005; Shet et al., 2003), Streptococcal pyrogenic exotoxin B (extracellular cysteine protease) (Kapur et al., 1994), streptococcal fibronectin-binding proteins, Combo vaccine (composed of Streptolysin O, SpyCEP and Spy0269) (Bensi et al., 2012), Pili (T antigen) (Koller et al., 2010), Streptococcal Heme Binding Protein (Shp) (Zhang et al., 2017), M-related protein (Mrp) (Courtney et al., 2017), but none of these candidates had entered clinical trials to date. The status of these vaccine candidates is summarized in Table 4.

Table 4. Current status of promising GAS non-M protein based vaccines

Vaccine candidates (antigen)	Current status	Significant findings	References
GAS carbohydrate	Preclinical studies	Conjugated to tetanus toxoid can protect mice from GAS challenge; no cross-reactivity was observed.	(Sabharwal et al., 2006)
GAS C5a peptidase (SCPA)	Preclinical studies	Highly conserved in all GAS serotypes; a virulence factor; not associated with autoimmune reactivity.	(Park and Cleary, 2005; Shet et al., 2003)
Fibronectin-Binding Protein	Preclinical studies	11 Fn-binding proteins were found, including protein F1 (PrtF1)/SfbI, protein F2 (PrtF2)/PFBP, FbaA (formerly Fba), FbaB, SfbII/serum opacity factor (SOF), SfbX, Fbp54, M1 protein, GAPDH/Plr, Shr and Scl1. These proteins have different coverage and	(Courtney et al., 2003; Kawabata et al., 2001; Schulze et al., 2006)

		some can elicit protective immune response.	
spy0469, spy1228 and spy1801	Preclinical studies	The antigens were tested for both antibody recognition and T cell responses in human adults and children; provide protection in a mice model.	(Mortensen et al., 2016)
M-related protein (Mrp)	Preclinical studies	The antisera opsonized GAS strains representing each Mrp family; also opsonized *emm* types not covered by the 30-valent M protein–based vaccine and can be combinated into the 30-valent vaccine improve the efficacy.	(Courtney et al., 2017)
Pili (T antigen)	Preclinical studies	Recombinant pilus proteins was shown to protect mice from GAS challenge; GAS FCT region displays considerable genetic diversity, with nine different FCT variants identified to date.	(Koller et al., 2010)
Combo vaccine (composed of Streptolysin O, SpyCEP and Spy0269)	Preclinical studies	Identified using 3 high throughput technologies; broad protection against a panel of four different GAS strains; SpyCEP was conjugated to J8 to improve the J8 efficacy.	(Bensi et al., 2012; Pandey et al., 2017; Zingaretti et al., 2010)
Streptococcal Heme Binding Protein (Shp)	Preclinical studies	Immunization elicited a robust IgG response, enhanced GAS clearance and reduced systemic dissemination.	(Zhang et al., 2017)

3. Issues worth of consideration during vaccinal design for Spn and GAS

3.1 Epidemiology--- essential parameter for both protein- and serotype-based strategy

The study of pneumococcal epidemiology is of particular importance because the efficacy varies by region, age and medical state of vaccinees. Pneumococcal proteins that are highly conserved and more widely distributed in human would be more likely to prevent Spn-related on a larger scale (Cornick et al., 2017). In addition, a recombinant multi-valent vaccine is recommended, preferably with adjuvants if needed. Similarly, overall and accurate epidemiological data is needed to optimize GAS vaccine design for better coverage and safety.

3.2 Whether Spn should be thoroughly eliminated is of discrete concern

Spn is a member of microbial commensals in human nasopharynx. Although colonization of Spn is the key step of bacterial invasion (Bogaert et al., 2004a), the balance of the microbiota in the respiratory tract plays an important role of immunity enhancement (Man et al., 2017). The serotype replacement is seen as the emergence of vaccine-excluded serotypes which becomes dominant following the use of PCVs. The "replacement phenomenon" was also thought to be observed as inverse correlate with other bacteria such as SA (Reiss-Mandel and Regev-Yochay, 2016), whilst whether the replacement by SA is a result of PCVs administration remains controversial (Bergenfelz and Hakansson, 2017; Fortanier et al., 2014; Reiss-Mandel and Regev-Yochay, 2016). Overall, it is paramount that the impact on pneumococcal carriage should be taken into consideration during the assessment on the efficacy of a vaccine, of which the original intention is to protect hosts from diseases rather than leading to possible side effects by unnecessary overtreatment. Although mucosal immunity is the best way to "shut out" Spn, vaccines that keep the colonized pneumococci below a pathogenic inoculum is suggested. Thus, a rationale endpoint setting is suggested during the efficacy assessment. Additionally, selectively targeting disease-causing factors of Spn could also be able to take the interests of the whole into account (McDaniel and Swiatlo, 2016), so as to achieve the goal of no interference to the asymptomatic pneumococcal carriage in human respiratory tract. As such, using genomic approach for identifying potential microbial targets can provide a comprehensive view to support rationale design of the future pneumococcal vaccines (Wizemann et al., 2001).

3.3 Using a systematic approach to optimize GAS vaccine design

Neither the multivalent vaccine nor the vaccine candidate based on a single conserved antigen can provide perfect coverage for all the serotypes, and serotype replacement might be developed under the selective pressure of vaccines. This prompts the optimization of vaccine design to increase the coverage and decrease the potential risk such as autoimmune responses. With an increasing number of antigens being discovered, a systematic approach can be employed to combine several highly conserved, immunogenic and non-autoimmune inducible antigen fragments to develop combination GAS vaccines which could potentially overcome the limitations of the current vaccine candidates.

Conclusions

As summarized in this review, there are a number of GAS vaccine candidates which have been proven to be promising. But the fact that must be faced is that there is no GAS vaccine currently on the market because of several impediments. This means there is no roadmap for GAS vaccine development. Unlike GAS, feasible Spn vaccines using novel tactics are much more fruitful, especially protein-based strategies which are considered the most prospective for developing next generation pneumococcal vaccines. As research moves along, the impediments are becoming largely solvable and the first GAS vaccine will come into being eventually.

Conflicts of interests

The authors declare no conflicts of interests.

Acknowledgements

Research in the lab of Hongren Wang is supported by a grant from Sichuan Provincial Science and Technology Support Project (2015SZ0178).

References

Alderson M., Malley, R., Anderson, P., Thompson, C., Morrison, R., Briles, D., King, J., Goldblatt, D., Green, N., Hural, J., *et al.* (2014). A phase 1 study to assess the safety, tolerability and immunogenicity of inactivated nonencapsulated streptococcus pneumoniae whole cell vaccine. In: Cripps A, editor. [Abstract ISSPD - 0121]. Pneumonia. Hyderabad; p. 94.

Alexander, J.E., Lock, R.A., Peeters, C.C., Poolman, J.T., Andrew, P.W., Mitchell, T.J., Hansman, D., and Paton, J.C. (1994). Immunization of mice with pneumolysin toxoid confers a significant degree of protection against at least nine serotypes of Streptococcus pneumoniae. Infect Immun 62, 5683-5688.

Andre, G.O., Converso, T.R., Politano, W.R., Ferraz, L.F., Ribeiro, M.L., Leite, L.C., and Darrieux, M. (2017). Role of Streptococcus pneumoniae Proteins in Evasion of Complement-Mediated Immunity. Front Microbiol 8, 224. doi:10.3389/fmicb.2017.00224

Babb, R., Chen, A., Hirst, T.R., Kara, E.E., McColl, S.R., Ogunniyi, A.D., Paton, J.C., and Alsharifi, M. (2016). Intranasal vaccination with gamma-irradiated Streptococcus pneumoniae whole-cell vaccine provides serotype-independent protection mediated by B-cells and innate IL-17 responses. Clin Sci (Lond) 130, 697-710. doi:10.1042/CS20150699

Batzloff, M.R., Hayman, W.A., Davies, M.R., Zeng, M., Pruksakorn, S., Brandt, E.R., and Good, M.F. (2003). Protection against group A streptococcus by immunization with J8-diphtheria toxoid: contribution of J8- and diphtheria toxoid-specific antibodies to protection. J Infect Dis 187, 1598-1608. doi:10.1086/374800

Beall, B., Facklam, R., and Thompson, T. (1996). Sequencing emm-specific PCR products for routine and accurate typing of group A streptococci. J Clin Microbiol 34, 953-958.

Bensi, G., Mora, M., Tuscano, G., Biagini, M., Chiarot, E., Bombaci, M., Capo, S., Falugi, F., Manetti, A.G., Donato, P., *et al.* (2012). Multi high-throughput approach for highly selective identification of vaccine candidates: the Group A Streptococcus case. Mol Cell Proteomics 11, M111 015693. doi:10.1074/mcp.M111.015693

Bergenfelz, C., and Hakansson, A.P. (2017). Streptococcus pneumoniae Otitis Media Pathogenesis and How It Informs Our Understanding of Vaccine Strategies. Curr Otorhinolaryngol Rep 5, 115-124. doi:10.1007/s40136-017-0152-6

Bergmann, S., and Hammerschmidt, S. (2006). Versatility of pneumococcal surface proteins. Microbiology 152, 295-303. doi:10.1099/mic.0.28610-0

Bessen, D., and Fischetti, V.A. (1988). Influence of intranasal immunization with synthetic peptides corresponding to conserved epitopes of M protein on mucosal colonization by group A streptococci. Infect Immun 56, 2666-2672.

Boccalini, S., Bechini, A., Gasparini, R., Panatto, D., Amicizia, D., and Bonanni, P. (2017). Economic studies applied to vaccines against invasive diseases: An updated budget impact analysis of age-based pneumococcal vaccination strategies in the elderly in Italy. Hum Vaccin Immunother 13, 417-422. doi:10.1080/21645515.2017.1264827

Bogaert, D., de Groot, R., and Hermans, P.W.M. (2004a). Streptococcus pneumoniae colonisation: the key to pneumococcal disease. The Lancet Infectious Diseases 4, 144-154. doi:10.1016/s1473-3099(04)00938-7

Bogaert, D., Hermans, P.W., Adrian, P.V., Rumke, H.C., and de Groot, R. (2004b). Pneumococcal vaccines: an update on current strategies. Vaccine 22, 2209-2220. doi:10.1016/j.vaccine.2003.11.038

Bridy-Pappas, A.E., Margolis, M.B., Center, K.J., and Isaacman, D.J. (2005). Streptococcus pneumoniae: description of the pathogen, disease epidemiology, treatment, and prevention. Pharmacotherapy 25, 1193-1212. doi:10.1592/phco.2005.25.9.1193

Briles, D.E., Hollingshead, S.K., King, J., Swift, A., Braun, P.A., Park, M.K., Ferguson, L.M., Nahm, M.H., and Nabors, G.S. (2000). Immunization of humans with recombinant pneumococcal surface protein A (rPspA) elicits antibodies that passively protect mice from fatal infection with Streptococcus pneumoniae bearing heterologous PspA. J Infect Dis 182, 1694-1701. doi:10.1086/317602

Brittan, J.L., Buckeridge, T.J., Finn, A., Kadioglu, A., and Jenkinson, H.F. (2012). Pneumococcal neuraminidase A: an essential upper airway colonization factor for Streptococcus pneumoniae. Mol Oral Microbiol 27, 270-283. doi:10.1111/j.2041-1014.2012.00658.x

Brown, J.S., Gilliland, S.M., and Holden, D.W. (2001a). A Streptococcus pneumoniae pathogenicity island encoding an ABC transporter involved in iron uptake and virulence. Mol Microbiol 40, 572-585.

Brown, J.S., Ogunniyi, A.D., Woodrow, M.C., Holden, D.W., and Paton, J.C. (2001b). Immunization with components of two iron uptake ABC transporters protects mice against systemic Streptococcus pneumoniae infection. Infect Immun 69, 6702-6706. doi:10.1128/IAI.69.11.6702-6706.2001

Camara, M., Boulnois, G.J., Andrew, P.W., and Mitchell, T.J. (1994). A neuraminidase from Streptococcus pneumoniae has the features of a surface protein. Infect Immun 62, 3688-3695.

Cao, J., Gong, Y., Cai, B., Feng, W., Wu, Y., Li, L., Zou, Y., Ying, B., and Wang, L. (2011). Modulation of human bronchial epithelial cells by pneumococcal choline binding protein A. Hum Immunol 72, 37-46. doi:10.1016/j.humimm.2010.10.007

Caro-Aguilar, I., Ottinger, E., Hepler, R.W., Nahas, D.D., Wu, C., Good, M.F., Batzloff, M., Joyce, J.G., Heinrichs, J.H., and Skinner, J.M. (2013). Immunogenicity in mice and non-human primates of the Group A Streptococcal J8 peptide vaccine candidate conjugated to CRM197. Hum Vaccin Immunother 9, 488-496.

Centers for Disease Control and Prevention (2015). Epidemiology and Prevention of Vaccine-Preventable Diseases. In The Pink Book: course textbook, Hamborsky J, Kroger A, Wolfe S, eds. (Washington D.C. Public Health Foundation). http://www.cdc.gov/vaccines/pubs/pinkbook/pneumo.html.

Chang, C. (2012). Cutting edge issues in rheumatic fever. Clin Rev Allergy Immunol 42, 213-237. doi:10.1007/s12016-011-8271-1

Chen, X., Li, N., Bi, S., Wang, X., and Wang, B. (2016). Co-Activation of Th17 and Antibody Responses Provides Efficient Protection against Mucosal Infection by Group A Streptococcus. PLoS One 11, e0168861. doi:10.1371/journal.pone.0168861

Cherazard, R., Epstein, M., Doan, T.L., Salim, T., Bharti, S., and Smith, M.A. (2017). Antimicrobial Resistant Streptococcus pneumoniae: Prevalence, Mechanisms, and Clinical Implications. Am J Ther 24, e361-e369. doi:10.1097/MJT.0000000000000551

Converso, T.R., Goulart, C., Rodriguez, D., Darrieux, M., and Leite, L.C. (2017a). Systemic immunization with rPotD reduces Streptococcus pneumoniae nasopharyngeal colonization in mice. Vaccine 35, 149-155. doi:10.1016/j.vaccine.2016.11.027

Converso, T.R., Goulart, C., Rodriguez, D., Darrieux, M., and Leite, L.C.C. (2017b). Rational selection of broadly cross-reactive family 2 PspA molecules for inclusion in chimeric pneumococcal vaccines. Microb Pathog 109, 233-238.

doi:10.1016/j.micpath.2017.06.004

Cornick, J.E., Tastan Bishop, O., Yalcin, F., Kiran, A.M., Kumwenda, B., Chaguza, C., Govindpershad, S., Ousmane, S., Senghore, M., du Plessis, M., *et al.* (2017). The global distribution and diversity of protein vaccine candidate antigens in the highly virulent Streptococcus pnuemoniae serotype 1. Vaccine 35, 972-980. doi:10.1016/j.vaccine.2016.12.037

Courtney, H.S., Hasty, D.L., and Dale, J.B. (2003). Serum opacity factor (SOF) of Streptococcus pyogenes evokes antibodies that opsonize homologous and heterologous SOF-positive serotypes of group A streptococci. Infect Immun 71, 5097-5103.

Courtney, H.S., Niedermeyer, S.E., Penfound, T.A., Hohn, C.M., Greeley, A., and Dale, J.B. (2017). Trivalent M-related protein as a component of next generation group A streptococcal vaccines. Clin Exp Vaccine Res 6, 45-49. doi:10.7774/cevr.2017.6.1.45

Cundell, D.R., Gerard, N.P., Gerard, C., Idanpaan-Heikkila, I., and Tuomanen, E.I. (1995). Streptococcus pneumoniae anchor to activated human cells by the receptor for platelet-activating factor. Nature 377, 435-438. doi:10.1038/377435a0

Cunningham, M.W. (2000). Pathogenesis of group A streptococcal infections. Clin Microbiol Rev 13, 470-511.

Dale, J.B. (1999). Multivalent group A streptococcal vaccine designed to optimize the immunogenicity of six tandem M protein fragments. Vaccine 17, 193-200.

Dale, J.B., Batzloff, M.R., Cleary, P.P., Courtney, H.S., Good, M.F., Grandi, G., Halperin, S., Margarit, I.Y., McNeil, S., Pandey, M., *et al.* (2016). Current Approaches to Group A Streptococcal Vaccine Development. In Streptococcus pyogenes : Basic Biology to Clinical Manifestations, J.J. Ferretti, D.L. Stevens, and V.A. Fischetti, eds. (Oklahoma City (OK)).

Dale, J.B., Penfound, T.A., Chiang, E.Y., and Walton, W.J. (2011). New 30-valent M protein-based vaccine evokes cross-opsonic antibodies against non-vaccine serotypes of group A streptococci. Vaccine 29, 8175-8178. doi:10.1016/j.vaccine.2011.09.005

Dale, J.B., Penfound, T.A., Tamboura, B., Sow, S.O., Nataro, J.P., Tapia, M., and Kotloff, K.L. (2013). Potential coverage of a multivalent M protein-based group A streptococcal vaccine. Vaccine 31, 1576-1581. doi:10.1016/j.vaccine.2013.01.019

Daniels, C.C., Rogers, P.D., and Shelton, C.M. (2016). A Review of Pneumococcal Vaccines: Current Polysaccharide Vaccine Recommendations and Future Protein Antigens. J Pediatr Pharmacol Ther 21, 27-35. doi:10.5863/1551-6776-21.1.27

Darrieux, M., Goulart, C., Briles, D., and Leite, L.C. (2015). Current status and perspectives on protein-based pneumococcal vaccines. Crit Rev Microbiol 41, 190-200. doi:10.3109/1040841X.2013.813902

Dave, S., Carmicle, S., Hammerschmidt, S., Pangburn, M.K., and McDaniel, L.S. (2004). Dual roles of PspC, a surface protein of Streptococcus pneumoniae, in binding human secretory IgA and factor H. J Immunol 173, 471-477.

De Amicis, K.M., Freschi de Barros, S., Alencar, R.E., Postol, E., Martins Cde, O., Arcuri, H.A., Goulart, C., Kalil, J., and Guilherme, L. (2014). Analysis of the coverage capacity of the StreptInCor candidate vaccine against Streptococcus pyogenes. Vaccine 32, 4104-4110. doi:10.1016/j.vaccine.2013.08.043

De Graeve, D., and Beutels, P. (2004). Economic aspects of pneumococcal pneumonia: a review of the literature. Pharmacoeconomics 22, 719-740.

Dieudonne-Vatran, A., Krentz, S., Blom, A.M., Meri, S., Henriques-Normark, B.,

Riesbeck, K., and Albiger, B. (2009). Clinical isolates of Streptococcus pneumoniae bind the complement inhibitor C4b-binding protein in a PspC allele-dependent fashion. J Immunol 182, 7865-7877. doi:10.4049/jimmunol.0802376

Dintilhac, A., Alloing, G., Granadel, C., and Claverys, J.P. (1997). Competence and virulence of Streptococcus pneumoniae: Adc and PsaA mutants exhibit a requirement for Zn and Mn resulting from inactivation of putative ABC metal permeases. Mol Microbiol 25, 727-739.

Drijkoningen, J.J., and Rohde, G.G. (2014). Pneumococcal infection in adults: burden of disease. Clin Microbiol Infect 20 Suppl 5, 45-51. doi:10.1111/1469-0691.12461

Entwisle, C., Hill, S., Pang, Y., Joachim, M., McIlgorm, A., Colaco, C., Goldblatt, D., De Gorguette D'Argoeuves, P., and Bailey, C. (2017). Safety and immunogenicity of a novel multiple antigen pneumococcal vaccine in adults: A Phase 1 randomised clinical trial. Vaccine 35, 7181-7186. doi:10.1016/j.vaccine.2017.10.076

Excler, J.L., and Kim, J.H. (2016). Accelerating the development of a group A Streptococcus vaccine: an urgent public health need. Clin Exp Vaccine Res 5, 101-107. doi:10.7774/cevr.2016.5.2.101

Fortanier, A.C., Venekamp, R.P., Boonacker, C.W., Hak, E., Schilder, A.G., Sanders, E.A., and Damoiseaux, R.A. (2014). Pneumococcal conjugate vaccines for preventing otitis media. Cochrane Database Syst Rev, CD001480. doi:10.1002/14651858.CD001480.pub4

Freschi de Barros, S., De Amicis, K.M., Alencar, R., Smeesters, P.R., Trunkel, A., Postol, E., Almeida Junior, J.N., Rossi, F., Pignatari, A.C., Kalil, J., *et al.* (2015). Streptococcus pyogenes strains in Sao Paulo, Brazil: molecular characterization as a basis for StreptInCor coverage capacity analysis. BMC Infect Dis 15, 308.

doi:10.1186/s12879-015-1052-3

Frey, S.E., Lottenbach, K.R., Hill, H., Blevins, T.P., Yu, Y., Zhang, Y., Brenneman, K.E., Kelly-Aehle, S.M., McDonald, C., Jansen, A., *et al.* (2013). A Phase I, dose-escalation trial in adults of three recombinant attenuated Salmonella Typhi vaccine vectors producing Streptococcus pneumoniae surface protein antigen PspA. Vaccine 31, 4874-4880. doi:10.1016/j.vaccine.2013.07.049

Gandhi, G.D., Krishnamoorthy, N., Motal, U.M.A., and Yacoub, M. (2017). Towards developing a vaccine for rheumatic heart disease. Glob Cardiol Sci Pract 2017, e201704. doi:10.21542/gcsp.2017.4

Gao, S., Zeng, L., Zhang, X., Wu, Y., Cui, J., Song, Z., Sun, X., Wang, H., Yin, Y., and Xu, W. (2016). Attenuated Streptococcus pneumoniae vaccine candidate SPY1 promotes dendritic cell activation and drives a Th1/Th17 response. Immunol Lett 179, 47-55. doi:10.1016/j.imlet.2016.08.008

Gentile, M.A., Melchiorre, S., Emolo, C., Moschioni, M., Gianfaldoni, C., Pancotto, L., Ferlenghi, I., Scarselli, M., Pansegrau, W., Veggi, D., *et al.* (2011). Structural and functional characterization of the Streptococcus pneumoniae RrgB pilus backbone D1 domain. J Biol Chem 286, 14588-14597. doi:10.1074/jbc.M110.202739

Ghaffar, K.A., Marasini, N., Giddam, A.K., Batzloff, M.R., Good, M.F., Skwarczynski, M., and Toth, I. (2016). Liposome-based intranasal delivery of lipopeptide vaccine candidates against group A streptococcus. Acta Biomater 41, 161-168. doi:10.1016/j.actbio.2016.04.012

Gianfaldoni, C., Censini, S., Hilleringmann, M., Moschioni, M., Facciotti, C., Pansegrau, W., Masignani, V., Covacci, A., Rappuoli, R., Barocchi, M.A., *et al.* (2007). Streptococcus pneumoniae pilus subunits protect mice against lethal challenge. Infect Immun 75, 1059-1062. doi:10.1128/IAI.01400-06

Ginsburg, A.S., Nahm, M.H., Khambaty, F.M., and Alderson, M.R. (2012). Issues and challenges in the development of pneumococcal protein vaccines. Expert Rev Vaccines 11, 279-285. doi:10.1586/erv.12.5

Godfroid, F., Hermand, P., Verlant, V., Denoel, P., and Poolman, J.T. (2011). Preclinical evaluation of the Pht proteins as potential cross-protective pneumococcal vaccine antigens. Infect Immun 79, 238-245. doi:10.1128/IAI.00378-10

Goulart, C., Rodriguez, D., Kanno, A.I., Lu, Y.J., Malley, R., and Leite, L.C. (2017). Recombinant BCG expressing a PspA-PdT fusion protein protects mice against pneumococcal lethal challenge in a prime-boost strategy. Vaccine 35, 1683-1691. doi:10.1016/j.vaccine.2017.02.029

Guerino, M.T., Postol, E., Demarchi, L.M., Martins, C.O., Mundel, L.R., Kalil, J., and Guilherme, L. (2011). HLA class II transgenic mice develop a safe and long lasting immune response against StreptInCor, an anti-group A streptococcus vaccine candidate. Vaccine 29, 8250-8256. doi:10.1016/j.vaccine.2011.08.113

Guilherme, L., Alba, M.P., Ferreira, F.M., Oshiro, S.E., Higa, F., Patarroyo, M.E., and Kalil, J. (2011). Anti-group A streptococcal vaccine epitope: structure, stability, and its ability to interact with HLA class II molecules. J Biol Chem 286, 6989-6998. doi:10.1074/jbc.M110.132118

Guilherme, L., Fae, K.C., Higa, F., Chaves, L., Oshiro, S.E., Freschi de Barros, S., Puschel, C., Juliano, M.A., Tanaka, A.C., Spina, G., *et al.* (2006). Towards a vaccine against rheumatic fever. Clin Dev Immunol 13, 125-132. doi:10.1080/17402520600877026

Guilherme, L., Postol, E., Ferreira, F.M., DeMarchi, L.M., and Kalil, J. (2013). StreptInCor: a model of anti-Streptococcus pyogenes vaccine reviewed. Auto Immun Highlights 4, 81-85. doi:10.1007/s13317-013-0053-8

Guilherme, L., Postol, E., Freschi de Barros, S., Higa, F., Alencar, R., Lastre, M., Zayas, C., Puschel, C.R., Silva, W.R., Sa-Rocha, L.C., *et al.* (2009). A vaccine against S. pyogenes: design and experimental immune response. Methods 49, 316-321. doi:10.1016/j.ymeth.2009.03.024

Hall, M.A., Stroop, S.D., Hu, M.C., Walls, M.A., Reddish, M.A., Burt, D.S., Lowell, G.H., and Dale, J.B. (2004). Intranasal immunization with multivalent group A streptococcal vaccines protects mice against intranasal challenge infections. Infect Immun 72, 2507-2512.

Hammerschmidt, S. (2006). Adherence molecules of pathogenic pneumococci. Curr Opin Microbiol 9, 12-20. doi:10.1016/j.mib.2005.11.001

Hanage, W.P. (2008). Serotype-specific problems associated with pneumococcal conjugate vaccination. Future Microbiol 3, 23-30. doi:10.2217/17460913.3.1.23

Hausdorff, W.P., Feikin, D.R., and Klugman, K.P. (2005). Epidemiological differences among pneumococcal serotypes. Lancet Infect Dis 5, 83-93. doi:10.1016/S1473-3099(05)01280-6

Heilmann, C. (1990). Human B and T lymphocyte responses to vaccination with pneumococcal polysaccharides. APMIS Suppl 15, 1-23.

Hendriksen, W.T., Bootsma, H.J., Estevao, S., Hoogenboezem, T., de Jong, A., de Groot, R., Kuipers, O.P., and Hermans, P.W. (2008). CodY of Streptococcus pneumoniae: link between nutritional gene regulation and colonization. J Bacteriol 190, 590-601. doi:10.1128/JB.00917-07

Hu, M.C., Walls, M.A., Stroop, S.D., Reddish, M.A., Beall, B., and Dale, J.B. (2002). Immunogenicity of a 26-valent group A streptococcal vaccine. Infect Immun 70, 2171-2177.

Huang, S.S., Johnson, K.M., Ray, G.T., Wroe, P., Lieu, T.A., Moore, M.R., Zell, E.R., Linder, J.A., Grijalva, C.G., Metlay, J.P., *et al.* (2011). Healthcare utilization and cost of pneumococcal disease in the United States. Vaccine 29, 3398-3412. doi:10.1016/j.vaccine.2011.02.088

Jambo, K.C., Sepako, E., Heyderman, R.S., and Gordon, S.B. (2010). Potential role for mucosally active vaccines against pneumococcal pneumonia. Trends Microbiol 18, 81-89. doi:10.1016/j.tim.2009.12.001

Jedrzejas, M.J. (2001). Pneumococcal virulence factors: structure and function. Microbiol Mol Biol Rev 65, 187-207 ; first page, table of contents. doi:10.1128/MMBR.65.2.187-207.2001

Josefsberg, J.O., and Buckland, B. (2012). Vaccine process technology. Biotechnol Bioeng 109, 1443-1460. doi:10.1002/bit.24493

Kadioglu, A., Weiser, J.N., Paton, J.C., and Andrew, P.W. (2008). The role of Streptococcus pneumoniae virulence factors in host respiratory colonization and disease. Nat Rev Microbiol 6, 288-301. doi:10.1038/nrmicro1871

Kapur, V., Maffei, J.T., Greer, R.S., Li, L.L., Adams, G.J., and Musser, J.M. (1994). Vaccination with streptococcal extracellular cysteine protease (interleukin-1 beta convertase) protects mice against challenge with heterologous group A streptococci. Microb Pathog 16, 443-450. doi:10.1006/mpat.1994.1044

Karthikeyan, G., and Guilherme, L. (2018). Acute rheumatic fever. The Lancet 392, 161-174. doi:10.1016/s0140-6736(18)30999-1

Kataoka, K., Fukuyama, Y., Briles, D.E., Miyake, T., and Fujihashi, K. (2017). Dendritic cell-targeting DNA-based nasal adjuvants for protective mucosal immunity to Streptococcus pneumoniae. Microbiol Immunol 61, 195-205. doi:10.1111/1348-0421.12487

Kaur, R., Casey, J.R., and Pichichero, M.E. (2016). Emerging Streptococcus pneumoniae Strains Colonizing the Nasopharynx in Children After 13-valent Pneumococcal Conjugate Vaccination in Comparison to the 7-valent Era, 2006-

2015. Pediatr Infect Dis J 35, 901-906. doi:10.1097/INF.0000000000001206

Kawabata, S., Kunitomo, E., Terao, Y., Nakagawa, I., Kikuchi, K., Totsuka, K., and Hamada, S. (2001). Systemic and mucosal immunizations with fibronectin-binding protein FBP54 induce protective immune responses against Streptococcus pyogenes challenge in mice. Infect Immun 69, 924-930. doi:10.1128/IAI.69.2.924-930.2001

Khan, M.N., and Pichichero, M.E. (2012). Vaccine candidates PhtD and PhtE of Streptococcus pneumoniae are adhesins that elicit functional antibodies in humans. Vaccine 30, 2900-2907. doi:10.1016/j.vaccine.2012.02.023

Khan, M.N., Sharma, S.K., Filkins, L.M., and Pichichero, M.E. (2012). PcpA of Streptococcus pneumoniae mediates adherence to nasopharyngeal and lung epithelial cells and elicits functional antibodies in humans. Microbes Infect 14, 1102-1110. doi:10.1016/j.micinf.2012.06.007

Khan, N., and Jan, A.T. (2017). Towards Identifying Protective B-Cell Epitopes: The PspA Story. Front Microbiol 8, 742. doi:10.3389/fmicb.2017.00742

Kim, E.H., Choi, S.Y., Kwon, M.K., Tran, T.D., Park, S.S., Lee, K.J., Bae, S.M., Briles, D.E., and Rhee, D.K. (2012). Streptococcus pneumoniae pep27 mutant as a live vaccine for serotype-independent protection in mice. Vaccine 30, 2008-2019. doi:10.1016/j.vaccine.2011.11.073

Kim, G.L., Choi, S.Y., Seon, S.H., Lee, S., Park, S.S., Song, J.Y., Briles, D.E., and Rhee, D.K. (2016). Pneumococcal pep27 mutant immunization stimulates cytokine secretion and confers long-term immunity with a wide range of protection, including against non-typeable strains. Vaccine 34, 6481-6492. doi:10.1016/j.vaccine.2016.10.071

Kim, G.L., Seon, S.H., and Rhee, D.K. (2017a). Pneumonia and Streptococcus pneumoniae vaccine. Arch Pharm Res. doi:10.1007/s12272-017-0933-y

Kim, G.L., Seon, S.H., and Rhee, D.K. (2017b). Pneumonia and Streptococcus pneumoniae vaccine. Arch Pharm Res 40, 885-893. doi:10.1007/s12272-017-0933-y

Koller, T., Manetti, A.G., Kreikemeyer, B., Lembke, C., Margarit, I., Grandi, G., and Podbielski, A. (2010). Typing of the pilus-protein-encoding FCT region and biofilm formation as novel parameters in epidemiological investigations of Streptococcus pyogenes isolates from various infection sites. J Med Microbiol 59, 442-452. doi:10.1099/jmm.0.013581-0

Kotloff, K.L., Corretti, M., Palmer, K., Campbell, J.D., Reddish, M.A., Hu, M.C., Wasserman, S.S., and Dale, J.B. (2004). Safety and immunogenicity of a recombinant multivalent group a streptococcal vaccine in healthy adults: phase 1 trial. JAMA 292, 709-715. doi:10.1001/jama.292.6.709

Kuipers, K., Jong, W.S.P., van der Gaast-de Jongh, C.E., Houben, D., van Opzeeland, F., Simonetti, E., van Selm, S., de Groot, R., Koenders, M.I., Azarian, T., *et al.* (2017). Th17-Mediated Cross Protection against Pneumococcal Carriage by Vaccination with a Variable Antigen. Infect Immun 85. doi:10.1128/IAI.00281-17

Lancefield, R.C. (1959). Persistence of type-specific antibodies in man following infection with group A streptococci. J Exp Med 110, 271-292.

Lancefield, R.C. (1962). Current knowledge of type-specific M antigens of group A streptococci. J Immunol 89, 307-313.

Lawrence, M.C., Pilling, P.A., Epa, V.C., Berry, A.M., Ogunniyi, A.D., and Paton, J.C. (1998). The crystal structure of pneumococcal surface antigen PsaA reveals a metal-binding site and a novel structure for a putative ABC-type binding protein. Structure 6, 1553-1561.

Laxminarayan, R., Duse, A., Wattal, C., Zaidi, A.K., Wertheim, H.F., Sumpradit,

N., Vlieghe, E., Hara, G.L., Gould, I.M., Goossens, H., *et al.* (2013). Antibiotic resistance-the need for global solutions. Lancet Infect Dis 13, 1057-1098. doi:10.1016/S1473-3099(13)70318-9

Laxminarayan, R., Matsoso, P., Pant, S., Brower, C., Rottingen, J.A., Klugman, K., and Davies, S. (2016). Access to effective antimicrobials: a worldwide challenge. Lancet 387, 168-175. doi:10.1016/S0140-6736(15)00474-2

LeMieux, J., Hava, D.L., Basset, A., and Camilli, A. (2006). RrgA and RrgB are components of a multisubunit pilus encoded by the Streptococcus pneumoniae rlrA pathogenicity islet. Infect Immun 74, 2453-2456. doi:10.1128/IAI.74.4.2453-2456.2006

Lipsitch, M., and Siber, G.R. (2016). How Can Vaccines Contribute to Solving the Antimicrobial Resistance Problem? MBio 7. doi:10.1128/mBio.00428-16

Loh, J.M.S., Lorenz, N., Tsai, C.J., Khemlani, A.H.J., and Proft, T. (2017). Mucosal vaccination with pili from Group A Streptococcus expressed on Lactococcus lactis generates protective immune responses. Sci Rep 7, 7174. doi:10.1038/s41598-017-07602-0

Lorenz, N., Loh, J.M., Moreland, N.J., and Proft, T. (2017). Development of a high-throughput opsonophagocytic assay for the determination of functional antibody activity against Streptococcus pyogenes using bioluminescence. J Microbiol Methods 134, 58-61. doi:10.1016/j.mimet.2017.01.010

Man, W.H., de Steenhuijsen Piters, W.A., and Bogaert, D. (2017). The microbiota of the respiratory tract: gatekeeper to respiratory health. Nat Rev Microbiol 15, 259-270. doi:10.1038/nrmicro.2017.14

Manzoor, I., Shafeeq, S., Kloosterman, T.G., and Kuipers, O.P. (2015). Co(2+)-dependent gene expression in Streptococcus pneumoniae: opposite effect of Mn(2+) and Co(2+) on the expression of the virulence genes psaBCA, pcpA, and prtA. Front Microbiol 6, 748. doi:10.3389/fmicb.2015.00748

Marasini, N., Ghaffar, K.A., Giddam, A.K., Batzloff, M.R., Good, M.F., Skwarczynski, M., and Toth, I. (2017). Highly Immunogenic Trimethyl Chitosan-based Delivery System for Intranasal Lipopeptide Vaccines against Group A Streptococcus. Curr Drug Deliv 14, 701-708. doi:10.2174/1567201813666160721141322

Marasini, N., Khalil, Z.G., Giddam, A.K., Ghaffar, K.A., Hussein, W.M., Capon, R.J., Batzloff, M.R., Good, M.F., Skwarczynski, M., and Toth, I. (2016). Lipid core peptide/poly(lactic-co-glycolic acid) as a highly potent intranasal vaccine delivery system against Group A streptococcus. Int J Pharm 513, 410-420. doi:10.1016/j.ijpharm.2016.09.057

McAllister, L.J., Tseng, H.J., Ogunniyi, A.D., Jennings, M.P., McEwan, A.G., and Paton, J.C. (2004). Molecular analysis of the psa permease complex of Streptococcus pneumoniae. Mol Microbiol 53, 889-901. doi:10.1111/j.1365-2958.2004.04164.x

McCool, T.L., Cate, T.R., Tuomanen, E.I., Adrian, P., Mitchell, T.J., and Weiser, J.N. (2003). Serum immunoglobulin G response to candidate vaccine antigens during experimental human pneumococcal colonization. Infect Immun 71, 5724-5732.

McDaniel, L.S., and Swiatlo, E. (2016). Should Pneumococcal Vaccines Eliminate Nasopharyngeal Colonization? MBio 7. doi:10.1128/mBio.00545-16

McMillan, D.J., Dreze, P.A., Vu, T., Bessen, D.E., Guglielmini, J., Steer, A.C., Carapetis, J.R., Van Melderen, L., Sriprakash, K.S., and Smeesters, P.R. (2013). Updated model of group A Streptococcus M proteins based on a comprehensive worldwide study. Clin Microbiol Infect 19, E222-229. doi:10.1111/1469-0691.12134

McNeil, S.A., Halperin, S.A., Langley, J.M., Smith, B., Warren, A., Sharratt, G.P., Baxendale, D.M., Reddish, M.A., Hu, M.C., Stroop, S.D., *et al.* (2005). Safety and immunogenicity of 26-valent group a streptococcus vaccine in healthy adult volunteers. Clin Infect Dis 41, 1114-1122. doi:10.1086/444458

Mellroth, P., Daniels, R., Eberhardt, A., Ronnlund, D., Blom, H., Widengren, J., Normark, S., and Henriques-Normark, B. (2012). LytA, major autolysin of Streptococcus pneumoniae, requires access to nascent peptidoglycan. J Biol Chem 287, 11018-11029. doi:10.1074/jbc.M111.318584

Miller, E., Andrews, N.J., Waight, P.A., Slack, M.P., and George, R.C. (2011). Herd immunity and serotype replacement 4 years after seven-valent pneumococcal conjugate vaccination in England and Wales: an observational cohort study. Lancet Infect Dis 11, 760-768. doi:10.1016/S1473-3099(11)70090-1

Minor, P.D. (2015). Live attenuated vaccines: Historical successes and current challenges. Virology 479-480, 379-392. doi:10.1016/j.virol.2015.03.032

Moffitt, K.L., and Malley, R. (2011). Next generation pneumococcal vaccines. Curr Opin Immunol 23, 407-413. doi:10.1016/j.coi.2011.04.002

Mortensen, R., Christensen, D., Hansen, L.B., Christensen, J.P., Andersen, P., and Dietrich, J. (2017). Local Th17/IgA immunity correlate with protection against intranasal infection with Streptococcus pyogenes. PLoS One 12, e0175707. doi:10.1371/journal.pone.0175707

Mortensen, R., Nissen, T.N., Fredslund, S., Rosenkrands, I., Christensen, J.P., Andersen, P., and Dietrich, J. (2016). Identifying protective Streptococcus pyogenes vaccine antigens recognized by both B and T cells in human adults and children. Sci Rep 6, 22030. doi:10.1038/srep22030

Mukerji, R., Mirza, S., Roche, A.M., Widener, R.W., Croney, C.M., Rhee, D.K., Weiser, J.N., Szalai, A.J., and Briles, D.E. (2012). Pneumococcal surface protein

A inhibits complement deposition on the pneumococcal surface by competing with the binding of C-reactive protein to cell-surface phosphocholine. J Immunol 189, 5327-5335. doi:10.4049/jimmunol.1201967

Nabors, G.S., Braun, P.A., Herrmann, D.J., Heise, M.L., Pyle, D.J., Gravenstein, S., Schilling, M., Ferguson, L.M., Hollingshead, S.K., Briles, D.E., *et al.* (2000). Immunization of healthy adults with a single recombinant pneumococcal surface protein A (PspA) variant stimulates broadly cross-reactive antibodies to heterologous PspA molecules. Vaccine 18, 1743-1754.

Navarre, W.W., and Schneewind, O. (1994). Proteolytic cleavage and cell wall anchoring at the LPXTG motif of surface proteins in gram-positive bacteria. Mol Microbiol 14, 115-121.

Nelson, A.L., Ries, J., Bagnoli, F., Dahlberg, S., Falker, S., Rounioja, S., Tschop, J., Morfeldt, E., Ferlenghi, I., Hilleringmann, M., *et al.* (2007). RrgA is a pilus-associated adhesin in Streptococcus pneumoniae. Mol Microbiol 66, 329-340. doi:10.1111/j.1365-2958.2007.05908.x

Nordstrom, T., Pandey, M., Calcutt, A., Powell, J., Phillips, Z.N., Yeung, G., Giddam, A.K., Shi, Y., Haselhorst, T., von Itzstein, M., *et al.* (2017). Enhancing Vaccine Efficacy by Engineering a Complex Synthetic Peptide To Become a Super Immunogen. J Immunol 199, 2794-2802. doi:10.4049/jimmunol.1700836

O'Brien, K.L., Wolfson, L.J., Watt, J.P., Henkle, E., Deloria-Knoll, M., McCall, N., Lee, E., Mulholland, K., Levine, O.S., Cherian, T., *et al.* (2009). Burden of disease caused by Streptococcus pneumoniae in children younger than 5 years: global estimates. Lancet 374, 893-902. doi:10.1016/S0140-6736(09)61204-6

Odutola, A., Ota, M.O.C., Antonio, M., Ogundare, E.O., Saidu, Y., Foster-Nyarko, E., Owiafe, P.K., Ceesay, F., Worwui, A., Idoko, O.T., *et al.* (2017). Efficacy of a novel, protein-based pneumococcal vaccine against nasopharyngeal carriage of

Streptococcus pneumoniae in infants: A phase 2, randomized, controlled, observer-blind study. Vaccine 35, 2531-2542. doi:10.1016/j.vaccine.2017.03.071

Pandey, M., Batzloff, M.R., and Good, M.F. (2009). Mechanism of protection induced by group A Streptococcus vaccine candidate J8-DT: contribution of B and T-cells towards protection. PLoS One 4, e5147. doi:10.1371/journal.pone.0005147

Pandey, M., Powell, J., Calcutt, A., Zaman, M., Phillips, Z.N., Ho, M.F., Batzloff, M.R., and Good, M.F. (2017). Physicochemical characterisation, immunogenicity and protective efficacy of a lead streptococcal vaccine: progress towards Phase I trial. Sci Rep 7, 13786. doi:10.1038/s41598-017-14157-7

Pandey, M., Wykes, M.N., Hartas, J., Good, M.F., and Batzloff, M.R. (2013). Long-term antibody memory induced by synthetic peptide vaccination is protective against Streptococcus pyogenes infection and is independent of memory T cell help. J Immunol 190, 2692-2701. doi:10.4049/jimmunol.1202333

Park, H.S., and Cleary, P.P. (2005). Active and passive intranasal immunizations with streptococcal surface protein C5a peptidase prevent infection of murine nasal mucosa-associated lymphoid tissue, a functional homologue of human tonsils. Infect Immun 73, 7878-7886. doi:10.1128/IAI.73.12.7878-7886.2005

Paterson, G.K., and Mitchell, T.J. (2006). The role of Streptococcus pneumoniae sortase A in colonisation and pathogenesis. Microbes Infect 8, 145-153. doi:10.1016/j.micinf.2005.06.009

Paton, J.C., and Ogunniyi, A.D. (2011). Evicting the pneumococcus from its nasopharyngeal lodgings. Cell Host Microbe 9, 89-91. doi:10.1016/j.chom.2011.01.013

Pettigrew, M.M., Fennie, K.P., York, M.P., Daniels, J., and Ghaffar, F. (2006). Variation in the presence of neuraminidase genes among Streptococcus pneumoniae isolates with identical sequence types. Infect Immun 74, 3360-3365. doi:10.1128/IAI.01442-05

Pichichero, M.E. (2017). Pneumococcal whole-cell and protein-based vaccines: changing the paradigm. Expert Rev Vaccines 16, 1181-1190. doi:10.1080/14760584.2017.1393335

Pichichero, M.E., Khan, M.N., and Xu, Q. (2016). Next generation protein based Streptococcus pneumoniae vaccines. Hum Vaccin Immunother 12, 194-205. doi:10.1080/21645515.2015.1052198

Plumptre, C.D., Ogunniyi, A.D., and Paton, J.C. (2013). Surface association of Pht proteins of Streptococcus pneumoniae. Infect Immun 81, 3644-3651. doi:10.1128/IAI.00562-13

Porchia, B.R., Bonanni, P., Bechini, A., Bonaccorsi, G., and Boccalini, S. (2017). Evaluating the costs and benefits of pneumococcal vaccination in adults. Expert Rev Vaccines 16, 93-107. doi:10.1080/14760584.2017.1242419

Postol, E., Alencar, R., Higa, F.T., Freschi de Barros, S., Demarchi, L.M., Kalil, J., and Guilherme, L. (2013). StreptInCor: a candidate vaccine epitope against S. pyogenes infections induces protection in outbred mice. PLoS One 8, e60969. doi:10.1371/journal.pone.0060969

Principi, N., and Esposito, S. (2018). Development of pneumococcal vaccines over the last 10 years. Expert Opin Biol Ther 18, 7-17. doi:10.1080/14712598.2018.1384462

Pruksakorn, S., Galbraith, A., Houghten, R.A., and Good, M.F. (1992). Conserved T and B cell epitopes on the M protein of group A streptococci. . J Immunol 149, 2729-2735.

Quin, L.R., Carmicle, S., Dave, S., Pangburn, M.K., Evenhuis, J.P., and McDaniel, L.S. (2005). In vivo binding of complement regulator factor H by Streptococcus pneumoniae. J Infect Dis 192, 1996-2003. doi:10.1086/497605

Ray, G.T. (2002). Pneumococcal conjugate vaccine: economic issues of the introduction of a new childhood vaccine. Expert Rev Vaccines 1, 65-74. doi:10.1586/14760584.1.1.65

Reiss-Mandel, A., and Regev-Yochay, G. (2016). Staphylococcus aureus and Streptococcus pneumoniae interaction and response to pneumococcal vaccination: Myth or reality? Hum Vaccin Immunother 12, 351-357. doi:10.1080/21645515.2015.1081321

Relf, W.A., Cooper, J., Brandt, E.R., Hayman, W.A., Anders, R.F., Pruksakorn, S., Currie, B., Saul, A., and Good, M.F. (1996). Mapping a conserved conformational epitope from the M protein of group A streptococci. Pept Res 9, 12-20.

Rigden, D.J., Galperin, M.Y., and Jedrzejas, M.J. (2003). Analysis of structure and function of putative surface-exposed proteins encoded in the Streptococcus pneumoniae genome: a bioinformatics-based approach to vaccine and drug design. Crit Rev Biochem Mol Biol 38, 143-168. doi:10.1080/713609215

Rioux, S., Neyt, C., Di Paolo, E., Turpin, L., Charland, N., Labbe, S., Mortier, M.C., Mitchell, T.J., Feron, C., Martin, D., *et al.* (2011). Transcriptional regulation, occurrence and putative role of the Pht family of Streptococcus pneumoniae. Microbiology 157, 336-348. doi:10.1099/mic.0.042184-0

Rosenow, C., Ryan, P., Weiser, J.N., Johnson, S., Fontan, P., Ortqvist, A., and Masure, H.R. (1997). Contribution of novel choline-binding proteins to adherence, colonization and immunogenicity of Streptococcus pneumoniae. Mol Microbiol 25, 819-829.

Sabharwal, H., Michon, F., Nelson, D., Dong, W., Fuchs, K., Manjarrez, R.C., Sarkar, A., Uitz, C., Viteri-Jackson, A., Suarez, R.S., *et al.* (2006). Group A streptococcus (GAS) carbohydrate as an immunogen for protection against GAS infection. J Infect Dis 193, 129-135. doi:10.1086/498618

Sampson, J.S., O'Connor, S.P., Stinson, A.R., Tharpe, J.A., and Russell, H. (1994). Cloning and nucleotide sequence analysis of psaA, the Streptococcus pneumoniae gene encoding a 37-kilodalton protein homologous to previously reported Streptococcus sp. adhesins. Infect Immun 62, 319-324.

Schodel, F., Moreland, N.J., Wittes, J.T., Mulholland, K., Frazer, I., Steer, A.C., Fraser, J.D., and Carapetis, J. (2017). Clinical development strategy for a candidate group A streptococcal vaccine. Vaccine 35, 2007-2014. doi:10.1016/j.vaccine.2017.02.060

Schulze, K., Ebensen, T., Chandrudu, S., Skwarczynski, M., Toth, I., Olive, C., and Guzman, C.A. (2017). Bivalent mucosal peptide vaccines administered using the LCP carrier system stimulate protective immune responses against Streptococcus pyogenes infection. Nanomedicine 13, 2463-2474. doi:10.1016/j.nano.2017.08.015

Schulze, K., Medina, E., and Guzman, C.A. (2006). Intranasal immunization with serum opacity factor (SOF) of Streptococcus pyogenes fails to protect mice against lethal mucosal challenge with a heterologous strain. Vaccine 24, 1446-1450. doi:10.1016/j.vaccine.2005.06.039

Seth, A., Kong, I.G., Lee, S.H., Yang, J.Y., Lee, Y.S., Kim, Y., Wibowo, N., Middelberg, A.P., Lua, L.H., and Kweon, M.N. (2016). Modular virus-like particles for sublingual vaccination against group A streptococcus. Vaccine 34, 6472-6480. doi:10.1016/j.vaccine.2016.11.008

Shaper, M., Hollingshead, S.K., Benjamin, W.H., Jr., and Briles, D.E. (2004). PspA protects Streptococcus pneumoniae from killing by apolactoferrin, and antibody to PspA enhances killing of pneumococci by apolactoferrin [corrected]. Infect Immun 72, 5031-5040. doi:10.1128/IAI.72.9.5031-5040.2004

Shapiro, E.D., Berg, A.T., Austrian, R., Schroeder, D., Parcells, V., Margolis, A., Adair, R.K., and Clemens, J.D. (1991). The protective efficacy of polyvalent pneumococcal polysaccharide vaccine. N Engl J Med 325, 1453-1460. doi:10.1056/NEJM199111213252101

Shelburne, S.A., Davenport, M.T., Keith, D.B., and Musser, J.M. (2008). The role of complex carbohydrate catabolism in the pathogenesis of invasive streptococci. Trends Microbiol 16, 318-325. doi:10.1016/j.tim.2008.04.002

Shet, A., Kaplan, E.L., Johnson, D.R., and Cleary, P.P. (2003). Immune response to group A streptococcal C5a peptidase in children: implications for vaccine development. J Infect Dis 188, 809-817. doi:10.1086/377700

Sims Sanyahumbi, A., Colquhoun, S., Wyber, R., and Carapetis, J.R. (2016). Global Disease Burden of Group A Streptococcus. In Streptococcus pyogenes : Basic Biology to Clinical Manifestations, J.J. Ferretti, D.L. Stevens, and V.A. Fischetti, eds. (Oklahoma City (OK)).

Singh, A., and Dutta, A.K. (2017). Pneumococcal Vaccines - How Many Serotypes are Enough? Indian J Pediatr. doi:10.1007/s12098-017-2449-3

Smit, P., Oberholzer, D., Hayden-Smith, S., Koornhof, H.J., and Hilleman, M.R. (1977). Protective efficacy of pneumococcal polysaccharide vaccines. JAMA 238, 2613-2616.

Speers, D.J., Levy, A., Gichamo, A., Eastwood, A., and Leung, M.J. (2017). M protein gene (emm type) analysis of group A Streptococcus isolates recovered during an acute glomerulonephritis outbreak in northern Western Australia. Pathology 49, 765-769. doi:10.1016/j.pathol.2017.09.001

Steer, A.C., Carapetis, J.R., Dale, J.B., Fraser, J.D., Good, M.F., Guilherme, L., Moreland, N.J., Mulholland, E.K., Schodel, F., and Smeesters, P.R. (2016). Status of research and development of vaccines for Streptococcus pyogenes. Vaccine

34, 2953-2958. doi:10.1016/j.vaccine.2016.03.073

Tandon, R., Sharma, M., Chandrashekhar, Y., Kotb, M., Yacoub, M.H., and Narula, J. (2013). Revisiting the pathogenesis of rheumatic fever and carditis. Nat Rev Cardiol 10, 171-177. doi:10.1038/nrcardio.2012.197

Telford, J.L., Barocchi, M.A., Margarit, I., Rappuoli, R., and Grandi, G. (2006). Pili in gram-positive pathogens. Nat Rev Microbiol 4, 509-519. doi:10.1038/nrmicro1443

Tilley, S.J., Orlova, E.V., Gilbert, R.J., Andrew, P.W., and Saibil, H.R. (2005). Structural basis of pore formation by the bacterial toxin pneumolysin. Cell 121, 247-256. doi:10.1016/j.cell.2005.02.033

Tseng, H.J., McEwan, A.G., Paton, J.C., and Jennings, M.P. (2002). Virulence of Streptococcus pneumoniae: PsaA mutants are hypersensitive to oxidative stress. Infect Immun 70, 1635-1639.

van der Poll, T., and Opal, S.M. (2009). Pathogenesis, treatment, and prevention of pneumococcal pneumonia. Lancet 374, 1543-1556. doi:10.1016/S0140-6736(09)61114-4

van Pee, K., Mulvihill, E., Muller, D.J., and Yildiz, O. (2016). Unraveling the Pore-Forming Steps of Pneumolysin from Streptococcus pneumoniae. Nano Lett 16, 7915-7924. doi:10.1021/acs.nanolett.6b04219

Vashishtha, A., and Fischetti, V.A. (1993). Surface-exposed conserved region of the streptococcal M protein induces antibodies cross-reactive with denatured forms of myosin. J Immunol 150, 4693-4701.

Vernatter, J., and Pirofski, L.A. (2013). Current concepts in host-microbe interaction leading to pneumococcal pneumonia. Curr Opin Infect Dis 26, 277-283. doi:10.1097/QCO.0b013e3283608419

Wang, S., Li, Y., Shi, H., Scarpellini, G., Torres-Escobar, A., Roland, K.L., and

Curtiss, R., 3rd (2010). Immune responses to recombinant pneumococcal PsaA antigen delivered by a live attenuated Salmonella vaccine. Infect Immun 78, 3258-3271. doi:10.1128/IAI.00176-10

WHO (2016). WHO | Pneumonia. World Health Organization; (fact sheet No 331). Available: http://www.who.int/mediacentre/factsheets/fs331/en/. Accessed 26 December 2017.

WHO (2017). WHO | WHO publishes list of bacteria for which new antibiotics are urgently needed. World Health Organization; (News release). Available: http://www.who.int/mediacentre/news/releases/2017/bacteria-antibiotics-needed/en/. Accessed 26 December 2017.

WHO Publication (2012). Pneumococcal vaccines WHO position paper – 2012 – Recommendations. Vaccine 30, 4717-4718. doi:https://doi.org/10.1016/j.vaccine.2012.04.093

Wizemann, T.M., Heinrichs, J.H., Adamou, J.E., Erwin, A.L., Kunsch, C., Choi, G.H., Barash, S.C., Rosen, C.A., Masure, H.R., Tuomanen, E., *et al.* (2001). Use of a whole genome approach to identify vaccine molecules affording protection against Streptococcus pneumoniae infection. Infect Immun 69, 1593-1598. doi:10.1128/IAI.69.3.1593-1598.2001

Wozniak, A., Garcia, P., Geoffroy, E.A., Aguirre, D.B., Gonzalez, S.A., Sarno, V.A., Dale, J.B., Salazar-Echegarai, F.J., Vera, A., Bueno, S.M., *et al.* (2014). A novel live vector group A streptococcal emm type 9 vaccine delivered intranasally protects mice against challenge infection with emm type 9 group A streptococci. Clin Vaccine Immunol 21, 1343-1349. doi:10.1128/CVI.00330-14

Xu, Q., Pryharski, K., and Pichichero, M.E. (2017). Trivalent pneumococcal protein vaccine protects against experimental acute otitis media caused by Streptococcus pneumoniae in an infant murine model. Vaccine 35, 337-344.

doi:10.1016/j.vaccine.2016.11.046

Zafar, M.A., Wang, Y., Hamaguchi, S., and Weiser, J.N. (2017). Host-to-Host Transmission of Streptococcus pneumoniae Is Driven by Its Inflammatory Toxin, Pneumolysin. Cell Host Microbe 21, 73-83. doi:10.1016/j.chom.2016.12.005

Zhang, J.R., Mostov, K.E., Lamm, M.E., Nanno, M., Shimida, S., Ohwaki, M., and Tuomanen, E. (2000). The polymeric immunoglobulin receptor translocates pneumococci across human nasopharyngeal epithelial cells. Cell 102, 827-837.

Zhang, X., Song, Y., Li, Y., Cai, M., Meng, Y., and Zhu, H. (2017). Immunization with Streptococcal Heme Binding Protein (Shp) Protects Mice Against Group A Streptococcus Infection. Adv Exp Med Biol 973, 115-124. doi:10.1007/5584_2016_198

Zingaretti, C., Falugi, F., Nardi-Dei, V., Pietrocola, G., Mariani, M., Liberatori, S., Gallotta, M., Tontini, M., Tani, C., Speziale, P., *et al.* (2010). Streptococcus pyogenes SpyCEP: a chemokine-inactivating protease with unique structural and biochemical features. FASEB J 24, 2839-2848. doi:10.1096/fj.09-145631

Zuhlke, L.J., Beaton, A., Engel, M.E., Hugo-Hamman, C.T., Karthikeyan, G., Katzenellenbogen, J.M., Ntusi, N., Ralph, A.P., Saxena, A., Smeesters, P.R., *et al.* (2017). Group A Streptococcus, Acute Rheumatic Fever and Rheumatic Heart Disease: Epidemiology and Clinical Considerations. Curr Treat Options Cardiovasc Med 19, 15. doi:10.1007/s11936-017-0513-y

www.ingramcontent.com/pod-product-compliance
Lightning Source LLC
LaVergne TN
LVHW070844210926
848482LV00016B/861
9781912530229